GeoPlanet: Earth and Planetary Sciences

For further volumes:
http://www.springer.com/series/8821

Paweł Rowiński
Editor

Experimental and Computational Solutions of Hydraulic Problems

32nd International School of Hydraulics

Springer

Editor
Paweł Rowiński
Institute of Geophysics
Polish Academy of Sciences
Warsaw
Poland

The GeoPlanet: Earth and Planetary Sciences Book Series is in part a continuation of Monographic Volumes of Publications of the Institute of Geophysics, Polish Academy of Sciences, the journal published since 1962 (http://pub.igf.edu.pl/index.php).

ISSN 2190-5193 ISSN 2190-5207 (electronic)
ISBN 978-3-642-30208-4 ISBN 978-3-642-30209-1 (eBook)
DOI 10.1007/978-3-642-30209-1
Springer Heidelberg New York Dordrecht London

Library of Congress Control Number: 2012955343

Printed on acid-free paper

Springer is part of Springer Science+Business Media (www.springer.com)

Series Editors

Advisory Board

Kon-Kee Liu
Institute of Hydrological
and Oceanic Sciences
National Central University Jhongli
Jhongli, Taiwan

Teresa Madeyska
Research Centre in Warsaw
Institute of Geological Sciences
Warszawa, Poland

Stanisław Massel
Institute of Oceanology
Polish Academy of Sciences
Sopot, Poland

Antonio Meloni
Instituto Nazionale di Geofisica
Rome, Italy

Evangelos Papathanassiou
Hellenic Centre for Marine Research
Anavissos, Greece

Kaja Pietsch
AGH University of Science and
Technology
Kraków, Poland

Dušan Plašienka
Prírodovedecká fakulta, UK
Univerzita Komenského
Bratislava, Slovakia

Barbara Popielawska
Space Research Centre
Polish Academy of Sciences
Warszawa, Poland

Tilman Spohn
Deutsches Zentrum für Luft-
und Raumfahrt
in der Helmholtz Gemeinschaft
Institut für Planetenforschung
Berlin, Germany

Krzysztof Stasiewicz
Swedish Institute of Space Physics
Uppsala, Sweden

Roman Teisseyre
Earth's Interior Dynamics Lab
Institute of Geophysics
Polish Academy of Sciences
Warszawa, Poland

Jacek Tronczynski
Laboratory of Biogeochemistry
of Organic Contaminants
IFREMER DCN_BE
Nantes, France

Steve Wallis
School of the Built Environment
Heriot-Watt University
Riccarton, Edinburgh
Scotland, UK

Wacław M. Zuberek
Department of Applied Geology
University of Silesia
Sosnowiec, Poland

Preface

This volume embraces a variety of research studies presented at the 32nd International School of Hydraulics *Experimental and computational solutions of hydraulic problems*. The papers that were accepted by the International Scientific Committee cover a wide range of subjects and research methods. They embody a mix of theory and practice, planning and reflection participation, and observation to provide the rich diversity of perspectives represented at the conference.

As is often the case, speakers had limited time in which to present their work at the conference and so the papers contained in this volume showcase the details of their research, approach, and outcomes. The authors represent a continuum from those with a record of research history in hydraulics up to those who have only very recently started to explore in this area, each with the benefits of their unique perspective. Particularly important to this volume is the invited comprehensive papers prepared by the internationally recognized indisputable authorities in particular fields of hydraulics. The keynote speaker at the meeting was Professor Donald W. Knight, whose research career over the years was associated with the Department of Civil Engineering of the University of Birmingham, United Kingdom. He prepared an excellent overview of hydraulic problems related to flooding. Professor Subhasish Dey from the Department of Civil Engineering, Indian Institute of Technology, India, together with S. K. Bose and O. Castro-Orgaz discussed the problems of hydrodynamics of undular free surface flows. Two Italian noble researchers highly enriched both the conference and the book: Professor Roberto Gaudio from the Soil Protection Department "Vincenzo Marone", University of Calabria (with coauthorship of S. Dey) with his contribution on the basic question of the universality of von Kármán's constant κ, and Professor Andrea Marion (with coauthorship of M. Tregnaghi) from the Department of Hydraulic, Maritime and Geotechnical Engineering, University of Padova, with his paper on novel methods of studying of incipient motion of sediment grains in open channels. A vision paper presenting research trends in the studies of aquatic ecosystems was provided by Professor Vladimir Nikora from the School of Engineering, University of Aberdeen, United Kingdom. The school has also ambition to raise very practical questions and the contribution of Professor Artur Radecki-Pawlik from the Department of Water Engineering, Agricultural University of Cracow in Poland went along such a line by discussing the issues related to construction and exploitation of rapid hydraulic structures in Polish rivers.

The International School of Hydraulics was also an occasion to celebrate **50 years** of pioneering work and role in shaping Polish environmental hydraulics by an extraordinary scientist, Professor Włodzimierz Czernuszenko, and this issue constitutes the subject of the paper of myself and Monika Kalinowska.

This volume contains 30 papers prepared by authors from 12 countries.

The 32nd International School of Hydraulics took place at the Palace Łochów located in a picturesque valley of Liwiec River, an hour away from the capital of Poland, Warsaw. The Łochów Palace is a nineteenth century Palace and park complex. The location, central part of Podlasie, is a historical gateway to the East, the place where many different religions and cultures converge. That beautiful surrounding and nice informal atmosphere made the School a unique scientific event.

Paweł M. Rowiński

Acknowledgments

I would like to thank all those who contributed to the 32nd International School of Hydraulics *Experimental and computational solutions of hydraulic problems* and the resulting volume. Great appreciation is therefore due to:

- All the invited speakers (Professors Donald W. Knight, Subhasish Dey, Roberto Gaudio, Andrea Marion, Vladimir Nikora, Artur Radecki-Pawlik) for delivering their excellent lectures and providing critical inputs and suggestions during the School.
- The Chairpersons of the sessions: Vladimir Nikora, Steve Wallis, Ian Guymer, Roberto Gaudio, Monika Kalinowska, Robert Bialik, Artur Radecki-Pawlik, Andrea Marion, Donald W. Knight, Subhasish Dey, and Jarosław Napiórkowski, who led them with insight, wisdom, and humor.
- The reviewers of the papers, particularly Robert Bialik, Włodzimierz Czernuszenko, Monika Kalinowska, Janusz Kubrak, Wojciech Majewski, Marek Mitosek, Jarosław Napiórkowski, and Steve Wallis for preparing their in-depth reports ensuring sufficient quality of all the contributions.
- The members of the Local Organizing Committee (Magdalena Mrokowska, Monika Kalinowska, Anna Łukanowska, and Anna Zdunek) who ensured that the process and the program remained as planned and did so effectively and with the necessary degree of flexibility; and who helped to make the School the success it was.
- Appreciation is also due to Anna Dziembowska for language editing of all the papers.

A significant factor in the success of the School was the support received from the following sponsors:

- Institute of Geophysics, Polish Academy of Sciences
- Polish Academy of Sciences and its Committee for Water Resources Management
- The International Association for Hydro-Environment Engineering and Research, IAHR
- GeoPlanet: Earth and Planetary Research Center of the Polish Academy of Sciences

Contents

Part II Computational Solutions of Hydraulic Problems

Pioneering Works in Polish Environmental Hydraulics: The Flavor of World-Class Science

50 Years of Work of Włodzimierz Czernuszenko

Paweł M. Rowiński and Monika B. Kalinowska

Abstract Achievements of a noteworthy Polish scientist, Professor Włodzimierz Czernuszenko, in the context of international cooperation are presented herein. He was one of the initiators of the in-depth research in the emerging field of environmental hydraulics. He is well-known for his studies on turbulence in open-channels, heat and mass transport in rivers and quite recently on solid particles' transport in rivers from the perspective of two-phase flows. This chapter was prepared on the occasion of the 50th anniversary of his scientific work.

1 Introduction

It is hard to point what is the beginning of the scientific discipline named environmental hydraulics or ecohydraulics. Although the subject attracted the interest of the mankind over thousands of years ago with, e.g., Aristotle (384–322 B.C.) and Archimedes (287–212 B.C.), the history of ecohydraulics as a separate discipline is rather new and may be counted in decades only. In Poland those decades coincide with the transformation of the political system and the development of science has obviously been correlated with those political changes. It so happened that in the last century a fundamental knowledge of flow in open channels was built often almost simultaneously—in the West on one side and in the Eastern and Central Europe on the other. The knowledge transfer particularly from the communist bloc to the western countries was rather limited and the scientific

P. M. Rowiński (✉) · M. B. Kalinowska
Institute of Geophysics, Polish Academy of Sciences, Ks. Janusza 64 01-452, Warszawa, Poland
e-mail: p.rowinski@igf.edu.pl

P. Rowiński (ed.), *Experimental and Computational Solutions of Hydraulic Problems*, GeoPlanet: Earth and Planetary Sciences, DOI: 10.1007/978-3-642-30209-1_1,

community in the West was usually not well informed about deep studies on this side of iron curtain, allowing for the better comprehension of the flow of water and matter in rivers and canals. Therefore, establishing links between those two scientific communities was extremely important for the benefit of our scientific discipline.

Professor Włodzimierz Czernuszenko, widely acknowledged as the father of Polish environmental hydraulics (as understood nowadays) and arguably the most influential Polish hydraulician at the time of the explosion of West-East collaboration… has celebrated **fifty years** of pioneering work and role in shaping Polish hydraulics. It is the first major retrospective to honor this extraordinary Polish scientist, one of the first to bring the flavor of world research into Polish environmental hydraulics community.

Professor Czernuszenko is one of the Polish most accomplished and highly acclaimed hydro-environmental scientists. His research and discoveries have influenced the study of transport processes and turbulence in open-channels, making it possible to understand the physics of those processes and, ultimately, to model those processes computationally. To those who know him only by professional reputation, Professor Czernuszenko is a scientist whose work opens new doors and lights new paths of discovery within environmental hydraulics. To those who know him personally, he is an amiable, soft-spoken man with a sly sense of humor, a man who is modest about his accomplishments and generous with his knowledge, advice, and encouragement.

Włodzimierz Czernuszenko has built an international reputation particularly with his works on turbulence in open-channels, heat and mass transport in rivers and quite recently on solid particles' transport in rivers from the perspective of two-phase flows. This chapter aims to briefly look into his selected research studies and to show their influence on the progress in the hydraulic engineering field. Modern hydraulics is no longer confined within the narrow boundaries of specific fields of thoughts—researchers must cross frontiers and Czernuszenko created an excellent example of such thinking.

2 River Turbulence

Flow in open channels is mainly in turbulent regime. Therefore, understanding of this ubiquitous nonlinear phenomenon is crucial for understanding practically all problems of environmental hydraulics. Studying of this phenomenon in respect to river flow had to draw on the achievements of all those research giants like Kolmogorov, Oboukhoff, Monin, Yaglom (in the East) and Hopf, Heisenberg, Taylor, Prandtl, von Karman (in the West) but also earlier works of Navier, Stokes and Reynolds. One cannot forget those who knew how to use that knowledge in broadening the understanding of river flow—Yalin, Grinvald, Grishanin, Nikora, Nezu, Nakagawa, Rodi, Knight and many others. Among them undoubtedly was Czernuszenko, although his early works were mostly known only in Poland.

Czernuszenko's early works of the 1980s were at the forefront of experimentation on turbulence in free surface flows. Although there was already quite a material gathered in international literature (e.g., Grinvald 1974; Imamomoto 1975; Raichlen 1967; Schuyf 1966) concerning turbulence characteristics in open channels but they all were of a very initial character and every well-performed experiment at that time was of extreme value. Early understanding of turbulence phenomenon referred to Richardson's energy cascade process and to the Kolmogorov theory of locally isotropic turbulence. The first experiments on open-channel turbulence were usually designed to confirm those theories and to find relevant scales of turbulence. The first experiments went along a well-established scheme which allowed for various comparison studies. In those studies turbulent flow velocity at any point was treated as a random variable. Obviously, the behavior of this velocity can be described by probability density function, $p(u)$. Velocity measurements provide a time series of values recorded at time instants at regular intervals Δt. The considerations were restricted to stationary and ergodic random processes. For such situation, the probability density function (pdf) is invariant with respect to time and only one sample record over a sufficiently long time interval is needed to define the pdf and relevant statistical characteristics. Thus, all information about the turbulent velocity at a point can be obtained from one time series. The pdf and the statistical moments depend only on the magnitudes of the velocities measured at one point in the flow and not on the sequence in which those values occur. On the contrary, relevant correlation functions and energy spectra depend on the sequence in which those magnitudes occur. Reynolds stresses depend on the simultaneously recorded velocities in two directions at a point, but similarly to the pdf function they are independent of the sequence of values. An analytical expression for the pdf for turbulent flows is not easy to establish. Nevertheless, for most practical (engineering) purposes, the function can be characterized by statistical moments of different orders that can be obtained relatively easily from experiments. All up to the fourth moment have usually been determined in experiments in open-channels.

The very first laboratory experiments in Poland allowing to determine the basic turbulence characteristics were carried out in the straight open channel (Fig. 1) of a width of 175 cm and a length of 16 m (Czernuszenko and Lebiecki 1980). Measurements were taken for Reynolds numbers ranging from 7,000 to 42,000. The technique was to measure the instantaneous longitudinal velocity values with use of micropropeller—those values were averaged over 1 s time intervals. Measurements were taken in numerous points allowing to obtain spatial distributions of basic characteristics, among them turbulence intensity, autocorrelation and structural functions as well as spectral distributions of energy. Those—from today perspective—very simple measurements allowed to determine the scales of turbulence: Kolmogorov's micro-scale characterizing "small eddies" in which energy dissipation and macroscale characterizing the greatest eddies occurring in the flow. Moreover, the measurements enabled the calculation of turbulent mixing coefficients, energy dissipation rate of eddies and range of eddies satisfying "5/3 Kolmogorov law". Czernuszenko further worked towards improvement of the

Fig. 1 Straight laboratory flume at the Institute of Meteorology and Water Management in the late 1970s

measuring techniques and ways of their interpretation and those studies encouraged him to go out of the laboratories to lead such measurements in the field. In the late eighties he organized field campaigns in Vistula River in the vicinity of Połaniec power station (three independent studies) (Czernuszenko and Lebiecki 1989) and a bit later in the Narew River in the reach behind warm water discharge from the electricity power station in Ostrołęka (Czernuszenko and Rowiński 1989). There were also studies performed in small streams and it gave an impulse to a long lasting collaboration with various groups of Prof. Vladimir Nikora (Fig. 2) at that time resulting in a common study of turbulence in (Nikora et al. 1993).

On top of providing invaluable data on turbulence structure in rivers, those studies were also extremely helpful for the verification of the fashionable jet models that were intensively built then. Together with gaining the experience in turbulence studies, Czernuszenko switched consistently to more and more complex situations and more difficult research questions. It is a common knowledge that rivers in flood are characterized by a compound cross-section (water overflows the banks to floodplains) and unsteadiness of the flow. The characteristics of turbulence are not well recognized in such a situation, even at present. In a compound channel one may suspect that additional flow resistance occurs due to an intensive momentum exchange between the deep main channel and the adjacent shallow floodplains. The flow structures that occur in rivers of a compound cross-section are very complex due to at least three mechanisms: boundary-generated turbulence, free shear layer turbulence and velocity fluctuations associated with perturbations in the longitudinal secondary flow cells. To make the considerations simpler, it is convenient to study the geometrical and temporal complexities in separation, i.e., to study the turbulence structure in a compound cross-section but at possibly steady state conditions and to additionally study turbulence under unsteady state condition. Czernuszenko has been recently involved in both of the above. Let us mention first his experimental investigations on the structure of turbulence in compound channels (Czernuszenko et al. 2007; Czernuszenko and Rowiński 2008a, b; Rowiński et al. 1998, 2002). Starting in the late 1990s, Czernuszenko and his collaborators initiated experimental studies in a compound channel. At the beginning, 1D experiments in a trapezoidal straight open channel with symmetrically complex cross-section with inclined side-walls were made (Rowiński et al. 1998, 2002). The originality of the approach consisted in simulating various hydraulic conditions by changing the roughness of floodplains. To achieve it, three series of experiments were performed—the first one was carried out in a smooth concrete channel (both main channel and floodplains), the second one was performed in a channel with rough floodplains and in the third test, plants (trees) growing on the floodplains were modelled by aluminum pipes placed on the floodplains. In that study it was confirmed that the vertical distributions of the local mean velocity can be satisfactorily described by logarithmic laws in two first variants of the experiment in the main channel and in the shallow part of it. The logarithmic law does not apply above sloping bank of the deeper section of the channel as well as in the experiment with high vegetation placed on the floodplains. Where the logarithmic law applies, the friction velocity and shear stress may be evaluated. In the central parts of the main channel and floodplains these values are in close agreement with the overall quantities obtained from the uniform flow formula. The model for the velocity distributions in the vegetated channel allows also for the evaluation of the friction velocity in such case. The longitudinal turbulence intensity variations in relations to depth turned out to be typical, i.e., the largest values occur at the channel bottom and the smallest ones at some short distance from the water surface. The intensity of turbulence increases significantly when the bed roughness increases and also at the interface between shallow and deep areas of the channel. Longitudinal sizes of the largest eddies were estimated with the use of autocorrelation functions and the hypothesis of "frozen turbulence". It was shown that in the floodplains these sizes are the smallest in the test with high vegetation (and the largest when the bed was smooth there). At some points the investigated spectral functions were characterized by the existence of an inertial subrange. They satisfied the Kolmogorov "-5/3" power law, which allowed for the use of the results of the theory of locally isotropic turbulence. In case of the vegetated channel, some additional energy supply except for the energy cascade is observed at the low frequencies, which is in agreement with the findings of other authors.

In the later studies, comprehensive measurements of *three-dimensional turbulent* velocities were carried out in the same channel (Czernuszenko et al. 2007; Czernuszenko and Rowiński 2008a, b). Tests were performed in a two-stage channel with smooth main channel bed made of concrete and rough floodplains

Fig. 2 Beginning of collaboration; Moldova, Kishinev, 1990; from the *left* W. Czernuszenko, V. Nikora, P. Rowiński

and sloping banks. Instantaneous velocities were measured with use of a three-component acoustic Doppler velocimeter. On top of determining traditional characteristics of turbulence, the main aim of the study was the recognition of structure of the Reynolds stresses in turbulent open channel flows. Particular attention has been paid to the bursting events such as ejections and sweeps. The bursting phenomenon occurs originally near the buffer layer and then shows a coherent or organized flow structure during its convection process. The probability density distributions of the turbulent velocities were measured at different distances from the bed in the main channel and also above inclined walls. In the main channel the lateral turbulent velocity is seen to follow the normal Gaussian distribution more closely than the remaining two components. Above the inclined walls all distributions turned out to have larger skewness. The probability density distributions of correlations between velocity fluctuations were also calculated. These distributions have long tails and sharp peaks and they fit the theoretical distributions very well. The structure of instantaneous Reynolds stresses was analyzed by quadrant technique with the arbitrarily chosen threshold level. It has been shown that the largest contribution to turbulent stresses comes from the second quadrant (ejection) and the fourth quadrant (sweep). The basic temporal characteristics for quadrant events like the average and maximum time for zero hole size have been determined in the study. Calculations of maximum duration time for all events reveal that those times are larger for even quadrants than for odd quadrants. The same channel was used much later for the studies of the migration of floating particles by means of an image processing technique (see Fig. 3). That work allowed to obtain Lagrangian characteristics in respect to the solid particles transported in a two-stage channel in turbulent regime (Rowiński et al. 2005). One

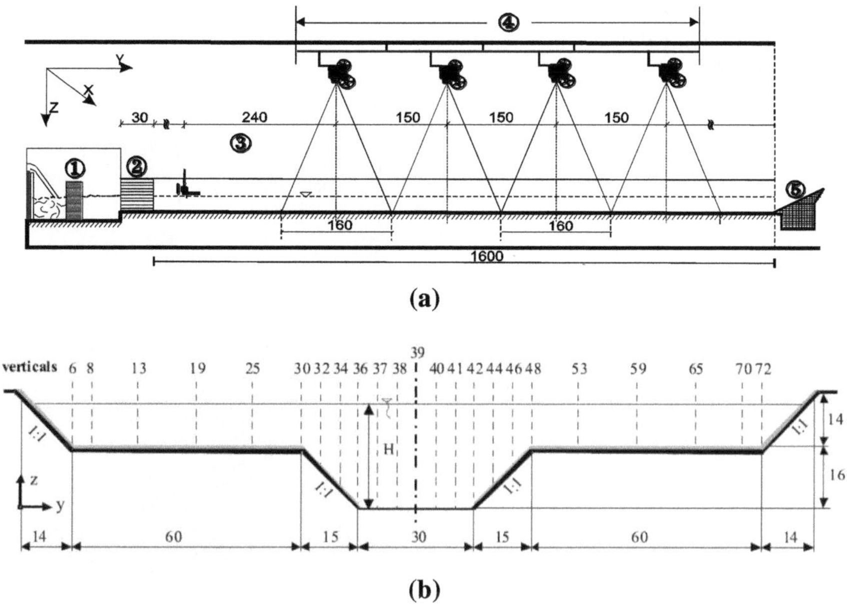

Fig. 3 **a** Scheme of the laboratory setup for the video tracking of particles in a compound channel (*1*—spillway; *2*—flow soothing pipes; *3*—particles' batcher; *4*—system of four digital cameras, *5*—net particles catcher). **b** Scheme of experimental cross-section (Rowiński et al. 1998, 2002)

should also note a very detailed 3D turbulence study that Czernuszenko performed together with Edward Holley (Fig. 4) for the data obtained in a rectangular flume channel (Fig. 5) at the University of Texas at Austin (Czernuszenko and Holley 2007). This is a marvelous, very detailed description of how to deal with such experiments and can with no doubts be recommended to students and young researchers—almost as a cook book for turbulence investigations.

Another issue related to hydraulic conditions under flood is associated with the flow unsteadiness. Czernuszenko's attempt was to fill the gap in interpreting the turbulence characteristics when the stationarity of the investigated random field $u(x, t)$ is violated. We do realize that most intensive transport processes in rivers occur during the passage of a flood wave. They strongly depend on turbulence characteristics so this challenge sooner or later had to be undertaken. An original study was designed to investigate the fluctuating velocity time series during the passage of two various flood waves generated on a lowland river (Wilga River) in central Poland and it was one of the first studies of that kind worldwide (Rowiński and Czernuszenko 1998a, b).

The statistical characteristics of the recorded time series were evaluated with the use of the discrete Fourier transform. It was observed that the horizontal turbulent intensities are larger in the rising branch than the ones in the falling branch of a flood wave. Surprisingly, the damping of turbulence is evident during

Fig. 4 Prof. Czernuszenko with Prof. E. Holley and his wife, Austin, 2002

Fig. 5 View of rectangular flume channel at the University of Texas at Austin (Czernuszenko and Holley 2007)

the passage of a flood wave and the degree of this damping is larger when unsteadiness parameter is larger. The studies performed during the passage of a flood wave led to yet another important result (Rowiński et al. 2000). A method of evaluation of time-dependent friction velocities was proposed and it made use of classic St Venant equations of motion. The key observation was that thus obtained friction velocities significantly exceeded the values obtained from the traditional uniform flow formula. This method gains more and more advocates nowadays (see, e.g., Shen and Diplas 2010; Mrokowska et al., this volume).

Experimental work was not the only approach that attracted Czernuszenko to work upon understanding of turbulence phenomenon. In 1994 a short monograph on modern mathematical models of the processes of transport and pollutant mixing in rivers was published (Czernuszenko and Rowiński 1994) and it turned out to be the first attempt in Poland to use two-equations turbulence models for open-channel flows. That study was also the first to discuss numerical problems with solving depth-averaged Reynolds equations in its k-ε version. A byproduct of that study was the ability to collaborate with leading research centers dealing with modeling of turbulence flows. Czernuszenko took special advantage of the gained experience in turbulence modeling a few years later. Together with his American colleagues he used a three-dimensional computational model, solving Reynolds equations with the k-ε turbulence closure to simulate the flow field in an open-channel near a side-discharge channel (Jia et al. 2002). It was a successful attempt showing the usefulness of the model for simulating 3D recirculating velocity field and it compared well with experimental results. In another paper (Czernuszenko and Rylov 2000), this time together with the researcher from Russia Alexey Rylov, he proposed his own model consisting in the generalization of the old Prandtl mixing-length hypothesis for 3D flows. That generalization consists in transition from scalar mixing length to mixing length second rank tensor. It turned out to be a very useful tool allowing for the calculation of the mean velocity distribution in non-homogeneous turbulent flows. This model has been used in different situations, also for compound channels (Czernuszenko 2001) and is successfully applied up to now (see Czernuszenko and Rylov, this volume).

His international reputation in modeling of open-channel flows led him to chair in transferring the state-of-the-art research and management tools and computational models in the areas of water resources and environmental engineering from National Center for Hydroscience and Engineering to Poland (see Altinakar et al. 2005) within a unique so-called US–Poland Technology Transfer Program financed by US Agency for International Development (Fig. 6). Actually with that institution—NCCHE in Mississippi (USA) Czernuszenko was specially connected—he has lectured there for almost 4 years starting in 1991.

There is one more modeling issue that should be noted as one of basic Czernuszenko's concerns. He has been always interested in searching for simple methods that could be useful for engineers. For example, when studying turbulence characteristics of open-channel flows Czernuszenko has not forgotten basic questions related to classic Prandtl velocity distributions. He has been keen to find relevant parametrization of the logarithmic law for the description of the velocity

Fig. 6 Ceremony of signing the *US–Poland Technology Transfer Agreement* in Warsaw, Poland, in November 5, 2002. From *left* to *right* (*back row*) Prof. M. Altinakar, Prof. Sam Wang, Prof. A. Clark, Prof. K. Rybicki, Prof. Z. Kaczmarek, Prof. W. Czernuszenko; (*front row*) Prof. C. Staton and Prof. B. Ney

profiles as a function of easily measurable variables (see, e.g., Franca and Czernuszenko 2006). As far as we know, this issue has been niggling him all the time and we expect solutions to this problem that will be beneficial for everybody in the field soon.

3 Mass Transport in Open-Channels

Rivers have always been the recipients of various kinds of pollutants resulting from human activities coming from domestic sources, industrial or agricultural effluents, or mining process waters. Over the last two centuries, since the age of industrialization, the massive increase of industrial production accompanied by the growth of large urban populations has led to severe water pollution problems in many rivers. Nowadays, the description and forecast of man's impact on water quality constitutes a key problem especially in well-developed countries. This problem receives an increasing attention of the public, politicians, decision makers nowadays. People have to answer how to achieve any given level and pattern of water quality, in particular watercourses and also how to mitigate the catastrophes that are inseparable elements of civilization. Czernuszenko has been always—throughout his career—part of the discussion and actually he was the initiator of numerous research trends in broadly understood water quality studies. Long lasting research in this subject matter of himself and of the entire team of the Institute of Geophysics PAS and strong international working tights led in 2005 to the preparation of the international monograph (Czernuszenko and Rowiński 2005) which was actually also a kick-off for intensive collaborative studies in the field that started afterwards. But for Czernuszenko all the story started in the 1960s.

In the second half of the twentieth century, as a result of the rapid development of industry, the need for environmental protection began to raise public awareness in Poland. Many new problems appeared in open channel hydraulics requiring a quick solution. One of the major problems was the description of mass transport of various types of substances discharged into surface waters affecting water quality. Meeting the needs of those times, Professor Czernuszenko put these problems in the mainstream of his scientific work. Most of his publications (including the first ones: Czernuszenko 1968, 1971, 1973a, c, d) and both his PhD thesis: "Mass transfer in the open channel flow" (Czernuszenko 1973b) and his habilitation based on the monograph "Dispersion of Pollutants in Rivers and Channels" (Czernuszenko 1983), all are dialing with them. Numerous of his works related to the mass transport processes (e.g., Czernuszenko 1983, 1987a, b, 1990, 2000a, b, 2001; Czernuszenko and Rowiński 1994) have become a great source of knowledge in the field of hydraulics. His publications (both in Polish and English) systematize the knowledge of pollutant transport in rivers and open channels. Complicated mathematical descriptions of the transport process are presented carefully and accurately, but in clear way, thus enabling to use them not only by scientists, but also by engineers and students. To all newcomers starting their adventure with the transport of pollutants (especially unfamiliar with the language of mathematics) the articles: Czernuszenko (2000a, b) (in Polish) or Czernuszenko (1990) (in English, part of the Encyclopaedia of Life Support Systems EOLSS), should be particularly recommended. The articles deal with the modeling of transport and mixing of different pollutants. Different substances which mix well with fluid medium (like mass, heat or electrical current) can be treated as pollutants and their concentration could be described by similar conservation equation. The first article cited above begins with the basic definitions and shows how to receive the equation and how to use it for the description of the concentration field in the water reservoir. The next two articles are an excellent continuation of the first one. Mass transport equations are derived there and then simplified (averaged) to two and one dimensional domains. Basic transport mechanisms, like molecular diffusion, turbulent diffusion, advection and dispersion are discussed in details with a special emphasis put on the differences among the molecular diffusion, turbulent diffusion and dispersion coefficients occurring in the derived equations. We all realize how important is to properly understand and use those coefficients in both scientific and practical applications but at the same time we still observe a big mess in the literature and user manuals of computer programs in this respect. Czernuszenko's works put some order in this field and his works constitute an excellent source of reference.

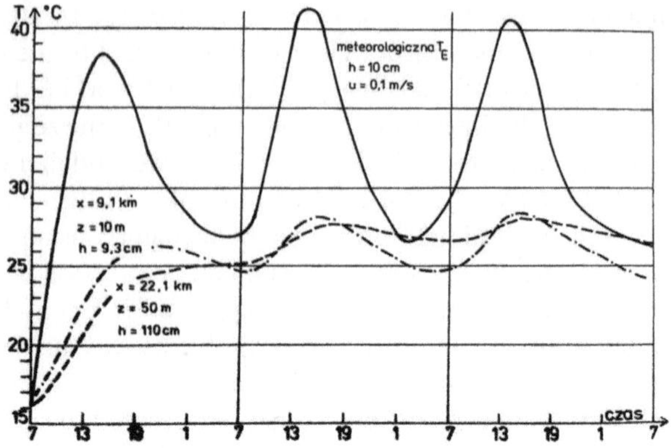

Fig. 7 Variation of natural temperature for 3-day water–air heat exchange, San River (Q = 28.7 m^3/s). Czernuszenko and Paradowska (1987)

One of the negative effects of industrial development have become thermal pollutions caused by discharges of heated water into rivers. From the works of Prof. Czernuszenko devoted to thermal pollution, particularly noteworthy is the paper: Czernuszenko and Paradowska (1987). That article deals with a two-dimensional spreading of thermal pollutants in a shallow river by means of a mathematical model taking into account the mechanism of convection, dispersion and heat exchange at the air–water (see Fig. 7) and soil–water interface. A substantial part of the paper is devoted to the comprehensive analysis of the air–water exchange with the description of all necessary empirical relationships. Also some examples of the applications of the presented mathematical model for rivers are discussed. Czernuszenko also took part in a series of experiments carried out on the Narew, Vistula and San Rivers, where three big Polish power plants: Ostrołęka, Połaniec and Stalowa Wola are located (see, e.g., Fig. 8).

River water is used for cooling purposes all over the world so the methods of analyses of the spread of a thermal plume are of extreme importance. The measurements of Czernuszenko and his collaborators included not only the cross sections, velocity profiles and two-dimensional (and sometimes even three-dimensional) temperature fields, but also the previously mentioned turbulence characteristics. The results are used and analyzed by Czernuszenko in several papers. He has also conducted a number of computational experiments. For example, the effect of the river bed profile on the temperature field was studied in details in: Czernuszenko and Paradowska (1987) and is presented fragmentarily in Fig. 9.

One of the major achievements of Prof. Czernuszenko was using the curvilinear (natural) coordinates system (Fig. 10) in the computations of spreading of pollutants in natural rivers. The lateral component of velocity greatly affects the mixing process, especially in natural, irregular, meandering rivers. At those times the measuring of this component or its calculation was extremely difficult and the use of the orthogonal curvilinear coordinate system allowed overcoming these difficulties (Czernuszenko 1986a, b, 1987a, b).

An important part of Czernuszenko's investigations were also related to the spread of pollutants in rivers over long distances. In such case, 1D approach is

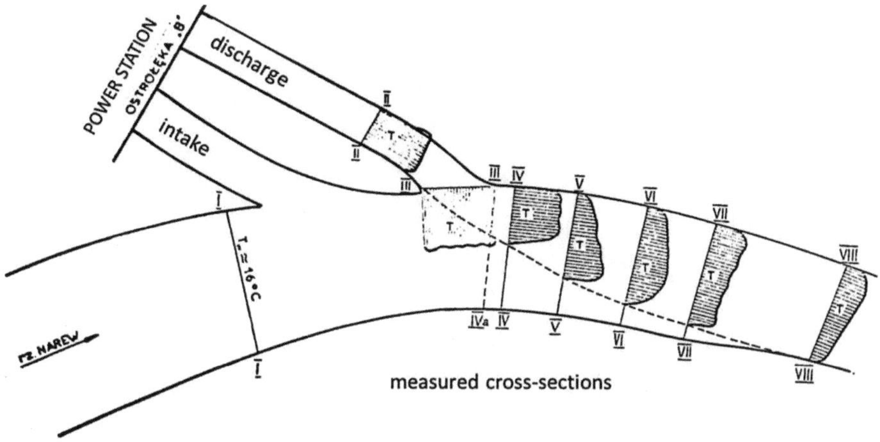

Fig. 8 Heated water jet behind the discharge from a thermal power plant in Ostrołęka (Narew River)—experimental results on the temperature field

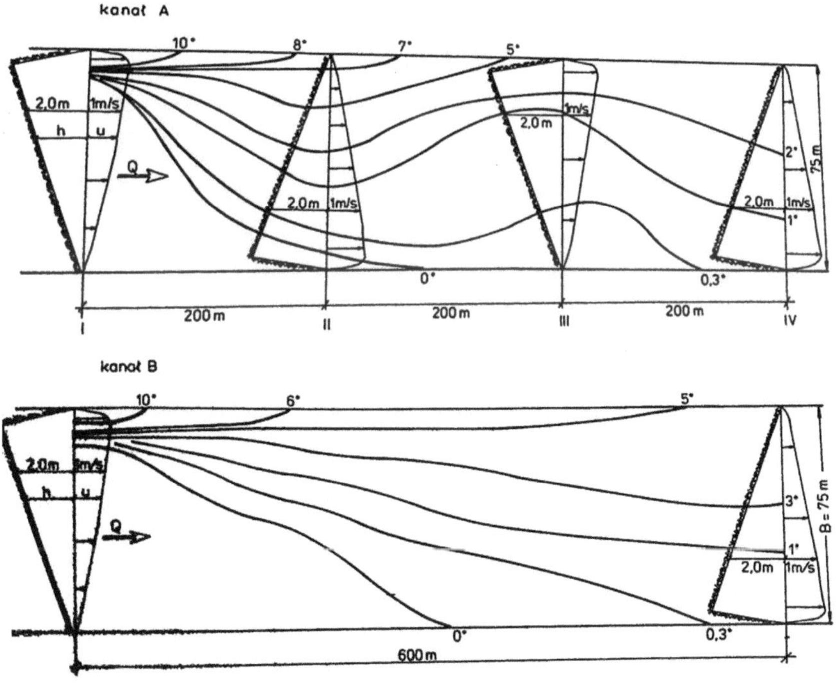

Fig. 9 Temperature distribution in two channels characterized by different bathymetries (Czernuszenko and Paradowska 1987)

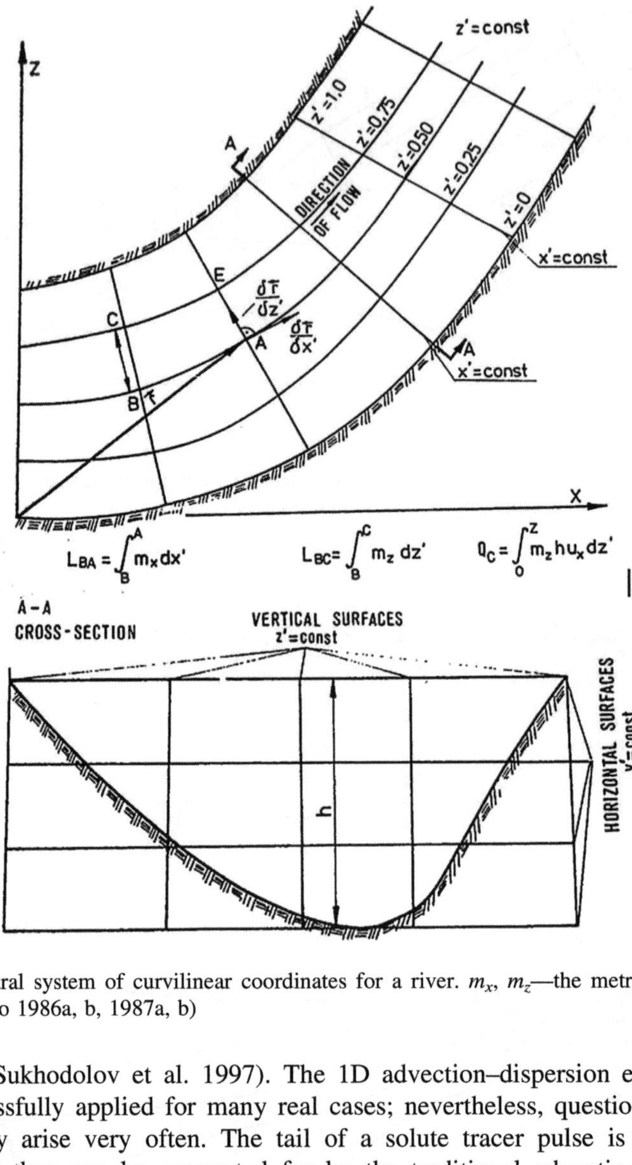

Fig. 10 Natural system of curvilinear coordinates for a river. m_x, m_z—the metric coefficients (Czernuszenko 1986a, b, 1987a, b)

sufficient (Sukhodolov et al. 1997). The 1D advection–dispersion equation has been successfully applied for many real cases; nevertheless, questions about its applicability arise very often. The tail of a solute tracer pulse is often more pronounced than can be accounted for by the traditional advection–dispersion model. A common method for simulating these long tails has been to allow for storage zones along the stream channel. These storage zones are assumed to be stagnant relative to the longitudinal flow of the stream and to obey a first-order mass transfer type of exchange relationship. Very often a quicker decrease of the concentration maximum than follows from traditional Fickian equation is observed. Also a nonlinear growth of the concentration distribution variance and dependence of the dispersion coefficient on time has been often manifested in experimental studies. The shift of the lower parts of temporal concentration distribution may be possibly explained by the dead-zone trapping and it motivated Czernuszenko to study a model taking into account this phenomenon. The dead-zone model has become increasingly popular for the calculation of the longitudinal dispersion of a solute in a river with irregular cross-sections and Czernuszenko's team was one of the first to study the model in details. This model is the reflection of the existence in the rivers of stagnant zones of water that are stationary relative to the faster moving waters near the center of the channel. The model results coincide well with experimental data (see Czernuszenko and Rowiński 1997; Czernuszenko et al. 1998).

Transport processes were treated by Czernuszenko from various perspectives. One way was to draw from the theory of two-phase flows and quite symbolically this is what ties the studies of the first doctoral dissertation that he supervised with the last one. At the beginning as an intermediate step towards the simulation of sediment transport and understanding of bed dynamics, Czernuszenko and Rowiński were interested in the determination of the paths of single grains in turbulent open channels or river flows (see, e.g., Rowiński and Czernuszenko 1999). In the early nineties the literature on sediment transport had led to major revisions in a view of turbulent transport of individual solid particles that was based on the equations proposed in the forties by Tchen. And the authors proposed a novel approach in comparison to other studies in literature. It is a common knowledge that the sediment grains may move in the form of rolling, saltation and suspension. The manner of their motion depends on many factors, like the position at which the particle initiates to move, its size, shape, density relative to the carrier fluid, as well as turbulent properties of the flow. In general, the way in which a particle moves is a direct effect of the role and significance of particular forces acting on a grain and also the variability of other solid particles concentration in its neighbourhood. It is expected that every mode of particles' movement should be described with the use of the same dynamic equations and this is what was achieved under supervision of Czernuszenko. When dealing with the particles contacting the bed, some supplementary model of a stochastic nature, responsible for the collision with solid boundary, was proposed. The proposed model included the term standing for the turbulent diffusion driven force and its main idea is based on the theory of turbulent flow of a dilute two-phase suspension derived from the first principles by Lee (1987), with whom Rowiński happened to collaborate that time. Czernuszenko was quick to realize the usefulness of the derived theory and he decided to look at it—the drift (diffusion) mechanism—from the perspective of mass conservation equation (Czernuszenko 1998). In sediment-laden flows with a concentration gradient the drift velocity arises as a result of nonsymmetrical turbulent fluctuations. Its existence is reflected in the proposed new model consisting in the mass conservation equation including three mechanisms—advection, diffusion and the drift.

This subject area came back in Czernuszenko's studies quite recently. Together with his last PhD student, Robert Bialik, he approached the problem again taking into account the new developments in other research centers. And as usually in

Professor's career he opened the gate to the new research path for a young scientist showing him the right way to progress his career. Among others, Czernuszenko and Bialik (2008) applied a scheme of particle–particle collision in a Lagrangian model of saltating grains in rivers. Within their studies the new and particularly fascinating results concerned the influence of the turbulence structure on the trajectories of saltating particles in open-channel flows. A special procedure based on the known spectra of turbulent kinetic energy has been designed to generate field velocities. The proposed approach allows to study the behaviour of saltating particles under the influence of turbulence for a broad range of the sizes of sediment particles. The analyzed distributions of dimensionless saltation lengths and heights show that with increase in turbulence intensities the particle movement becomes more unpredictable and the analyzed dimensional saltation lengths and heights significantly change depending on particle Reynolds number.

All the methodology initiated by Czernuszenko is well established now and it is in quick progress and Czernuszenko all the time gladly shares his skills and knowledge with younger collaborators.

4 The End

Professor Włodzimierz Czernuszenko is a scientist of the highest level—creating a legacy of hydro-environment research with grateful friends and researchers both in Poland and all over the world. He has tackled the scientific problems on a broad front, working in an interdisciplinary field that is difficult to penetrate, and he has been the leading figure in the advances made there. He made pioneering contributions in a number of fields of hydraulics. He has the unique ability to recognize and exploit connection between phenomena in different fields of hydro-physics. With all his skills he has played an important role in the development of science and engineering in Poland, particularly in the time of transformation to a new democratic system. Yet he is modest about it all and we believe this is the dominant feature of His character. This chapter is by no means a biographical sketch; rather it bears testimony to his role on the development of environmental hydraulics, the work that is continued by himself and a group of people who have been privileged to work with Him.

Acknowledgments This work was partly supported by grant No. N N306 658140 from National Science Centre, Poland.

References

Altinakar MS, Czernuszenko W, Rowiński PM, Wang SSY (2005) Computational modeling for the development of sustainable water-resources systems in Poland, US–Poland Technology Transfer Program, Monographic Volume, Publications of the Institute of Geophysics, vol E-5(387). Polish Academy of Sciences, Warsaw

Bialik RJ, Czernuszenko W (2008) Particle–particle collision for Lagrangian modeling of saltating grain: theoretical background. In: Rowiński PM (ed) Hydraulics methods for catastrophes: floods, droughts, environmental disasters, Monographic Volume, Publications of the Institute of Geophysics, vol E-10(406), Polish Academy of Sciences, Warsaw, pp 29–37
Czernuszenko W(1968) Study of diffusion in two-dimensional flow in open channel (in Polish). Wiadomości służby hydrologicznej 4(3–4)
Czernuszenko W (1971) The process of diffusion at the channel's wall with turbulent flow (in Polish). Wiadomości służby hydrologicznej i meteorologicznej 19(1):19–27
Czernuszenko W (1973a) Lateral diffusion of particles floating in the open channel (in Polish). Archiwum Hydrotechniki 20(1):21–32
Czernuszenko W (1973b) Mass transfer in the open channel flow (in Polish). PhD thesis, Warsaw University of Technology
Czernuszenko W (1973c) Mixing processes in wide open channels. J Hydrol 19(3):259–268. doi:10.1016/0022-1694(73)90084-X
Czernuszenko W (1973d) Transversal diffusion in the open channel. In: Proceedings of the XV Congress IAHR, Istanbul
Czernuszenko W (1983) Dispersion of pollutants in rivers and channels (in Polish). Technical report, IMGW
Czernuszenko W (1986a) Diffusion and dispersion in open flow (in Polish). Archiwum Hydrotechniki XXXIII(3):229–245
Czernuszenko W (1986b) Natural mixing in rivers (in Polish). Archiwum Hydrotechniki XXXIII(1–2):22–44
Czernuszenko W (1987a) Dispersion coefficient identification. In: Proceedings of the XXII congress IAHR, Lausanne, Ecole Polytechnique Fdrale de Lausanne, Switzerland
Czernuszenko W (1987b) Dispersion of pollutants in rivers. Hydrol Sci J 32(1):33–42
Czernuszenko W (1990) Dispersion of pollutants in flowing surface water. In: Encyclopedia of fluid mechanics, surface and groundwater flow phenomena, vol 10, Gulf Publishing Company, Houston, London, Paris, chap 4, pp 119–168
Czernuszenko W (1998) The drift velocity concept for sediment-laden flows. J Hydraul Eng 124(10):1026–1033
Czernuszenko W (2000a) Modelling principles of the transport of pollutants (in Polish). Przegląd Geofizyczny XLV(1):33–42
Czernuszenko W (2000b) Transport of pollutants in open channels (in Polish). Przegląd Geofizyczny XLV(2):150–193
Czernuszenko W (2001) Quality models of river waters. Monographic volume, Publications of the Institute of Geophysics, vol E-2(325), Polish Academy of Sciences, Warsaw, pp 53–70
Czernuszenko W, Holley E (2007) Open-channel turbulence measurements with a three-components acoustics doppler velocimeter. In: Rowiński PM (ed) Transport phenomena in hydraulics. Monographic volume, Publications of the Institute of Geophysics, vol E-7(401). Polish Academy of Sciences, Warsaw, pp 49–79
Czernuszenko W, Lebiecki P (1980) Turbulent characteristics of stream in open channel (in Polish). Archiwum Hydrotechniki 27(1):19–38
Czernuszenko W, Lebiecki P (1989) Turbulence in the river flow (in Polish). Archiwum Hydrotechniki 36(1–2):17–34
Czernuszenko W, Paradowska E (1987) Two-dimensional model for water temperature of river (in Polish). Przegląd Geofizyczny XXXII(2):131–145
Czernuszenko W, Rowiński PM (1989) Measurements of turbulence in heated water jet. In: Proceedings of 23rd IAHR congress, Ottawa, pp A71–A76
Czernuszenko W, Rowiński P (1994) Modern mathematical models of turbulent flows and pollutant spreading in rivers (in Polish), Monografie Komitetu Gospodarki Wodnej Polskiej Akademii Nauk, vol 6. Oficyna Wydawnicza Politechniki Warszawskiej, Warsaw
Czernuszenko W, Rowiński PM (1997) Properties of the dead-zone model of longitudinal dispersion in rivers. J Hydraul Res 35(4):491–504

Czernuszenko W, Rowiński PM (2005) Water quality hazards and dispersion of pollutants. Springer, USA
Czernuszenko W, Rowiński PM (2008a) Reynolds stresses in a compound open channel flow—flume experiments. In: Altinakar MS, Kokpinar MA, Aydin I, Cokgor S, Kirkgoz S (eds) Proceedings of the international conference of fluvial hydraulics, Rive Flow 2008, pp 289–297
Czernuszenko W, Rowiński PM (2008b) Shear stress statistics in a compound channel. Arch Hydro-Eng Environ Mech 55(1–2):3–27
Czernuszenko W, Rylov A (2000) A generalization of prandtl's model for 3d open channel flows. J Hydraul Res 38(2):133–139
Czernuszenko W, Rowiński PM, Sukhodolov A (1998) Experimental and numerical validation of dead-zone model for longitudinal dispersion in rivers. J Hydraul Res 36(2):269–280
Czernuszenko W, Kozioł A, Rowiński PM (2007) Measurements of 3d turbulence structure in a compound channel. Arch Hydro-Eng Environ Mech 54(1):3–21
Franca MJ, Czernuszenko W (2006) Equivalent velocity profile for turbulent flows over gravel riverbeds. In: Ferreira RML, Alves ECTL, Leal JGAB, Cardoso AH (eds) Proceedings of the international conference of fluvial hydraulics, Rive Flow 2006, Braunschweig, Taylor & Francis Group, London, pp 189–197
Grinvald DI (1974) Turbulence of open channel flows (in Russian). Hydrometeoizdat
Imamomoto H (1975) Universal representation of turbulence characteristics in free surface shear flow. In: Proceedings of 16th congress IAHR, Sao Paolo
Jia Y, Czernuszenko W, Wang SSY (2002) Simulation of three-dimension side discharge into an open channel. Arch Hydro-Eng Environ Mech 49(1):3–21
Lee SL (1987) A unified theory on particle transport in a turbulent dilute two-phase suspension flow—II. Int J Multiph Flow 13(1):137–144
Nikora VI, Czernuszenko W, Sukhodolov A, Rowiński PM (1993) Structure of turbulence in small rivers. In: Wang SY (ed) Advances in hydro-sciences and -engineering, vol 1. Washington, pp 1110–1115
Raichlen F (1967) Some turbulence measurements in water. J Eng Mech Div 93(2):73–97
Rowiński PM, Czernuszenko W (1998a) Experimental study of river turbulence under unsteady conditions. Acta Geophysica Polonica XLVI(4):462–480
Rowiński PM, Czernuszenko W (1998b) Friction and turbulent velocities in a river under unsteady conditions. In: Proceedings of the 3rd international conference on hydroscience and engineering, Cottbus/Berlin, Germany, http://www.bauinf.tucottbus.de/ICHE98/proceedings
Rowiński PM, Czernuszenko W (1999) Modeling of sand grain paths in a turbulent open channel flow. In: Proceedings of the 28th IAHR congress, CD-ROM, Graz, Austria
Rowiński PM, Czernuszenko W, Kozioł A, Kuśmierczuk K, Kubrak J (1998) Longitudinal turbulence characteristics in a compound channel under various roughness conditions. In: Proceedings of the 3rd international conference on hydroscience and engineering, Cottbus/Berlin, Germany
Rowiński PM, Czernuszenko W, Pretre JM (2000) Time-dependent shear velocities in channel routing. Hydrol Sci J 45(6):881–895
Rowiński PM, Czernuszenko W, Kozioł A, Kubrak J (2002) Properties of a streamwise turbulent flow field in an open two-stage channel. Arch Hydro-Eng Environ Mech 49(2):37–57
Rowiński PM, Czernuszenko W, Krukowski M (2005) Migration of floating particles in a compound channel. In: Czernuszenko W, Rowiński PM (eds) Water quality hazards and dispersion of pollutants. Springer, USA, pp 121–141
Schuyf JP (1966) The measurement of turbulent velocity fluctuations with a propeller-type current meter. J Hydraul Res 4(2):37–54. doi:10.1080/00221686609500005
Shen YM, Diplas P (2010) Modeling unsteady flow characteristics for hydropeaking operations and their implications on fish habitat. J Hydraul Eng 136(12):1053–1066
Sukhodolov A, Nikora VI, Rowiński PM, Czernuszenko W (1997) A case study of longitudinal dispersion in small lowland rivers. Water Environ Res 69(7):1246–1253

Hydraulic Problems in Flooding: From Data to Theory and from Theory to Practice

Donald Knight

Abstract The value of integrating mathematical modelling with experimental work in both the laboratory and field is illustrated through the development of a software tool that deals with key practical issues related to rivers in flood. The Conveyance Estimation System software (www.river-conveyance.net) is aimed primarily at estimating the stage-discharge relationship, the distribution of depth-averaged velocity and boundary shear stress across channels of any prismatic shape for both inbank and overbank flows. The practical problems in obtaining data and the theoretical issues in identifying relevant flow parameters for stream wise and planform vorticity, turbulence shear stresses and frictional resistance are highlighted. The significance of these and their relevance to other hydraulics problems are noted. The issues involved in moving from data to theory (or vice-versa), then to practical application, are described in general terms, beginning with how to develop a model as a research tool, testing it against different data sets, through to using the model in practice with embedded tools. The tools deal with uncertainties in estimates and give guidance on roughness coefficients in natural channels.

Keywords Rivers · Floods · Modelling · Software · Roughness · Vorticity · Turbulence

D. Knight (✉)
Department of Civil Engineering, The University of Birmingham,
Edgbaston, Birmingham B15 2TT, UK
e-mail: d.w.knight@bham.ac.uk

P. Rowiński (ed.), *Experimental and Computational Solutions of Hydraulic Problems*, GeoPlanet: Earth and Planetary Sciences, DOI: 10.1007/978-3-642-30209-1_2,

1 Introduction

Experimental and computational approaches are frequently used in hydraulics to solve certain types of practical problem that are not amenable to a single approach. When used together, they offer an instructive way of dealing with flow-related issues in a fundamental manner on account of using actual data, theoretical tools, as well as numerical models that may be used for further analysis and comparative studies with other data. Although many engineers and scientists utilize and maybe rely on numerical models to solve many of their problems, there is an increasing need to go back to experimental data, not least because most 3-D mathematical simulations rely heavily on empirical information for key coefficients and, in some cases, cannot be validated properly due to lack of knowledge concerning complex turbulence phenomena that occur in those cases requiring investigation.

Whatever the general approach that may be adopted to solve a particular problem, the process itself needs to be appreciated and understood. Although the focus of this article concerns flooding, the various steps in blending experimental data, theoretical knowledge and modelling know-how to achieve a particular goal are described herein in general terms in order to show that whatever the problem is, the principles and process of moving from data to theory, and then from theory into practice, are relevant to whatever problem is being tackled.

Many authors have written about modelling flow in rivers, e.g. Anderson et al. (1996), Ashworth et al. (1996), Chang (1988), Ikeda and McEwan (2009), Knight (1996, 2008), Knight et al. (2009, 2010a) and Nakato and Ettema (1996). A strong theme to emerge is that a hybrid model, involving both physical and mathematical models, is not a thing of the past but still relevant as demands for precision and authenticity in numerical models increase. High quality graphical output from a CFD model may look convincing to some but is no substitute for actual data, with all its attendant shortcomings, cost implications and effort.

2 What are Some of the Problems in Modelling Flows in Rivers?

Defining the issues involved in solving a problem is often a necessary step in arriving at a solution. In the case of flooding, there are a number of particular issues that set it apart from other hydraulic problems and these need to be recognised and appreciated during any model calibration. Four of these are now described briefly.

2.1 High Discharges

Flood hydrology deals with extreme events that occur infrequently and involve high discharges. These combine to make it difficult to monitor floods effectively, since they need to be anticipated and suitable preparations of equipment and personnel made beforehand. Moreover, it is not just the water levels, or water surface slopes, that need to be measured in these extreme events, but also velocity distributions and turbulence structure as well. Measuring these are difficult enough under normal circumstances, and considerably more so under extreme conditions. Consequently, acquiring high quality data on the flow field, flow resistance and turbulence, in sufficient temporal and spatial detail, for subsequent use in a model, is not without difficulty. As a result, the seemingly simple hydraulic task of extending a rating relationship for flows outside the observed range at a specific gauging station, is not straightforward, as well documented by Ramsbottom and Whitlow (2003). Some practical and theoretical issues on stage-discharge relationships are dealt with in ISO 1100-2 (2010) and Knight (2006a), respectively.

2.2 Channel Geometry and Roughness

Assuming a stable river cross-section and ignoring sediment issues, floods introduce one particular complicating feature related to the geometry of the cross-section which must be recognised before any model calibration takes place with respect to roughness. Moderate flood discharges, typically with a return period of 2–3 years, will generally cause a river to flow overbank, with the water inundating any adjoining floodplains. Although the precise determination of bankfull flow is not easy, as illustrated by Navratil et al. (2004), it is commonly used as a parameter in geomorphological studies. At higher discharges, the river and its floodplains will both convey flow as part of the natural alluvial process. The continuum from inbank, through bankfull, to overbank flow conditions needs to be appreciated in order to understand the effect on key phenomena such as resistance, dispersion and wave speed. The discontinuity in cross-sectional shape at the bankfull stage, where the sudden increase in wetted perimeter, without a corresponding equivalent increase in area, causes the hydraulic radius to decrease rapidly as the flow goes just overbank. Since R (=A/P) is used widely in most open channel problems involving inbank flows, the uncritical use of R when dealing with overbank flow become problematic. Figures 1 and 2 illustrate the effect of this on resistance coefficients for a reach of the River Severn at Montford bridge, where the value of R decreases rapidly from around 4 to 2 m at the bankfull stage. The overall roughness does not in reality decrease, only R does, leading to an apparent decrease in the overall Manning n, as shown in Fig. 1. Likewise, Fig. 2 shows a dramatic decrease in the value of f by a factor of 2, as well as an interesting looped pattern, all due to this effect. The departure from the customary Moody type diagram of f v Re for various k_s/4R, makes any attempt at trying to determine a Nikuradse k_s value from these data pointless, unless it is done with due care.

A further feature is also revealed in Figs. 1 and 2, and shown more explicitly in Figs. 3 and 4. A distinction should be made between overall friction factors (using

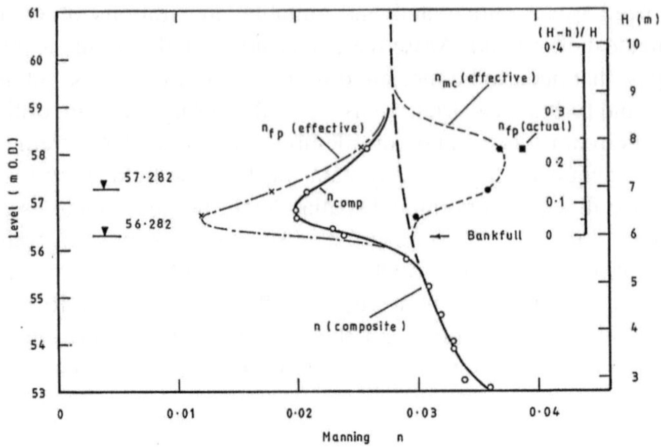

Fig. 1 Variation of overall and zonal Manning's n values with depth for overbank flow in the River Severn at Montford bridge (after Knight et al. 1989, 2010a)

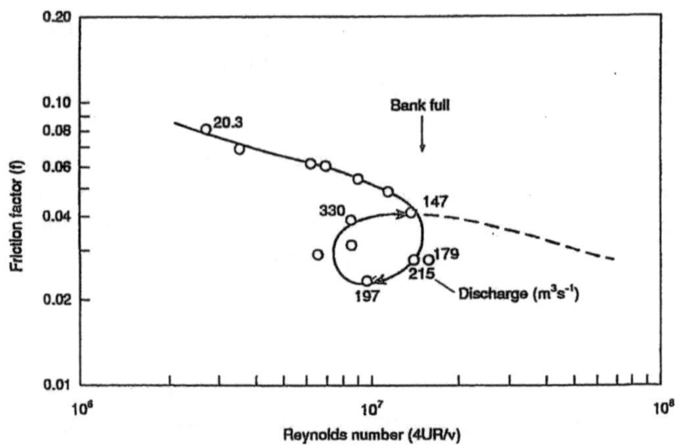

Fig. 2 Variation of overall Darcy-Weisbach resistance coefficient, f, with Reynolds number for River Severn at Montford bridge for discharges from 20.3 to 330 m^3s^{-1}, showing transition from in bank to overbank flows (bank full, $Q_b = 170\ m^3s^{-1}$) (after Knight 2006a)

cross-section parameter values), zonal friction factors (using sub-area values) and local friction factors (using depth-averaged velocity based values). The lateral variation in local friction factor ($f = 8\tau_b/(\rho U_d^2)$ shown in Fig. 3, is based on measured velocities and boundary shear stresses taken from a laboratory study of overbank flow. When several series are plotted in the same f v Re manner as in Figs. 2, 4 reveals a herring-bone pattern of curves for both overall and local values for each floodplain width (B/b value). The reduction in f values observed in the field data of Fig. 2 can then be understood in terms of the reduction in R and the inadvisability of using a simple resistance law. Figure 5 shows the ratio between

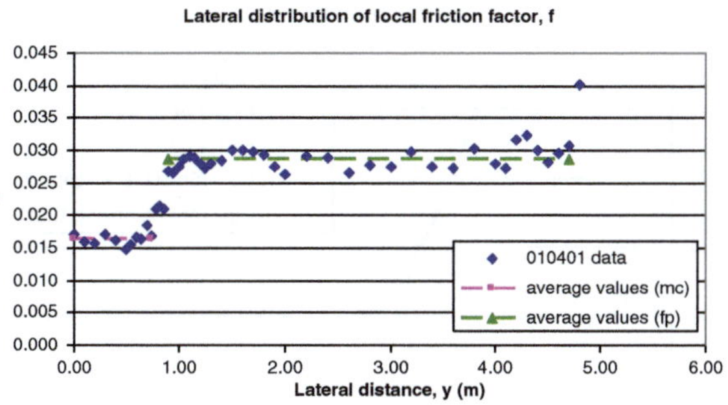

Fig. 3 Lateral variation of local friction factors: main channel (mc) to floodplain (fp)

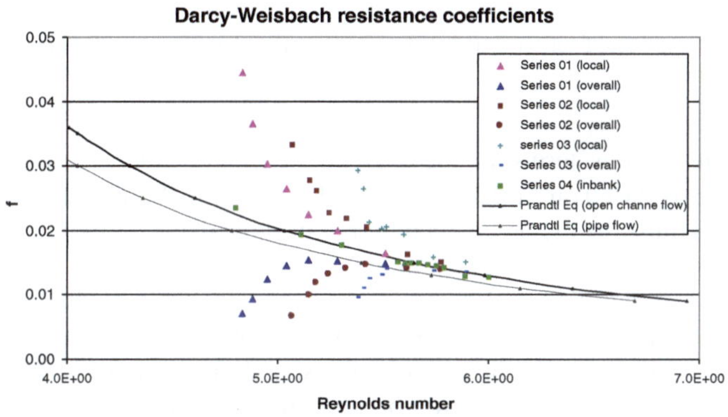

Fig. 4 Overall and local friction factors for FCF data (Series 01–04)

the flood plain and main channel friction factors, deduced from plots similar to Fig. 3 for the same overbank experimental series (01–03) shown in Fig. 4. These all follow a similar trend, with f_{fp}/f_{mc} increasing as the relative depth, Dr, decreases.

The role of roughness and use of resistance coefficients when calibrating river models is another issue worthy of reflection and mature knowledge, as indicated by Morvan et al. (2008) and McGahey et al. (2009). Specialist knowledge is also required when dealing with the additional resistance that may arise from either sediment bed forms (flow related) or drag forces on different types of vegetation.

2.3 Unsteadiness in Flow

The nature of floods means that data must be collected under unsteady flow conditions, making measurements at sufficiently comprehensive temporal and spatial details difficult. Figure 6, taken from Knight (1981), shows the measured

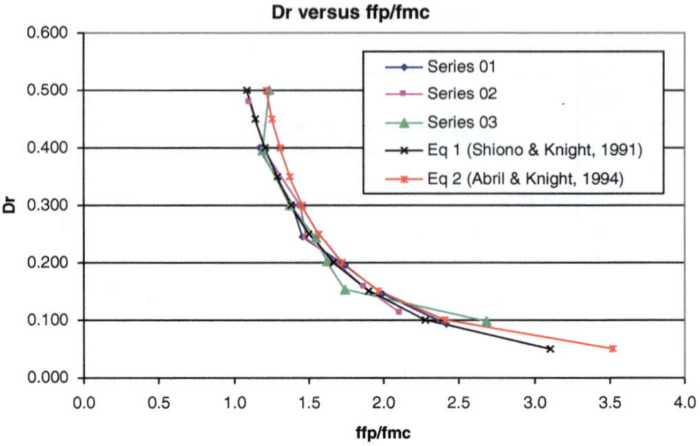

Fig. 5 Variation of local friction factors between the main channel and a floodplain with relative depth, *Dr*

values of the various terms in the 1-D St Venant equations, used in a model calibration study based on a tidal estuary, as described by Wallis and Knight (1984). This serves to illustrate the precision required in slope data required to subsequently determine resistance coefficients accurately. Validation issues in 1-D flood routing models, using either the St Venant or Variable Parameter Muskingum-Cunge (VPMC) equations, are dealt with by Knight (2006b) and Tang et al. (2001).

Figure 7, taken from Sellin and van Beesten (2004), shows a typical looped resistance relationship for vegetation on a floodplain, where the resistance is seen to be less during the flood recession (falling limb), due to the vegetation being flattened by the flood in its first progress overbank when water initially inundates the floodplain (rising limb). See Sellin and van Beesten (2004) and Knight (1981) for further details of the data acquisition and resistance analysis, together with Gunawan et al. (2010), for information concerning the River Blackwater studies that have continued investigating hydraulic resistance, turbulence, particle image velocimetry (PIV) and the modelling of floodplain vegetation.

2.4 Data for Model Calibration

The previous three sections serve to illustrate the care that needs to be taken in calibrating any model, even with respect to a single parameter, such as resistance. This is further explored in Sects. 3 and 4 where data acquisition is related to the philosophy and technical objectives of the model. As is often the case, a model calibration may change as further data become available, or understanding of the original data and maybe other related data improves. An example of this may be seen in Fig. 5, where the original equation proposed in 1991 by Shiono and Knight (1991), relating the ratio of local friction factors, laterally averaged along the

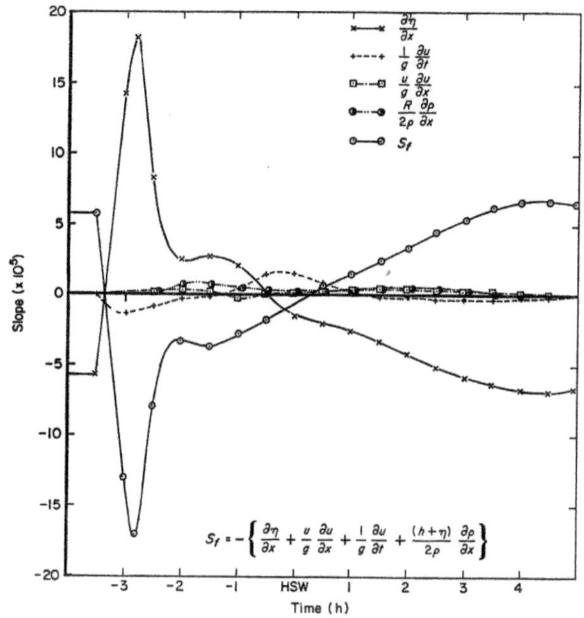

Fig. 6 Resistance data for Conwy estuary showing terms in the 1-D St Venant eq. (after Knight 1981)

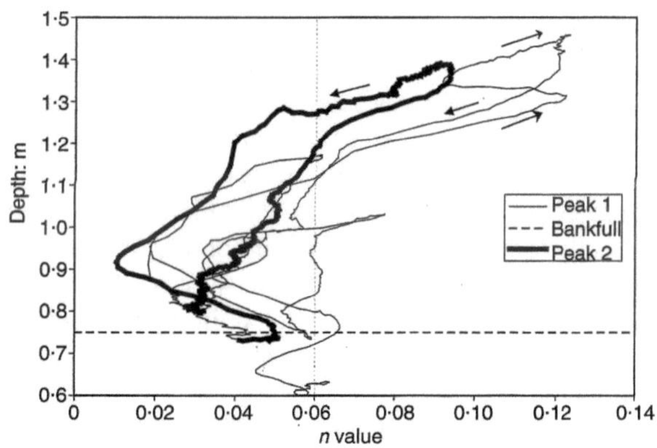

Fig. 7 Looped resistance relationships for a two-stage channel with vegetated floodplains (after Sellin and van Beesten 2004)

wetted perimeters of the main channel and floodplain, with the relative depth, *Dr* (=floodplain depth/main channel depth), was later examined in 1996 by Knight and Abril (1996). This was subsequently refined in 2004 by extensive testing against all the FCF data, as shown by Abril and Knight (2004). Atabay and Knight (2006) have continued this review still further, summarising analysis of many other data sets for overbank flow. More recently, Sun and Shiono (2009) and Knight et al. (2010b) have extended the modelling to include the effects of vegetation.

3 General Approach to Solving Problems

3.1 Defining the Problem

Defining what the fundamental issues are in relation to a particular problem is well worth attempting. Even if the mathematical equations cannot either be formulated or solved, it is always valuable to see where the actual points are that assumptions have to made about a physical process, or use made of a borrowed piece of theory, to close the equations for solution. It also helps to avoid going down well worn tracks, however commonly used by others, and to see afresh where the key difficulties arise. Of course, it is sensible to review the literature critically, as the same problem may already have been solved and a wealth of data exist, but without a grasp of the basic physics and underlying mathematics, such a review will be less valuable than it might have been without this preliminary step.

3.2 Acquiring Data

One usually discovers that for a particular problem requiring solution, there is not enough data, or in some cases none at all. The enterprise of acquiring sufficient data with which to solve the problem is then another process needing careful consideration. Measurements in the natural environment at full scale are clearly desirable, but expensive and rarely comprehensive enough, whereas laboratory based experiments lack all the complexity occurring in natural rivers, though are more controllable and can provide high quality data. It is always good practice to actually assess oneself the quality of any primary data, using ones own experience, data mining techniques and other devices for sorting out good data from poor.

It is too commonly accepted in hydrometry that data acquisition is about measuring what is easy, rather than what is really required for use in modern computational models. For example, velocity and water levels are relatively easy to monitor via ultrasonics and ADCP. However, turbulence, secondary flows and boundary shear stress are much more difficult, as they require measuring accurately temporal variations in water surface slope, Reynolds stresses and flow structure. Measuring the migration rate of bed forms or dispersion parameters, where gradient terms have to be obtained very precisely, bring similar difficulties.

The design of new apparatus, or modifications to existing facilities, requires consideration as to its purpose, likely errors and the limits of all measuring equipment. To illustrate this, consider how uniform flow was established for overbank flow at the specified relative depths in Figs. 4 and 5. A preliminary series of experiments was undertaken in which for a given channel geometry and discharge, the tailgate was adjusted to give 4 longitudinal water surfaces profiles (two M1 and two M2 profiles close to the estimated uniform depth). The mean water surface slope and depth were then plotted against tailgate level and the tailgate setting which gave a mean water surface slope equal to the bed slope interpolated from the graph. The normal depth was also interpolated. This procedure was repeated for around 20 discharges to obtain a smooth stage-discharge curve. The particular depths corresponding to the required relative depths, *Dr*, then gave the discharges for each main experiment in which detailed velocity, shear stress and turbulence measurements were to be made. Finally, during each main experiment, the water surface slope was again measured to ensure that uniform flow had in fact been achieved. Only occasionally were very minor adjustments made to the tail-gate to ensure precise uniformity. All flumes and wind tunnels were also designed to be long enough to ensure fully developed flow, with full secondary flow development necessary for all boundary shear stress and turbulence measurements. Post processing checks were run daily to ensure compliance with <1–2 % error in water surface slopes, $<2 \sim 3$ % in integrated local velocities to match with the input discharge (Q) and $<3 \sim 5$ % in boundary shear to match with the total shear stress ($\rho g R S_o$). Experiments were repeated if these errors were ever exceeded. See Knight and Demetriou (1983) and Knight and Shiono (1990) for further details.

Large scale experiments usually involve collaborative work at national or international level to defray costs and to maximise technical expertise. Monographs, technical reports and books are then a useful source of information, as illustrated by Ikeda and Parker (1989), Ikeda and McEwan (2009) and RIBAMOD (1999). Websites also make the acquisition of data easier task than it once was, as shown by the list of those on flooding cited in Knight and Samuels (2007). See also www.flowdata@bham.ac.uk for the FCF data. General topics on flooding are discussed by Bronstert (2006), Knight and Shamseldin (2006) and Knight et al. (2006).

3.3 Recognising Physical and Theoretical Concepts

The first step in the solution of a problem is to identify the physical and theoretical concepts involved, which are also pre-requisites before acquiring any data. For novel research work, it may not be known a priori what types of flow mechanisms are actually involved and what should be effectively measured. Experiments are usually undertaken with some pre-conceived objective in mind, focusing on some general concept, or investigating in detail some parameter of particular interest. Thus the 'definition of the problem' and 'acquiring data' are linked to the recognition of 'concepts' in a fundamental way as illustrated in Fig. 8.

It is also possible with certain types of simple practical problem, especially those undertake routinely, to proceed to solve it directly by using a pre-prepared model. For many research problems however, the construction of a model is often one of the objectives and therefore a final step in the process, not the first. It is then debatable where one starts—is it with data, concept or equations? In reality each is important, as illustrated elsewhere by Knight (2008), and discussed further in Sect. 4, and 5, describing the construction, testing and use of models generally.

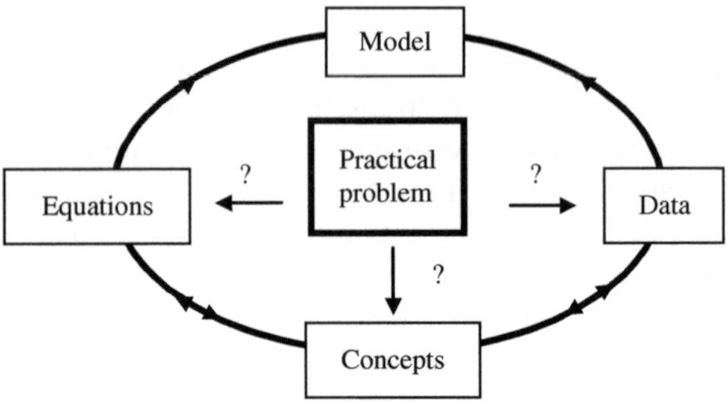

Fig. 8 Solving a practical problem–where to start?

Another way of looking at solving problems is to regard it as an art in applying the many topics within theoretical fluid mechanics to the particular problem under consideration. This is illustrated for river engineering in Fig. 9, by way of an example. Knowledge and understanding have to be gained from both 'river banks', one marked 'theoretical fluid mechanics' and the other 'practical problems'. Concepts, data, equations and ideas may arise in one area that may be equally applicable to other areas and assist in the development of a model to solve the particular problem in mind. Figure 9 is explained further in Chap. 6 of Knight et al. (2010a).

4 Constructing a Model

In order to illustrate the steps involved in the construction of a model, beginning with concepts, to development of the model, through to finally applying the model to solve practical problems, the example of a simple lateral distribution model is used. The process took many years, involved wide collaborative experimental work, extensive data analysis and a leading professional software company. Appendix 1 gives the background to this particular model, referred to as the Shiono & Knight model (SKM). For further details, see Shiono and Knight (1988, 1991), Abril and Knight (2004) and Knight et al. (2010a). The three key parameters are now considered and the rationale behind their adoption explained.

The philosophy behind resistance coefficients was based on distinguishing between the use of the section-mean velocity, U_A, the zonal velocity U_z, the depth-mean velocity, U_d, and any local near-bed velocity, u, used in a 'law of the wall' turbulence model in various friction factors. This leads to the important distinction between 'global', 'zonal' and 'local' friction factors used in 1-D, 2-D & 3-D river models, as introduced in Sect. 2.2 and shown in Fig. 1. In equation form:

$$\underset{\text{(global)}}{\tau_o = \left(\frac{f}{8}\right)\rho\, U_A^2;} \quad \underset{\text{(zonal/sub-area)}}{\tau_z = \left(\frac{f_z}{8}\right)\rho U_z^2;} \quad \underset{\text{(local/depth-averaged)}}{\tau_b = \left(\frac{f_b}{8}\right)\rho U_d^2\,;} \quad \underset{\text{(turbulence)}}{\tau_b = \left(\frac{f_t}{8}\right)\rho u^2} \qquad (1)$$

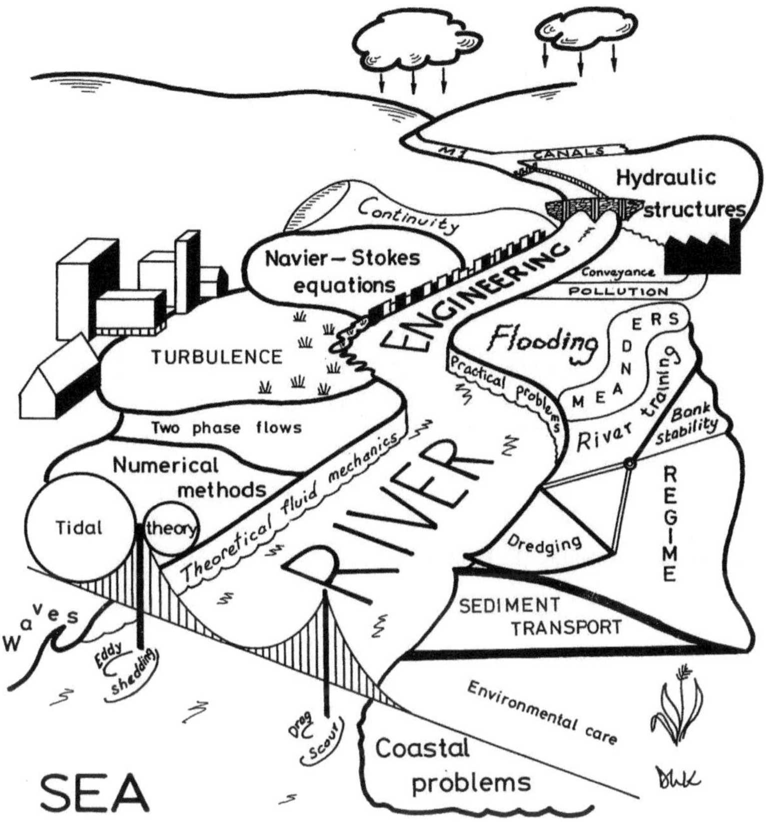

Fig. 9 The art and science of river engineering (after Knight) [reproduced from Nakato and Ettema (1996), p. 448]

In the SKM approach the third option is used, with the shear stress on the bed assumed to be in the same stream wise direction as U_d. This is valid for moderately straight channels where secondary flows cause differences of less than 4°. The relationship between local resistance coefficients in different parts of a compound channel turned out to be surprisingly easy to define for overbank flow with floodplains, since the bed shear stress, τ_b, and depth-mean velocity, U_d, given by

$$U_d = \frac{1}{H}\int_0^H U\,dz; \tag{2}$$

are then linked by Eq. (15a) in Appendix 1. As already illustrated in Figs. 3–5, the experimental data from the Flood Channel facility (FCF) showed that the flow adjusts itself so that f is sensibly constant along certain sections of the wetted

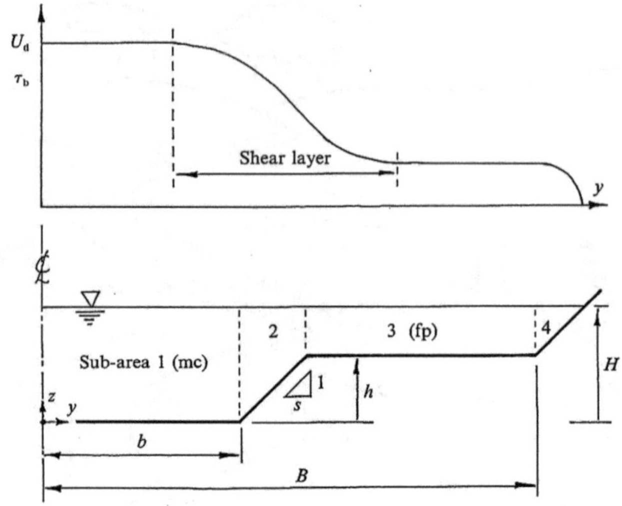

Fig. 10 Flood channel facility (FCF) notation

perimeter for flat floodplains and walls parallel to the main channel. Analysis of the FCF data, as shown in Fig. 5, suggested an equation of the form

$$\frac{f}{f_{mc}} = 0.669 + 0.331 Dr^{-0.719} \tag{3}$$

where *Dr* is defined as the relative depth, the ratio between floodplain depth and main channel depth, defined as (H–h)/H using the notation in Fig. 10, or by H_1/H in Appendix 2, where H is the main channel depth, H(y) or $\xi(y)$ the local depth on any element with a side slope, and (H–h) or H_1 the depth on the floodplain. Figures 11, and 12 show one typical set of U_d and τ_b distributions, with model predictions.

The eddy viscosity was found to follow a similar trend, with higher values on the floodplain. A similar type of equation was formulated through using data, as

$$\frac{\lambda}{\lambda_{mc}} = -0.20 + 1.20 Dr^{-1.44} \tag{4}$$

where λ is the dimensionless eddy viscosity, defined by Eq. (14b) and $U_* = (\tau_b/\rho)^{1/2}$= shear velocity. The intent behind Eqs. (3) and (4) was to make it only necessary to estimate the values of the two coefficients f_{mc} and λ_{mc} in the main channel, making calibration easier for the model user.

The relationship for the coefficient, Γ, defined by Eq. (17), was obtained by measuring all the terms in Eq. (12). The lateral variation of the depth-averaged Reynolds stresses, $\overline{\tau}_{yx}$, is shown in Fig. 13 for series 02 and the lateral variation of apparent shear stress, $\left(\rho\overline{UV}\right)_d$, in Fig. 14. The latter indicates that within certain

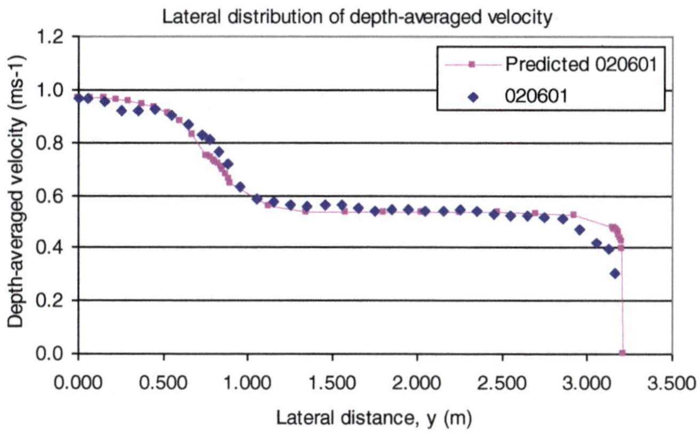

Fig. 11 Measured and predicted U_{d} v y

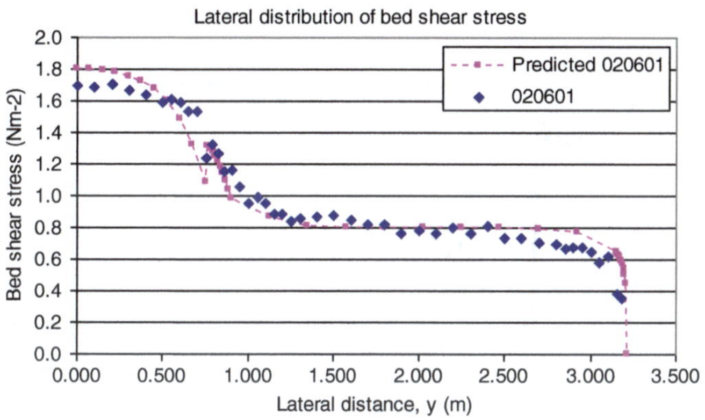

Fig. 12 Measured and predicted τ_{b} v y

zones, the gradient of the secondary flow term was constant, allowing constant values of Γ to be assigned to the 4 regions shown in Figs. 10 and 14.

The model was thus consistent with some of the dominant mechanisms, known to be important in overbank flow by using experimental data from the FCF, a large scale facility. The theoretical concepts were later developed further, by linking boundary shear stress and secondary flow, giving finally two simple equations:

$$\Gamma^*_{mc} = \frac{\Gamma_{mc}}{H} = 0.15\rho g S_o \quad \text{and} \quad \Gamma^*_{fp} = \frac{\Gamma_{fp}}{(H-h)} = -0.25\rho g S_o \tag{5}$$

where $\Gamma^*= \Gamma$/local depth in the region, as used by Abril and Knight (2004). These equations were subsequently re-examined in the light of more data and the need to consider stream wise and planform vorticity separately, see Omran et al. (2008a). Flows in non-prismatic channels with either skew or diverging/converging

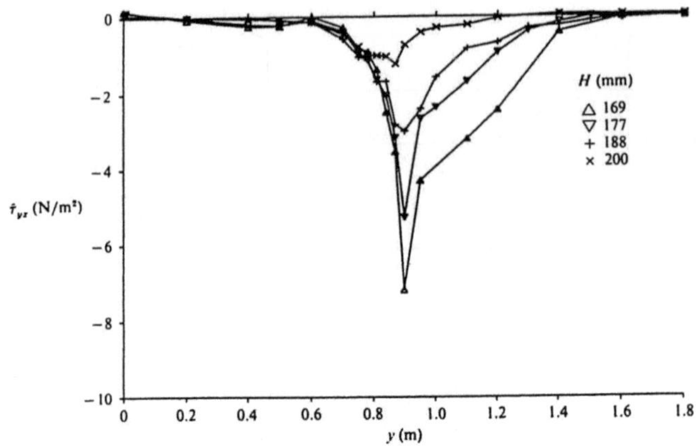

Fig. 13 Lateral variation of depth-averaged Reynolds stress, $\overline{\tau}_{yx}$, for different depths, H, in FCF Series 02

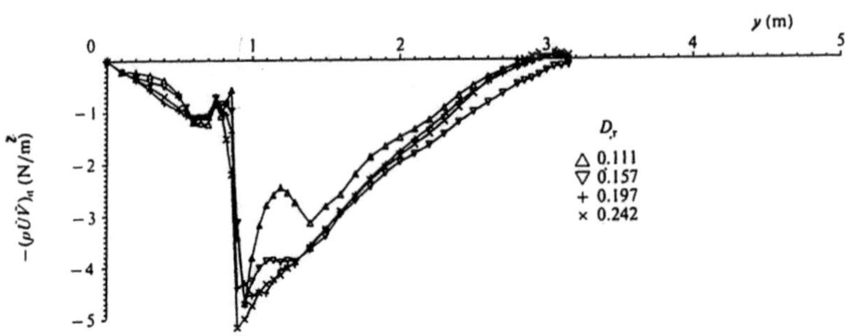

Fig. 14 Lateral variation of apparent shear stress, $\left(\rho\overline{UV}\right)_d$, for $Dr = 0.111$–0.242 in FCF Series 02

floodplains were also examined. Some details of these studies may be found in Chlebek and Knight (2008), Chlebek et al. (2010) and Rezaei and Knight (2009, 2011).

5 Testing a Theoretical Model

5.1 Overall Integrity

The refinement of any model often takes place during an extended period of testing, using data from sources other than those used in its original development,

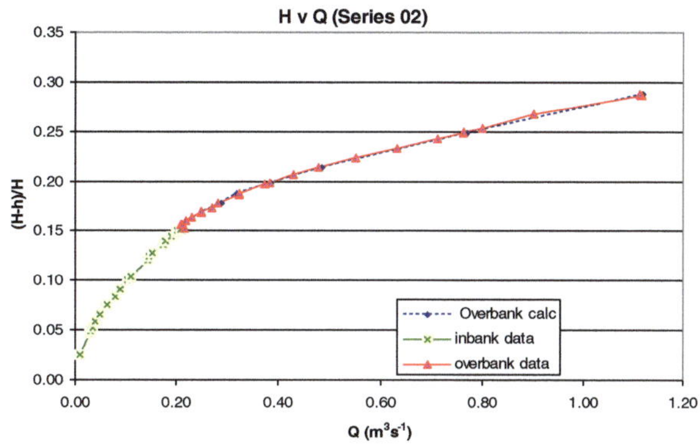

Fig. 15 *H* v *Q* simulation (FCF series 02)

in order to test its generality. There is also usually some focus on other key parameters within the model, not examined in the original model development stage, but which subsequently are considered to be significant.

In testing this model, many of the FCF experiments were simulated with other options on the choice of the 3 coefficients for each panel, not necessarily basing them on Eqs. (3)–(5). In order to assess the results systematically, 6 physical outputs were selected in each numerical experiment. Initially, for given inputs of depth, number of panels and the 3 coefficients, given by Eqs. (3)–(5) for each panel, the 6 resulting outputs were examined. The 6 criteria were: the total discharge, *Q*, the total shear force, *SF*, the lateral distribution of velocity U_d v *y*, the lateral distribution of τ_b around the wetted perimeter, the %*Q* in each panel and the %*SF* on each panel wetted perimeter. Graphs on the behaviour of the apparent shear forces (ASF) on vertical, inclined and horizontal interfaces, expressed as percentages of the total shear force as $\%SF_V$, $\%SF_I$ and $\%SF_H$, respectively, were also obtained. One set of results for this entire process is shown in Figs. 15, 16, 17, 18, 19, 20, 21, 22.

In addition, the differences between the experimental and simulated results were tabulated, as shown for one experiment (020601) in Table 1. These helped to identify where the panel coefficients might need some very minor adjustment from those determined by Eqs. (3)–(5). In most cases it was minimal, which is not surprising, since Eqs. (3)–(5) were based on the average best fit through the same data. Figures 23, 24, 25, 26, 27, 28, 29, 30, 31 illustrate another set of results, using data from a different source, that of the experiments conducted by Knight and Demetriou (1983).

The main parameters, shown in bold in Table 1, account for the bulk of the discharge or shear force. It should be noted that the large 'errors' shown above for the sloping sidewall elements, especially panel 4, are due to using only 4 panels to schematize the whole cross-section, as well as the poor simulation at the flood plain edge, as shown in Figs. 11 and 12. This case has been deliberately chosen to illustrate this point. The 'error' can be reduced by simply using more panels.

Fig. 16 $\%Q_i$ v *Dr*

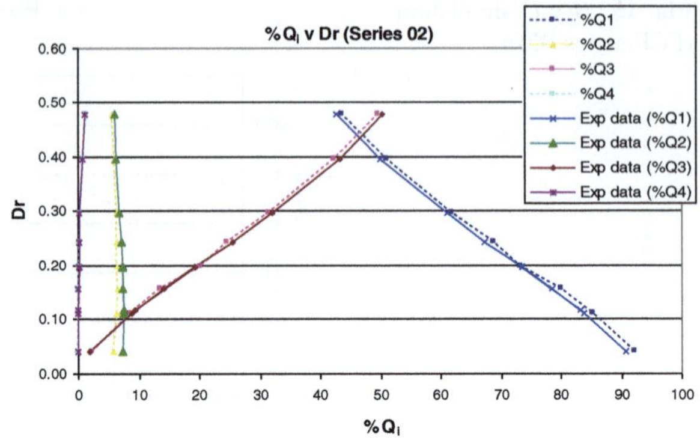

Fig. 17 $\%Q_{mc}$ and $\%Q_{fp}$ v *Dr*

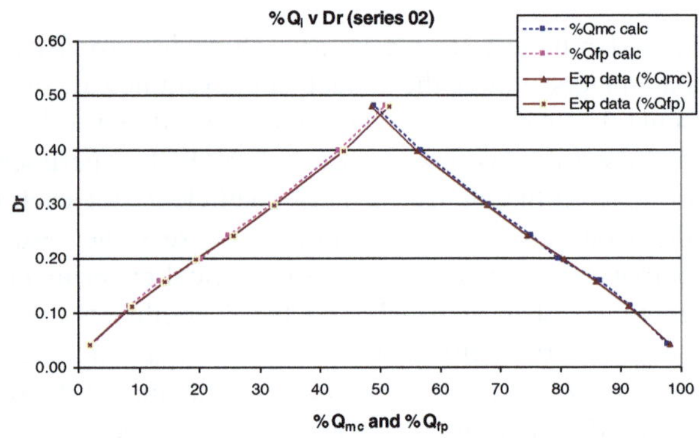

Fig. 18 $\%SF_i$ v *Dr*

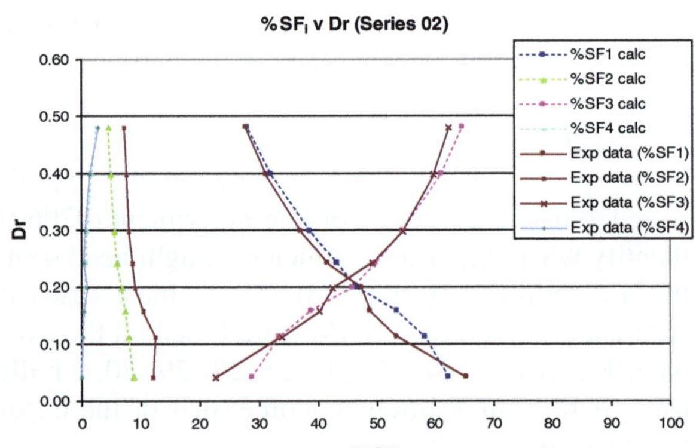

The simulations using the check validation data, shown in Figs. 23–31, exhibit similar trends to those shown in Figs. 15–22. The local friction factors shown in Fig. 28 are sensibly constant for the mc and fp, the total discharge and division of flow between zones are well simulated by the model, as are %SF values on various boundary elements and distributions of U_d and τ_b shown in Figs. 30 and 31. Zonal discharges and $\%SF_i$ are specifically modelled in Knight and Tang (2008).

Fig. 19 $\%SF_{mc}$ and $\%SF_{fp}$ v *Dr*

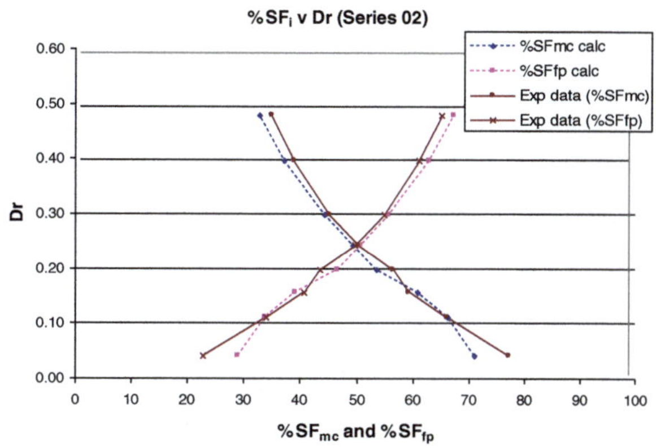

Fig. 20 $\%ASF_v$ v *Dr*

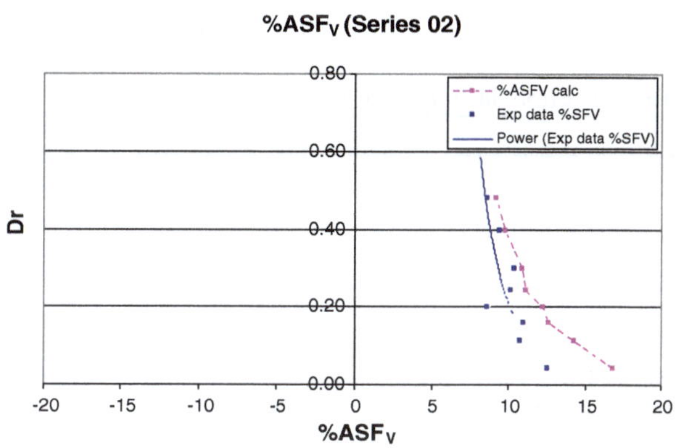

Fig. 21 $\%ASF_I$ v *Dr*

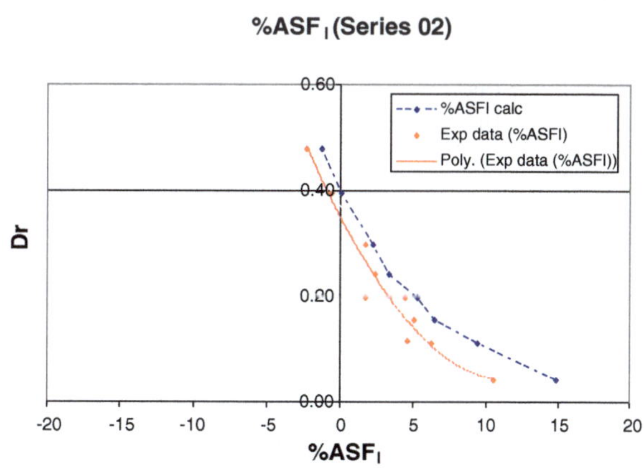

The shear forces on the internal division lines that are designated in Fig. 23 and shown in Figs. 20–22 and 29, are reasonably well simulated in trend, if not in precise detail. It should be recognised that these are particularly sensitive to the

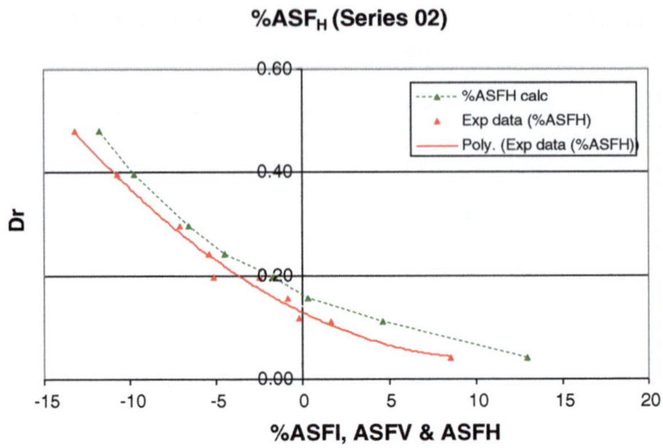

Fig. 22 %ASF$_H$ v *Dr*

Table 1 Errors in simulation for FCF experiment 020601

	Parameter	Error (%)
	Q total	−0.11
	SF total	1.80
Panel 1	Q_1	0.43
Panel 2	Q_2	−7.40
Panel 3	Q_3	−0.02
Panel 4	Q_4	51.98
Panels 1 & 2 (mc)	Q_{mc}	−0.35
Panels 3 & 4 (fp)	Q_{fp}	0.39
Panel 1	SF_1	0.89
Panel 2	SF_2	−3.69
Panel 3	SF_3	1.63
Panel 4	SF_4	194.78
Panels 1 & 2 (mc)	SF_{mc}	0.09
Panels 3 & 4 (fp)	SF_{fp}	3.40

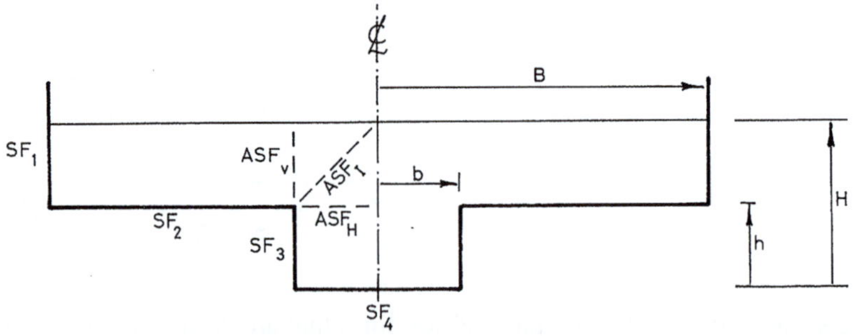

Fig. 23 Notation for apparent shear forces, ASF, on vertical, inclined and horizontal interfaces in a rectangular compound channel

Fig. 24 H v Q (Series DWK3)

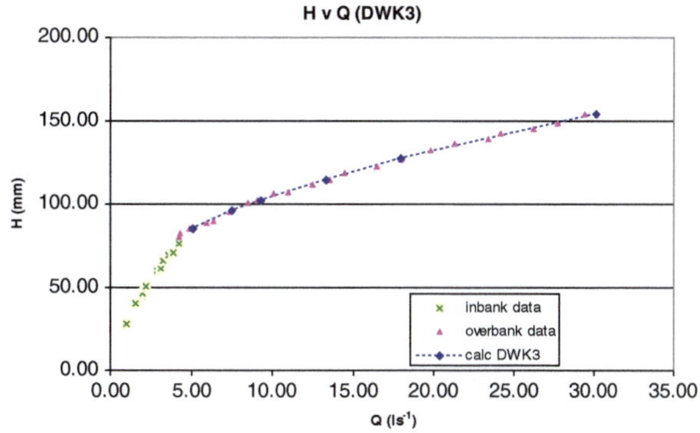

Fig. 25 $\%Q_{mc}$ v *Dr*

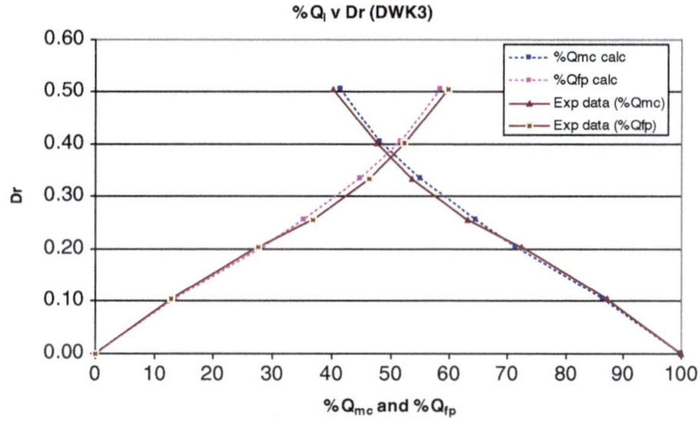

Fig. 26 $\%SF_i$ v *Dr*

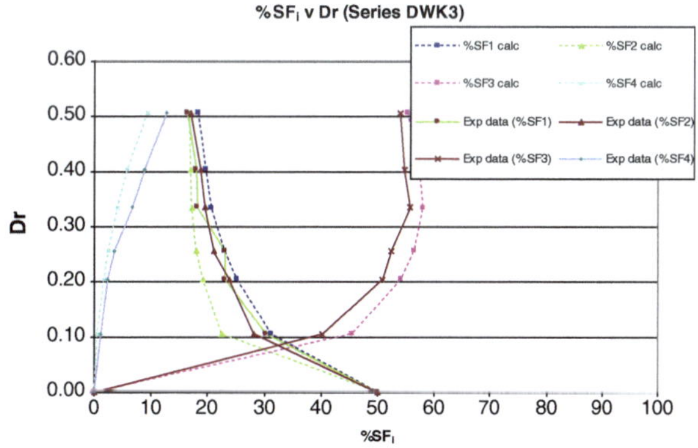

$\%SF_i$ values, and are in reality not linear, as sketched in Fig. 23. Recent work by de Cacqueray et al. (2009) and Ansari et al. (2011), using data from Knight et al. (1994) and 3-D CFD models, reveals the complex nature of these division lines, which are approximately linear in only certain regions and cases.

Fig. 27 %SF v Dr

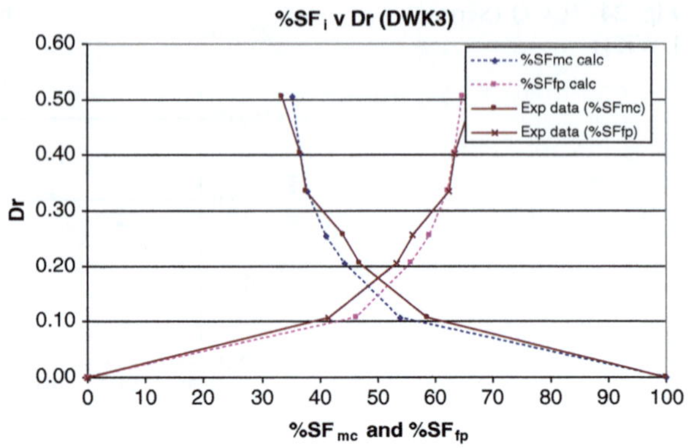

Fig. 28 f_{fp}/f_{mc} v y

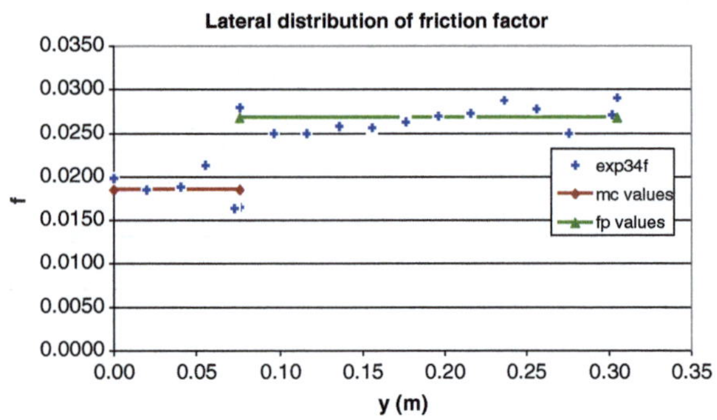

Fig. 29 %ASF_H v Dr

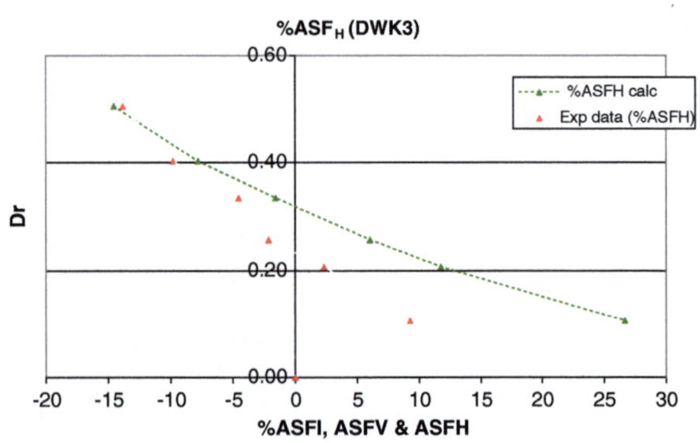

5.2 Number of Panels

The simulations shown through Figs. 11–31 have highlighted the use of data to provide equations for each of the three parameters, f, λ and Γ to be adopted in each panel for flows in trapezoidal and rectangular compound channels. The number of

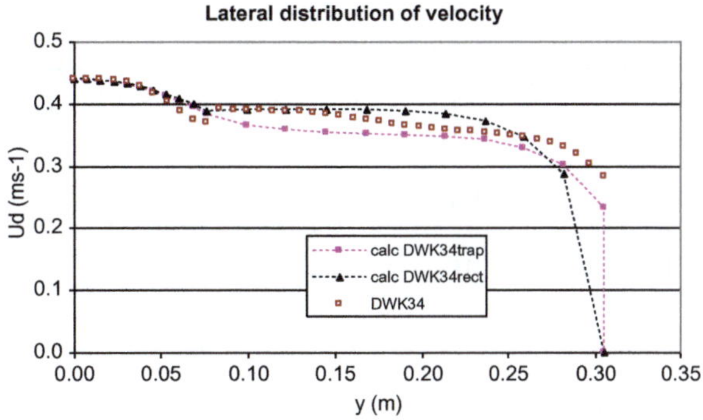

Fig. 30 Calculated and experimental distributions of U_d (DWK34)

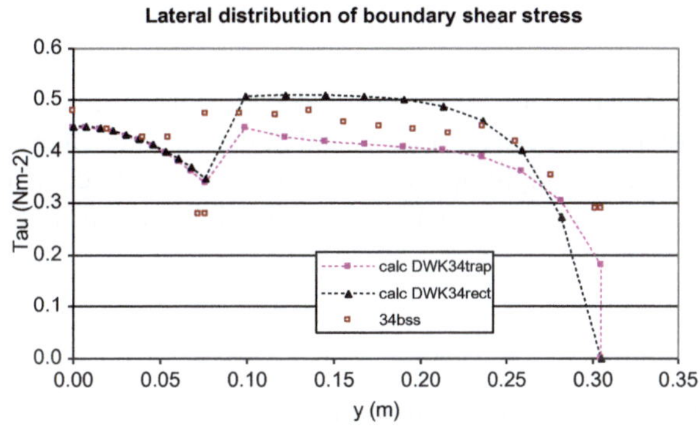

Fig. 31 Calculated and experimental distributions of boundary shear stress (DWK34)

panels has been deliberately kept to a minimum of 4 here, in order to demonstrate that not many are needed to get a quite reasonable representation of overbank flow in a trapezoidal compound channel. In principle, more could have been chosen and a better simulation would have resulted. However, Table 1 has shown that the errors are mostly small enough to allow for a 4 panel representation of the cross-section. Furthermore, the main results, such as the predicted stage-discharge curve, the division of flow between the main river channel and the floodplains, the lateral distributions of velocity and boundary shear stress, are all adequate enough for most practical purposes. Under normal circumstance, 4–6 panels are more than adequate to simulate well most inbank flows in trapezoidal channels, even where secondary flows effects may be strong, as shown by Knight et al. (2007). Overbank flows will require more panels for good simulations.

5.3 Boundary Conditions

As well as the appropriate choice of coefficients for resistance, turbulence and secondary flow within each panel, and the selection of the number of panels to

Table 2 Possible boundary conditions for internal walls in rectangular compound channels

Form	U_d or q continuity	U_d gradient or unit force continuity	Notes
[A]	$U_d^{(1)} = U_d^{(2)}$	$\left(\phi \frac{\partial U_d^2}{\partial y}\right)_{y=b}^{(1)} = \left(\phi \frac{\partial U_d^2}{\partial y}\right)_{y=b}^{(2)} - h\tau_w$	$\phi = \frac{1}{2}\rho\lambda H^2\sqrt{f/8}$ $\tau_w = \rho f_w U_d^2(\mathrm{y} = \mathrm{b})/8$
[B]	$U_d^{(1)} = U_d^{(2)}$	$\left(\mu \frac{\partial U_d}{\partial y}\right)_{y=b}^{(1)} = \left(\phi\mu \frac{\partial U_d}{\partial y}\right)_{y=b}^{(2)}$	$\mu = H^2\lambda\sqrt{f}$ with an adjustment factor φ
[C]	$[HU_d]^{(1)} = [HU_d]^{(2)}$	$\left(\frac{\partial U_d}{\partial y}\right)_{y=b}^{(1)} = \left(\frac{\partial U_d}{\partial y}\right)_{y=b}^{(2)}$	
[D]	$U_d^{(1)} = U_d^{(2)}$	$\left(\mu \frac{\partial U_d}{\partial y}\right)_{y=b}^{(1)} = \left(\mu \frac{\partial U_d}{\partial y}\right)_{y=b}^{(2)}$	$\mu = H^2\lambda\sqrt{f}$
[E]	$[U_d]^{(1)} = [U_d]^{(2)}$	$\left(\frac{\partial U_d}{\partial y}\right)_{y=b}^{(1)} = \left(\frac{\partial U_d}{\partial y}\right)_{y=b}^{(2)}$	

schematize the river cross section, care needs to be taken over the boundary conditions between panels and at the channel edges. It may seem obvious that the following might apply at $y = 0$ (the centreline), if the flow is symmetric, and at $y = B$ (the remote edge of the floodplain), but in a depth-averaged model,

$$\left.\frac{\partial U_d^{(1)}}{\partial y}\right|_{y=0} = 0 \text{ (centerline)}; \quad U_d^{(2)}\big|_{y=B} = 0 \text{ (floodplain edge)} \tag{6}$$

the velocity and shear force conditions at any vertical wall of a floodplain edge ($y = B$) are in conflict. There are also difficulties in specifying boundary conditions at vertical internal walls, an issue explored in detail by Omran et al. (2008b) and Tang and Knight (2008). To illustrate this point, Table 2 lists 5 boundary conditions that might be used for an internal vertical interface between the main channel and the floodplains in a rectangular compound channel. These were investigated by Tang against a wide selection of data in preparing the Tang and Knight (2008) paper. The results of applying them are presented for a hypothetical case in Fig. 32, where bc [A] and bc [B] are shown to give identical results, provided φ is chosen appropriately.

From this analysis it was clear that boundary condition [A] with the following relationship for the continuity of unit force was technically the most suitable:

$$\left(H\overline{\tau}_{yx}\right)_{y=b}^{(i)} + h\tau_W = \left(H\overline{\tau}_{yx}\right)_{y=b}^{(i+1)} \tag{7}$$

$$\left(\varphi \frac{\partial U_d^2}{\partial y}\right)_{y=b}^{(1)} = \left(\varphi \frac{\partial U_d^2}{\partial y}\right)_{y=b}^{(2)} - h\tau_w \tag{8}$$

$$\varphi = \frac{1}{2}\rho\lambda H^2\sqrt{f/8} \text{ and } \tau_w = \rho f_w U_d^2\,(\mathrm{y} = \mathrm{b})\,/\,8 \tag{9}$$

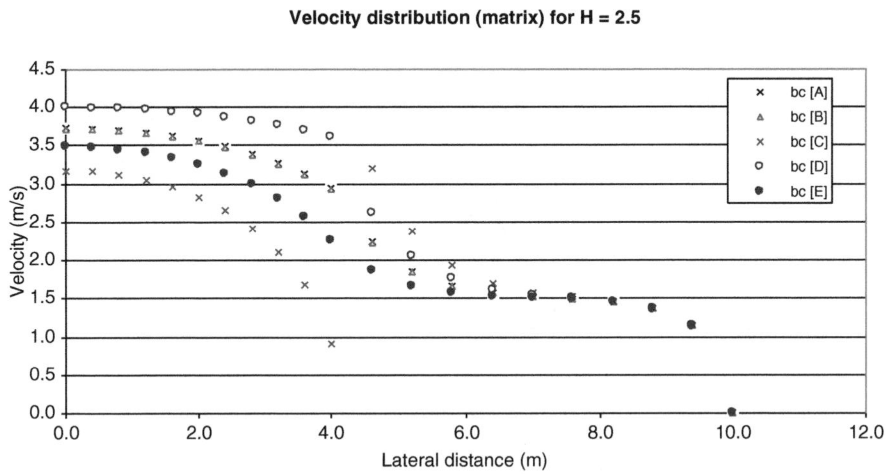

Fig. 32 Effect of different boundary conditions on U_d for a symmetric rectangular compound channel for $H = 2.5$ m ($S_o = 0.001$, $b = 4$ m, $B = 10$ m, $h = 2$ m; $f_1 = f_w = 0.01$ & $f_2 = 0.02$; $\lambda_1 = 0.01$ & $\lambda_2 = 0.2$; $\Gamma_1 = 1.0$ & $\Gamma_2 = -0.75$)

Because bc [A] suffers from the drawback that the wall shear stress needs to be known, the approximation is made that the friction factor in the main channel and the depth-averaged velocity at the interface can be used to estimate the wall shear force. In the subsequent model simulations, this was achieved by assuming the main channel friction factor could be used in (9) to determine the wall shear force.

Thus, bc [A] was subsequently used in a 2 panel simulation of flow in the same hypothetical case as before and checked for several depths using a 3 panel trapezoidal simulation with a very steep internal wall (s = 0.001), as shown in Fig. 33.

The results for one of these simulations are shown in Fig. 34 and the matrix equations are given in Appendix 2. The schematisation in Fig. 33 should be compared with those already presented in Figs. 10 and 23. Figure 34 shows that the simulations in a 2 panel rectangular channel using bc [A] agree well with those with a steep internal wall

The balance between satisfying both $U_d = 0$ and the shear wall shear force criterion at $y = B$, was investigated for simple channels by Chlebek and Knight (2006). For inbank flows this is more straightforward problem to solve since the bed shear force may be determined by integration of the bed shear stress distribution and then subtracted from the total shear force ($=\rho gAS_o$) to obtain the wall shear force directly. For compound channels this is not possible, and alternative methods have to be found, as illustrated for an internal vertical wall.

Finally, there is the issue of establishing an automatic testing procedure for optimum parameter values and hence the solutions. This was undertaken, several years later, using multi-objective evolutionary algorithms, as shown by Sharifi et al. (2008, 2009). The results of these optimisation techniques were then compare the corresponding results undertaken using visual inspection of the various output

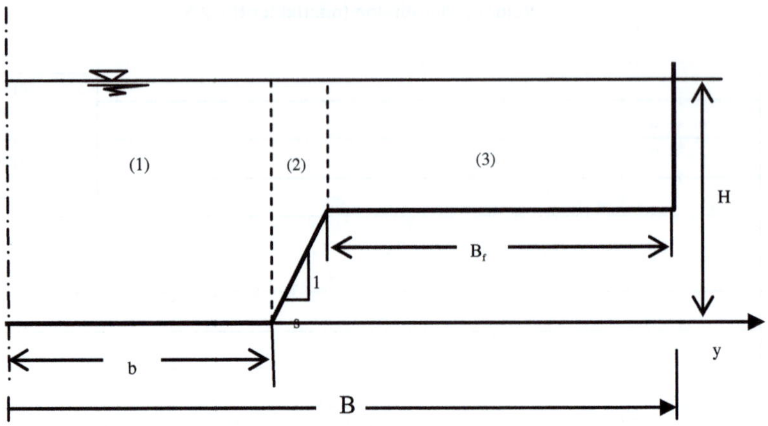

Fig. 33 Symmetric compound channel with a very steep internal wall

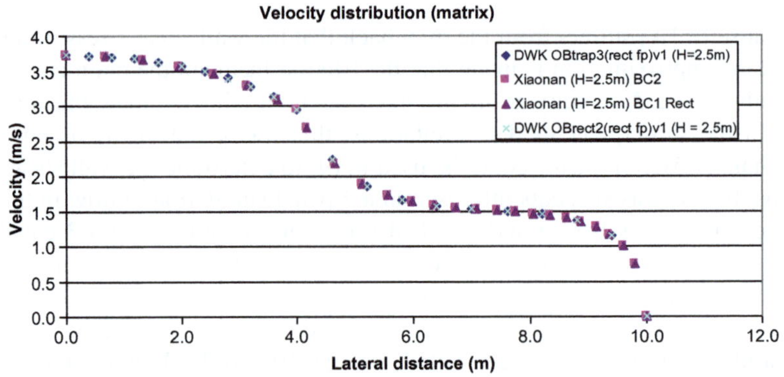

Fig. 34 Comparison between U_d distributions for a trapezoidal compound channel with a nearly vertical internal wall with that for a rectangular compound channel with bc [A] for $H = 2.5$ m ($S_o = 0.001$, $b = 4$ m, $B = 10$ m, $h = 2$ m; $f_1 = f_2 = 0.01$ & $f_3 = 0.02$; $\lambda_1 = \lambda_2 = 0.1$ & $\lambda_3 = 0.2$; $\Gamma_1 = \Gamma_2 = 1.0$ & $\Gamma_3 = -0.75$)

graphs, such as Figs. 11, 12 and 15–33, and tabular formats for errors, such as indicated by Table 1. Despite some advantages in numerical assessments, the focus on a single determinant for a multi-objective function may cause one to overlook some important physical feature in either model or data.

6 Using a Model in Practice

The last stage in many research processes is to bring the fruits of the work to both academic colleagues and practising engineers and scientists. The former is often achieved via journal papers and conferences, whereas the latter may be via user-friendly software that encapsulates the essence of the research and applies it effectively to real problems faced by those in practice.

The many studies undertaken in the Flood Channel Facility (FCF) at HR Wallingford between 1985 and 1995, led to the scientific basis of the Conveyance Estimation System and Afflux Estimation System (CES-AES) described fully on the website www.river-conveyance.net and in a companion book by Knight et al. (2010a) which covers the key scientific issues and applies the CES to many practical river problems. The experimental programme included many large scale experimental studies of inbank and overbank flows in straight, skewed and sinuous channels, with a variety of geometries and roughness conditions. Some studies were also undertaken with a mobile bed, with the sediment being re-circulated in both the straight and meandering channel cases.

Figure 35 shows a comparison between measured and predicted flows, based on the CES software and data from a wide range of rivers. Further details are available in McGahey et al. (2008) and Knight et al. (2010a). The CES contains a roughness advisor, a conveyance generator, an uncertainty estimator, a backwater module and an afflux estimator for flow through brides and culverts. The conveyance calculation is based on a depth-integration of the Reynolds-Averaged Navier–Stokes (RANS) equations and is heavily built on the concepts in the Shiono & Knight Method (SKM) described earlier. A finite-element approximation is made to the CES equations, expressed in terms of the discharge per unit width, q, rather than U_{d} ($=q/H$), but one is readily converted via Eq. (10) to the other to give the lateral variation in U_{d} as illustrated previously in several examples.

$$q = \int_0^H udz = HU_d \tag{10}$$

Figure 36 illustrates the use of the CES to back-calculate the overall roughness of a river from its constituent panel values, using two mountain rivers in which boulders with a d_{90} of around 2.0 m form the main roughness. The CES software can handle up to 200 panels per river cross-section, making it sufficiently flexible to deal with natural geometries in sufficient detail for most practical purposes.

Channels with a low sinuosity can also be simulated with the CES, but overbank flow in fully developed meandering channels produces more complex flow structures. Experimental studies that highlight these structures for overbank flow in meandering channels have been undertaken by Ikeda and Parker (1989), Sellin et al. (1993), Rameshwaran and Willetts (1999), Shiono and Muto (1998) and Ikeda et al. (2002) and Fukuoka et al. (2009).

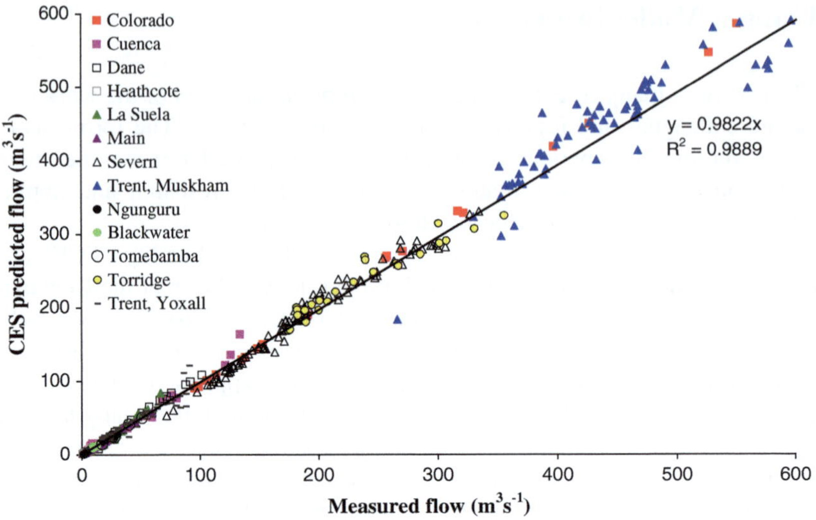

Fig. 35 Predicted flows (using CES) compared with measured flows

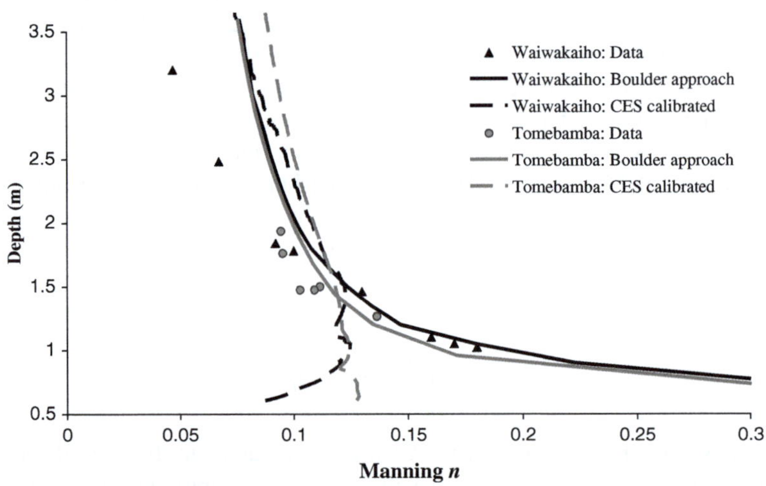

Fig. 36 Back-calculated Manning *n* values for the rivers Waiwakailo and Tomebamba (mountain rivers in New Zealand and Ecuador)

7 Concluding Remarks

The various steps in solving a hydraulics problem, in this case that of overbank flow, have been traced from concepts and data through to a finished model product. The whole process has taken a number of years from start to finish, and is now entering a stage of reflection, before any upgrades are made to the software, arising out of continuing research and further gains in understanding of the relevant phenomena. It has been an instructive exercise in how to obtain 'experimental and computational solutions of hydraulic problems', which is the title of this book.

An attempt has been made to write briefly about the general approach to solving problems in Sect. 3, with Figs. 8 and 9 describing the art of river engineering.

As someone wrote in a recent Editorial of the Water Management Journal, we might aim to "Measure more, and model less; think more, and publish less". The first couplet seeks to redress the current trend to only rely on numerical modelling, and the second is a plea that I have singularly failed to comply with, with yet another document to add to the burden on diligent academics and readers!

Appendix 1: Governing Equations Used in the Shiono & Knight Model

In a prismatic open channel, the equation for the stream wise component of momentum on a fluid element in a steady flow may be combined with the continuity equation to give:

$$\rho\left[\frac{\partial UV}{\partial y}+\frac{\partial UW}{\partial z}\right]=\rho gS+\frac{\partial}{\partial y}(-\rho\overline{uv})+\frac{\partial}{\partial z}(-\rho\overline{uw}) \tag{11}$$

where U, V, W are the mean velocity components in the x (stream wise), y (lateral) and z (normal to bed, but also nearly the vertical) directions, respectively, u, v, w are turbulent fluctuations of velocity with respect to the mean, ρ is the density of water, and S is the bed slope gradient ($S = \sin\theta$).

The depth-mean-averaged momentum equation can be obtained by integrating (11) over the water depth, H, provided $W(H) = W(0) = 0$, as given by Shiono and Knight (1991):

$$\frac{\partial\left[H(\rho UV)_{\mathrm{d}}\right]}{\partial y}=\rho gHS+\frac{\partial H\bar{\tau}_{yx}}{\partial y}-\tau_b\sqrt{1+\frac{1}{s^2}} \tag{12}$$

in which τ_b is the bed shear stress, s is the side slope (1:s = vertical: horizontal), and

$$(\rho UV)_{\mathrm{d}}=\frac{1}{H}\int_0^H(\rho UV)dz \quad \text{and} \quad \bar{\tau}_{yx}=\frac{1}{H}\int_0^H(-\rho uv)dz \tag{13}$$

An analytical solution for (12) may be obtained based on the commonly used eddy viscosity assumptions, given as follows:

$$\bar{\tau}_{yx} = \rho\bar{\varepsilon}_{yx}\frac{\partial U_\mathrm{d}}{\partial y} \tag{14a}$$

and

$$\bar{\varepsilon}_{yx} = \lambda U_* H \tag{14b}$$

where λ is the dimensionless eddy viscosity coefficient and U_* (= $\tau_b/\rho^{1/2}$) is the local shear velocity. Using the Darcy-Weisbach friction coefficient, *f*, relating the local boundary shear stress, τb, with the depth-mean velocity, *Ud*, by the customary relationship

$$\tau_b = \rho\frac{f}{8}U_d^2 \tag{15a}$$

or

$$U_* = \sqrt{\frac{f}{8}}U_d \tag{15b}$$

Then, substituting (14a) and (15a) into (12) yields:

$$\rho gHS - \rho\frac{f}{8}U_b^2\sqrt{1+\frac{1}{s^2}} + \frac{\partial}{\partial y}\left[\rho\lambda H^2\sqrt{\frac{f}{8}}U_d\frac{\partial U_d}{\partial y}\right] = \frac{\partial}{\partial y}\left[H(\rho UV)_\mathrm{d}\right] \tag{16}$$

Experimental results show that the shear stress due to secondary flow, (ρUV)d, decreases approximately linearly either side of a maximum value occurring at the edge of the main channel and the floodplain. The lateral gradient of the secondary flow force per unit length of the channel may therefore be written as

$$\frac{\partial}{\partial y}\left[H(\rho UV)_\mathrm{d}\right] = \Gamma \tag{17}$$

where Γ is a dimensionless secondary flow parameter (different for different flow regions). An analytical solution to (16) for the lateral distribution of depth-mean velocity has been obtained by Shiono and Knight (1988, 1991), as follows:

<1> For a sub-area with a constant water depth *H*, the analytic U_d distribution has the form:

$$U_d = \left[A_1 e^{\gamma y} + A_2 e^{-\gamma y} + k\right]^{1/2} \tag{18}$$

where

$$k = \frac{8gSH}{f}(1-\beta) \tag{19}$$

$$\gamma = \sqrt{\frac{2}{\lambda}}\left(\frac{f}{8}\right)^{\frac{1}{4}}\frac{1}{H} \tag{20}$$

$$\beta = \frac{\Gamma}{\rho g H S} \tag{21}$$

A_1 & A_2 can be determined by considering the relevant boundary conditions.

<2> For a sub-area with a channel side slope of 1: s, the U_d distribution is written in the form

$$U_d = \left[A_3\xi^{\alpha} + A_4\xi^{-\alpha-1} + \omega\xi + \eta\right]^{1/2} \tag{22}$$

where

$$\alpha = -\frac{1}{2} + \frac{1}{2}\sqrt{1 + \frac{s(1+s^2)^{\frac{1}{2}}}{\lambda}(8f)^{\frac{1}{2}}} \tag{23}$$

$$\omega = \frac{gS}{\frac{(1+s^2)^{\frac{1}{2}}}{s}\left(\frac{f}{8}\right) - \frac{\lambda}{s^2}\left(\frac{f}{8}\right)^{\frac{1}{2}}} \tag{24}$$

$$\eta = -\frac{\Gamma}{\frac{(1+s^2)^{\frac{1}{2}}}{s}\rho\left(\frac{f}{8}\right)} \tag{25}$$

and ξ is the local depth, given here for one side sloping element as

$$\xi = H - \frac{y-b}{s} \tag{26}$$

In a similar way to obtaining A_1 & A_2, the coefficients A_3 & A_4 can be determined by considering the relevant boundary conditions.

Where there are discontinuities in the roughness distribution across the section, it is important to alter the velocity gradient boundary condition between panels, such that Eq. (27) is satisfied, as in these cases $\mu \neq 1.0$. Based on an approximation of the exact force balance, linearly varying the value of *f* within each panel, maintaining the mean value,

$$\left(\mu\frac{\partial U_d}{\partial y}\right)^{(i)} = \left(\mu\frac{\partial U_d}{\partial y}\right)^{(i+1)} \text{ with } \mu = \lambda\sqrt{f} \tag{27}$$

aids smoothing of the τ_b distributions. Otherwise, τ_b varies in a saw-tooth pattern in direct response to lateral changes in *f* between panels, since U_d is the same for both panels at the interface. This arises because of the relationship between τ_b and depth-averaged velocity, given by (15a). See Knight et al. (2007).

Appendix 2: Matrix Equations for 3 Types of Overbank Flow in Rectangular & Trapezoidal Channels

2 Panel Rectangular Compound Channel (OBrect2)

$$\begin{bmatrix} 1 & -1 & 0 & 0 \\ 0 & 0 & e^{\gamma_2 B} & e^{-\gamma_2 B} \\ e^{\gamma_1 b} & e^{-\gamma_1 b} & -e^{\gamma_2 b} & -e^{-\gamma_2 b} \\ (\phi_1\gamma_1 + \rho f_w h/8)e^{\gamma_1 b} & -(\phi_1\gamma_1 - \rho f_w h/8)e^{-\gamma_1 b} & -\phi_2\gamma_2 e^{\gamma_2 b} & \phi_2\gamma_2 e^{-\gamma_2 b} \end{bmatrix} \times \begin{bmatrix} A_1 \\ A_2 \\ A_3 \\ A_4 \end{bmatrix} = \begin{bmatrix} 0 \\ -k_2 \\ k_2 - k_1 \\ -\rho f_w h k_1/8 \end{bmatrix}$$

3 Panel Rectangular Compound Channel with a Steep Internal Main Channel Wall (OBtrap3)

$$\begin{bmatrix} 1 & -1 & 0 & 0 & 0 & 0 \\ 0 & 0 & 0 & 0 & e^{\gamma_3 B} & e^{-\gamma_3 B} \\ e^{\gamma_1 b} & e^{-\gamma_1 b} & -H^{\alpha_2} & -H^{-(\alpha_2+1)} & 0 & 0 \\ \mu_1\gamma_1 s_2 e^{\gamma_1 b} & -\mu_1\gamma_1 s_2 e^{-\gamma_1 b} & \mu_2\alpha_2 H^{(\alpha_2-1)} & -\mu_2(\alpha_2+1)H^{-(\alpha_2+2)} & 0 & 0 \\ 0 & 0 & H_1^{\alpha_2} & H_1^{-(\alpha_2+1)} & -e^{\gamma_3 b_1} & -e^{-\gamma_3 b_1} \\ 0 & 0 & \mu_2\alpha_2 H_1^{(\alpha_2-1)} & -\mu_2(\alpha_2+1)H_1^{-(\alpha_2+2)} & \mu_3\gamma_3 s_2 e^{\gamma_3 b_1} & -\mu_3\gamma_3 s_2 e^{-\gamma_3 b_1} \end{bmatrix} \times \begin{bmatrix} A_1 \\ A_2 \\ A_3 \\ A_4 \\ A_5 \\ A_6 \end{bmatrix} = \begin{bmatrix} 0 \\ -k_3 \\ \omega_2 H + \eta_2 - k_1 \\ -\mu_2\omega_2 \\ k_3 - \omega_2 H_1 - \eta_2 \\ -\mu_2\omega_2 \end{bmatrix}$$

4 Panel Trapezoidal Compound Channel (Obtrap4)

$$\begin{bmatrix} 1 & -1 & 0 & 0 & 0 & 0 & 0 & 0 \\ 0 & 0 & 0 & 0 & 0 & 0 & 0 & 1 \\ e^{\gamma_1 b} & e^{-\gamma_1 b} & -H^{\alpha_2} & -H^{-(\alpha_2+1)} & 0 & 0 & 0 & 0 \\ \mu_1\gamma_1 s_2 e^{\gamma_1 b} & -\mu_1\gamma_1 s_2 e^{-\gamma_1 b} & \mu_2\alpha_2 H^{\alpha_2-1} & -\mu_2(\alpha_2+1)H^{-(\alpha_2+2)} & 0 & 0 & 0 & 0 \\ 0 & 0 & H_1^{\alpha_2} & H_1^{-(\alpha_2+1)} & -e^{\gamma_3 b_1} & -e^{-\gamma_3 b_1} & 0 & 0 \\ 0 & 0 & \mu_2\alpha_2 H_1^{\alpha_2-1} & -\mu_2(\alpha_2+1)H_1^{-(\alpha_2+2)} & \mu_3\gamma_3 s_2 e^{\gamma_3 b_1} & -\mu_3\gamma_3 s_2 e^{-\gamma_3 b_1} & 0 & 0 \\ 0 & 0 & 0 & 0 & e^{\gamma_3 B} & e^{-\gamma_3 B} & -H_1^{\alpha_4} & 0 \\ 0 & 0 & 0 & 0 & \mu_3\gamma_3 s_4 e^{\gamma_3 B} & -\mu_3\gamma_3 s_4 e^{-\gamma_3 B} & \mu_4\alpha_4 H_1^{\alpha_4-1} & 0 \end{bmatrix}$$

$$\times \begin{bmatrix} A_1 \\ A_2 \\ A_3 \\ A_4 \\ A_5 \\ A_6 \\ A_7 \\ A_8 \end{bmatrix} = \begin{bmatrix} 0 \\ 0 \\ \omega_2 H + \eta_2 - k_1 \\ -\mu_2\omega_2 \\ k_3 - \omega_2 H_1 - \eta_2 \\ -\mu_2\omega_2 \\ \omega_4 H_1 + \eta_4 - k_3 \\ -\mu_4\omega_4 \end{bmatrix}$$

References

Abril JB, Knight DW (2004) Stage-discharge prediction for rivers in flood applying a depth-averaged model. J Hydraul Res IAHR 42(6):616–629

Anderson MG, Walling DES, Bates PD (1996) Floodplain processes. Wiley, Chichester, pp 1–658

Ansari K, Morvan HP, Hargreaves DM (2011) Numerical investigation into secondary currents and wall shear in trapezoidal channels. J Hydraul Eng ASCE 137(4):1–9 (April)

Ashworth PJ, Bennett SJ, Best JL, McLelland SJ (1996) Coherent flow structures in open channels. J Wiley, Chichester, pp 1–733

Atabay S, Knight DW (2006) 1-D modelling of conveyance, boundary shear and sediment transport in overbank flow. J Hydraul Res IAHR 44(6):739–754

Bronstert A (2006) The effects of climate change on flooding. In: Knight DW, Shamseldin AY (eds) Chapter 4 in river basin modelling for flood risk mitigation. Taylor & Francis, NY, pp 77–91

Chang HH (1988) Fluvial processes in river engineering. Wiley, NY, pp 1–432

Chlebek J, Knight DW (2006) A new perspective on sidewall correction procedures, based on SKM modeling. In: Alves F, Cardoso L (eds) RiverFlow 2006, vol 1. Taylor & Francis, Lisbon, pp 135–144

Chlebek J, Knight DW (2008) Observations on flow in channels with skewed floodplains. In: Altinakar MS, Kokpinar MA, Aydin I, Cokgar S, Kirkgoz S (eds) RiverFlow 2008. Taylor & Francis, Cesme, pp 519–527

Chlebek J, Bousmar D, Knight, DW, Sterling M (2010) A comparison of overbank flow conditions in skewed and converging/diverging channels. In: Dittrich A, Koll K, Aberle J, Geisenhainer P (eds) Riverflow 2010, Proceedings of the international conference on fluvial hydraulics, vol I. Braunschweig, Germany, 8–10 Sept, Bundesanstalt fur Wasserbau (BAW), Karlsruhe, Germany, pp 503–511

de Cacqueray N, Hargreaves DM, Morvan HP (2009) A computational study of shear stress in smooth rectangular channels. J Hydraul Res IAHR 47(1):50–57

Fukuoka S, Watanabe A, Wormleaton PR (2009) Design considerations. In: Ikeda S, McEwan I (eds) Flow and sediment transport in compound channels: the experience of Japanese and UK research. IAHR monograph, pp 235–304

Gunawan B, Sterling M, Tang X, Knight DW (2010) Measuring and modelling flow structures in a small river. In: Dittrich A, Koll K, Aberle J, Geisenhainer P (eds) Riverflow 2010, proceedings of the international conference on fluvial hydraulics, vol I. Braunschweig, Germany, 8–10 Sept, Bundesanstalt fur Wasserbau (BAW), Karlsruhe, Germany, Keynote address, pp 179–186

Ikeda S, McEwan I (2009) Flow and sediment transport in compound channels: the experience of Japanese and UK research. IAHR monograph, pp 1–320

Ikeda S, Parker G (1989) River meandering. American geophysical union, water resources monograph, vol 12. AGU, Washington, pp 1–485

Ikeda S, Kawamura K, Toda Y, Kasuya I (2002) Quasi-three dimensional computation and laboratory tests on flow in curved open channels. In: Bousmar D, Zech Y (eds) Proceedings of RiverFlow 2002, vol 1, Balkema, Louvain La-Neuve, Belgium, pp 233–245

ISO 1100-2 (2010) Hydrometry-measurement of liquid flow in open channels-Part 2: Determination of the stage-discharge relationship. International Standards Organisation, 3rd edn. ISO 1100-2, pp 1–28

Knight DW (1981) Some field measurements concerned with the behaviour of resistance coefficients in a tidal channel. Estuarine, coastal and shelf science, vol 12. Academic, London, pp 303–322

Knight DW (1996) Issues and directions in river mechanics–closure of sessions 2, 3 and 5. In: Nakato T, Ettema R (eds) Issues and directions in hydraulics. An Iowa Hydraulics Colloquium in honour of Professor John F Kennedy, Iowa Institute of Hydraulic Research, Iowa, USA, Balkema, pp 435–462

Knight DW (2006a) River flood hydraulics: theoretical issues and stage-discharge relationships. In: Knight DW, Shamseldin AY (eds) Chapter 17 in river basin modelling for flood risk mitigation. Taylor & Francis, Chichester, pp 301–334

Knight DW (2006b) River flood hydraulics: calibration issues in one-dimensional flood routing models. In: Knight DW, Shamseldin AY (eds) Chapter 18 in river basin modelling for flood risk mitigation. Taylor & Francis, Chichester, pp 335–385

Knight DW (2008) Modelling overbank flows in rivers—data, concepts, models and calibration. In: Garcia-Navarro P, Playan E (eds) Chapter 1 in numerical modelling of hydrodynamics for water resources. Taylor & Francis, Chichester, pp 3–23

Knight DW and Abril B (1996) Refined calibration of a depth-averaged model for turbulent flow in a compound channel. In: Proceedings of institution of civil engineers, water, maritime and energy division, London, vol 118, issue 3. Paper no. 11017, pp 151–159

Knight DW, Demetriou JD (1983) Flood plain and main channel flow interaction. J Hydraul Eng ASCE 109(8):1073–1092

Knight DW and Samuels PG (2007) Examples of recent floods in Europe. J Disas Res 2(3):190–199 (Fuji Technology Press, Tokyo, Japan)

Knight DW, Shamseldin A (2006) River basin modelling for flood risk mitigation. Taylor & Francis, Chichester, pp 1–607

Knight DW, Shiono K (1990) Turbulence measurements in a shear layer region of a compound channel. J Hydraul Res IAHR 28(2):175–196 (Discussion in IAHR J 1991, 29(2):259–276)

Knight DW, Tang X (2008) Zonal discharges and boundary shear in prismatic channels. In: Proceedings of the institution of civil engineers, London, vol 161. J Eng Comput Mech, EM2, pp 59–68

Knight DW, Shiono K, Pirt J (1989) Prediction of depth mean velocity and discharge in natural rivers with overbank flow. In: Falconer RA, Goodwin P, Matthew RGS (eds) Proceedings of

international conference on hydraulic and environmental modelling of coastal, estuarine and river waters, Gower Technical, University of Bradford, Paper 38, pp 419–428
Knight DW, Yuen KWH, Alhamid AAI (1994) Boundary shear stress distributions in open channel flow. In: Beven K, Chatwin P, Millbank J (eds) Physical mechanisms of mixing and transport in the environment. Wiley, NY, Chapter 4, pp 51–87
Knight DW, Cao S, Liao H, Samuels PG, Wright NG, Liu X, Tominaga A (2006) Floods—are we prepared? J Disas Res 1(2):325–333 (Fuji Technology Press, Tokyo, Japan)
Knight DW, Omran M, Tang X (2007) Modelling depth-averaged velocity and boundary shear in trapezoidal channels with secondary flows. J Hydraul Eng ASCE 133(1):39-47
Knight DW, Aya S, Ikeda S, Nezu I, Shiono K (2009) Flow structure. In: Ikeda S, McEwan IK (eds) Chapter 2 in Flow and sediment transport in compound channels: the experience of Japanese and UK research, IAHR monograph, pp 1–320
Knight DW, McGahey C, Lamb R, Samuels PG (2010) Practical channel hydraulics—roughness, conveyance and afflux. CRC Press/Taylor & Francis, NY, pp 1–354
Knight DW, Tang X, Sterling, M, Shiono K, McGahey C (2010b) Solving open channel flow problems with a simple lateral distribution model. In: Dittrich A, Koll K, Aberle J, Geisenhainer P (eds) Riverflow 2010, proceedings of the international conference on fluvial hydraulics, Braunschweig, Germany, 8–10 Sept, Bundesanstalt fur Wasserbau (BAW), Karlsruhe, Germany, Keynote address, vol I, pp 41–48
McGahey C, Samuels P G, Knight DW and O'Hare MT (2008) Estimating river flow capacity in practice. J Flood Risk Manage 1(1):23–33
McGahey C, Knight DW, Samuels PG (2009) Advice, methods and tools for estimating channel roughness. In: Proceedings of the institution of civil engineers, water management, London, vol 162, issue WM6, pp 353–362
Morvan H, Knight DW, Wright NG, Tang X, Crossley M (2008) The concept of roughness in fluvial hydraulics and its formulation in 1-D, 2-D & 3-D numerical simulation models. J Hydraul Res IAHR 46(2):191–208
Nakato T, Ettema R (1996) Issues and directions in hydraulics. Balkema, pp 1–495
Navratil O, Albert MB, Gresillon JM (2004) Using a 1D steady flow model to compare field determination methods of bank-full stage. In: Carravetta G, and Morte D (eds) Proceedings of river flow 2004, Naples, June, Balkema, pp 155–161
Omran M, Knight DW, Beaman F, Morvan H (2008a) Modelling equivalent secondary current cells in rectangular channels. In: Altinakar MS, Kokpinar MA, Aydin I, Cokgar S, Kirkgoz S (eds) RiverFlow 2008, Cesme, Turkey, vol 1, pp 75–82
Omran M, Atabay S, Knight DW, Seckin G (2008b) Boundary conditions for a depth-averaged flow model in overbank flow. In: Altinakar MS, Kokpinar MA, Aydin I, Cokgar S, Kirkgoz S (eds) RiverFlow 2008, Cesme, Turkey, vol 1, pp 485–492
Rameshwaran R, Willetts BB (1999) Conveyance prediction for meandering two-stage channel flows. In: Proceedings of institution of civil engineers, water, maritime & energy, vol 136, pp 153–166
Ramsbottom DM, Whitlow CD (2003) Extension of rating curves at gauging stations: best practice guidance manual. R&D Manual W6-061/M, the environment agency, Bristol, UK, pp 1–254
Rezaei B and Knight DW (2009) Application of the Shiono and Knight Method in compound channels with non-prismatic floodplains. J Hydraul Res IAHR 47(6):716–726
Rezaei B, Knight DW (2011) Overbank flow in compound channels with non-prismatic floodplains. J Hydraul Eng ASCE 137(8): 815–824
RIBAMOD (1999) River basin modelling and flood mitigation: concerted action. In: Proceedings of the final workshop, February 1998, The European Commission, EUR 18287 EN, EC, pp 1–404
Sellin RHJ, van Beesten DP (2004) Conveyance of a managed vegetated two-stage river channel. J Water Manage, Inst Civil Eng 157(1):21–33
Sellin RHJ, Ervine AE, Willetts BB (1993) Behaviour of meandering two-stage channels. Water Maritime Energy Proc Inst Civil Eng 101(2):99–111

Sharifi S, Knight DW and Sterling M (2008) Modelling flow using SKM and a multi-objective evolutionary algorithm. In: Altinakar MS, Kokpinar MA, Aydin I, Cokgar S, Kirkgoz S (eds) RiverFlow 2008, Cesme, Turkey, vol 3, pp 2149–2158

Sharifi S, Knight DW and Sterling M (2009) A novel application of a multi-objective evolutionary algorithm in open channel flow modeling. J Hydroinf, IWA Publishing, vol 11(1), pp 31–50

Shiono K and Knight DW (1988) Two-dimensional analytical solution for a compound channel. In: Proceedings of 3rd international symposium on refined flow modelling and turbulence measurements, Tokyo, Japan, pp 503–510

Shiono K, Knight DW (1991) Turbulent open channel flows with variable depth across the channel. J Fluid Mech 222:617–646 and vol 231: Oct. 693

Shiono K, Muto Y (1998) Complex flow mechanisms in compound meandering channels with overbank flow. J Fluid Mech 376:221–261

Sun X, Shiono K (2009) Flow resistance of one-line emergent vegetation along the floodplain edge of a compound channel. Advs Water Resour 32:430–438

Tang X, Knight DW, Samuels PG (2001) Wave speed-discharge relationship from cross-section survey. In: Proceedings of institution of civil engineers, water and maritime engineering, London, vol 148. June, Issue 2, pp 81–96

Tang X, Knight DW (2008) Lateral depth-averaged velocity distributions and bed shear in rectangular compound channels. J Hydraul Eng ASCE, vol 134, No. 9, September, pp 1337–1342

Wallis SG, Knight DW (1984) Calibration studies concerning a one-dimensional numerical tidal model with particular reference to resistance coefficients. Estuarine, coastal and shelf science, vol 19. Academic, London, pp 541–562

Hydrodynamics of Undular Free Surface Flows

Subhasish Dey, Sujit K. Bose and Oscar Castro-Orgaz

Abstract In this study, the hydrodynamics of three types of steady undular free surface flow problems are theoretically analyzed. These flows are governed by the steady-state equations of conservation of mass and momentum developed by Bose and Dey in 2007 (J Hydraul Eng 133:1074–1079) and 2009 (Phys Rev E 80:036304). The first type of flow is an undular hydraulic jump on a plane smooth boundary when the approach flow Froude number marginally exceeds its critical value of unity. Here, it is established that the undular hydraulic jump phenomenon can be treated as an instability of the free surface flow being described by the instability principle. The results reveal that the threshold of an undular hydraulic jump is represented by a monotonic increase of approach flow Froude number with boundary inclination. The elevation of the undular free surface increases, as the boundary inclination increases. However, the amplitude of the free surface waves decreases with downstream distance. The second type of flow is that over a submerged hemi-cylindrical boss (inverted semi-circular section) placed on the channel bottom with its horizontal diameter lateral to the flow direction. The steady-state flow analysis shows that there is a drop in the free surface elevation on the downstream end of the cylinder with an undular free surface profile. The third

S. Dey (✉)
Department of Civil Engineering, Indian Institute of Technology, Kharagpur,
West Bengal 721302, India
e-mail: sdey@iitkgp.ac.in

S. K. Bose
Centre for Theoretical Studies, Indian Institute of Technology, Kharagpur,
West Bengal 721302, India
e-mail: sujitkbose@yahoo.com

O. Castro-Orgaz
Instituto de Agricultura Sostenible, Consejo Superior de Investigaciones Cientificas,
Finca Alameda del Obispo E-14080 Cordoba, Spain
e-mail: oscarcastro@ias.csic.es

P. Rowiński (ed.), *Experimental and Computational Solutions of Hydraulic Problems*,
GeoPlanet: Earth and Planetary Sciences, DOI: 10.1007/978-3-642-30209-1_3,

type of flow is that over a continuous sinusoidal boundary of a channel. A numerical experiment shows that there exists a phase lag between the free surface profile and the boundary profile. Also, it is revealed that if the flow depth is reduced, an accumulation of heaved wave occurs indicating an onset of chocking.

1 Introduction

Free surface profiles of open channel flows can exhibit considerable undulations, depending on the flow and the rigid boundary conditions. For example, an undular hydraulic jump can occur when the approach flow Froude number (F_0) slightly exceeds its critical value of unity. The free surface undulations, in this type of jump, result in an increase in mean free surface elevation, as F_0 becomes slightly greater than its critical value. It is therefore formally termed a *jump phenomenon*. Undulations of the channel boundary that can induce undular free surface profiles are also common in field conditions with complex bathymetry. Here, the undulations of the channel boundaries are considered as (1) isolated one in the form of a submerged hemi-cylindrical boss (inverted semi-circular section) and (2) continuous sinusoidal boundary profile in the streamwise direction. It may be noted that the fluid motion in the theory of such types of flows may be assumed as a steady-state, although the time development phenomenon can also be possible in case of loose boundary streams (Bose and Dey 2009).

The progress in hydrodynamic theories related to undular hydraulic jumps has so far been achieved with varying success. Primarily, it is attributed to the fact that an inviscid solution for the equations of fluid motion does not exist, even though the viscosity effect is quite weak (Benjamin and Lighthill 1954). Fawer (1937) was the first to treat an undular hydraulic jump problem by applying the potential flow theory as a series of cnoidal waves that existed following the first wave crest, but a transition from supercritical to subcritical flow was not considered. He determined the main undular flow characteristics by conducting the experiments. Applying the potential flow theory, Benjamin and Lighthill (1954) and Mandrup-Andersen (1978) hypothesized the first wave crest portion by a solitary wave, which included the transition from supercritical to subcritical flow, and the downstream flow portion by a series of subcritical cnoidal waves. Mandrup-Andersen (1978) adopted Boussinesq-type energy equation in the analysis. Iwasa (1955) also used the potential flow theory to analyze undular hydraulic jumps. He suggested a connection of the solitary and cnoidal waves at the location of the critical point $F_0 = 1$. On the other hand, Fawer (1937) and Mandrup-Andersen (1978) preferred a subcritical point ahead of the first wave crest. It is a traditional theoretical approach to have continuity at the meeting point of the solitary and the cnoidal wave portions to define the entire undular hydraulic jump profile. In this way, it ultimately takes into account the transitional flow from supercritical to subcritical flow induced by the boundary resistance (Marchi 1963; Hager and Hutter 1984a, b).

A failure on this account insisted Johnson (1972) to put forward a viscous flow theory pertaining to an asymptotic analysis for the approach flow Froude number close to unity, having high Reynolds numbers. His basic finding was a steady-state interpretation of the well-known *Korteweg-de Vries-Burgers* equation governing the perturbation on the free surface profiles. Inasmuch as the turbulence is involved in free surface flows, Kaufmann (1934) attempted by considering the effects of small-tube curvature and a boundary resistance coefficient. He found an exponentially decaying harmonic function for the free surface profiles. However, the solution obtained by him was not in conformity with the observations by Böβ (1927). In this regard, an improved theory was provided by Mandrup-Anderson (1978) and Montes and Chanson (1998) utilizing Boussinesq-type energy equation. Montes (1998), on the other hand, applied perturbation analysis to treat the turbulent flow equations. He found that the streamwise velocity obeys the 1/7th power law of the wall. Hager and Hutter (1984b) used an improved Boussinesq-type energy equation obtained from the Euler equations in natural coordinates of the flow net. They accounted for the energy dissipation by merging upstream and downstream solutions at a suitably chosen position on the undular hydraulic jump profiles. Applying the logarithmic law of the wall, Grillhofer and Schneider (2003) gave an asymptotic analysis of the turbulent flow equations. They solved a third order ordinary-differential equation for the free surface profiles leading to an undular hydraulic jump. Castro-Orgaz (2010) analyzed the influence of boundary resistance on the undular hydraulic jump by applying the turbulent boundary layer method for an oscillating pressure gradient. He put forward a simplified method to figure out the oscillatory boundary layer characteristics in a weakly undular hydraulic jump under a steadily changing pressure gradient from a positive to a negative and vice versa. Besides, number of good experimental studies detailing the flow characteristics of undular hydraulic jumps are reported elsewhere (Chanson 1993, 1995, 2000; Chanson and Montes 1995; Reinauer and Hager 1995; Ohtsu et al. 2001, 2003; Gotoh et al. 2005). In this study, we treat the phenomenon of the undular hydraulic jump in the context of the instability of flow giving rise to undulations in the free surface profiles of the subcritical flow zone.

Theoretical approaches to flows over isolated-submerged obstacles are limited to the geometry of an obstacle. Flow over a circular or rectangular cylinder placed (across the flow direction) on the channel bottom was mainly studied in the context of the wake flow behind the two-dimensional obstacles (Good and Joubert 1968; Sforza and Mons 1970; Dimaczek et al. 1989; Schulte and Rouvé 1986; Durão et al. 1991; Akoz and Kirkgoz 2009; Akoz et al. 2010). Here, we consider a hemi-cylindrical boss (inverted semi-circular section with its horizontal diameter in contact with the channel bottom) placed on the boundary as an obstacle with fillets at the base in order to make a smooth passage of the limiting streamlines near the base. In the context of classical hydrodynamics of an ideal fluid, the flow characteristics over a submerged hemi-cylindrical boss with axis lateral to the flow direction in a fluid of infinite depth can be obtained from the flow past a full cylinder (Lamb 1932). But when the viscous effect is taken into account, the solution is likely to be of completely different in nature.

Further, the undular channel boundaries of practical interest, such as flow over a natural dunal bed-form, are continuous. There exist a number of theoretical approaches, even though numerical techniques have widely been used to solve the problem (Zhaoshun and Zhan 1989; Patel et al. 1991; Nakayama and Sakio 2002; Tsai and Chou 2008). Henderson (1964) used the standard backwater approach, while Iwasa and Kennedy (1968) used Boussinesq-type energy equations in a boundary-fitted system of reference (Castro-Orgaz and Dey 2009), thereby accounting for the channel boundary curvature in the governing equations. Benjamin (1959), who studied the shear flow over a wavy boundary to find out normal and tangential stresses, also calculated the phase lag due to an effect of the boundary resistance. Motzfeld (1937) and Hsu and Kennedy (1971) observed a phase lag in the boundary pressure distribution relative to the sinusoidal boundary. Mizumura (1995) gave the detailed difference of the free surface profiles in the supercritical and subcritical flows over rigid wavy boundaries by using the potential flow theory. Here, we consider a continuous sinusoidal boundary for the analytical tractability. The elementary solution of the problem for the ideal fluid flow is available (Lamb 1932), but its viscous counterpart seems to have received little attention.

The aim of the study is, therefore, to provide a theoretical analysis for the aforementioned three types of flows by using a single governing equation, thereby providing a unified and generalized treatment of the undular free surface flows. The analysis is based on the Reynolds averaged Navier-Stokes (henceforth RANS) equations given by Bose and Dey (2007, 2009), where the 1D equations of conservation of mass and momentum for the 2D free surface flows on sloping and undular boundaries were systematically derived. In the next section, the generalized third-order ordinary differential equation is derived as the basic theory for the undular free surface profiles; and then, in the subsequent sections, the aforementioned types of flows are analyzed in the light of the developed theory.

2 Governing Equations of Undular Free Surface Flows

Considering a convective accelerated flow over a rigid undular boundary, whose mean boundary is inclined to an angle β with the horizontal (Fig. 1), the flow is essentially governed by the equations of continuity and momentum in the direction of the flow. Bose and Dey (2007, 2009) obtained the basic equation of fluid motion from RANS equations considering the streamwise velocity to preserve 1/7th power law of wall. With the x-axis along the mean level of the boundary and the y-axis perpendicular to it, the equation of continuity is

$$\frac{\partial}{\partial t}(\eta - h) + \frac{\partial}{\partial x}[(\eta - h)U] = 0 \tag{1}$$

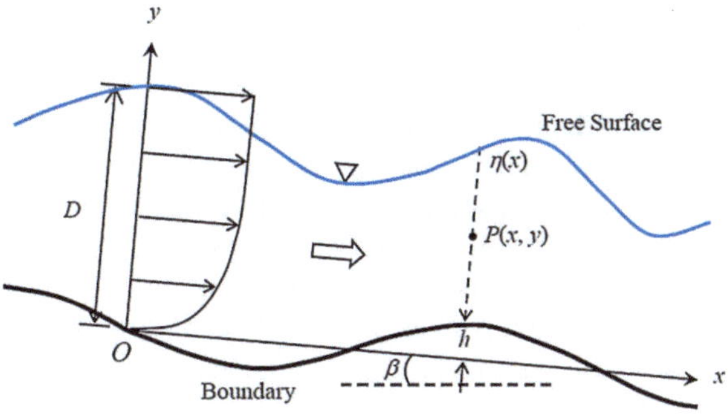

Fig. 1 Definition sketch of curvilinear flow over an undular boundary

where t is the time; $U(x, t)$ is the depth-averaged streamwise velocity, that is $(\eta - h)^{-1} \int_h^\eta \bar{u}(x, y, t)dy$; $h(x)$ is the elevation of the boundary profile from the mean level of the boundary or the x-axis; $\bar{u}$ is the time-averaged streamwise velocity at point $P(x, y)$; and $\eta(x, t)$ is the free surface elevation from the mean level of the boundary or the x-axis. The momentum equation is obtained as

$$\frac{\partial U}{\partial t} + U\frac{\partial U}{\partial x} + \frac{2}{5(\eta - h)} \cdot \frac{\partial}{\partial x}\left[(\eta - h)^2 U^2 \left(\frac{\partial^2 \eta}{\partial x^2} + \frac{7}{16} \cdot \frac{\partial^2 h}{\partial x^2}\right)\right] - \frac{7}{22}(\eta - h)^2 \frac{\partial^3 U}{\partial t \partial x^2}$$
$$+ g\cos\beta \frac{\partial \eta}{\partial x} - g\sin\beta + \frac{\tau_0}{\rho(\eta - h)} = 0 \tag{2}$$

where g is the gravitational acceleration; ρ is the mass density of fluid; and τ_0 is the boundary shear stress. According to the Manning equation, τ_0 is given by $\rho g n^2 U^2 / (\eta - h)^{1/3}$; where n is the Manning roughness coefficient.

The types of flows studied here are in steady-state, and accordingly, $\partial(\cdot)/\partial t = 0$. Equation (1) thus yields

$$(\eta - h)U = q \tag{3}$$

where q is the constant discharge per unit width across the section of interest. Equation (2) then under the same assumption becomes

$$\frac{d^3\eta}{dx^3} + \frac{5}{2}\left[\frac{g(\eta - h)\cos\beta}{q^2} - \frac{1}{(\eta - h)^2}\right]\frac{d\eta}{dx} + \frac{7}{16} \cdot \frac{d^3 h}{dx^3} + \frac{5}{2(\eta - h)^2} \cdot \frac{dh}{dx}$$
$$+ \frac{5}{2q^2}\left[\frac{\tau_0}{\rho} - g(\eta - h)\sin\beta\right] = 0 \tag{4}$$

Equation (4) determines the steady-state free surface profiles, when the undular boundary profile $h(x)$, inclination β and discharge q are known parameters. Next, we proceed to consider the specific cases.

3 Undular Hydraulic Jump over a Plane Sloping Boundary

Figure 2 depicts a schematic diagram of an undular hydraulic jump over a plane sloping boundary, where a supercritical approach flow changes to a subcritical over a short transitional reach. In this case, as the boundary is considered to be a plane, $h(x)$ is zero. It is well-known that as soon as the approach flow Froude number exceeds its critical value of unity by a small magnitude, the phenomenon of hydraulic jump takes place with an undular free surface profile in the subcritical flow zone. Therefore, Eq. (4) applies as

$$\frac{d^3\eta}{dx^3}+\frac{5}{2}\left(\frac{g\eta\cos\beta}{q^2}-\frac{1}{\eta^2}\right)\frac{d\eta}{dx}+\frac{5}{2q^2}\left(\frac{\tau_0}{\rho}-g\eta\sin\beta\right)=0 \tag{5}$$

where τ_0 is $\rho g n^2 U^2/\eta^{1/3}$. Here, the phenomenon of undular hydraulic jump is treated as an instability in the flow governed by Eq. (5). Let the flow depth be D at the origin O (Fig. 2). In the surroundings of O, Eq. (5) with $\eta = D$ yields

$$\tau_{0D} = \rho g D\sin\beta \tag{6}$$

where τ_{0D} is $\rho g n^2 q^2/D^{7/3}$. For an unstable solution of Eq. (5), let $\eta(x >> 0) = D(1 + \eta_1)$. Then, we can write to the first order approximation as

$$\tau_0 = \frac{\rho g n^2 q^2}{D^{7/3}}\left(1-\frac{7}{3}\eta_1\right) = \rho g D\sin\beta\left(1-\frac{7}{3}\eta_1\right) \tag{7}$$

and Eq. (5) becomes

$$\frac{d^3\eta_1}{dx^3}+\frac{5}{2}\left(\frac{gD\cos\beta}{q^2}-\frac{1}{D^2}\right)\frac{d\eta_1}{dx}-\frac{25}{3}\cdot\frac{g\sin\beta}{q^2}\eta_1=0 \tag{8}$$

Also, introducing flow Froude number F_0, that is, $q^2 = F_0^2 g D^3$, Eq. (8) becomes

$$\frac{d^3\eta_1}{dx^3}+\frac{5}{2D^2}\left(\frac{\cos\beta}{F_0^2}-1\right)\frac{d\eta_1}{dx}-\frac{25}{3}\cdot\frac{\sin\beta}{F_0^2 D^3}\eta_1=0 \tag{9}$$

Equation (9) is a third-order linear ordinary differential equation. Its solution is of the form $\eta_1 = E\exp(\lambda x)$, having a constant E and a coefficient λ. Inserting it into Eq. (9), λ must satisfy the cubic equation. Then

Fig. 2 Schematic of an undular hydraulic jump on a sloping boundary

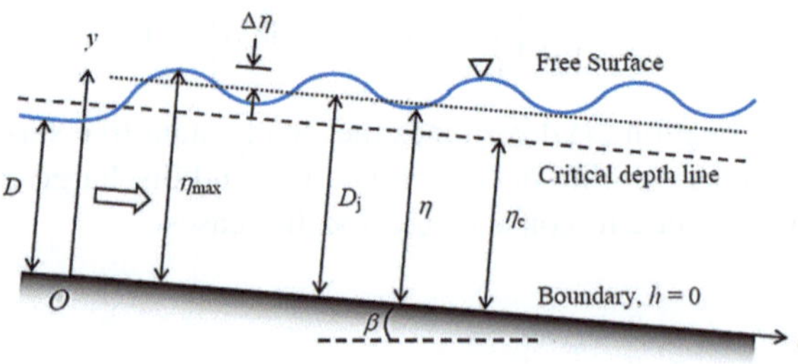

$$(D\lambda)^3+\frac{5}{2}\left(\frac{\cos\beta}{F_0^2}-1\right)D\lambda-\frac{25}{3}\cdot\frac{1}{F_0^2}\sin\beta=0 \tag{10}$$

Since the boundary inclination β is usually very small, the three roots of Eq. (10) are approximated as

$$D\lambda\approx\frac{10}{3}\cdot\frac{\sin\beta}{F_0^2-\cos\beta}\ \text{and}\ \frac{5}{3}\cdot\frac{\sin\beta}{F_0^2-\cos\beta}\pm\left[\frac{5}{2}\cdot\left(1-\frac{\cos\beta}{F_0^2}\right)\right]^{0.5} \tag{11}$$

When $F_0^2<\cos\beta$, the real parts of $D\lambda$ are all negative, referring to the decaying solutions. It indicates that there is no instability in the solution due to an exponential decrease in flow depth. Alternatively, when $F_0^2>\cos\beta$, all the roots are real and two of them are positive. It suggests an instability in fluid motion, resulting in an undular hydraulic jump, as observed in experiments.

More specifically, all the three roots of Eq. (10) are real when $\hat{q}^3+\hat{r}^2<0$ [see Abramowitz and Stegun (1972), p 17]; where

$$\hat{q}=\frac{5}{6}\cdot\left(\frac{\cos\beta}{F_0^2}-1\right)\ \text{and}\ \hat{r}=\frac{25}{6}\cdot\frac{\sin\beta}{F_0^2} \tag{12}$$

The condition for real roots on simplification is as follows:

$$\left(\cos\beta-F_0^2\right)^3+30F_0^2\sin^2\beta<0 \tag{13}$$

Equation (13) is satisfied if F_0^2 to some extent exceeds $\cos\beta$, as $\sin^2\beta$ is very small. The condition for an undular hydraulic jump to occur can therefore be defined by plotting a curve F_0 versus β, as given in Fig. 3, obtained by equating the left hand side of Eq. (13) to zero. The curve defines the necessary hydraulic condition for the formation of an undular hydraulic jump. It is evident that the threshold of an undular hydraulic jump corresponds to a monotonic increase of F_0 with β. A negative value of the left hand side of Eq. (13) refers to the zone of the upper side of the curve. It is therefore the instability zone resulting in an undular hydraulic jump. Experimental data of Gotoh et al. (2005) for undular hydraulic jumps on sloping boundaries are also shown in Fig. 3. It confirms that the experimental data plots lie on the zone of the upper side of the threshold curve.

For plotting the free surface profiles, Eq. (5) is expressed in normalized form as

$$\frac{d^3\hat{\eta}}{d\hat{x}^3}+\frac{5}{2}\left(\frac{\hat{\eta}}{F_0^2}\cos\beta-\frac{1}{\hat{\eta}^2}\right)\frac{d\hat{\eta}}{d\hat{x}}+\frac{5}{2}\left(\frac{\Omega}{\hat{\eta}^{2/3}}-\frac{\hat{\eta}}{F_0^2}\sin\beta\right)=0 \tag{14}$$

where $\hat{\eta}$ is η/D; $\hat{x}$ is x/D; and Ω is $gn^2/D^{1/3}$, that is termed a resistance parameter.

The numerical integration of Eq. (14), which is a third-order differential equation, is performed by using the phase variables to convert Eq. (14) into three first-order differential equations. The initial conditions are given as: at $\hat{x}=0$, $\hat{\eta}=1$, $d\hat{\eta}/d\hat{x}=0$ and $d^2\hat{\eta}/d\hat{x}^2=0.01$ (say) a small value. The value of $\Omega=0.004$ that is relevant for a smooth boundary is considered for the computation. Equation (14) is

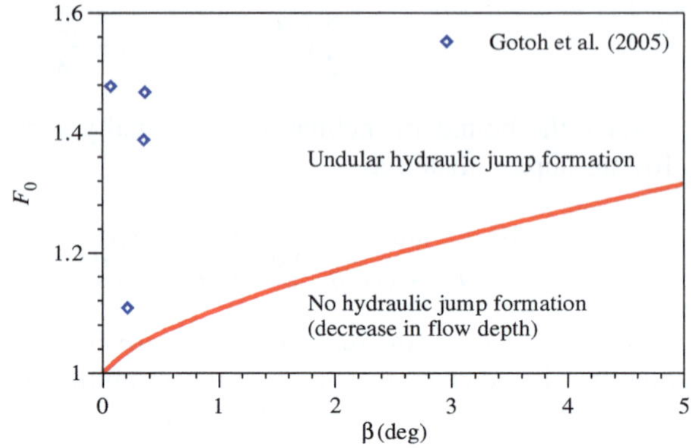

Fig. 3 F_0 versus β as threshold condition for the formation of an undular hydraulic jump

solved for the given values of F_0 and β by the Runge–Kutta method. Figure 4a and b present the computed free surface profiles ($\hat{\eta}$ versus $\hat{x}$) of undular hydraulic jumps for $F_0 = 1.11$ and 1.39 corresponding to $\tan\beta = 0.00355$ and 0.00588, respectively. The values of F_0 and $\tan\beta$ are chosen in such a way so that a direct comparison with the experimental data of Gotoh et al. (2005) is possible. Note that their experimental conditions were same as F_0 and $\tan\,\beta$ values in Fig. 4a and b having $B/D = 9$ and 10.4 for the former and latter pairs of F_0 and $\tan\beta$, respectively, where B is the channel width. The computed free surface profiles in general agree well with the experimental data, although a slight discrepancy is apparent in the third wave

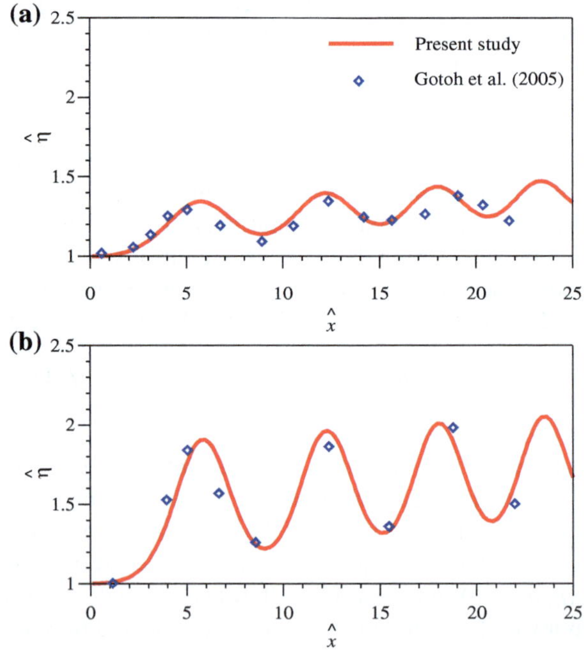

Fig. 4 Normalized profiles ($\hat{\eta}$ versus $\hat{x}$) of undular hydraulic jumps for (**a**) $F_0 = 1.11$ and $\tan\,\beta = 0.00355$ and (**b**) $F_0 = 1.39$ and $\tan\,\beta = 0.00588$ overlapping experimental data of Gotoh et al. (2005)

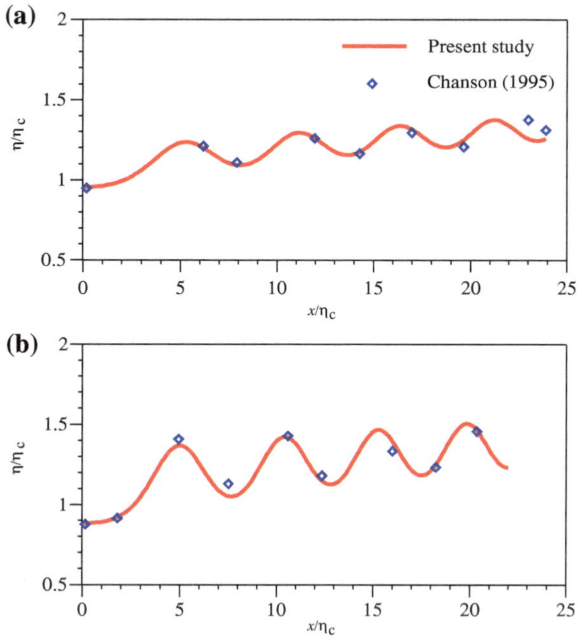

Fig. 5 Normalized profiles ($\hat{\eta}$ versus $\hat{x}$) of undular hydraulic jumps for (**a**) $F_0 = 1.07$ and $\tan\beta = 0.00433$ and (**b**) $F_0 = 1.21$ and $\tan\beta = 0.00567$ overlapping experimental data of Chanson (1995)

portion. The discrepancy may be attributed to the sidewall effect on an experimental flume that could induce lateral instability resulting in shifting of the consecutive crests of the undulations at the free surface. Besides, the free surface profiles, for $F_0 > 1.2$, are 3D with shockwaves (Reinauer and Hager 1995; Chanson and Montes 1995), which could be another source for the deviation of the prediction from the experimental data.

However, more comparative results are provided in Fig. 5a and b for $F_0 = 1.07$ and 1.21 corresponding to $\tan\beta = 0.00433$ and 0.00567, respectively, where the experimental data of Chanson (1995) are compared. His experimental conditions were same as F_0 and $\tan\beta$ values in Fig. 5a and b having $\eta_c/B = 0.454$ for both the experiments; where η_c is the critical flow depth, that is $DF_0^{2/3}$. The computed curves that are shown as η/η_c versus x/η_c according to the experimental data plotted by Chanson (1995) are in agreement with the experimental data.

Figure 6 displays the theoretically obtained normalized free surface profiles ($\hat{\eta}$ versus $\hat{x}$) of undular hydraulic jumps for $\beta = 1$, 2 and 3° and $F_0 = 1.3$. The undular free surface profiles are elevated progressively with an increase in boundary inclination β; and the amplitude of the free surface waves decreases with distance x becoming a flat surface at far downstream.

In Fig. 7, the variation of computed normalized elevation of first wave crest $\hat{\eta}_m$ ($= \eta_{max}/D$; where η_{max} is the elevation of first wave crest) with F_0 for $\beta = 0°$ is shown. The $\hat{\eta}_m$ increases with an increase in F_0. It is pertinent to mention that the curves ($\hat{\eta}_m$ versus F_0) are not apparently different for the small variation of β; such as $\beta = 0°$, arctan(1/1,000) and arctan(1/163), which correspond to the horizontal,

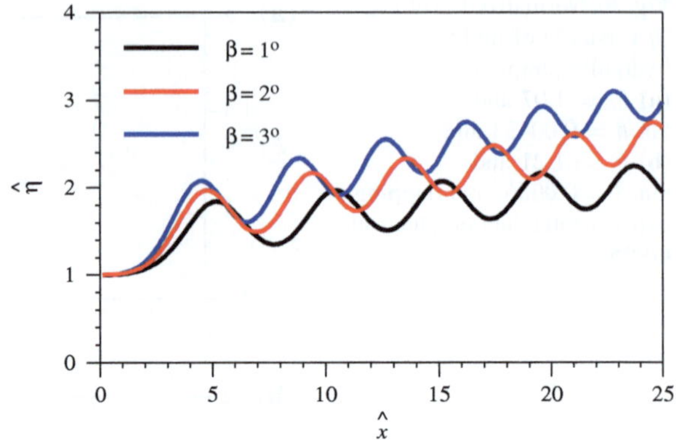

Fig. 6 Normalized profiles ($\hat{\eta}$ versus $\hat{x}$) of undular hydraulic jumps for $\beta = 1$, 2 and 3° and $F_0 = 1.3$

mild and steep slopes, respectively. The curve appears to be somewhat linear and is satisfactorily in agreement with the experimental data of Gotoh et al. (2005), even though a departure is noticeable for the higher values of F_0. In reality, for higher values of F_0, the hydraulic jump phenomenon in the experiments does not belong to the category of an undular hydraulic jump. Figure 8 presents dependency of the normalized first wave height $\Delta\eta/\eta_c$ on F_0; where $\Delta\eta$ is the first wave height with respect to the conjugate flow depth D_j [= $0.5D(\sqrt{1 + 8F_0^2} - 1)$] of the free hydraulic jump. The $\Delta\eta/\eta_c$ varies monotonically with an increase in F_0. There is some agreement between the computed curve of the present study and that obtained from the Boussinesq equation developed by Mandrup-Anderson (1978) for $F_0 < 1.4$, but a disagreement is always prevalent for the higher values of F_0, as the reason has already been stated.

Figure 9 presents the variation of normalized first wave length $\Delta L/\eta_c$ with F_0; where ΔL is the first wave length, that is, the distance between the first two wave crests. The $\Delta L/\eta_c$ diminishes with F_0 becoming almost independent of F_0 for $F_0 > 3$, where in fact the undular hydraulic jump phenomenon hardly exists. Due to the backwater effects in the experimental flume induced by the tailgate or the cross flows induced by the sidewalls, a reduction in wave length of undular

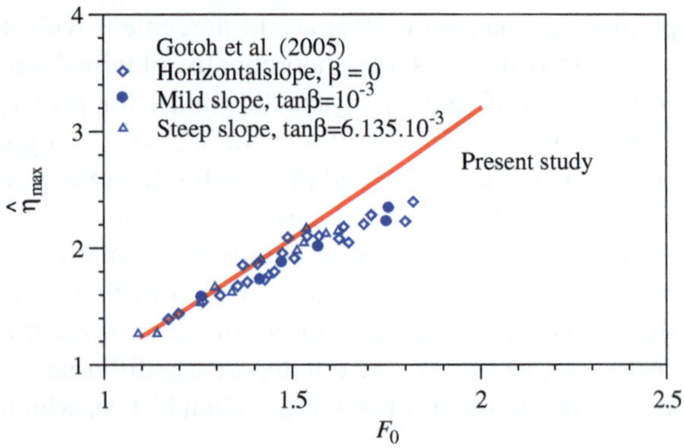

Fig. 7 Normalized elevation of first wave crest $\hat{\eta}_{max}$ as a function of F_0 and a comparison with the experimental data of Gotoh et al. (2005)

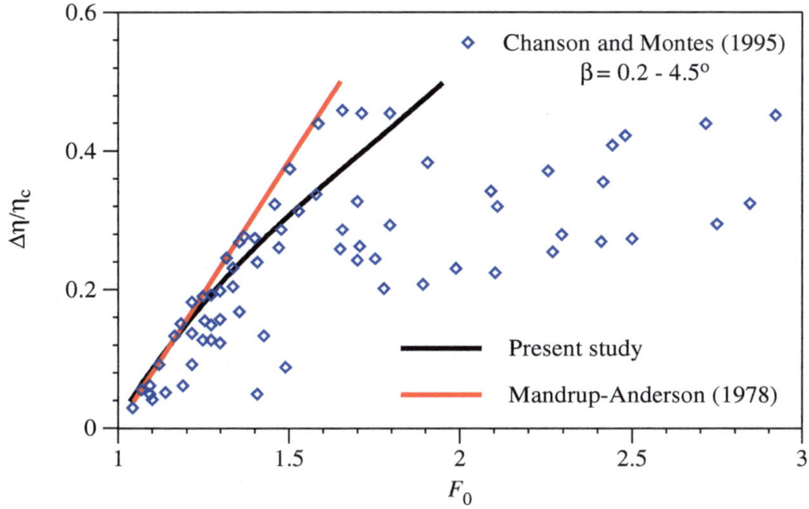

Fig. 8 Normalized first wave height $\Delta\eta/\eta_c$ as a function of F_0 and a comparison with the experimental data of Chanson and Montes (1995)

hydraulic jumps is usually prevalent. It is therefore reflected in the computed curve, $\Delta L/\eta_c$ versus F_0, as it is compared with the experimental data of Chanson and Montes (1995) in Fig. 9. However, the use of an empirical factor of 0.77 enables to bring down the computed curve over the experimental data plots. Note that the factor 0.77 shows the degree of discrepancy that exists between the theoretical curve and the experimental data of Chanson and Montes (1995). Nevertheless, there is also a possibility that the present theory overestimates the experimental data on first wave length.

4 Flow over a Hemi-Cylindrical Boss

Flow over a submerged hemi-cylindrical boss (inverted semi-circular section), having radius a placed on the channel bottom lateral to the flow direction with its horizontal diameter in contact with the bottom, is considered here (Fig. 10). The approach flow depth D is greater than the radius a of the boss. In order to have a gradual turn of the limiting streamlines at the base of the boss, it is assumed that the fillets of width $a\varepsilon$ on both sides of the base of the boss are attached. Here, ε is a factor. For this type of flow, Eq. (4) applies where $\beta = 0$ and $\tau_0 = \rho g n^2 U^2/(\eta - h)^{1/3}$, so that

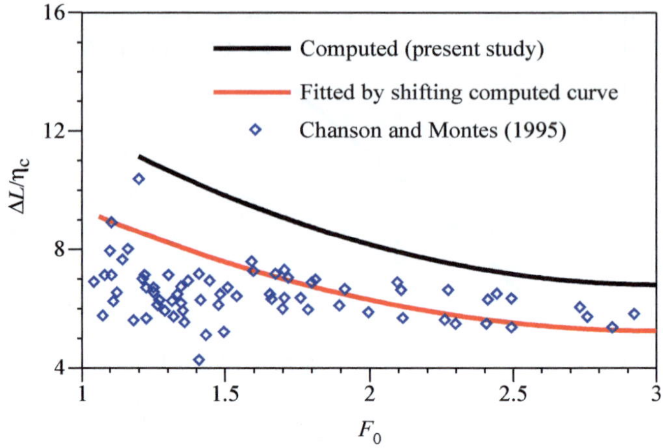

Fig. 9 Normalized length of first wave $\Delta L/\eta_c$ as a function of F_0 and a comparison with the experimental data of Chanson and Montes (1995)

$$\frac{d^3\eta}{dx^3}+\frac{5}{2}\left[\frac{g(\eta-h)}{q^2}-\frac{1}{(\eta-h)^2}\right]\frac{d\eta}{dx}+\frac{7}{16}\cdot\frac{d^3h}{dx^3}+\frac{5}{2(\eta-h)^2}\cdot\frac{dh}{dx}+\frac{5}{2}\cdot\frac{gn^2}{(\eta-h)^{7/3}}=0 \tag{15}$$

By setting $\tilde{\eta}=(\hat{\eta}-1)/\alpha$; $\alpha=a/D$; $\tilde{x}=x/a$; and $\tilde{h}=h/a$, Eq. (15) can be rewritten in normalized form as

$$\frac{d^3\tilde{\eta}}{d\tilde{x}^3}+\frac{5\alpha^2}{2}\left\{\frac{1+\alpha(\tilde{\eta}-\tilde{h})}{F_0^2}-\frac{1}{[1+\alpha(\tilde{\eta}-\tilde{h})]^2}\right\}\frac{d\tilde{\eta}}{d\tilde{x}}+\frac{7}{16}\cdot\frac{d^3\tilde{h}}{d\tilde{x}^3}+\frac{5\alpha^2}{2}\cdot\frac{1}{[1+\alpha(\tilde{\eta}-\tilde{h})]^2}\frac{d\tilde{h}}{d\tilde{x}}+\frac{5\Omega\alpha^2}{2}\cdot\frac{1}{[1+\alpha(\tilde{\eta}-\tilde{h})]^{7/3}}=0 \tag{16}$$

where $\tilde{x}^2+\tilde{h}^2=1$ on the surface of the boss, so that $d\tilde{h}/d\tilde{x}=-\tilde{x}/(1-\tilde{x}^2)^{1/2}$ and $d^3\tilde{h}/d\tilde{x}^3=-3\tilde{x}/(1-\tilde{x}^2)^{5/2}$. Thus, for $|\tilde{x}|\le 1-\varepsilon$ (surface of the boss), writing $f=1+\alpha[\tilde{\eta}-(1-\tilde{x}^2)^{1/2}]$ yields

Fig. 10 Schematic of flow over a hemi-cylindrical boss

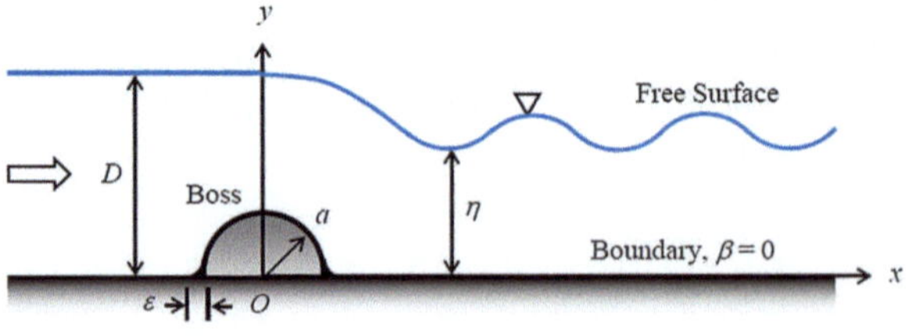

$$\frac{d^3\tilde{\eta}}{d\tilde{x}^3}+\frac{5\alpha^2}{2}\left(\frac{f}{F_0^2}-\frac{1}{f^2}\right)\frac{d\tilde{\eta}}{d\tilde{x}}-\frac{21}{26}\cdot\frac{\tilde{x}}{(1-\tilde{x}^2)^{5/2}}-\frac{5\alpha^2}{2}\cdot\frac{\tilde{x}}{f^2(1-\tilde{x}^2)^{1/2}} \\ +\frac{5\Omega\alpha^2}{2}\cdot\frac{1}{f^{7/3}}=0 \tag{17}$$

In the left fillet region $|\tilde{x}+1|<\varepsilon$, if $f_{-1}=1+\alpha[\tilde{\eta}-m(\tilde{x}-1-\varepsilon)]$, where m is the slope given by $m=0.5[(2-\varepsilon)\varepsilon]^{1/2}/\varepsilon$, then Eq. (17) becomes

$$\frac{d^3\tilde{\eta}}{d\tilde{x}^3}+\frac{5\alpha^2}{2}\left(\frac{f_{-1}}{F_0^2}-\frac{1}{f_{-1}^2}\right)\frac{d\tilde{\eta}}{d\tilde{x}}+\frac{5\alpha^2 m}{2}\cdot\frac{1}{f_{-1}^2}+\frac{5\Omega\alpha^2}{2}\cdot\frac{1}{f_{-1}^{7/3}}=0 \tag{18}$$

In the right fillet region $|\tilde{x}-1|<\varepsilon$, similarly, if $f_{+1}=1+\alpha[\tilde{\eta}+m(\tilde{x}-1-\varepsilon)]$, then Eq. (17) becomes

$$\frac{d^3\tilde{\eta}}{d\tilde{x}^3}+\frac{5\alpha^2}{2}\left(\frac{f_{+1}}{F_0^2}-\frac{1}{f_{+1}^2}\right)\frac{d\tilde{\eta}}{d\tilde{x}}-\frac{5\alpha^2 m}{2}\cdot\frac{1}{f_{+1}^2}+\frac{5\Omega\alpha^2}{2}\cdot\frac{1}{f_{+1}^{7/3}}=0 \tag{19}$$

Beyond the fillet region $1+\varepsilon>\tilde{x}>-(1+\varepsilon)$, $\hat{h}=0$, and Eq. (17) simply reduces to

$$\frac{d^3\tilde{\eta}}{d\tilde{x}^3}+\frac{5\alpha^2}{2}\left(\frac{f}{F_0^2}-\frac{1}{f^2}\right)\frac{d\tilde{\eta}}{d\tilde{x}}+\frac{5\Omega\alpha^2}{2}\cdot\frac{1}{f^{7/3}}=0 \tag{20}$$

where $f=1+\alpha\tilde{\eta}$.

Equations (17)–(20) are third-order ordinary differential equation and can be cast as systems of first order differential equations that can be integrated by Runge-Kutta method (Bose 2009). For numerical computation, it is assumed that the approach flow Froude number is $F_0=0.2$ and the fillet dimension $\varepsilon=0.1$. The values of α are selected as 1/4 and 1/3; and the computational results are shown as normalized free surface profiles ($\tilde{\eta}$ versus $\tilde{x}$) in Fig. 11. For both the cases, there is a reduction in mean free surface elevation that exhibits undular profiles extending downstream. A close observation reveals that the amplitude of the waves diminishes with an increase in downstream distance $\tilde{x}$. However, the reduction in mean free surface elevation and the amplitude of the waves increase with an increase in α.

5 Flow over a Sinusoidal Boundary

In Fig. 12, a unidirectional flow over a sinusoidal boundary of small amplitude is considered. The x-axis is set through the mean boundary level, which is a horizontal line, and the origin O is set conveniently on this axis. The y-axis is therefore vertically upwards. Due to the consideration of gradual variation of the boundary undulation, the maximum boundary perturbation $|h|$ is small compared to the wave

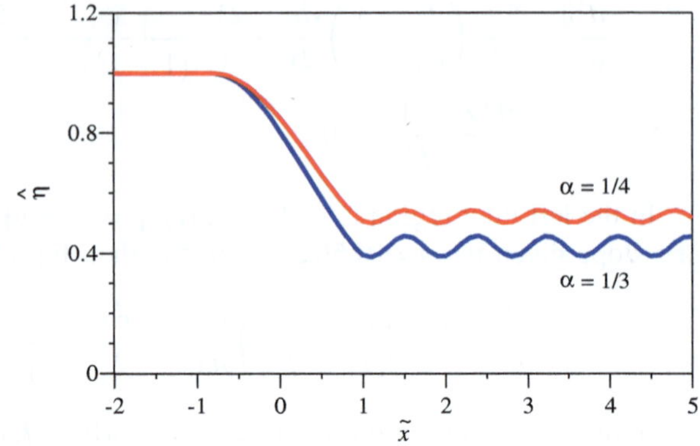

Fig. 11 Normalized free surface profiles ($\hat{\eta}$ versus $\tilde{x}$) for flow over bosses having $\alpha = 1/4$ and $1/3$

length, and its streamwise gradient is therefore $|\partial h/\partial x| << 1$. Likewise, maximum free surface perturbation $|\eta|$ must be small and hence $|\partial \eta/\partial x| << 1$. The boundary has a sinusoidal form given by $h = A \sin(kx)$; where A is the amplitude; and k is the wave number. Then, introducing the normalized quantities as $\eta_0 = (\hat{\eta} - 1)/\delta$; $\delta = A/D$; $\xi = kx$; and $\sigma = kD$, Eq. (4) for the free surface profile becomes

$$\frac{d^3\eta_0}{d\xi^3} + \frac{5}{2\sigma^2}\left(\frac{1}{f_s F_0^2} - \frac{1}{f_s^2}\right)\frac{d\eta_0}{d\xi} - \frac{7}{16}\cos\xi + \frac{5}{2\sigma^2}\cdot\frac{1}{f_s^2}\cos\xi + \frac{5\Omega}{2\delta\sigma^3}\cdot\frac{1}{f_s^{7/3}} = 0 \quad (21)$$

where $f_s = 1 + \delta(\eta_0 - \sin\xi)$. The quantities δ and σ can be interpreted as normalized amplitude and wave number relative to the mean flow depth D.

The solution of Eq. (21) must be periodic with a period 2π. Thus, to compute the wavy pattern of the free surface, we can simulate the solution of the differential equation, Eq. (21), subject to the periodicity of elevation η_0, slope $d\eta_0/d\xi$ and approximate curvature $d^2\eta_0/d\xi^2$. A typical numerical experiment has been conducted for the values of $\delta = 0.1$, $\sigma = 13$, $F_0 = 0.2$ and $\Omega = 0.004$. Equation (21) was then solved by the Runge-Kutta method (Bose 2009); and periodicity of the solution was checked for five wavelengths of the boundary. A satisfactory solution was obtained by taking the suitable initial values of $\eta_0 = 0.8$, $d\eta_0/d\xi = -0.71$ and $d^2\eta_0/d\xi^2 = -0.002$ at the origin $\xi = 0$ by trial. The role of initial values is to

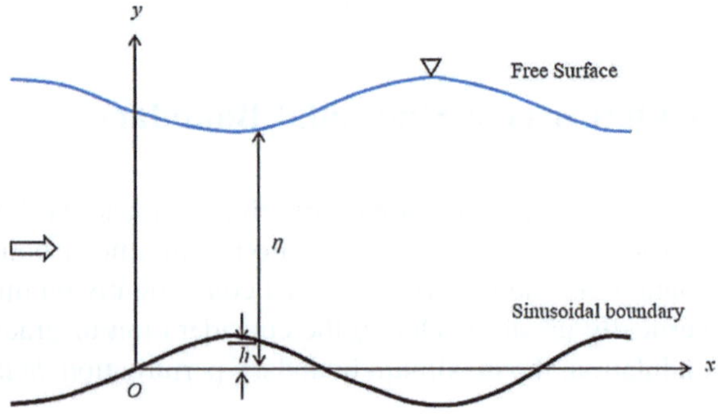

Fig. 12 Schematic of flow over a sinusoidal boundary

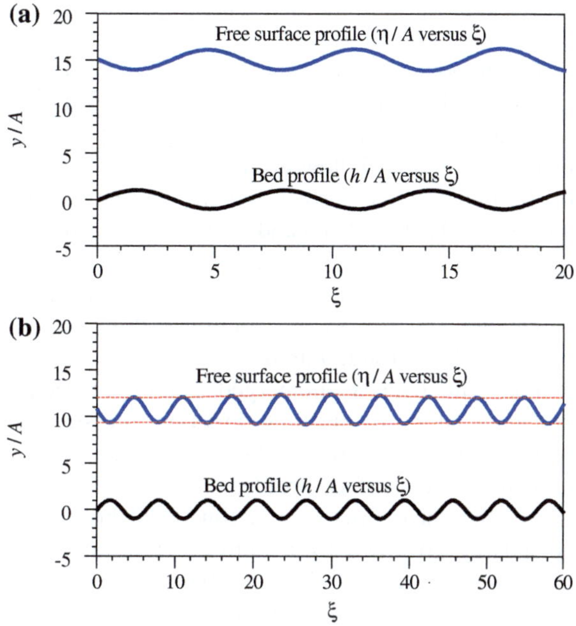

Fig. 13 Normalized free surface profiles (y/A versus ξ) for flow over sinusoidal boundaries: (**a**) for $\delta = 0.1$, $\sigma = 13$, $F_0 = 0.2$ and $\Omega = 0.004$; and (**b**) for $\delta = 0.1$, $\sigma = 9.5$, $F_0 = 0.2$ and $\Omega = 0.004$

initiate the computation for the periodical type of solution of the differential equation. The computed wavy free surface profile is shown in Fig. 13a. It is evident that there is a spatial phase lag between the wavy free surface and the sinusoidal boundary. It is the normalized distance of a wave crest of free surface from that of the nearest boundary crest. The lag is found to be approximately 3.

In another numerical experiment, the value of σ was reduced keeping the other parameters to remain unchanged. It implies that the mean flow depth D is reduced and made closer to the wavelength of the boundary profile. Having taken $\sigma = 9.5$, it was observed that the periodicity of 2π could not be attained. The initial values that yielded closest to the periodicity were found to be $\eta_0 = 0.8$, $d\eta_0/d\xi = -1.41$ and $d^2\eta_0/d\xi^2 = -0.003$. The profile of the free surface is plotted in Fig. 13b, where the peaks of the waves definitely show periodic groups of waves in a heaving motion. However, the effect becomes more pronounced for smaller values of σ. This is an interesting feature of the free surface profile for shallow flow depths over a sinusoidal boundary.

6 Conclusions

A theory of undular free surface flows has been developed. The theory is based on the RANS and the time-averaged continuity equations. For the analysis, the time-averaged streamwise velocity and the boundary resistance are assumed to follow the 1/7th power law and the Manning equation, respectively, as proposed by Bose and Dey (2007, 2009).

It is hypothesized that the undular hydraulic jump phenomenon can be treated by the instability principle of a third-order differential equation. The threshold of an undular hydraulic jump is illustrated by a curve that represents a monotonic increase of approach flow Froude number with boundary inclination. The elevation of the undular free surface increases progressively, as the boundary inclination increases. On the other hand, the amplitude of the free surface waves decreases with an increase in downstream distance. The elevation of first wave crest gets higher with an increase in approach flow Froude number. Also, the first wave height with respect to the conjugate flow depth increases with an increase in approach flow Froude number.

For flow over a submerged hemi-cylindrical boss placed on the channel boundary lateral to the flow direction, the steady-state flow analysis shows that there is a drop in the free surface elevation on the downstream of the cylinder with an undular free surface profile.

For the shear flow over a sinusoidal boundary, the free surface elevation is adequately governed by a third-order ordinary differential equation. Numerical experiments revealed that the free surface profile lags the boundary profile. Additionally, it has been observed that when the flow depth decreases, an accumulation of heaved waves in the free surface is formed.

This study, therefore, presents a generalized unified treatment of shear flows with a wavy free surface profile of interest both in hydraulic structures, e.g., undular jumps on a flat bottomed channel, and in geophysical processes, e.g., flow over a natural dunal bed-form.

References

Abramowitz M, Stegun IA (1972) Handbook of mathematical functions with formulas, graphs, and mathematical tables. Dover, New York

Akoz MS, Kirkgoz MS (2009) Numerical and experimental analyses of the flow around a horizontal wall-mounted circular cylinder. Trans Can Soc Mech Eng 33(2):189–215

Akoz MS, Sahin B, Akilli H (2010) Flow characteristic of the horizontal cylinder placed on the plane boundary. Flow Meas Instrum 21:476–487

Benjamin TB (1959) Shearing flow over a wavy boundary. J Fluid Mech 6:161–205

Benjamin TB, Lighthill MJ (1954) On cnoidal waves and bores. Proc R Soc Lond A 224:448–460

Bose SK (2009) Numeric computing in FORTRAN. Narosa, New Delhi

Bose SK, Dey S (2007) Curvilinear flow profiles based on Reynolds averaging. J Hydraul Eng 133(9):1074–1079

Bose SK, Dey S (2009) Reynolds averaged theory of turbulent shear flow over undulating beds and formation of sand waves. Phys Rev E 80:036304

Böβ P (1927) Berechnung der wasserspiegellage. Forschungsarbeiten VDI 284

Castro-Orgaz O (2010) Weakly undular hydraulic jump: effects of friction. J Hydraul Res 48(4):453–465

Castro-Orgaz O, Dey S (2009) One-dimensional channel flow equations with curvature revisited. J Hydraul Res 47(2):157–166

Chanson H (1993) Characteristics of undular hydraulic jumps. Research report CE146, Department of Civil Engineering, University of Queensland, Brisbane, Australia
Chanson H (1995) Flow characteristics of undular hydraulic jumps: comparison with near-critical flows. Research report CH45/95, Department of Civil Engineering, University of Queensland, Brisbane, Australia
Chanson H (2000) Boundary shear stress measurements in undular flows: application to standing wave bed forms. Water Resour Res 36(10):3063–3076
Chanson H, Montes JS (1995) Characteristics of undular hydraulic jumps: experimental apparatus and flow patterns. J Hydraul Eng 121(2):129–144
Dimaczek G, Tropea C, Wang AB (1989) Turbulent flow over two-dimensional, surface mounted obstacles: plane and axisymmetric geometries. In: Proceedings of the second European turbulence conference, Berlin, Germany, pp 114–121
Durão DFG, Gouveia PST, Pereira JCF (1991) Velocity characteristics of the flow around a square cross section cylinder placed near a channel wall. Exp Fluids 11:341–350
Fawer C (1937) Etude de quelques écoulements permanents à filets courbes. PhD thesis, Université de Lausanne, La Concorde, Lausanne, Switzerland (in French)
Good MC, Joubert PN (1968) The form drag of two-dimensional bluff plates immersed in the turbulent boundary layers. J Fluid Mech 31:547–582
Gotoh H, Yasuda Y, Ohtsu I (2005) Effect of channel slope on flow characteristics of undular hydraulic jumps. Trans Ecol Environ 83:33–42
Grillhofer W, Schneider W (2003) The undular hydraulic jump in turbulent open channel flow at large Reynolds numbers. Phys Fluids 15(3):730–735
Hager WH, Hutter K (1984a) Approximate treatment of plane channel flow. Acta Mech 51(3–4):31–48
Hager WH, Hutter K (1984b) On pseudo-uniform flow in open channel hydraulics. Acta Mech 53(3–4):183–200
Henderson FM (1964) Steady flow in sinusoidally varying channels. In: Proceedings of the first Australasian fluid mechanics conference, pp 51–67
Hsu ST, Kennedy JF (1971) Turbulent flow in wavy pipes. J Fluid Mech 47:481–502
Iwasa Y (1955) Undular jump and its limiting conditions for existence. In: Proceedings of the fifth Japan national congress for applied mechanics, Japan, vol II-14, pp 315–319
Iwasa Y, Kennedy JF (1968) Free surface shear flow over a wavy bed. J Hydraul Div 94(2):431–454
Johnson RS (1972) Shallow water waves on a viscous fluid—the undular bore. Phys Fluids 15:1693–1699
Kaufmann K (1934) Hydromechanik II. Springer, Berlin
Lamb H (1932) Hydrodynamics. Cambridge University Press, Cambridge
Mandrup-Andersen V (1978) Undular hydraulic jump. J Hydraul Div 104(8):1185–1188
Marchi E (1963) Contributo allo studio del risalto ondulato. Giornale del Genio Civile 101(9):466–476
Mizumura K (1995) Free-surface profile of open-channel flow with wavy boundary. J Hydraul Eng 121(7):533–539
Montes JS (1998) Hydraulics of open channel flow. ASCE Press, Reston
Montes JS, Chanson H (1998) Characteristics of undular hydraulic jumps: Results and calculations. J Hydraul Eng 124(2):192–205
Motzfeld H (1937) Die turbulente stromung an welligen wanden. Zeitschrift für Angewandte Mathematik und Mechanik 17:193–212
Nakayama A, Sakio K (2002) Simulation of flows over wavy rough boundaries. Annual research briefs, Center for Turbulence Research, Kobe University, Japan
Ohtsu I, Yasuda Y, Gotoh H (2001) Hydraulic condition for undular jump formations. J Hydraul Res 39(2):203–209
Ohtsu I, Yasuda Y, Gotoh H (2003) Flow conditions of undular hydraulic jumps in horizontal rectangular channels. J Hydraul Eng 129(12):948–955

Patel VC, Chon JT, Yoon JY (1991) Turbulent flow in a channel with a wavy wall. J Fluid Mech 113(4):579–586
Reinauer R, Hager WH (1995) Non-breaking undular hydraulic jump. J Hydraul Res 33(5):1–16
Schulte H, Rouvé G (1986) Turbulent structures in separated flow. In: Proceedings of the third international symposium on application of laser anemometry to fluid mechanics, Lisbon, Portugal, pp 2.2
Sforza PM, Mons RF (1970) Wall-wake: flow behind a leading edge obstacle. AIAA J 8(2):2162–2167
Tsai Y-P, Chou C–C (2008) Investigations of turbulent flow over solid wavy boundary. Chung Hua J Sci Eng 6(2):9–14
Zhaoshun Z, Zhan C (1989) Numerical study of turbulent flows over wavy boundaries. Acta Mech Sin 5(3):197–204

Evidence of Non-Universality of von Kármán's κ

Roberto Gaudio and Subhasish Dey

Abstract A notable universal feature of wall-bounded turbulent flows is the universal logarithmic law of the wall deduced by Theodore von Kármán. This law of the wall describes how time-averaged streamwise velocity changes with distance from the wall. Despite the law of the wall having a universal von Kármán constant $\kappa = 0.41$ that governs the slope of the log-law velocity profile, as commonly known over a period of about 80 years, in fluvial streams there are a number of instances of the non-universality of κ. To be specific, it behaves as a variable in flows with low relative submergence, or where there is bed-load and/or suspended-load sediment transport. This article focuses on the aspect of non-universality of κ by inviting various open questions relating to future research directions.

1 Introduction

The *von Kármán constant* κ relates the time-averaged streamwise velocity $u(y)$ profile along the vertical axis y in a wall-bounded shear flow to the shear stress at the bed surface, y being the elevation above the wall. The universal logarithmic law of time-averaged streamwise velocity profile for the inner scaling

R. Gaudio (✉)
Dipartimento di Difesa del Suolo "V. Marone", Università della Calabria,
87036 Rende (CS), Italy
e-mail: gaudio@unical.it

S. Dey
Department of Civil Engineering, Indian Institute of Technology,
Kharagpur 721302, West Bengal, India
e-mail: sdey@iitkgp.ac.in

P. Rowiński (ed.), *Experimental and Computational Solutions of Hydraulic Problems*,
GeoPlanet: Earth and Planetary Sciences, DOI: 10.1007/978-3-642-30209-1_4,

region in the classical turbulent boundary layer discovered by Theodore von Kármán in 1930 reads as follows (von Kármán 1930):

$$\frac{u(y)}{u_\tau} = \frac{1}{\kappa}\ln\left(\frac{y}{y_0}\right) \tag{1}$$

where u_τ is the shear velocity and y_0 is the zero-velocity level. In the classical theory of wall-law, the coefficient κ of the logarithmic term, termed the *von Kármán constant*, is considered to have a universal value of 0.41. The zero-velocity level y_0 is dependent on smooth, transitional and rough flow regimes governed by the shear Reynolds number R_* ($=u_\tau k_s/\upsilon$; where k_s is the Nikuradse equivalent sand roughness and υ is the kinematic viscosity of fluid) and/or k_s. The expressions of y_0 for different flow regimes are (van Rijn 1993):

$$y_0(R_* \leq 3) = \frac{\upsilon}{9.1u_\tau} \quad \text{for smooth regime} \tag{2a}$$

$$y_0(3 < R_* \leq 70) = \frac{\upsilon}{9.1\,u_\tau} + \frac{k_s}{30} \quad \text{for transition regime} \tag{2b}$$

$$y_0(R_* > 70) = \frac{k_s}{30} \quad \text{for rough regime} \tag{2c}$$

In another way, the von Kármán constant κ can be defined as the ratio of the mixing-length to the vertical distance y from the wall. There have been some attempts to compute κ mathematically for an idealized flow over a smooth rigid wall or for homogeneous turbulence. For instance, for an idealized flow, Long et al. (1993) analyzed the turbulent flow in smooth pipes with an eddy viscosity closure, obtaining $\kappa = 0.408 \pm 0.004$. Lo et al. (2005) estimated von Kármán's κ for wall-bounded turbulence based on the similarity with homogeneous constant-shear turbulence, obtaining $\kappa \approx 0.42$. Using an energy argument and the mathematical symmetry for an idealized flow, Hughes (2007) argued that κ is equal to 0.414 for smooth wall conditions.

Some considerations were made to set the u-profiles over rough beds and for low submergence by fixing the y_0 value. Koll (2006) described the parameterization of the u-profiles over rough beds. She identified three layers that include the roughness layer, in which the u-profile is influenced by wall roughness; the log-law layer, in which the log-law velocity profile exits; and the outer layer, which covers up to the free surface. The first two layers form the wall region. The u-profile holds, in the log-law layer, as $u/u_\tau = \kappa^{-1}\ln[(y-d)/(y_R-d)] + (u_R/u_\tau)$, where d is the zero-plane displacement height and u_R is the velocity at the top of the roughness layer y_R. Nikora et al. (2000) put forward a concept that the large-scale turbulent eddies are in the order of the zero-plane displacement height d, where the mixing length l [$=\kappa(y-d)$] vanishes. The parameters κ, d, y_R and u_R/u_τ can be predicted from the fitting of $u(y)$-profiles using experimental data. Koll (2006) showed that κ decreases with an increase in d. In addition, for low submergence, the wakes produced by the roughness layer reduce severely or eliminate the log-law and the outer layers, implying that κ does not hold. Explicitly, the evaluation of κ from the log-law loses its significance if the roughness layer destroys the log-law or the outer layer. Considering the roughness layer is extended up to $4k_s$ on the top of the roughness elements, the non-existence of the log-law layer is found up to $h \approx 4k_s$, where h is the flow depth. As a consequence, κ loses its implication and cannot be evaluated.

Despite serious efforts over a number of decades, consideration of the von Kármán's κ as having a universal value of 0.41 or a different value in flows over mobile sediment beds remains uncertain. In general, the basic problem of flows over a sediment bed can be stated in a deceptively simple way: given the sediment characteristics, flow rate and bed slope, what are the probable flow depth and the amount of sediment transported? Even for the simplest case of a two-dimensional channel bed with a uniform sediment size, a general solution can only be presented with estimates involving a high degree of uncertainty, as much of the intricacy lies in the consideration of whether κ is universal. Advances in flow measurement technology and progresses in understanding of the turbulence phenomena in shear flows within near-bed logarithmic law layer have inspired a recent research trend that may add to a more satisfactory response to the basic questions. Moreover, this topic has attracted the attention not only of hydraulicians but also of sedimentologists, with potentially constructive results and contributions being published in leading journals, reports and monographs not essentially familiar to the fluid engineering communities. In the literature, there remain many examples of the non-universality of von Kármán's κ in flows over sediment beds. For instance, von Kármán's κ behaves as a variable in flows with low relative submergence S_r $(=h/k_s)$ and where there is bed-load and/or suspended-load transport. Therefore, to organize the discussion, the problem may be divided into three categories (Gaudio et al. 2010), namely, flows with (1) low relative submergence, (2) bed-load transport and (3) suspended-load transport. Nevertheless, the real-world flow problems in natural streams involve such a huge number of factors, including temporally and spatially varying flows; and, on the other hand, laboratory experiments and theoretical investigations have almost customarily chosen to study simplified cases in the hope of clarifying essential mechanisms.

2 Flows with Low Submergence

Rand (1953) was the first to conduct a flume experiment to study the effect of relative submergence on von Kármán's κ with artificial bottom roughness (63 mm square bars spaced at 25.3 mm). He observed a value of $\kappa = 0.3$ for $S_r = 3.3$. Bayazit (1976) reported a gradually increasing κ (greater than 0.4) as the relative submergence S_r decreases below 2.5. They evaluated κ by fitting the velocity data to the logarithmic law having the position of virtual bed level at 0.33 k_s below the roughness crest for all experiments. Pokrajac et al. (2006) reanalyzed the experimental data of Bayazit (1976) with the roughness crest shear velocity instead of the bed shear velocity and obtained modified values of κ. However, the values of κ remained greater than 0.41 for $S_r < 2.5$. Using the experimental data of Dittrich and Hammann de Salazar (1993); Dittrich and Koll (1997); Koll (2002) and Koll (2006) showed that κ is non-universal and depends on both the irregularity of the surface and S_r. These parameters influence the formation and the evolution of the turbulence coherent structures, and in turn the velocity gradient (Kirkbride 1993; Sirovich and Karlsson 1997). Koll (2006) observed that κ approaches a value of 0.4 for regular surfaces and large values of S_r. It decreases significantly down to 0.2 if the bed roughness becomes large enough in order to decrease the flow depth relative to the roughness height. The value of κ reaches a minimum value within the range $4 < S_r < 7$. Koll (2006) commented that the non-monotonic dependency of κ on S_r could not be explained and a further investigation would be required. Cooper (2006) also observed the values of κ primarily being lower than 0.41 for lower values of S_r in the range $4 < S_r < 13$ (see Fig. 1). Here, a mean variation of κ with S_r is inferred (shown by the solid gray line), although the data plots exhibit considerable scatter. It suggests that κ reaches the minimum value 0.25 at $S_r = 5.5$, becoming universal for $S_r > 15$. The curve κ versus S_r is sagging in nature within $3.5 < S_r < 15$. Now the questions concerned with the experiments and the estimations of the non-universality of κ are: (1) if Nikuradse's equivalent sand roughness k_s used for scaling h were the same as those obtained from the velocity profiles, and (2) if the roughness layer influences the value of κ [see Nikora et al. (2000)]. The latter is the major decisive factor towards non-universality of κ, while the former provides the scale up to which κ is a function of S_r. Additionally, hyporheic exchange is often controlled by subsurface advection driven by the interaction of the fluid flow with sedimentary pore water, since macro-rough beds

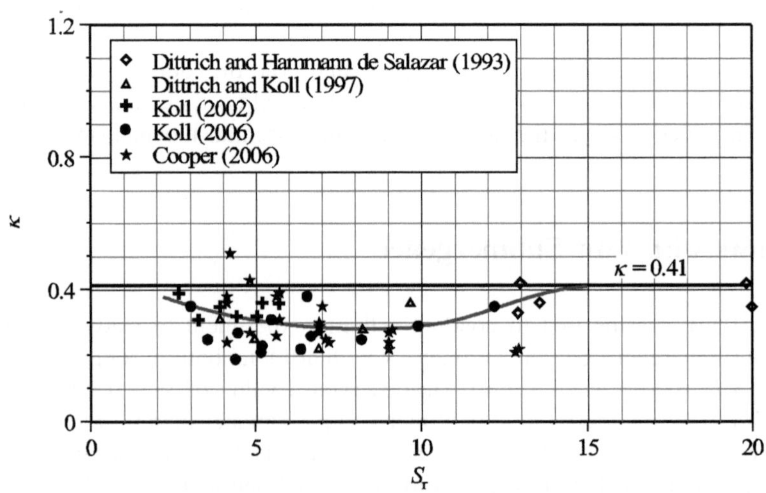

Fig. 1 Dependency of von Kármán's κ with relative submergence S_r (modified from Gaudio et al. 2010)

such as gravel-beds are permeable in general (Packman et al. 2004). The nature and magnitude of the induced hyporheic exchange flow influences the main stream by changing the velocity and Reynolds shear stress profiles and possibly also κ.

3 Flows with Bed-Load Transport

Gust and Southard (1983) observed a decrease in κ with an increase in bed-load transport. They suggested that, after a transitional regime related to the entrainment threshold of sediments, κ adjusted to a constant value of 0.32 $\pm$ 0.04 for all the experiments with bed-load transport, in which the transport rate varied by a factor of 10. This implies that the uncertainties associated with the measured bed-load transport rate had a minimal effect on the result that κ is reduced by 25 % from its universal value. Best et al. (1997) used a phase Doppler anemometer to differentiate the characteristics of the fluid from those of the sediment particles to quantify the influence of the sediment on the carrier fluid turbulence. They observed that the average value of κ was 0.385 in presence of bed-load transport.

Nikora and Goring (2000) reported a study on the characteristics of turbulent structure of high Reynolds number in quasi-two-dimensional flows with fixed and weakly mobile gravel beds. Three sets of measurements were recorded with an acoustic Doppler velocimeter in an irrigation field canal. Measurements were taken first with a weakly mobile-bed flow (WMBF) and then repeated with a fixed-bed flow (FBF). Nikora and Goring (2000) obtained $\kappa \approx 0.29$ for the WMBF, significantly smaller than $\kappa \approx 0.4$ for the FBF. They argued that the adjustment of the virtual bed level would have achieved for $\kappa \approx 0.4$ with the WMBF if the virtual bed level were shifted 3 cm upwards. Since such a shift is physically unjustifiable, it implies that the differences of κ values between the WMBF and the FBF are possible because of the effect of bed-load transport.

Bennett and Bridge (1995); Nikora and Goring (1999) and Gallagher et al. (1999) also revealed an appreciable decrease in κ under bed-load transport. Nikora and Goring (1999) imagined that the reduction in κ might reflect the special turbulence characteristics within a rather narrow range of the Shields parameter when the bed shear stress is approximately equal to the critical shear stress. In Nikora and Goring (2000), the drag reduction effects were expressed as decreased values of κ. The general concept is that the drag reduction prevails when the spacing between turbulent bursting events increases in comparison to the spacing in flows with no sediment (Tiederman et al. 1985). However, it is revealed that κ reduces when spanwise (lateral) spacing between bursting events increases, while streamwise spacing remains unchanged (Hetsroni et al. 1997). Nikora and Goring (2000) found that the streamwise spacing between bursting events was approximately the same for both the WMBF and the FBF, referring to an increase in spanwise spacing for the WMBF.

Dey and Raikar (2007) reported the laboratory experimental results on the turbulent flow characteristics measured by an acoustic Doppler velocimeter. The primary endeavor was to investigate the response of the turbulent flow field, with a zero-pressure gradient, to the uniform gravel beds at the near-threshold of motion. Tests were performed with a reasonably wide range of relative submergence h/d_{50}, where d_{50} is the median size of gravel being considered as k_s. They observed that the variation of the mixing-length is considerably linear with the elevation above the bed within the inner-layer, whose thickness was considered as 0.23 times the boundary layer thickness, and von Kármán's κ was obtained as 0.35. It is pertinent to mention that in some experiments the ratios of open channel flume width B to flow depth h were less than 6 and therefore secondary flows could have influenced the measured flow data.

Recently, Gaudio et al. (2011) performed laboratory tests in a narrow flume with bed-load feeding on an artificially roughened fixed bed and considered the log-law layer velocity measurement data ($y/h \leq 0.20$) using a Pitot-Prandtl tube. They found a decrease in local velocities and κ, and an increase in bed friction factor with volumetric concentration C_b of bed-load transport; specifically, κ varied from 0.3 (at maximum C_b) to 0.4 (clear-water conditions). They also presented the following empirical relationship (obtained using multiple regression), describing the variation of κ with C_b and relative submergence, S_r ($=R_b/d_{50}$, where R_b = bed hydraulic radius):

$$\kappa = 5.19 \times 10^{-4} \frac{S_r^{0.123}}{C_b^{0.731}} \tag{3}$$

which is valid within the ranges of $52.5 \leq R_b/d_{50} \leq 67.81$ and $2.38 \times 10^{-4} \leq C_b \leq 3.23 \times 10^{-4}$, with coefficient of multiple correlation of 0.99 and average relative error of 1.2 %. The variation of κ with C_b is shown in Fig. 2, where κ decreases as C_b increases. Further, Dey et al. (2012) fitted a log-low for mobile-bed flows to obtain $\kappa = 0.37$ for bed-load weight discharge per unit width $g_b = 2 \times 10^{-3}$–7×10^{-3} kg/(ms) with $S_r = h/d_{50}$. Table 1 furnishes a summary of the results for κ in flows with bed-load transport. It is evident that the data available are so limited that the variation of κ with g_b is not so specific, although it has been well-recognized that κ-values with bed-load are less than the universal value 0.41.

4 Flows with Suspended-Load Transport

Also in flows with suspended sediment load, von Kármán's κ has been a long disputed parameter in the data analysis of the logarithmic velocity profile fittings. During 1946–1961, pioneering research was conducted by Vanoni (1946); Einstein and Chien (1955) and Elata and Ippen (1961) to examine the effect of suspended sediment concentration C_v on the velocity profile. One of the key findings was that κ decreases with an increase in C_v, although Coleman (1981, 1986) expressed a strong dissatisfaction on this issue. He believed that Vanoni's finding was an

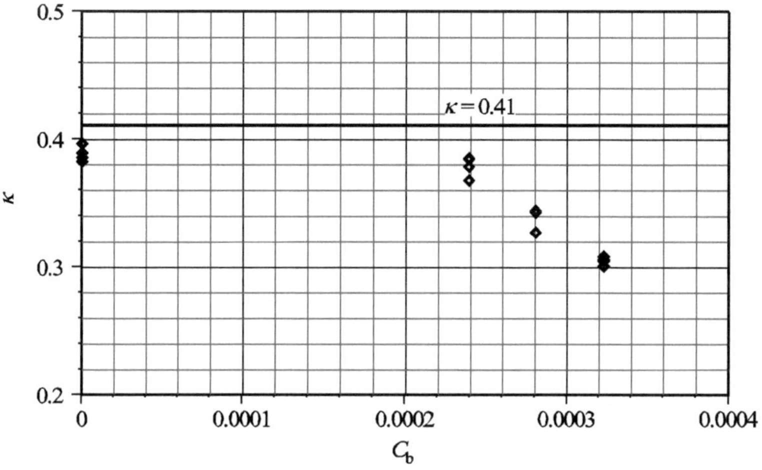

Fig. 2 von Kármán's κ as a function of bed-load sediment concentration C_b

artifact of the erroneous technique of evaluating κ that was generally accepted at the time when he did this work. According to Coleman, the experimental studies by Einstein and Chien (1955) were also unacceptable because the measurements were taken only over the lower 40 % of the flow depth in the experimental channel that they used. Thus, it was difficult to determine the boundary layer thickness, the maximum velocity or any general information about the flow. The experiments by Elata and Ippen (1961) were done using virtually neutrally buoyant polystyrene particles to simulate the sediment suspension. The velocity profiles were presented in a velocity defect form, and the apparent decrease of κ was due to its incorrect evaluation. However, Coleman (1981) used the wake-law to study velocity profiles in sediment-laden flows and suggested that the wake coefficient rather than κ is affected by the existence of sediment suspension. The wake-law that describes the logarithmic velocity profile only in the vicinity of the bed has the same value of κ as that in clear-water flow ($\kappa = 0.41$). Lyn (1986) noted that the use of pure log-law profile leads to a decreased value of κ with C_v, while the use of log-wake-law leads to a universal one. Cioffi and Gallerano (1991) determined experimentally the velocity and sediment concentration profiles on a mobile flat bed using the phase difference method; without assessing κ in the inner region, they only verified that the measured velocity profiles were reasonably interpolated with $\kappa = 0.4$ for $y/h < 0.15$. Cellino and Graf (1999) carried out an experimental study to investigate the influence of suspended sediments in the flow under non-capacity and capacity conditions. They accepted $\kappa = 0.4$ as a reasonable value for describing the velocity profiles through the velocity defect law. After Muste (2002), a constant value for $\kappa = 0.41$ would be appropriate for the sediment-laden flows involving only small C_v (smaller than 0.05). The aforementioned discussion is given on the investigators who opposed the changed value of κ from its universal value.

Table 1 Summary of experimental data and results for κ in flows with bed-load transport (modified from Gaudio et al. 2010)

Source	d_{50} (mm)	S_r	B/h	R_*	F	Mode of transport/type of bed/sediment feeding	g_b (g/s/m)	κ
Gust and Southard (1983)	0.16	625	6	–	0.2–0.28	Bed-load/mobile bed/no feeding ($y/h \leq 0.2$)	0.0015–0.015	0.32 ± 12.5 %
Best et al. (1997)	0.22	261	5.22	8.9	0.76–0.78	Bed-load/fixed bed/with feeding	9–22	0.385
Nikora and Goring (2000)	6.4	166	7.57–10.08	429	0.64–1.09	Bed-load/mobile bed/no feeding ($y/h \leq 0.2$)	13.8	0.29 ± 10.3 %
Dey and Raikar (2007)	4.1–14.25	6.74–54.15	4.05–11.84	210–1573	0.17–0.38	Bed-load/mobile bed/near threshold ($y/h \leq 0.23$)	12.3–90	0.35 ± 0.86 %
Gaudio et al. (2011)	1	52.5–67.81	2.9–3.6	101–120	0.98–1.01	Bed-load/fixed bed/with feeding ($y/h \leq 0.2$)	33.4–64.9	0.3–0.39 ± 10.7 %
Dey et al. (2012)	0.95, 2.6, 4.1	29.2–158.0	6.3–34.2	63–508	0.55–0.77	Bed-load/fixed bed/with feeding ($y/h \leq 0.2$)	2–7	0.35–0.42

F $U/(gh)^{0.5}$, U mean flow velocity, and g gravitational acceleration

On the other hand, other investigators believed that κ is non-universal in sediment-laden flows. Besides Vanoni (1946); Einstein and Chien (1955) and Elata and Ippen (1961), many other researchers during the 1960 and 1970s reported that κ diminishes as C_v increases, primarily owing to increasing gradient of the velocity profiles in presence of sediment suspension (Vanoni and Nomicos 1960; Hino 1963; Paintal and Garde 1964; Bohlen 1969; Ippen 1971). In the 1980s, Nouh (1989) reported that the variation of κ with average concentration C_{av} of suspended sediments in straight open channel flows depends on the flow Reynolds number $Re = 4Uh/\upsilon$. As C_{av} increases, κ decreases in flows for $Re < 7 \times 10^5$, but increases in flows for $Re > 7 \times 10^5$. The variation of κ with C_{av} is insignificant in flows with Re equal to about the critical value $Re_c = 7 \times 10^5$. Nouh (1989) hypothesized that the turbulence level close to the boundary decreases for $Re < Re_c$ and increases for $Re > Re_c$, as C_{av} increases. In addition, he hypothesized that the turbulence level close to the boundary in sediment-laden flows increases as C_{av} decreases and Re increases, with a rate depending on the values of C_{av} and Re. For $Re < Re_c$, C_{av} affects the turbulence level more than Re, and vice versa for $Re > Re_c$. These two effects are balanced for $Re \approx Re_c$. Nouh (1989) also observed that the increased values of κ in flows with fine suspended sediments are larger than those in flows with relatively coarse suspended sediments, and also those in flows with high C_{av} are larger than those in flows with low C_{av}. In fact, he explained that, for a given C_{av}, coarse suspended sediments produce a larger reduction in turbulence level than fine suspended sediments. In a clear-water flow, universal value of κ is invariant of Re (for $4 \times 10^5 < Re < 2 \times 10^6$). Wang and Qian (1992) showed that, in the lower-flow region, the values of κ in sediment-laden flows are less than those in clear-water flows. Guo and Julien (2001) argued that the reduction of κ in sediment-laden flows is governed by C_{av} and the mass density gradient given by the Richardson number. In presence of sediment suspension, Wang et al. (2001) modified von Kármán's κ_p for the log-law and κ_w for the wake-law separately. For the log-law, they proposed $\kappa_p = 2.08\kappa/(\Delta u^+\kappa + 2.08)$, where Δu^+ is $(u_c - u_s)/u_\tau$, and u_c and u_s are the velocities at $y/h = 0.05$ in clear-water and sediment-laden flows, respectively. It is apparent that κ_p decreases with an increase in Δu^+. For the wake-law, they obtained an average value of κ_w as 0.346 with a relative error equal to 229 %. Nezu and Azuma (2004) conducted simultaneous measurements of both the suspended particles and water in particle-laden flows by means of a discriminator particle-tracking velocimeter leading to the observation that κ decreases with an increase in C_{av}.

In fact, the investigators supporting the universality of κ also in sediment-laden flows referred to Coleman's (1981, 1986) data and analysis. Owing to the available evidence that κ varies in the presence of suspended sediments, it seems to be necessary to reanalyze Coleman's dataset in order to verify the proposed constancy of κ. By adopting the velocity defect law as in the original work, the dimensionless data $(u_{max} - u)/u_\tau$ were plotted as a function of y/δ in a semi-log graph, where δ is the boundary layer thickness, that is, the distance above the bed at which the time-averaged streamwise velocity reaches its maximum value, u_{max}. The slopes $-\kappa^{-1}$ were evaluated from the regression analysis fitting straight lines within the log-law

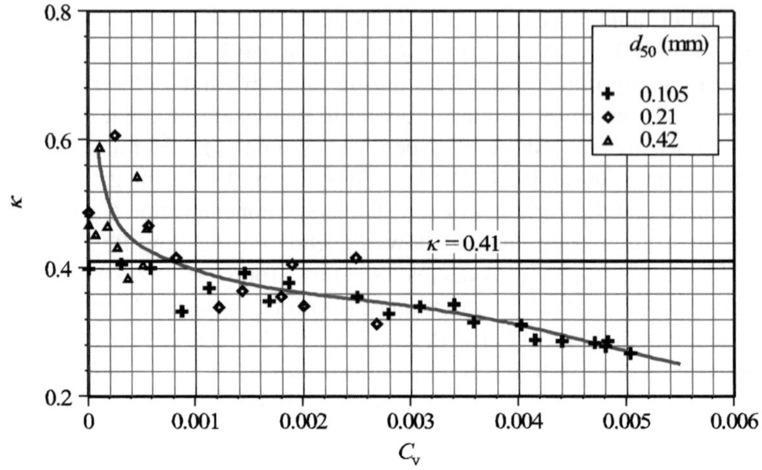

Fig. 3 Variation of von Kármán's κ with suspended sediment concentration C_v obtained by reanalyzing the data of Coleman (1981, 1986) (modified from Gaudio et al. 2010)

layer in the inner flow region ($y \leq 0.15\ \delta$). We obtained values of κ different from those given in Coleman (1981, 1986). It therefore invites potential uncertainty in the analysis of Coleman. Figure 3 presents the data plots of the reanalyzed κ values as a function of C_v. The mean trend shown by the solid gray line gives progressively diminishing values of κ with increase in C_v. In general, for $C_v > 0.0008$, the values of κ are less than its clear-water value 0.41.

5 Conclusions

It is evident that von Kármán's κ remains non-universal in flows with either a low relative submergence or a bed-load and/or suspended-load transport. Reasons for the variability of κ are recognized as follows. Sediment particles during the bed-load transport interact with both the flow that accelerates them and the rough bed that decelerates them (Gyr and Schmid 1997). Collisions cause momentum extraction from the mean flow in the near-bed region, resulting in reduction of local streamwise velocities and increase of velocity gradients and turbulence intensities (Owen 1964; Smith and McLean 1977). Plotting measured data on a semi-logarithmic plane and applying a regression fitting of a straight line according to Eq. (1), the slope $a = u_\tau/\kappa$ is determined and thus κ can be easily obtained. The increment of a and the substantial constancy of u_τ leads to a decrease in κ. Analogous considerations can be made for the decreasing κ with suspended load.

In low submergence, κ was found to depend on the irregularity of the surface and the relative submergence, which influence the formation and the expansion of the coherent turbulent flow structures, and thus the velocity gradient. Nevertheless, the cause of the non-monotonic dependency of κ on S_r remains unexplained.

Although an attempt is made to provide possible variations of κ with relative submergence and suspended sediment concentration, there remain a number of unanswered questions suggesting that the variations are rather *ad hoc*. On the other hand, in case of bed-load transport, it was not even possible to provide any solid relationship between κ and the bed-load transport rate, although Gaudio et al. (2011) seems to have conducted the only study providing an empirical relationship. However, it is established that the values of κ are lower than its universal value 0.41. The von Kármán κ is given by l/y; where l is the Prandtl mixing length equal to $[-\overline{u'v'}/(du/dy)^2]^{0.5}$, u' and v' being the fluctuations of the streamwise and vertical velocity components, u and v, respectively. Therefore, κ is dependent on the Reynolds shear stress relative to the mass density of fluid, $-\overline{u'v'}$, and the gradient of the velocity profile, du/dy. Hence, the increased values of κ depend on the prevailing effect between turbulence intensity and velocity gradient. For instance, if turbulence intensity increases and/or velocity gradient decreases, then κ increases and vice versa. Thus, in order to assess the non-universality of κ, future researchers have to keep an eye on the variations of these two parameters with the vertical distance within the logarithmic law layer.

Acknowledgments The Taylor and Francis Group is kindly acknowledged for having granted free permission of reuse of the content of the chapter Gaudio et al. (2010) (www.tandfonline.com).

References

Bayazit M (1976) Free surface flow in a channel of large relative roughness. J Hydraul Res 14(2):115–126

Bennett SJ, Bridge JS (1995) The geometry and dynamics of low-relief bed forms in heterogeneous sediment in a laboratory channel, and their relationship to water flow and sediment transport. J Sediment Res A65:29–39

Best J, Bennett S, Bridge J, Leeder M (1997) Turbulence modulation and particle velocities over flat sand beds at low transport rates. J Hydraul Eng 123(12):1118–1129

Bohlen WF (1969) Hotwire anemometer study of turbulence in open-channel flows transporting neutrally buoyant particles. Report number 69-1, Experimental sedimentology laboratory, Department of earth and planetary sciences, Massachusetts institute of technology, Cambridge, Mass, USA

Cellino M, Graf WH (1999) Sediment-laden flow in open-channels under noncapacity and capacity conditions. J Hydraul Eng 125(5):455–462

Cioffi F, Gallerano F (1991) Velocity and concentration profiles of solid particles in a channel with movable and erodible bed. J Hydraul Res 29(3):387–401

Coleman NL (1981) Velocity profiles with suspended sediment. J Hydraul Res 19(3):211–227

Coleman NL (1986) Effects of suspended sediment on the open-channel velocity distribution. Water Resour Res 22(10):1377–1384

Cooper JR (2006) Spatially-induced momentum transfer over water-worked gravel beds. PhD thesis, The University of Sheffield, Sheffield, UK

Dey S, Raikar RV (2007) Characteristics of loose rough boundary streams at near-threshold. J Hydraul Eng 133(3):288–304
Dey S, Das R, Gaudio R, Bose SK (2012) Turbulence in mobile-bed streams. Acta Geophysica (in press)
Dittrich A, Hammann de Salazar K (1993) Bed instability caused by clear water flow. Final report of project Eroslope (EV5 V-CT92-0179), Institute for hydraulic engineering, Braunschweig technical university, Germany
Dittrich A, Koll K (1997) Velocity field and resistance of flow over rough surface with large and small relative submergence. Int J Sedim Res 12(3):21–33
Einstein HA, Chien N (1955) Effects of heavy sediment concentration near the bed on velocity and sediment distribution. MRD sediment series report number 8, University of California, Berkeley, US Army Corps of Engineers, Missouri Division, St. Louis, MO, USA
Elata C, Ippen AT (1961) The dynamics of open channel flow with suspensions of neutrally buoyant particles. Technical report number 45, Massachusetts institute of technology, Boston, MA, USA
Gallagher M, McEwan I, Nikora V (1999) The changing structure of turbulence over a self-stabilising sediment bed. Internal report number 21, Department of Engineering, University of Aberdeen, Aberdeen, UK
Gaudio R, Miglio A, Calomino F (2011) Friction factor and von Kármán's κ in open channels with bed-load. J Hydraul Res 49(2):239–247
Gaudio R, Miglio A, Dey S (2010) Non-universality of von Kármán's κ in fluvial streams. J Hydraul Res 48(5):658–663
Guo J, Julien PY (2001) Turbulent velocity profiles in sediment-laden flows. J Hydraul Res 39(1):11–23
Gust G, Southard JB (1983) Effects of weak bed load on the universal law of the wall. J Geophys Res 88(C10):5939–5952
Gyr A, Schmid A (1997) Turbulent flows over smooth erodible sand beds in flumes. J Hydraul Res 35(4):525–544
Hetsroni G, Zakin JL, Mosyak A (1997) Low-speed streaks in drag-reduced turbulent flow. Phys Fluids 9(8):2397–2404
Hino M (1963) Turbulent flow with suspended particles. J Hydraul Div 89(HY4):161–185
Hughes RL (2007) A mathematical determination of von Kármán's constant, κ. J Hydraul Res 45(4):563–566
Ippen AT (1971) A new look at sedimentation in turbulent streams. J Boston Soc Civil Eng 58(3):131–163
Kirkbride A (1993) Observations of the influence of bed roughness on turbulence structure in depth limited flows over gravel beds. In: Clifford NJ, French JR, Hardisty J (eds.) Turbulence: perspectives on flow and sediment transport, Wiley, Chichester, pp 185–196
Koll K (2002) Feststofftransport und Geschwindigkeitsverteilung in Raugerinnen. Karlsruhe University, Fak. f. Bauingenieur- und Vermessungswesen, Diss. v. 12.07.2002, http://www.ubka.uni-karlsruhe.de/cgibin/
Koll K (2006) Parameterisation of the vertical velocity profile in the wall region over rough surfaces. In: Ferreira RML, Alves ECTL, Leal JGAB, Cardoso AH (eds.) River flow 2006, Taylor & Francis, London, Proceedings of international conference of fluvial hydraulics, Lisbon, Portugal, pp 163–171
Lo TS, L'vov VS, Pomyalov A, Procaccia I (2005) Estimating von Kármán's constant from homogeneous turbulence. Europhys Lett 72(6):943–949
Long CE, Wiberg PL, Nowell ARM (1993) Evaluation of von Kármán's constant from integral flow parameters. J Hydraul Eng 119(10):1182–1190
Lyn DA (1986) Turbulence and turbulent transport in sediment-laden open-channel flows. Report number KH-R-49, In: Keck WM (ed.) Laboratory of hydraulic and water resources, California Institute of Technology, Pasadena, CA, USA
Muste M (2002) Sources of bias errors in flume experiments on suspended-sediment transport. J Hydraul Res 40(6):695–708

Nezu I, Azuma R (2004) Turbulence characteristics and interaction between particles and fluid in particle-laden open channel flows. J Hydraul Eng 130(10):988–1001
Nikora VI, Goring DG (1999) Effects of bed mobility on turbulence structure. NIWA Internal report number 48, NIWA, Christchurch, New Zealand
Nikora V, Goring D (2000) Flow turbulence over fixed and weakly mobile gravel beds. J Hydraul Eng 126(9):679–690
Nikora V, Goring D, McEwan I, Griffiths G (2000) Spatially averaged open-channel flow over rough bed. J Hydraul Eng 127(2):123–133
Nouh M (1989) The von-Kármán coefficient in sediment laden flow. J Hydraul Res 27(4):477–499
Owen PR (1964) Saltation of uniform grains in air. J Fluid Mech 20:225–242
Packman AI, Salehin M, Zaramella M (2004) Hyporheic exchange with gravel beds: Basic hydrodynamic interactions and bedform-induced advective flows. J Hydraul Eng 130(7):647–656
Paintal AS, Garde RJ (1964) Discussion of 'suspended transportation mechanics: suspension of sediment'. J Hydraul Div 90(HY4):257–265
Pokrajac D, Finnigan JJ, Manes C, McEwan I, Nikora V (2006) On the definition of the shear velocity in rough bed open channel flows. In: Ferreira RML, Alves ECTL, Leal JGAB, Cardoso AH (eds.) River flow 2006, Taylor & Francis, London, UK, Proceedings of international conference of fluvial hydraulics, Lisbon, Portugal pp 88–96
Rand W (1953) Discussion of 'some effects of suspended sediment on flow characteristics'. In: Proceedings of fifth hydraulics conference, Bulletin 34, State University of Iowa, Iowa City, Iowa, USA pp 156–158
Sirovich L, Karlsson S (1997) Turbulent drag reduction by passive mechanisms. Nature 388:753–755
Smith JD, McLean SR (1977) Spatially averaged flow over a wavy surface. J Geophys Res 82(12):1735–1746
Tiederman WG, Luchik TS, Bogard DG (1985) Wall layer structure and drag reduction. J Fluid Mech 156:419–437
Vanoni VA (1946) Transportation of suspended sediment by water. Trans Am Soc Civil Eng 111:67–133
Vanoni VA, Nomicos GN (1960) Resistance properties of sediment laden stream. Trans Am Soc Civil Eng 125:1140–1175
van Rijn LC (1993) Principles of sediment transport in rivers, estuaries and coastal seas. Aqua Publications, The Netherlands
von Kármán T (1930) Mechanische ähnlichkeit and turbulenz. Nachrichten der Adademie der Wissenchaften Göttingen, Mathematisch-Physikali-Sche Klasse 58–76
Wang X, Qian N (1992) Velocity profiles of sediment-laden flow. Int J Sedim Res 7(1):27–58
Wang X, Wang ZY, Yu M, Li D (2001) Velocity profile of sediment suspensions and comparison of log-law and wake-law. J Hydraul Res 39(2):211–217

Nezu I, Azuma R (2004) Turbulence characteristics and interaction between particles and fluid in particle-laden open channel flows. J Hydraul Eng 130(10):988–1001
Nikora VI, Goring DG (1999) Effects of bed mobility on turbulence structure. NIWA internal report number 48, NIWA, Christchurch, New Zealand
Nikora V, Goring D (2000) Flow turbulence over fixed and weakly mobile gravel beds. J Hydraul Eng 126(9):679–690
Nikora V, Goring D, McEwan I, Griffiths G (2001) Spatially averaged open-channel flow over rough bed. J Hydraul Eng 127(2):123–133
[illegible]
[illegible]
[illegible]
[illegible]
[illegible]
[illegible]

A New Theoretical Framework to Model Incipient Motion of Sediment Grains and Implications for the Use of Modern Experimental Techniques

Andrea Marion and Matteo Tregnaghi

Abstract The entrainment of sediments in rivers is recognized to exhibit an intermittent nature, hence incipient motion is inherently a random process that requires an appropriate stochastic description. The effect of near-bed turbulence on grain entrainment and the variation in stability of randomly configured bed particles due to local surface heterogeneity are included into a probabilistic framework based on a concept first proposed by Grass. Bedload transport tests were carried out in a flume where sediment movement was monitored using a three-camera 3D PIV system. Simultaneous grain motion and flow velocity measurements were made on a plane located slightly above and parallel to the sediment bed. Detailed statistical velocity information was acquired to model the velocity distribution at the bed level accounting for the probabilistic distribution of particle exposures. This was combined with the probabilistic distribution of grain resistance to motion, which was obtained from discrete particle modeling (DPM) simulations. The analysis provides detailed insight, in terms of grain dynamics, into the physical aspects that determine the initiation of movement, and the stochastic equations of incipient motion are derived. The key feature of the proposed analysis is the potential of including into the model as much statistical information as one can obtain from experimental observations based on state-of-the-art flow measurement techniques and from the use of numerical simulations performed with discrete particle models.

A. Marion (✉) · M. Tregnaghi
Department of Industrial Engineering, University of Padua,
Via Marzolo 9,
35131 Padua, Italy
e-mail: p.rowinski@igf.edu.pl

P. Rowiński (ed.), *Experimental and Computational Solutions of Hydraulic Problems*, GeoPlanet: Earth and Planetary Sciences, DOI: 10.1007/978-3-642-30209-1_5,

1 Introduction

Sediment dynamics in natural water bodies has been the subject of research over several decades. Initial interest on sediment transport was motivated by predicting channel morphology; later, the development of tools for natural risk assessment drove research activity. More recently, interest in sediment dynamics is emerging within the engineering and ecological science community dealing with the prediction of river water quality (e.g., Battin et al. 2003; Bottacin et al. 2009). Since the work of Einstein (1942) and studies by Grass (1970) and Paintal (1971), researchers have promoted a probabilistic view of entrainment and transport processes, whilst others have still used the deterministic threshold concept proposed by Shields (1936). To date a major limitation of probabilistic models was the lack of data at the scale of individual processes to test these models. Grain scale observations have only recently been possible as instrumental capabilities have improved (e.g., Papanicolaou et al. 1999; Schmeeckle et al. 2007; Bottacin-Busolin et al. 2008).

The initial movement of sediment by a turbulent flow remains one of the most complex problems in river hydraulics. It is quite well accepted that, when the hydrodynamic forces over the bed exceeds a certain critical value, particle motion begins. From this definition, one would suppose that it may be fairly simple to estimate the hydraulic conditions at incipient motion. However, for turbulent flows the beginning of movement of sediment is still an open issue as no single criterion to define the threshold condition has been universally accepted (e.g., Buffington and Montgomery 1997). Because both turbulent fluctuations and individual grain characteristics are randomly distributed, the process of sediment entrainment has a random and discontinuous character; therefore it requires an appropriate stochastic interpretation. Currently, the difficulty with defining criteria that adequately capture these features is a major problem.

Recently, several contributors have developed theoretical formulations that link the 'pickup probability' or 'probability of entrainment' to both turbulent flow conditions and bed texture properties that characterize the resistance to motion of sediments (e.g., Cheng and Chiew 1998; Papanicolaou et al. 2002; Wu and Chou 2003; Wu and Yang 2004). In these studies, the near-bed velocity U_f (or the fluid shear stress T_f) is represented by a cumulative distribution F_{Uf} with probability density function f_{Uf}, and that the probability of entrainment P_E can be expressed as

$$P_E = P(U_f > u_f = u_g) = \int_{u_g}^{\infty} f_{Uf}(u)\mathrm{d}u = 1 - F_{Uf}(u_g) \qquad (1)$$

where u_g denotes the threshold velocity or critical velocity for grain entrainment (equivalent to the critical shear stress τ_g for individual particles). Different notations for random variables and the single occurrence of such variables, respectively, are accounted for herein by conventional use of capital and lower-case letters. Being U_g a random variable represented by a cumulative distribution F_{Ug} with probability density function f_{Ug}, the mean entrainment probability is the expected value of Eq. (1) over the full range of U_g

$$R_E = \int_0^\infty f_{Ug}(u) \cdot [1 - F_{Uf}(u)] \mathrm{d}u = \int_{-\infty}^\infty f_{Uf}(u) \cdot F_{Ug}(u) \mathrm{d}u \qquad (2)$$

where $F_{Ug}(u_f) = P(U_g < u_g = u_f)$, and the notation R_E stands for risk of entrainment (e.g., Lopez and Garcia 2001; McEwan et al. 2004). Solutions for the entrainment probability in the form of Eq. (2) are here referred to as following the 'Grass approach' (Grass 1970).

In this chapter, grain scale observations are reported on the correlation between the near-bed velocity field and the initial movement of sediments. Based on this evidence, a general derivation of Eq. (2) is provided along with the analysis of the conditions that applies to the present model. The use of shear stress variables T_f and T_g is here preferred to the velocity-based (U_f and U_g, respectively) probability definition as (a) it is consistent with the original approach suggested by Grass; (b) grain dislodgment relies on the fundamental concept of balance of forces (or moments); (c) this renders the application of the model comparable with conventional analysis and measurements of grain entrainment and incipient motion conditions. The resulting equations link fluid shear and grain exposure by accounting for the statistical distribution of the surface grain elevations, Z_g. The model is combined with detailed statistical information of the near-bed flow-field and bed texture, and compared with low-rate sediment transport experimental data. This is achieved through a series of laboratory experiments based on particle imaging velocimetry (PIV) techniques, and using numerical simulations performed with discrete particles models (DPM).

2 Experimental Observations

2.1 3D Flow-Field Measurements

The experiments were carried out in a tilting flume equipped with state-of-the-art monitoring equipment that enabled high resolution velocity measurement of free surface flows over rough boundaries and sediment fluxes at high spatial and temporal resolutions. Measurements were made approximately at $z_{laser} = 10$ mm above and parallel to the original sediment bed to avoid interference between the flow-field measurement and the bed-load transport. A fixed bed made of gravel glued on a plastic surface covered the upstream end of the flume for a distance of 1.50 m; this ensured the uniform development of a stable upstream boundary layer for all experiments. The remaining part of the flume was filled with loose natural river gravel with an average diameter $d_{50} = 5.0$ mm and standard deviation $\sigma_g = 1.3$. Particular care was taken when placing the sediment in order to produce a well mixed flat sediment deposit with no appreciable bed forms along this reach of the flume. A uniform flow was obtained by adjusting a weir placed at the downstream end of the flume. Flow velocity and grain motion measurements were taken simultaneously during each test.

An area of 220 by 35 mm^2 located on the centerline of the flume was selected for measurement. This area was viewed by the three cameras placed vertically above the water surface. The area size was selected to provide velocity data at a suitable spatial scale, and the area was large enough so that enough grain motions could be observed in an experiment so that statistically significant data on grain entrainment could be obtained. Two cameras had a green filter and the other operated under white light conditions. The cameras with the green filter were used to obtain flow field information, whilst the other camera was used to gather data on grain entrainment, depositions and movements. They collected pairs of images with the time step between the images $\Delta t_p = 1$ ms and the frequency of the collection of image pairs $fr = 45$ Hz. The images recorded the "instantaneous" positions of the flow tracing particles. In order to measure the spatial flow velocity pattern, seeding particles were fed into the flow with an average size of $d_p = 200$ μm and density $\rho_p = 995$ kg/m^3. Camera-1 was focused on the sediment bed surface; it was illuminated by a white strobe light and captured an image of the bed at the starting time of the second pair images recorded by cameras-2 and -3. Bed images were also collected at a 45-Hz frequency. This setup provided velocity data sets with a spatial resolution that was comparable to the grain scale. This was done so that velocities at particular locations and times could be associated with the movement of individual grains.

Data from 20 tests carried out at different slopes but with similar water depths are reported. A number of tests was repeated to optimize data measurements. Experiments were run with increasing levels of boundary shear stress, ranging from critical conditions to about twice the threshold of grain motion, so that the distribution of velocities and linked grain motions could be determined as the system moved away from the threshold of motion to conditions of well developed transport. The sediment distributions had a similar average grain size ensuring that all the experiments were carried out with a similar bed shear stress range. Full details of the experiments are reported in Tregnaghi et al. (2010).

2.2 Link Between Near-Bed Velocity Field and Grain Motion

The experiments described in the present paper were focused on the observations of the movement of single grains from deposits of natural gravel particles and the relevance of the local flow field. They were therefore designed so that grain entrainments could be clearly identified and simultaneous measurements of the near-bed local flow field could be analyzed before any grain had been entrained.

Flow-field images were used to obtain the three velocity components, u, v and w, and the relevant temporal fluctuations, u', v' and w'. The temporal trends of the flow field velocities in the proximity of the moved grains were collated from the flow field data and used by linking the locations of grain motion with the local velocity patterns, e.g., streamwise velocity (Fig. 1a). Quadrant analysis is one of the most effective methods that best classifies burst events and quantifies the differences in turbulent structures in open-channel flow. This analysis has been

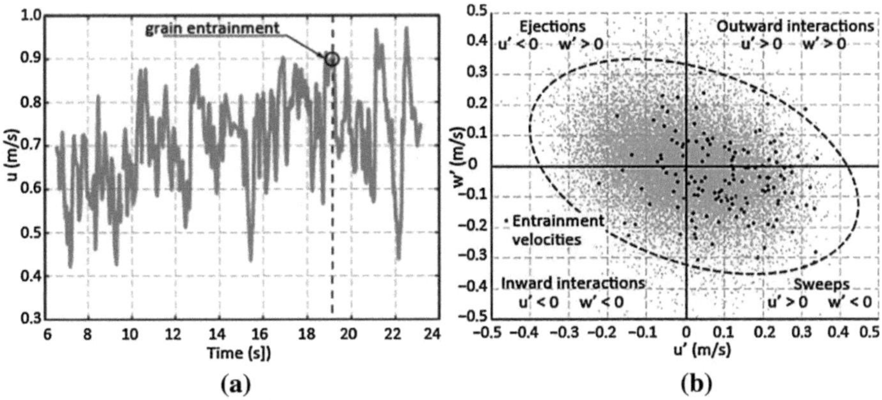

Fig. 1 **a** Velocity timeseries before grain entrainment. **b** Quadrant analysis

applied to the current dataset. Although it is commonly believed that ejections ($u' < 0$, $w' > 0$) and sweeps ($u' > 0$, $w' < 0$) are primarily responsible for the entrainment of sediment since they both contribute positively to the Reynolds stress, recent studies showed that periods of significant motion were found to correspond at those times when the near-bed streamwise velocity is high, with little correlation to the cross product of the velocity fluctuations (e.g., Nelson et al. 1995; Papanicolaou et al. 2001). Figure 1b shows that sweeps and outward interactions ($u' > 0$, $w' > 0$) move more sediment than ejections and inward interactions, indicating they are individually more effective. During sweeps, downward flow enhances transport by advecting fluid with high velocities toward the bed; during outward interactions, upward flow tends to move particles away from the bed. This means that a strong correlation exists with large u' as particle motion is dominated by drag and lift forces, while Reynolds stresses were found to have weak correlations with grain entrainment.

Moving on from these observations, data analysis was focused on time periods >10s before grains were entrained. The pdf of such velocity samples were investigated (see Fig. 2a) and a 'statistical match' with an entrainment event was assumed to occur if the flow velocity fell above the 85th percentile of the relevant cumulative distribution (see Fig. 2b). Results have indicated that about 70 % of entrainment events were associated to peak values of the streamwise velocity u. For 20 % of cases, the streamwise velocity was found around average or below, but they were associated to high fluctuations of lateral or vertical components, v and w. This evidence suggests a strong link between grain motion and local drag with some combination of lift or lateral destabilizing forces, as supported by direct observations of bed images during the first instants of the movement. The remaining part of the sample was not found to be significantly associated to any flow signature. However, in some cases closer inspection of data suggests that these grains were possibly put in motion by other particles passing through their neighborhood, or leaving their position few diameters upstream, thus reducing the sheltering effect. One goal

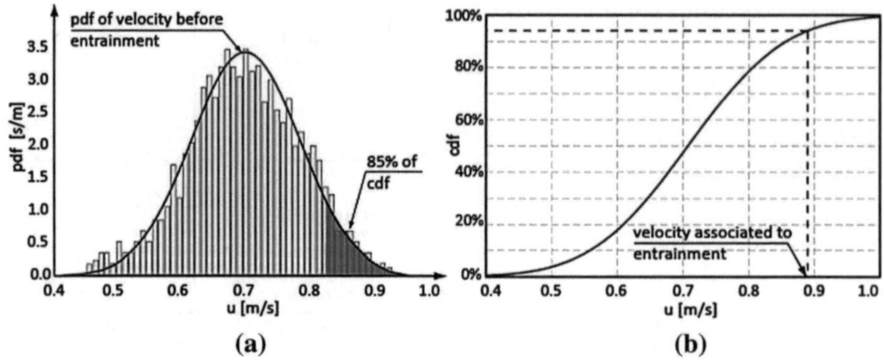

Fig. 2 **a** Distribution of velocities before grain entrainment. **b** Probability of the entrainment velocity

of these experiments was clearly identifying the most effective physical mechanisms leading to the entrainment of particles. The analysis performed has demonstrated that the local near bed velocity, rather than shear stress can be representative of the physical process affecting entrainment, and this working assumption will be used in the analysis described in the next section.

3 Derivation of Probability Density Functions

3.1 Fluid Shear Stress

In the literature the term shear stress usually refers to the space and time averaged fluid force exerted on an area of the bed. Based on the evidence that a strong link exists between grain motion and the instantaneous streamwise velocity, the pressure distribution on the particle resulting from the flow around its surface can be accounted for by a sufficient parameterization of the drag (and lift) term. As a consequence, it is the turbulence fluctuations in the near-bed velocity that give rise to fluctuations in the forces on sediment grains. The entrainment mechanism is formulated in terms of force balance, where the fluid force is represented by the hydrodynamic drag force. The fluid shear stress for an individual grain can then be obtained from the ratio of its destabilizing force F_D to the particle surface exposed to the flow, as follows

$$\tau_f = \frac{4F_D}{\pi d^2} = \frac{1}{2}\rho C_D \cdot \omega_f \cdot u_f\left|u_f\right| \tag{3}$$

where ω_f is the ratio of the exposed area to the cross-section of the particle, ρ the fluid density, and C_D the drag coefficient. The shear stress that one calculates for

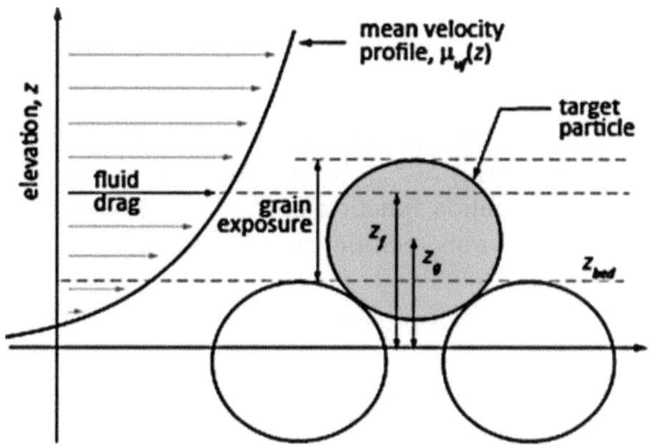

Fig. 3 Geometrical definition of the fluid drag on particle at $z = z_g$ (longitudinal section)

a single grain is therefore equivalent to the 'drag stress' described by Buffington and Montgomery (1999), which is derived from the instantaneous streamwise velocity rather than the instantaneous Reynolds stress. Equation (3) possibly describes the simplest formulation of a static equilibrium of forces, which allows a statistical description by using univariate distributions.

Figure 3 shows the geometrical definition of the fluid drag on a 'target' particle located at the elevation $z = z_g$, where z_g is a certain occurrence of the random variable Z_g having cumulative distribution F_{Zg} and probability density function f_{Zg}. The fluid is assumed to flow over a flat surface composed of uniformly-sized spherical particles, whose top elevation is located at $z = z_{bed}$. A number of studies reported that zero-velocity level was commonly taken as 0.2–0.3d below the tops of the bed grains, z_{bed} (e.g., Bridge and Bennet 1992). Determining the position of the fluid drag, z_f, presents some difficulties as it depends on the local velocity profile and the height of the grain above the bed level. Values of z_f are commonly taken as 0.5–0.7d with respect to the reference level and measured normal to the bed surface, as this is the average position of the most exposed particles that rest in an interstice formed by the top layer particles. Here, the elevation of fluid drag and the exposure of the particle to the fluid are estimated as functions of the target particle position and characterized as random variables. The distribution of bed elevations is assumed to follow a near-Gaussian distribution (e.g., Kirchner et al. 1990; Nikora et al. 1998), i.e., $f_{Zg} \sim N(\mu_{zg}, \sigma_{zg})$, where $\mu_{zg} = (z_{bed} - d/2)$ is the mean particle elevation, and σ_{zg} is the standard deviation of the distribution.

The mean velocity vertical distribution in the wall region of the turbulent flow over a hydraulically rough bed can be described as

$$\mu^*_{uf} = g_1(z^*) = \frac{1}{\kappa}\ln(z^*) + C \tag{4}$$

where the star notation denotes dimensionless quantities, whose scales are the shear velocity u_s and the particle size d. The fluid drag z^*_f computed over the exposed grain surface $A_f = \omega_f \pi d^2/4$ is

$$z_f^* = g_2\left(z_g^*\right) = \frac{\int_{A_f} z^* \cdot g_1^2(z^*)\mathrm{d}A}{\int_{A_f} g_1^2(z^*)\mathrm{d}A} \tag{5}$$

Equation (5) represents a monotonic function that holds for $z_g^* \geq (z_{bed}^* - 1/2)$, otherwise $g_2(z_g^*) = z_{bed}^*$ by definition. Being $F_{Zf}(z_f^*) = P(g_2(Z_g^*) < z_f^*)$ the cumulative distribution function of Z_f^*, if the function g_2 is invertible and monotonic, then the previous relation can be extended to obtain

$$F_{Zf}\left(z_f^*\right) = P\left(Z_g^* < g_2^{-1}\left(z_f^*\right)\right) = F_{Zg}\left(g_2^{-1}\left(z_f^*\right)\right) \tag{6}$$

where the notation g^{-1} stands for inverse function. Differentiating both sides of Eq. (6) with respect to z_f^* results in

$$\frac{\mathrm{d}F_{Zf}\left(z_f^*\right)}{\mathrm{d}z_f^*} = \left|\frac{\mathrm{d}g_2^{-1}}{\mathrm{d}z_f^*}\right| \cdot f_{Zg}\left(g_2^{-1}\left(z_f^*\right)\right) \tag{7}$$

where the distribution function $F_{Zf}(z_f^*)$ is discontinuous in $z_f^* = z_{bed}^*$. Therefore, the probability density function $f_{Zf}(z_f^*)$ includes an impulse term beside the conventional derivative, i.e.,

$$f_{Zf}\left(z_f^*\right) = \frac{1}{\left|g_2'\right|} \cdot f_{Zg}\left(g_2^{-1}\left(z_f^*\right)\right) + \Gamma_z \tag{8}$$

where Γ_z reflects the discontinuity of the cumulative distribution for $z_f^* = z_{bed}^*$ and the notation g' stands for derivative of g. Figure 4a shows both the probabilistic description of the drag force elevation z_f^*, and the relation between $g_2(z_g^*)$ and $f_{Zf}(z_f^*)$. Following a similar reasoning, the exposed area of the target particle can be obtained as a function of z_g^* as

$$\omega_f = g_3\left(z_g^*\right) = 1 - \frac{\arccos\left(2z_g^*\right)}{\pi} + \frac{z_g^*\sqrt{1-\left(2z_g^*\right)^2}}{\pi} \tag{9}$$

and the probability density function of the particle exposure $f_{\Omega f}(\omega_f)$ is given as

$$f_{\Omega f}\left(\omega_f\right) = \frac{1}{\left|g_3'\right|} \cdot f_{Zg}\left(g_3^{-1}\left(\omega_f\right)\right) + \Gamma_\omega \tag{10}$$

where the impulse term Γ_ω reflects the discontinuity of the cumulative distribution $F_{\Omega f}(\omega_f)$ at the lower and upper boundaries $\omega_f = 0$ and $\omega_f = 1$, respectively, as illustrated in Fig. 4b.

Assume that a normal distribution describes satisfactorily well the streamwise velocity fluctuations in the turbulent boundary layer, and that the moments of the distribution are known functions of the vertical distance from the bed, i.e. $f_{Uf}(u_f^*|z^*) \sim N(\mu_{uf}^*, \sigma_{uf}^*)$. From the transformation rules of random variables, the probability density function of T_f^* can then be expressed as

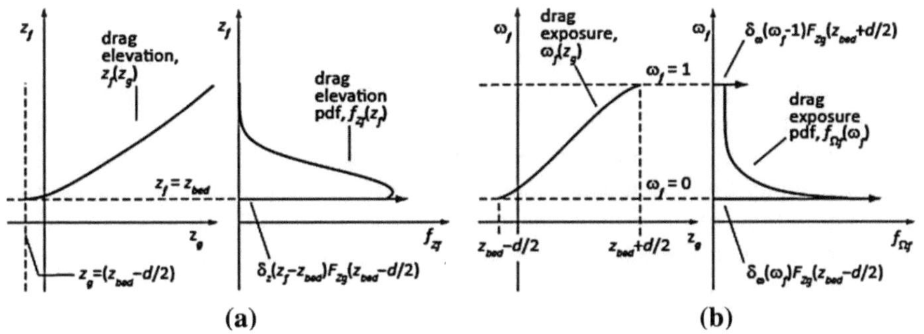

Fig. 4 **a** Fluid drag elevation function and pdf. **b** Particle exposure function and pdf

$$f_{\mathrm{T}f}\left(\tau_f^*\right) = \int_{-\infty}^{\infty} f_{\mathrm{T}f}\left(\tau_f^* \middle| z_g^*\right) \mathrm{d}z_g^* \tag{11}$$

where the conditional probability density function of T_f^* given $Z_g^* = z_g^*$ is

$$f_{\mathrm{T}f}\left(\tau_f^* \middle| z_g^*\right) = \frac{1}{\omega_f\sqrt{K\left|\tau_f^*\right|}} \cdot f_{Uf}\left(\frac{\tau_f^*}{\sqrt{K\omega_f\left|\tau_f^*\right|}} \middle| \mu_{uf}^*, \sigma_{uf}^*\right) + \Gamma_z \tag{12}$$

and $K = 0.5C_D(u_s)^2/(g\Delta d)$ is a dimensionless coefficient, γ is the acceleration due to gravity, and Δ the relative submerged density of sediments. Here, the impulse term Γ_z accounts for the probability of fully sheltering conditions for particles located below the mean bed level, while the former term involves partial exposure conditions for $0 < \omega_f < 1$, i.e., $(z_{bed}^* - 1/2) < z_g^* < (z_{bed}^* + 1/2)$, and full exposure conditions for $\omega_f = 1$, i.e., $z_g^* \geq (z_{bed}^* + 1/2)$. Equation (11) can then be introduced in Eq. (2) to obtain the risk of entrainment, R_E.

3.2 *Critical Shear Stress*

McEwan and Heald (2001) proposed a computational approach to calculate the form of the critical entrainment shear stress distribution for spheres under the action of a constant, uniform flow. They identified the contributions made by the supporting arrangement (or pocket friction angle) and the sheltering effects by recognizing that two grains may have the same critical entrainment shear stress but would actually be subjected to quite different fluid forces because of differences in the geometry of the bed upstream. These differences were then accounted for by empirically adjusting the critical entrainment force for individual particles. The authors suggested that *discrete particle modeling provides an additional* *means to evaluate these distributions. In the longer term, this may lead to refinements in probabilistic models and, consequently, to improvements in sediment transport predictions.* To our knowledge, to date this suggestion has remained unheeded, as probabilistic models have usually assumed the distributions of critical entrainment shear stress and fluid shear stress with no clear experimental evidence supporting the assumptions used to define the probability distributions. The distributions of the critical shear stress, $f_{\mathrm{T}g}(\tau_g^*)$, for different grain sizes were found to fall onto the same curve and to a first approximation they could be considered as lognormal, i.e.,

$$f_{\mathrm{T}g}\left(\tau_g^*\right) \sim logN\left(\mu_{\tau g}^*, \sigma_{\tau g}^*\right) \tag{13}$$

where the mean and standard deviation of the distribution for the sheltering case were found to be $\mu_{\tau g}^* = 0.87$ and $\sigma_{\tau g}^* = 2.77$. The expected value of the random variable T_f^* is different from the classical critical shear stress that can be computed using the Shields diagram, as $\mu_{\tau g}^*$ is associated with the instability of more bed particles than those implied by the critical shear stress (the corresponding fraction by weight entrained is roughly 0.5%).

A similar approach was developed by Cheng et al. (2003), who carried out flume experiments to investigate directly the distribution of the distance between particles on a flat bed comprised of uniformly sized spheres. The variation of the critical shear stress, T_g^*, was theoretically demonstrated to be approximately a narrow-banded random process with probability density function, $f_{\mathrm{T}g}$, following the Rayleigh distribution

$$F_{\mathrm{T}g}\left(\tau_g^*\right) = 1 - \exp\left(-\frac{\pi\tau_g^*}{4\left(\mu_{\tau g}^*\right)^2}\right) \tag{14}$$

For $\tau_g^* = \mu_{\tau g}^*$ the cumulative distribution $F_{\mathrm{T}g} = 54$ %, indicating that slightly more than half of the bed particles are likely to be entrained by the flow. Assuming that the classical incipient condition for grain motion corresponds approximately to the probability $F_{\mathrm{T}g} = 0.5 \div 1.0$ % leads to $\tau_{0c}^* \approx 0.1\ \mu_{\tau g}^*$, similarly to the DPM-based results.

The discrete particle model originally developed by by McEwan and Heald (2001) was used in the present analysis, such that the random contribution of the several geometrical factors involved in the definition of the critical shear stress distribution could be included into the model. Figure 5 shows the comparison between the probability density functions of the drag stress, obtained with Eq. (11), and the critical shear stress, obtained with Eq. (13). According to the Grass method, the larger the overlapping of the shear stress distributions, the higher the probability that grains are entrained. As recognized in previous experimental studies, the hydrodynamic forces on the bed do not follow a normal distribution, rather they appear to be better approximated by an exponential density function (e.g., Jimenez 1998).

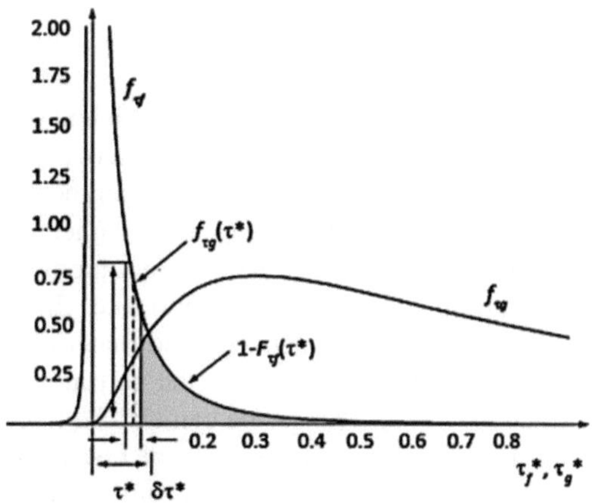

Fig. 5 *Pdfs* of drag and critical stress based on the Grass joint-probability approach

4 Comparison Of Experimental and Numerical Observations

Beside the analysis of the features of the near-bed turbulent flow that cause the initiation of motion, modeling sediment transport requires to gain information on the statistics of grain movement once particles are entrained. Einstein (1942) first introduced the notion that the number of bed particles eroded per unit time is related to the probability that a stationary particle is entrained by the fluid over a 'characteristic time', T_E. This represents the theoretical starting point to move from a probabilistic model of grain entrainment to a physically-based formulation of sediment transport. Most sediment transport is formed from a parameterization of Einstein step-length model

$$q_s = \lambda_s \cdot E_s \tag{15}$$

in which q_s is the transport rate per unit width, E_s is the entrainment (or pick-up) rate and λ_s is the mean particle step length. In this view of fluvial transport, particles are assumed to be transported in a discontinuous series of steps (each consisting of multiple hops or saltations) which are punctuated by rest periods.

Following Dancey et al. (2002), if R_E is the probability that an individual grain is being displaced, given that an entrainment 'event' has occurred or, in other words, given that a 'probabilistic experiment' has been performed, then the average mass of grains that is entrained per unit area over a time interval Δt is given by

$$m_s(\Delta t) = n_s(\rho_s V_s) \cdot \frac{R_E}{T_E} \cdot \Delta t \tag{16}$$

where n_s is the number of available sediment particles distributed over the unit area; V_s is the volume of the particles; T_E is the timescale of an entrainment event, corresponding to the average period between occurrences of the events. Recognizing that $E_s = m_s(\Delta t)/\Delta t$ and that number of grains per unit area is inversely proportional to the particle area yields

$$E_s = (1 - p_s)(\alpha_s \rho_s d) \cdot \frac{R_E}{T_E} \Rightarrow E_s \propto \frac{R_E}{T_E} \quad (17)$$

where $\alpha_s = V_s/(d \cdot A_s)$ is a shape coefficient equal to 2/3 for spherical particles, with $A_s = (1 - p_s)/n_s$ being the surface area occupied by a single grain and p_s the surface porosity of the sediment deposit. It can be drawn that an entrainment event with a characteristic timescale T_E (and a characteristic spatial scale) must be defined so that the probability or risk of entrainment can be scaled with experimental observations of sediment transport rate. Einstein defined the characteristic time as the time of exchange between bed material and bedload, which is proportional to the ratio of the grain size to the settling velocity of the particle. According to Paintal (1971), the timescale can also be interpreted as the time through which the particle moves from its position of rest to occupy a new position of deposit.

The probabilistic distributions of the hop lengths and travel times or 'hop times' of the entrained grains were determined based on image analysis of data collected during the experiments described in Sect. 2.1. The data identified the path of single entrained grains, including velocity, length, and orientation. From Fig. 6a it can be observed that about 50 % of particles have travel times shorter than about 1.0–1.5 s, with longer travel times being about 6–8 s. For similar hydraulic and transport conditions, Wu and Yang (2004) reported that the travel time of individual particles was less than about 4 s for most particles, with more than 80 % of the data having travel times shorter than 2 s. Following the original Einstein approach, the characteristic timescale T_E is assumed to approximate the mean travel times of the particles entrained by the flow.

Hop lengths were found to follow approximately an exponential distribution (see Fig. 6b). Comparing distributions from experiments having different boundary conditions, no evident trends were identified. The mean hop length was found approximately equal to 40–50 mm, corresponding roughly to 8–10 grain diameters, which is in agreement with previous studies on bedload transport at low transport stages. Einstein (1942) assumed the step distance of bed load movement

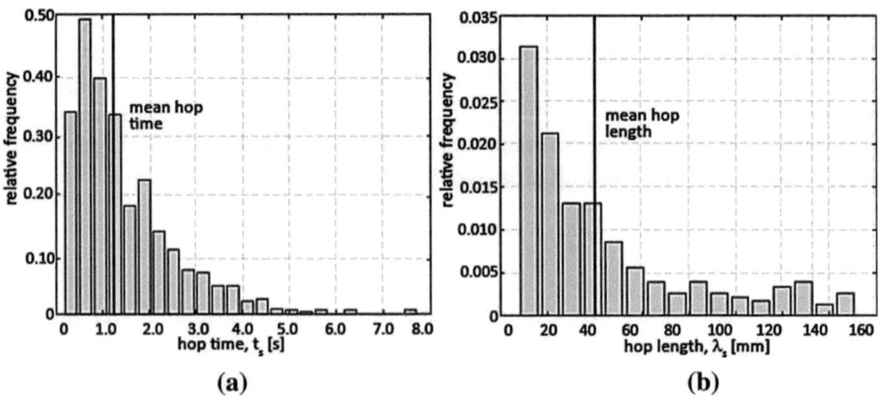

Fig. 6 **a** Experimental distribution of hop lengths. **b** Experimental distribution of travel times

to be proportional to the particle diameter, i.e., $\lambda_s \propto d$. Drake et al. (1988) found that displacement lengths increase with the grain size, while mean step lengths were observed approximately equal to 15 particle diameters. Papanicolaou et al. (1999) obtained similar results as they used glass beads to evaluate the particle displacement distance and the characteristics of the particle motion. They found that the travelling distance in the longitudinal direction was between one bead diameter and ten bead diameters. For low ratios of the boundary shear stress to the critical boundary shear stress (Bottacin et al. 2008) concluded that the vast majority of the displacements have a length lower than 50 mm (10 grain diameters).

These data combined with the entrainment data are used to examine experimentally the derived probabilistic formulation of entrainment risk given by Eq. (2). Assuming that an event of entrainment and the length that a grain covers once it is entrained are statistically independent, the total mass transport rate in bed load per unit length can be written as

$$q_s = \int E_s \cdot \lambda f(\lambda) d\lambda \tag{18}$$

where $f(\lambda)$ is the probability density function of step lengths. Introducing Eq. (17) into Eq. (18) and assuming $\alpha_s(1 - p_s) \approx 0.8 \div 0.9$ leads to the following expression of the volumetric sediment transport per unit length and yields

$$q_s = \frac{2}{3}(\rho_s d) \cdot \lambda_s \frac{R_E}{T_E} \tag{19}$$

Noting that $\lambda_s \sim 0.5 \div 1.0\ 10^{-1}$ m; $d \sim 5.0 \cdot 10^{-3}$m; $T_E \sim 1.0$ s; $R \sim 0.5 \div 1.0\ 10^{-3}$ leads to $q_s \sim 5 \div 9$ g/ms. This compares favourably with sediment transport rate measurements collected through visual inspection of the time series of bed images for tests T5, T9 and T11. Figure 7 shows the comparison between experimental and numerical observations, and the prediction of the experimental transport rates using an empirical formula commonly used in river engineering.

Fig. 7 Comparison of experimental and numerical observations

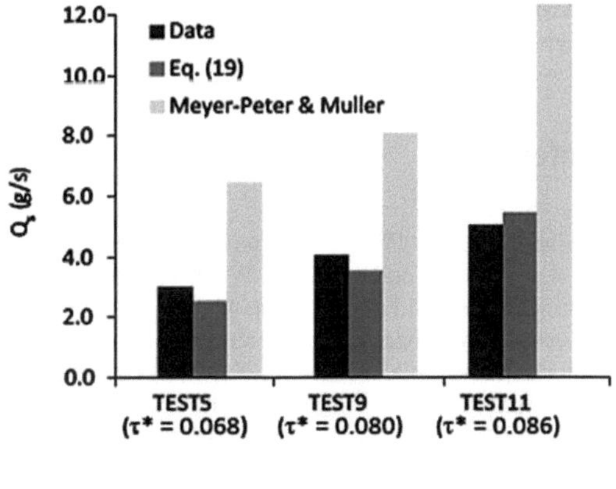

5 Conclusions

Despite the knowledge gained from intensive research on sediment incipient motion, a definite solution to this complicated problem is not yet available. It is believed that the development of entrainment models derived from a well-founded stochastic approach will be an important contribution to the prediction of sediment transport. To date, limitations to this development mainly depend on (a) a poor understanding of the interdependent random processes that lead to the incipient motion of sediments; (b) the lack of detailed statistical information of the variables involved in the entrainment process. Following on the original Grass theory, it is suggested that the bed-load entrainment can be predicted from the joint probability distribution of the instantaneous fluid shear stress and the local critical shear stress. The approach is founded on the assumption that the contributions of the near-bed turbulence and the variability of the resisting forces arising from highly disordered grain arrangements, respectively, can be disjoined and modeled separately.

The need for such disjoined description stems not only from the demand of more simplified analytical solutions, but it is also motivated by the evidence that the different nature of the random processes involved requires different and specifically-developed tools of analysis and investigation. As a consequence, the potential is shown of including into the model as much statistical information as required by the expected level of accuracy for the sediment transport prediction. This can be achieved by enhancing the description of the near-bed turbulence with the rapid advancement of particle imaging velocimetry (PIV) techniques. On the other hand, the use of numerical simulations performed with discrete particles models (DPM) can provide detailed representations of random bed geometries and, consequently, reliable distributions of the critical shear stress. By using such information it was possible to show that a probabilistically derived sediment entrainment threshold is significantly different from that of the traditional deterministic methods. The probabilistic method derived is general and can be applied in situations in which detailed knowledge of the bed geometry is available.

Acknowledgments Modeling and data analysis was carried out within the Project "PARTS: Probabilistic Assessment of the Retention and Transport of Sediments and Associated Pollutants in Rivers", funded by the EU Research Executive Agency via an Intra-European Fellowship to Dr Tregnaghi under a Marie Curie action funding scheme.

References

Bridge JS, Bennet JW (1992) A model for the entrainment and transport of sediment grains of mixed sizes, shapes, and densities. Water Resour Res 28(2):337–363. doi:10.1029/91WR02570

Battin TJ, Kaplan LA, Newbold JD, Hansen CME (2003) Contributions of microbial biofilms to ecosystem processes in stream mesocosms. Nature 426:439–441. doi:10.1038/nature02152

Bottacin-Busolin A, Singer G, Zaramella M, Battin TJ, Marion A (2009) Effects of streambed morphology and biofilm growth on the transient storage of solutes. Environ Sci Technol 43:7337–7342. doi:10.1021/es900852w

Bottacin-Busolin A, Tait SJ, Marion A, Chegini A, Tregnaghi M (2008) Probabilistic description of grain resistance from simultaneous flow field and grain motion measurements. Water Resour Res 44:W09419. doi:10.1029/2007WR006224

Buffington JR, Montgomery DR (1997) A systematic analysis of eight decades of incipient motion studies, with special reference to gravel-bedded rivers. Water Resour Res 33(8):1993–2029. doi:10.1029/96WR03190

Buffington JM, Montgomery DR (1999) Effects of sediment supply on surface textures of gravel-bed rivers. Water Resour Res 35:3523–3530. doi:1029/1999WR900138

Cheng NS, Chiew YN (1998) Pickup probability for sediment entrainment. J Hydraul Eng 124(2):232–235. doi:10.1061/(ASCE)0733-9399(1998)124:2(232)

Cheng NS, Law AWK, Lim SY (2003) Probability distribution of bed particleinstability. Advances in Water Resources 26(4):427–433. doi: 10.1016/S0309-1708(02)00184-7.

Dancey CL, Diplas P, Papanicolaou AN, Bala M (2002) Probability of individual grain movement and threshold condition. J Hydraul Eng 128(12):1069–1075. doi:10.1061/(ASCE)0733-9429(2002)128:12(1069)

Drake TG, Shreve RL, Dietrich WE, Whiting PJ, Leopold L (1988) Bedload transport of fine gravel observed by motion picture. J Fluid Mech 192:2193–2217. doi:10.1017/S0022112088001831

Einstein HA (1942) Formulas for the transportation of bed load. Trans ASCE 107:561–597

Grass AJ (1970) Initial instability of fine sand. J Hydraul Div ASCE 96(3):619–632

Jimenez J (1998) Turbulent velocity fluctuations need not be gaussian. J Fluid Mech 376:139–147. doi:10.1017/S0022112098002432

Kirchner JW, Dietricht WE, Iseya F, Ikeda H (1990) The variability of critical shear stress, friction angle, and grain protrusion in water-worked sediments. Sedimentology 37:647–672. doi:10.1111/j.1365-3091.1990.tb00627.x

Lopez F, Garcia MH (2001) Risk of sediment erosion and suspension in turbulent flows. J Hydraul Eng 127(3):231–235. doi:10.1061/(ASCE)0733-9429(2001)127:3(231)

McEwan IK, Heald JGC (2001) Discrete particle modelling of entrainment from flat uniformly sized sediment beds. J Hydraul Eng 127(7):588–597. doi:10.1061/(ASCE)0733-9429(2001)127:7(588)

McEwan IK, Soressen M, Heald JGC, Tait SJ, Cunningham G, Goring D, Willetts BB (2004) Probabilistic modelling of bed-load composition. J Hydraul Eng 130(2):129–140. doi:10.1061/(ASCE)0733-9429(2004)130:2(129)

Nelson JM, Shreve RL, McLean SR, Drake TG (1995) Role of near bed turbulence structure in bed load transport and bed mechanics. Water Resour Res 31(8):2071–2086. doi:10.1029/95WR00976

Nikora V, Goring DK, Biggs BJF (1998) On gravel-bed roughness characterization. Water Resour Res 34(3):515–527. doi:0043-1397/98/98WR-02886

Paintal AS (1971) A stochastic model of bed-load transport. J. Hydraul Res 9(4):527–554. doi:10.1080/00221687109500371

Papanicolaou AN, Diplas P, Balakrishnan M, Dancey CL (1999) Computer vision techniques for tracking bedload movement. J Comput Civil Eng 13(2):71–80. doi:10.1061/(ASCE)0887-3801(1999)13:2(71)

Papanicolaou A, Diplas P, Dancey C, Balakrishnan M (2001) Surface roughness effects in near-bed turbulence: implications to sediment transport. J Eng Mech 127(3):211–218. doi:10.1061/(ASCE)0733-9399(2001)127:3(211)

Papanicolaou A, Diplas P, Evaggelopoulos N, Fotopoulos S (2002) Stochastic incipient motion criterion for spheres under various bed packing conditions. J Hydraul Eng 128(4):369–380. doi:10.1061/(ASCE)0733-9429(2002)128:4(369)

Schmeeckle MW, Nelson JM, Shreve RL (2007) Forces on stationary particles in near-bed turbulent flows. J Geophys Res 112:F02003. doi:10.1029/2006JF000536

Shields A (1936) Application of the similarity principles and turbulence research to bedload movement, Report 167. California Institute of Technology, Pasadena, (translated from German)

Tregnaghi M, Busolin-Bottacin A, Tait S.J, Marion A (2010) 3D near-bed flow field measurements at low sediment transport rates. Paper presented at river flow 2010, International conference on fluvial hydraulics, Braunschweig, Germany, 8–10 Sept
Wu FC, Chou YJ (2003) Rolling and lifting probabilities for sediment entrainment. J Hydraul Eng 129(2):110–119. doi:10.1061/(ASCE)0733-9429(2003)129:2(110)
Wu FC, Yang KH (2004) A stochastic partial transport model for mixed-size sediment. Application to assessment of fractional mobility. Water Resour Res 40:W04501. doi:10.1029/2003WR00225

On Using Artificial Rapid Hydraulic Structures (RHS) Within Mountain Stream Channels: Some Exploitation and Hydraulic Problems

Artur Radecki-Pawlik

Abstract The chapter describes rapid hydraulic structure (RHS) with increased roughness: one of the technical solutions intended to maintain river and stream beds in a good condition and at the same time ensure the development of a braided channel and contribute to the regeneration of channel bar structures, creating suitable living conditions for macrobenthos, and enabling unconstrained fish migration without additional fish passes. The chapter also discusses the problem of rebuilding the existing water straight drop structure in Brenna on the Brennica River (Polish Carpathians) which was changed into the rapid hydraulic structure. The technical project was set up in 1988 and finished in the same year. The structure was rebuilt in the field in early autumn 1990. The author of the chapter was a co-designer of the hydraulic structure.

Keywords Rapid hydraulic structure · Straight drop structure · Roughness · Rebuilding · Energy dissipating · Artificial roughness · Water discharge · Water velocity

1 Introduction

For the past years, member states across Europe have been implementing Water Framework Directive (WFD) of the European Union. There has been a prolonged debate, especially in Poland, among designers, river managers, fishermen and

A. Radecki-Pawlik (✉)
Department of Hydraulic Engineering and Geotechnics, Faculty of Environmental Engineering and Land Surveying, University of Agriculture in Kraków, Aleja Adama Mickiewicza 21, 30-059 Kraków, Poland
e-mail: rmradeck@cyf-kr.edu.pl

P. Rowiński (ed.), *Experimental and Computational Solutions of Hydraulic Problems*, GeoPlanet: Earth and Planetary Sciences, DOI: 10.1007/978-3-642-30209-1_6,

biologists on the application of all hydraulic structures and engineering methods meant to preserve river beds in best possible condition, and at the same time maintain a close to nature river morphology while preventing river bed erosion and providing flood protection. Along many sections of the Carpathian rivers, river bed systems are still affected by not always justified technical river regulation and therefore it is necessary to recognize which mountain river training structures can be accommodated in a mountain river fluvial system and positively affect the biological life of both macrobenthos and fish (Korpak et al. 2008; Wyżga et al. 2009). Field studies were thus undertaken to examine artificial rapid hydraulic structures (RHS) that were already existing within the mountain channels and later to use the results of such investigations for modelling RHS to improve their construction in terms of their hydraulics and impact on river environment, especially for fish and macroinvertebrates.

This chapter describes rapid hydraulic structure (RHS) with increased roughness: one of the technical solutions intended to maintain river and stream beds in a good condition and at the same time ensure the development of a braided channel and contribute to the regeneration of channel bar structures, creating suitable living conditions for macrobenthos, and enabling unconstrained fish migration without additional fish passes. At the same time, RHS is one of the river training structures which prevent river bed against erosion, stabilise bed slope and help to keep river channels stable under flooding conditions.

The chapter also discusses the problem of rebuilding the existing water straight drop structure in Brenna on the Brennica River (Polish Carpathians) which was changed into the rapid hydraulic structure. The technical project was set up in 1988 and finished in the same year. The structure was rebuilt in the field in early autumn 1990. Artificial roughness of a slope plate of the rapid hydraulic structure was reached by placing cobbles into it. The dimension of cobbles was calculated by applying different methods and the optimum value for that dimension was chosen. The author of the chapter was a co-designer of the rapid hydraulic structure.

2 Artificial Rapid Hydraulic Structure with Increased Roughness

While planning stream training, one should take into consideration a compromise between natural environment requirements and human interference (Ślizowski and Radecki-Pawlik 1996). A natural stream with overgrown banks usually remains in a state of dynamic and biological equilibrium. Disturbing the natural state of the stream entails decreasing the roughness and thus disturbing the dynamic equilibrium. The stabilisation of mountain stream channels, characterised by steep longitudinal slopes, considerable changeability of water stage, flash floods and massive bed-load transport, can be obtained by a construction of stages of falls or weirs in order to create a given critical slope of the stream bed. Such measures unfavourably influence the natural environment. In this context, the best solution to the problem of river channel protection from the impact of flash flows by the reduction of the longitudinal slope of the river seems to be the application of rapid hydraulic structures with increased roughness (Niel 1960; Hartung and Scheurlein 1970; Knauss 1980; Żelazo 1992; Ślizowski and Radecki-Pawlik 1996, 1998, 1999, 2000; Ślizowski et al. 1997, 2008; Radecki-Pawlik and Ślizowski 1998; Radecki-Pawlik 2006a, b). These structures enable the migration of fish and benthic macroinvertebrates (benthos), lead to water oxidation and blend into the landscape (Kłonowska-Olejnik and Radecki-Pawlik 2000; Kłonowska-Olejnik et al. 1999, 2006). The pools formed, the presence of which is justified by hydrodynamics of the flow, should be preserved at the sections between the rapid hydraulic structures. Stones of different sizes should be placed in the river bed, creating a shelter for fish and other living organisms (Ślizowski et al. 2008). Such shelters should also be located along river banks. The proposed solutions meet the ecological requirements—blending into landscape—as well as those connected with the stabilisation of the stream channel.

The proper choice of stone sizes and their positioning on the rapid hydraulic structure significantly influence the efficiency of the rapid and its integration with the natural environment. The dimensions of stones in the slope plate of the rapid in relation to the computational velocity and the slope of the rapid are presented in Table 1 (Knauss 1980). Velocity over RHS could be measured (Photo 6—later in the text) for better understanding the hydraulics of structures and finally to improve their construction towards environmental requirements given by Water Framework Directive EU.

The choice of stones on the rapid hydraulic structure plate can be made with the help of the chart suggested by Ślizowski et al. (1997). In this case, the dimensions of stones are determined by the specific discharge on the rapid (q) and the height of the rapid (H) (Fig. 1).

Photographs 1, 2, 3 beneath present examples of rapid hydraulic structures with increased roughness built in Polish rivers and congruent with the principles of the appropriate maintenance of river beds in the close to nature condition. In Photograph 2, note the concentration of flow on the slope plate of the rapid hydraulic structure, ensuring the passage of fish even at low water stages, while Photographs 3, and 5 and 6 depict the apparently carelessly placed crushed rock, playing an important role in the reduction of erosion of the river bed. The designed rapid

Table 1 Dimensions of stones on the rapid hydraulic plate versus the plate slope and water velocity, according to Knauss (1980)

Water velocity v for RHS inclination 1:8 (m s^{-1})	Water velocity v for RHS inclination 1:10 (m s^{-1})	Water velocity v for RHS inclination 1:15 (m s^{-1})	Stone dimension (m)
2.50	2.70	3.70	0.6
4.60	4.90	5.80	0.8
7.00	7.60	8.90	1.2

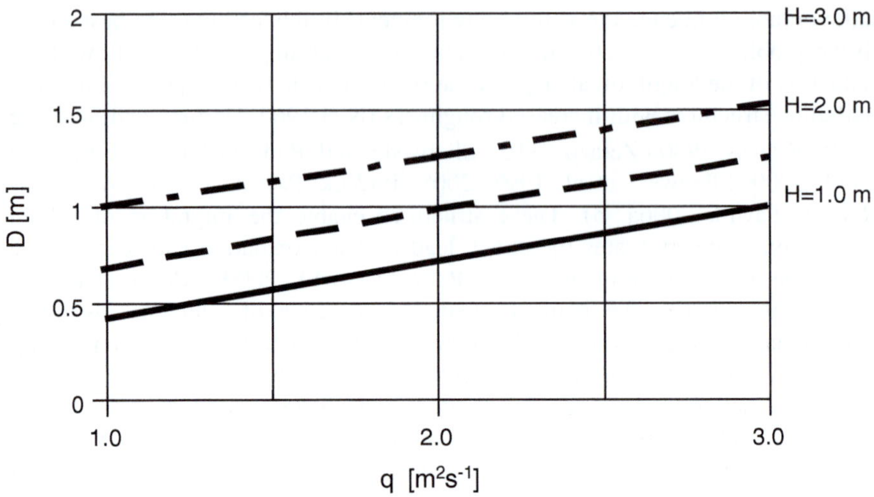

Fig. 1 The graph suggesting the choice of dimension of stones (D) on the RHS plate depending on the specific discharge (q), and the RHS height (H)

Photo 1 A RHS in the Raba River. Photo K. Plesinski

Photo 2 A RHS in the Porębianka Stream. Look at the concentration of water discharge in the middle of the RHS plate. Photo K. Plesinski

hydraulic structures better fulfil their ecological function of ensuring the fish and macrobenthos passage if the slope of the rapid plate is not very steep. The optimum is 1:10, but taking into consideration local conditions, steeper inclinations of slopes can be allowed, to the maximum of 1:5 (Photo 4).

The rapid hydraulic structures built in the Poniczanka and the Sanoczek Streams are made of crushed stones arranged side by side, enabling the reduction of water surface slope as well as local strengthening of the river bed. The form of the structure in the river channel is the closest to nature.

The rapids constructed in the Porębianka stream are stone platforms, enabling, at the achieved river bed slope, an unconstrained fish and macrobenthos passage and recreating channel bars up- and downstream of the structures. It is manifested by the restoration of the natural, braided-meandering river channel pattern. Additionally, the rapid built in the Porębianka stream has an overflow part which converts into a concentrated channel passing along the slope plate of the structure. It causes the concentration of the water stream and in consequence guarantees good conditions for fish passage even at low water stages.

Photo 3 A RHS in the Poniczanka Stream. Photo K. Plesinski

3 The Rapid Hydraulic Structure on the Brennica River in the Beskid Slaski Mts

3.1 Polish Carpathians

When rebuilding any engineering construction we often struggle with many more problems than in the case of the new structures. It is obvious that sometime there is no need to reassemble a structure and limited reconstruction work is sufficient to repair it, especially when its construction corpus is not seriously damaged. Hydraulic structures work in special conditions (within the water) thus a careful way of thinking is needed when designing, building and reconstructing them. Application of research results into engineering works seems to be very beneficial from practical point of view. These two ways of thinking, based on scientific and engineering experience, have been combined in Brenna project in the Polish Carpathians s during the repair works of the straight drop spillway on the Brennica River (Radecki-Pawlik 1993). The structure, situated in 7 + 964 km along the Brennica River (Fig. 2), is of a great local importance because it is the last hydraulic structure in a chain of other hydraulic structures of a river training works built upstream. It means that it supports all other water sills and bank revetments upstream.

Photo 4 A RHS in the Jachówka-Bienkowka Stream. Photo K. Plesinski

As has been noticed during field-expert visit some parts of the 7 + 964 km straight drop weir had been damaged. Figure 3 shows the severe damage to the downstream-sill as well as the floor of the energy dissipating pool (a silting basin). The conditions of the two side-walls along the energy dissipating pool had also been very poor and needed repairs.

It has been noticed after 19 years after the construction of the weir that a river bed was lowered about 2 m downstream of the structure. The main reason for that was probably an illegal mining procedure of gravel from the river bed leading to bed degradation. It also seemed to be the main reason for the damage to the hydraulic structure (Radecki-Pawlik and Wójcik 1987). Because of good conditions of the straight drop wall of the weir as well as upstream revetments, the new concept of the repair of the structure was applied, based on the existing local conditions. The existing structure was to be rebuilt into a rapid hydraulic structure (spillway) with an artificial roughness on its slope chute plate to reduce the energy of flowing water. The artificial roughness was achieved using stones from the river bed. It was advised to fix stones into the slope plate of the structure.

Using the above-mentioned method to determine the stone dimension which might be put into the apron of the RHS and the other experiences—"Hydroprojekt" design office instructions and Polish standards (Radecki-Pawlik and Wójcik, 1987)—the technical project of the rapid hydraulics structure was proposed. The theoretical plot of a rapid hydraulic structure and the stones fixed in chess-like manner to the chute plate is shown in Fig. 4.

Photo 5 A RHS in the Sanoczek Stream. Photo A. Radecki-Pawlik

Precise information about hydrological data is required to rebuild any hydraulic structure in a river. Listed below are some of the main water discharges for a rapid hydraulic structure cross-section in the Brennica River: NNQ (low flow) is 0.033 m^3s^{-1}, SNQ (mean-low flow) is 0.21 m^3s^{-1}, SRQ (mean flow) is 1.58 m^3s^{-1}, SWQ (mean-high flow) is 57.20 m^3s^{-1}, Q-50 % is 32.00 m^3s^{-1}, and finally Q-5 % is 173.00 m^3s^{-1}. Discharges shown here are intended for the designing purposes. The way of calculating the discharges are in accordance with the Polish standards and were calculated by IMGW (Water Management Authority).

Hydraulics calculation has been carried out using Q-50 and Q-5 % discharge values as design discharge values, according to the Polish standards. Calculations have been carried out in three ways: for straight drop spillway structure (Chow 1959), for RHS structure ("HYDROPROJEKT"—1980) and finally using USBR stilling basin II method (Dziewoński 1973). These calculation methods are well-known and commonly used by engineers. Finally, the USBR method was applied. As a result, the structure presented in Photos 7, 8, 9 was finally designed.

Hydraulics calculations involving artificial roughness (stones on a slope plate) indicated that the silting basin length (calculated following usual USBR sets of equations; see http://en.wikipedia.org/wiki/United_States_Bureau_of_Reclamation) needed to be reduced by 30 % to decrease velocity of water flowing through the

Photo 6 Velocity wading measurements along the RHS in the Sanoczek Stream. Photo K. Huta

structure. This reduction in water velocity was possible by using the artificial roughness of a slope chute plate.

4 Summary

The current concept of river and stream training makes it possible to use technical measures only where it is necessary, for example, for the protection of road bridges or flood protection of large population centres. In places where a river constitutes no threat, its broad flood area should remain unchanged. One should also enable an unconstrained development of morphological processes forming pool and riffle sequences, and deposition of mid-channel bars. It is unnecessary to protect meadows and wastelands by means of very costly irrigation works; on the contrary, these lands should be treated as flood zones, where water can stagnate thus increasing valley retention. In some regions, the government should strive to buy out agricultural lands, in other, the issue of building permits for flood areas should be restricted. Hydraulic engineers should be encouraged to use "close to nature" techniques in the design and execution of river training structures, such as rapid hydraulic structures with increased roughness presented in this work. The readers

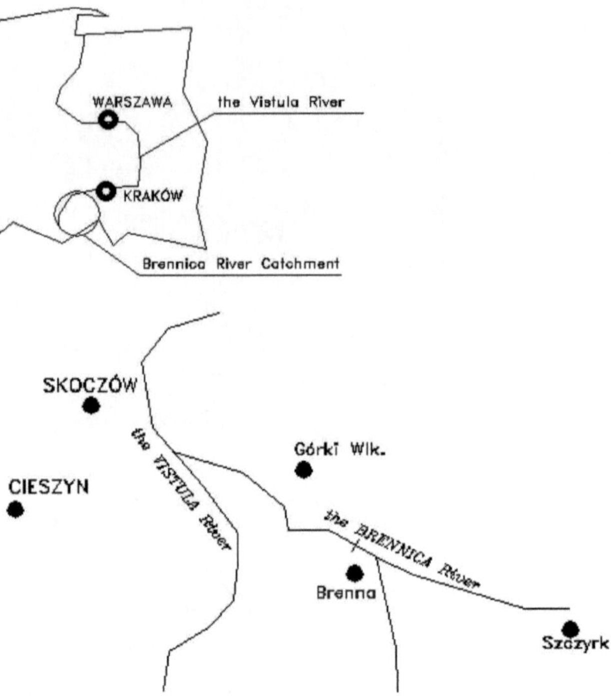

Fig. 2 The Brenna River catchment

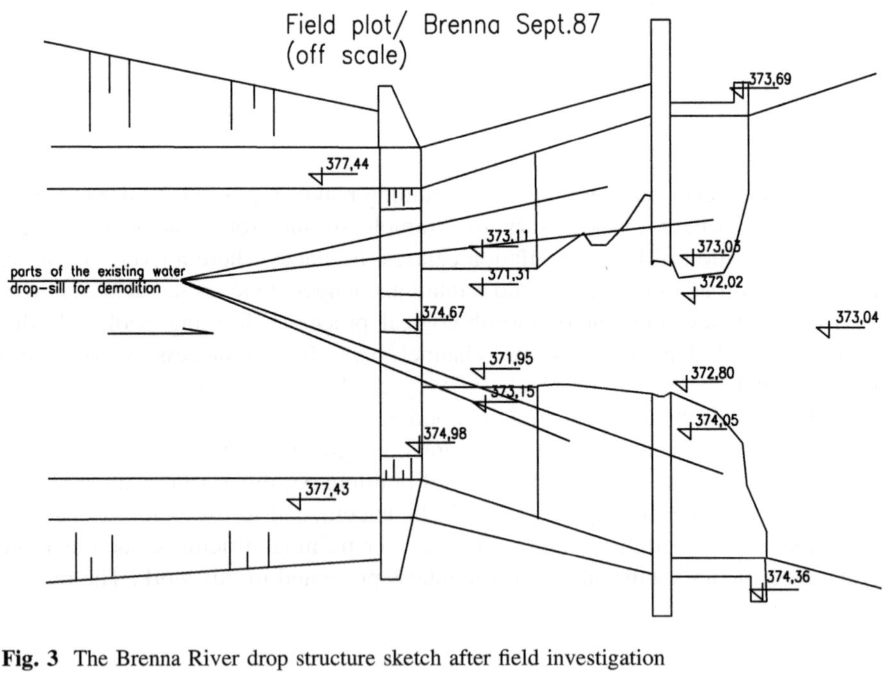

Fig. 3 The Brenna River drop structure sketch after field investigation

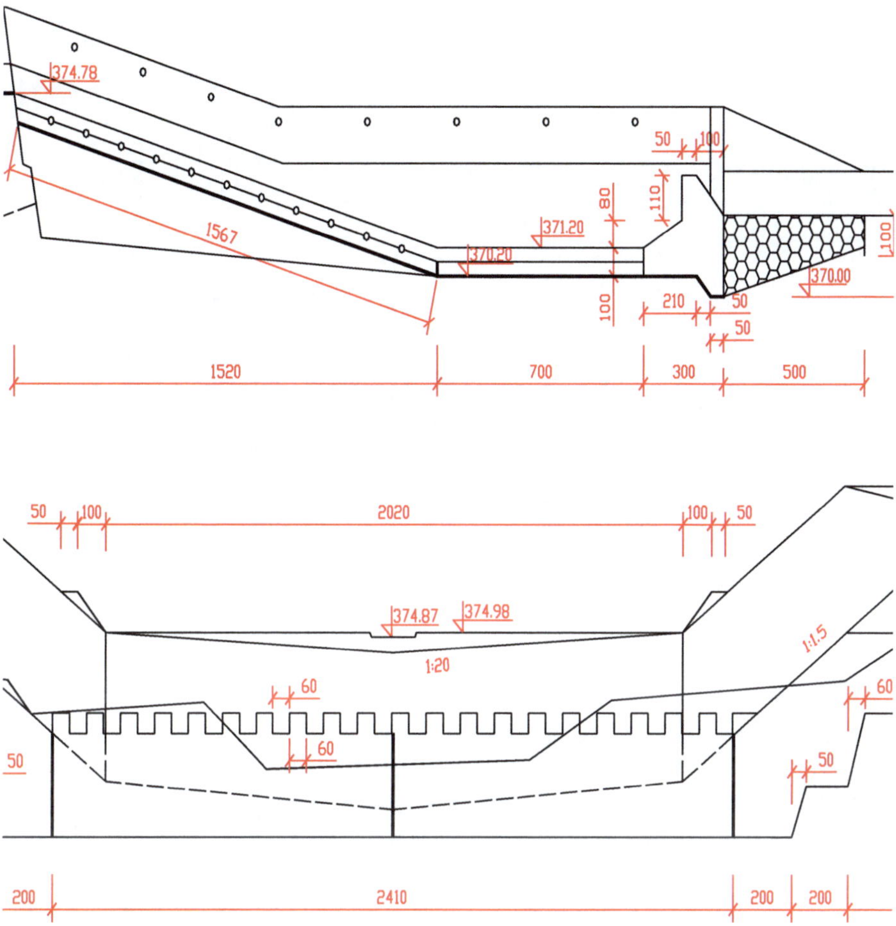

Fig. 4 The Brenna River RHS designing plots

who are more interested in the issue described in the chapter are kindly requested to become acquainted with the monograph of Korpak et al. (2008), where the problems concerning anthropogenic changes of the Carpathian river beds and the use of many different hydraulic structures have been described in detail.

As far as the Brenna RHS is concerned, it is important to combine the results of scientific research and designing technique when building any hydraulic structure. Such a situation appeared in the Brennica River site where the existing straight-drop hydraulic structure had been rebuilt. The artificial roughness on the slope rapid plate of the rapid hydraulic structure which replaced the previously existing weir was used to reduce the energy of the stream water flume. That roughness was reached by placing cobbles into the slope rapid plate along its length. Some formulae were presented and used to determine the dimension of cobbles to be

Photo 7 The RHS in the Brenna River. Photo A. Radecki-Pawlik

Photo 8 The RHS in the Brenna River. Photo A. Radecki-Pawlik

used. As a consequence, the length of the silting basin pool of the rapid hydraulic structure was reduced. To reach the artificial roughness of a slope plate the cobbles were taken out from the river bed and the hydraulic structure is therefore very well fitted to the local environment (Ratomski 1992; Ślizowski 1993; Ślizowski and

Photo 9 The RHS in the Brenna River, a close look at the steep apron with artificial roughness. Photo A. Radecki-Pawlik

Radecki-Pawlik 1996). Thus, the object seems to work like a natural rapid in a stream. Such a solution combines then engineering and environmental needs. The project was finished in 1988. The rebuilding of the structure in the field was finished in early autumn 1990. For designers as well as for the users of any structure, a so-called natural test of an object is important. Such a "natural test" took place on 5th of August 1991 when the water discharge of a range Q-50 % was recorded. It was 33.8 m^3/sec (close to the competent flood). During that time, the structure was working properly and no damage occurred. The authors of the project hope that such designing solutions would be applied in the future when working in similar conditions.

References

Chow VT (1959) Open channel hydraulics. McGraw-Hill, NY

Dziewoński Z (1973) Rolnicze zbiorniki retencyjne (Agricultural water reservoirs), PWN, Warszawa (in Polish)

Hartung F, Scheurlein H (1970) Design of overflow rock dams. Beitrag no. 36, Talsperenkongres, Montreal

Kłonowska-Olejnik M, Radecki-Pawlik A (2000) Zróżnicowanie mikro-siedliskowe makrobezkręgowców dennych w obrębie łach korytowych potoku górskiego o dnie żwirowym (The micro-habitat difference in water benthos structures within the region of gravel point bars). XVIII Zjazd Hydrobiologów Polskich "Szacunek dla wody", 4–8 Sept 2000, Polskie Towarzystwo Hydrobiologiczne, Białystok, pp 119–120 (in Polish)

Kłonowska-Olejnik M, Fiałkowski W, Radecki-Pawlik A (1999) Wpływ zabudowy żłobem kamiennym na zespoły makrobentosu potoku Białego w Pienińskim Parku Narodowym (The influence of concrete channel lining on macrobentos of the Biały Stream, Pieniny National Park, Ogólnopolska Konf. Nauk. nt.: Współczesne Kierunki Badań Hydrobiologicznych, 22–24 Sept 1999, Supraśl k/Białegostoku, pp 115–116 (in Polish)

Kłonowska-Olejnik M, Radecki-Pawlik A, Skalski T (2006) Wybrane metody biologiczne w ocenie stanu ekosystemu uregulowanego cieku górskiego (potok Młynne, Gorce) (Chosen biological methods used in assessment of of ecosystems of engineered river channels). XX Zjazd Hydrobiologów Polskich, 5–8 września 2006, Uniwersytet Mikołaja Kopernika, Torun (UMK), p 124 (in Polish)

Knauss J (1980) Drsne skluzy (Weirs and rapids). Vodni Hospodarstvi, rada A C 1, Praha (in Czech)

Korpak J, Krzemień K, Radecki-Pawlik A (2008) Wpływ czynników antropogenicznych na zmiany koryt cieków karpackich (Influence of anthropogenic factors on changes of Carpathian river channels). Infrastruktura i Ekologia Terenów Wiejskich. Tech. Inf. Wsi, PAN Oddz. w Krakowie, 4, Monografia, p 88 (in Polish)

Niel A (1960) Über die wernichtung kinetischer Energie durch niedere Geffellastufen. Wien. Österreische Wasserwirtshaft 4:5

Radecki-Pawlik A (1993) Stopień—bystrze w Brennej na rzece Brennicy jako przykład wariantu remontu istniejącego stopnia klasycznego (Rapid hydraulic structure on the Brenna River as an example of typical rebuilding engineering works on the drop hydraulic structure. I Krajowa Konferencja Naukowa z Udziałem Gości Zagranicznych nt. "Bezpieczeństwo i trwałość budowli wodnych", Wrocław—Rydzyna (in Polish)

Radecki-Pawlik A (2006a) Podstawy hydrogeomorfologii cieków górskich dla inżynierów środowiska, biologów, geografów i ekologów (Principles of hydromorphology of mountain streams and rivers, for engineers, geographers and biologists). Monografia, AR-UJ, Kraków-Warszawa, Instytut Nauk o Środowisku Uniwersytetu Jagiellońskiego w Krakowie, BEL-Studio—Warszawa, p 131 (in Polish)

Radecki-Pawlik A (2006b) Bystrza o zwiększonej szorstkości jako budowle regulacyjne bliskie naturze a formacje korytowe rzeki górskiej (Rapid hydraulic structures used as nature friendly hydraulic structures similar to natural gravel bed forms). In: Kłonowska-Olejnik M, Fiałkowski W (eds) Zastosowanie hydrologii w badaniach biologicznych wód płynących, Instytut Nauk o Środowisku, Uniwersytet Jagielloński, Kraków, pp 66–67 (in Polish)

Radecki-Pawlik A, Ślizowski R (1998) Bystrze o zwiększonej szorstkości jako element ekologicznej stabilizacji potoków górskich (Rapid hydraulic structures as ecological elements stabilizing mountain river channels). Przegląd Naukowy, SGGW, Warszawa, vol 15. pp 153–162 (in Polish)

Radecki-Pawlik A, Wójcik A (1987) Remont budowli regulacyjnych na rzece Brennicy w km. Od 7+964 do 8+215—projekt techniczny (Rebuilding of a river engineering system on the Brennica River along km 7+964 to 8+215—professional project) CBSiPBW Hydroprojekt o/ Kraków (in Polish)

Ratomski J (1992) Proekologiczna zabudowa rzek i potoków górskich (Pro-ecological river engineering works). XII Ogólnopolska Szkoła Hydrauliki, Międzyzdroje, 21–25 września. Materiały Szkoły s. pp 59–68 (in Polish)

Ślizowski R (1993) Bystrza o zwiększonej szorstkości jako element zabudowy potoków górskich (Rapid hydraulic structures as an element of river engineering works), Rozpr. hab. nr 181, Zesz. Nauk. AR w Krakowie (in Polish)

Ślizowski R, Radecki-Pawlik A (1996) Badania bystrza o zwiększonej szorstkości na potoku Brennica w miejscowości Brenna (On some investigations of rapid hydraulic structures—the Brennica River in Brenna, Zesz. Nauk. AR w Krakowie, vol 306. pp 71–77 (in Polish)

Ślizowski R, Radecki-Pawlik A (1998) Bystrze z kamienia naturalnego jako przykład nowoczesnej, ekologicznej budowli regulacyjnej w zlewniach rzecznych (Rapid hydraulic structure with artificial roughness as an example of a model river training structure). Przegląd Naukowy, SGGW, Warszawa, vol 15. pp 162–170 (in Polish)

Ślizowski R, Radecki-Pawlik A (1999) Rozmycie nieumocnionego dna poniżej bystrzy o zwiększonej szorstkości (Scouring problems downstream of the rapid hydraulic structures). Zesz. Nauk. AR w Krakowie, vol 19. pp 115–123 (in Polish)

Ślizowski R, Radecki-Pawlik A (2000) Distribution of maximum velocities and Froude numbers on a rapid hydraulic structure apron. Electron J Pol Agric Univ (EJPAU) 3(1):1–13

Ślizowski R, Radecki-Pawlik A, N'Famara Sambou (1997) Badania porównawcze bystrzy z kamienia naturalnego—elementu ekologicznej zabudowy potoków górskich—z bystrzami betonowymi typu Peterki (Investigtions of rapid hydraulic structures and their comparison with the Peterka-like hydraulic structures, Zesz. Nauk. AR w Krakowie, 321, pp 33–41 (in Polish)

Ślizowski R, Radecki-Pawlik A, Huta K (2008) Analiza wybranych parametrów hydrodynamicznych na bystrzu o zwiększonej szorstkości na potoku Sanoczek (Analysis of some hydraulic parameters of a rapid hydraulic structure on the Sanoczek stream. Infrastruktura i Ekologia Terenów Wiejskich. Tech. Inf. Wsi, PAN Oddz. w Krakowie, 2, pp 47–58 (in Polish)

Wyzga B, Amirowicz A, Radecki-Pawlik A, Zawiejska J (2009) Hydromorphological conditions, potential fish habitats and the fish community in a mountain river subjected to variable human impacts, the Czarny Dunajec. Pol Carpathians River Res Appl 25(5):499–659

Żelazo J (1992) Badania prędkości i oporów przepływu w naturalnych korytach rzek nizinnych (Water velocity and flow friction investigations in natural lowland river channels). Rozprawy naukowe i monografie, Warszawa, SGGW (in Polish)

[illegible] spatial [illegible]
[illegible] R, Radecki-Pawlik A [illegible] (in Polish)
[illegible] (in Polish)
[illegible]

Part I
Experimental Solutions of Hydraulic Problems

Scale Model Study of Propeller Induced Scour Development

Peter Geisenhainer and Jochen Aberle

Abstract This chapter describes the experimental setup and preliminary results of a scale model study (1:16) aiming at the investigation of the temporal evolution of propeller jet induced scours. Time series of scour depth and geometry from experiments carried out with the stern of a barge equipped with two different propulsion systems, a Kaplan propeller in Kort nozzle with double rudder and a Wageningen B-series propeller with central rudder, respectively, are presented. The time series are used to validate the experimental methodology and to show the reproducibility of the experiments. The influence of the different propulsion systems on scour depth and geometry is discussed. The preliminary results are used to highlight the need for a more accurate description of the near bed propeller jet flow field.

1 Introduction

Inland water transport represents an economical and energy-efficient way for transportation but can also impact the ecological and morphological character and functions of waterways (e.g., PIANC 2003). The development of adequate management strategies for inland waterways taking into account ecological,

P. Geisenhainer (✉) · J. Aberle
Leichtweiß-Institut für Wasserbau, Technische Universität Braunschweig,
Beethovenstr. 51a 38106 Braunschweig, Germany
e-mail: p.geisenhainer@tu-bs.de

J. Aberle
Department of Hydraulic and Environmental Engineering,
The Norwegian University of Science and Technology,
Norway, S. P. Andersens vei 5 N-7491 Trondheim, Norway

P. Rowiński (ed.), *Experimental and Computational Solutions of Hydraulic Problems*, GeoPlanet: Earth and Planetary Sciences, DOI: 10.1007/978-3-642-30209-1_7,

hydrological, economical and political needs and boundary conditions, respectively, requires the proper understanding of vessel related impacts on waterways. These impacts result from drawdown, ship-induced primary and secondary wave systems, return flow, and propeller wash (e.g., BAW 2005) and have increased from the beginning of last century until today due to increases in ship size and motor power (e.g., Spitzer et al. 2012a, b).

Especially, the fast-moving turbulent propeller induced water jet can cause severe damage to the bed, banks, and quay structures, particularly in maneuvering situations (e.g., Flamm 1914; Kempf 1939; FAS 1974; Verhey 1983; Oebius 1984; BAW 2005). The investigation of the scouring action of a propeller jet requires the knowledge of the complex velocity field in the propeller wash (see Lam et al. 2011 for a review) which depends on many boundary conditions, such as stern shape, engine power, propeller characteristics, number of propellers, rudder arrangement, underkeel clearance, draught, etc. (e.g., Fuehrer and Römisch 1977; Blaauw and van de Kaa 1978; Berger et al. 1981; Verhey 1983; Hamill et al. 1995; Oebius 2000; BAW 2005).

Most of the existing approaches for the calculation of the propeller jet velocity field are based on the axial momentum theory predicting the velocity distribution in relation to the efflux velocity u_0 (maximum time-averaged velocity along the initial propeller plane; Lam et al. 2011). Subdividing the flow field into a zone of flow establishment and a zone of established flow (e.g., Blaauw and van de Kaa 1978), the corresponding approaches allow for the calculation of the velocity $u_{x,r}$ in an infinitive wide and deep water body at any distance to the propeller plane x and jet axis r as a function of u_0 (e.g., FAS 1972).

However, in real world situations, the dispersion of the jet is hindered by the water surface, channel bottom as well as side walls and influenced by the stern shape and rudder arrangement (e.g., BAW 2005). For example, a rudder behind a propeller splits the jet into two streams with one stream being directed towards the bottom (e.g., Gebers 1911; Hamill et al. 2001) leading to distinct differences in the near bed flow field compared to the situation without a central rudder. As a consequence, BAW (2005) suggested the classification of the propeller jet dispersion using four standard situations (SS). SS 1 and SS 2 describe the situation in the absence of lateral and longitudinal obstacles hindering jet dispersion with (SS2) and without (SS1) splitting of the jet at a rudder (Fig. 1), respectively.

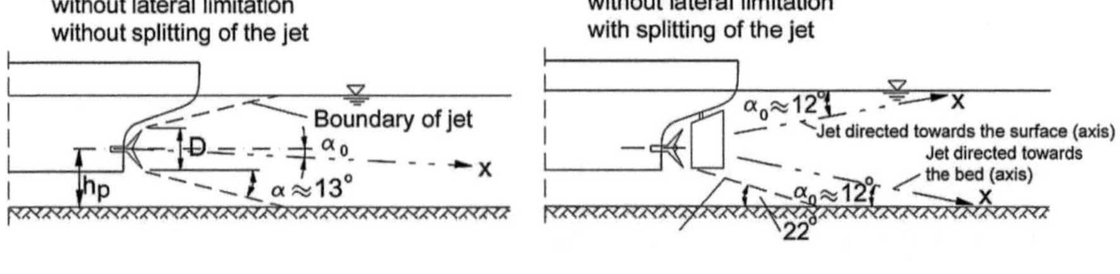

Fig. 1 Standard situations 1 and 2 for jet dispersion according to BAW (2005)

Similarly, SS 3 and SS 4 describe the situations when jet dispersion is affected by quay structures (not considered here).

According to Fig. 1, the main axis of the propeller jet is inclined towards the bed at an angle of $\alpha_0 = 2.5°$ for SS 1 (note that $\alpha_0 = 0°$ for a ducted propeller or a vessel with tunnel stern) and by $\alpha_0 = 12.5°$ for the situation with jet splitting. The outer boundaries of the jet (for which $u_{x,r}/u_{x,max} = 0.1$) are inclined at an angle of $\alpha = 13°$ for SS 1 and $\alpha = 22°$ for SS 2. Using the inclination angles of the main jet axis and the velocity distribution resulting from the axial momentum theory according to Blaauw and van de Kaa (1978), it can be shown that the maximum bed velocity occurs at a distance to the propeller axis of $x_s = h_p/\tan(11.6°)$ for SS 1 and $x_s = h_p/\tan(17.9°)$ for SS 2, respectively (LWI 2007). The corresponding maximum bed velocity $u_{b,max}$ can be expressed via the following simplified general expression (e.g., Fuehrer et al. 1981; BAW 2005):

$$\frac{u_{b,Max}}{u_0} = E\left(\frac{h_P}{D_P}\right)^{-1} \tag{1}$$

where h_p defines the height of propeller axis above the bed and E is a constant depending on rudder arrangement and vessel type. E-values reported in the literature for different situations and vessel types are summarized in Table 1. The variety of these values shows that the stern shape and/or presence of a rudder have a significant impact on jet dispersion and hence on $u_{b,max}$.

The dependency of E and hence $u_{b,max}$ on vessel and propulsion system characteristics hampers the determination of bed shear stress which can, e.g., be estimated via the expression recommended by Blaauw and van de Kaa (1978) $\tau = 0.5\rho C_f u_{b,max}^2$ (where ρ denotes the fluid density and C_f a friction coefficient depending on bed material properties). Thus, according to this expression, τ depends on barge characteristics even for situations where h_p, D_p and u_0 are identical. This is of particular importance for the accurate determination of scour depths associated with the propeller jet attack of the bed (e.g., Aberle and Söhngen 2008).

Table 1 Summary of E-coefficients reported in the literature (maneuvering situation)

E	Case	Source
0.21	Free propeller (unbounded jet)	Maynord (1998), LWI (2007)
0.25	Barge with tunnel stern and twin rudders	Fuehrer et al. (1981)
0.27	SS 1, unducted propeller	LWI (2007)
0.30	Propeller in Kort nozzle (unbounded jet)	Maynord (1998), LWI (2007)
0.38	SS 1, ducted propeller	LWI (2007)
0.42	Slender sterns without central rudder	BAW (2005)
0.47	Ducted single screw, trawler, central rudder	Fuehrer et al. (1981)
0.51	SS 2, unducted propeller	LWI (2007)
0.58	Twin screw vessel with Kort nozzles, central rudder and tunnel stern	Maynord (1998)
0.71	Slender sterns with central rudder	BAW (2005)

In fact, most of the existing approaches for the determination of scour depth contain h_p, D_p and u_0 to parameterize the propeller jet attack of the bed (e.g., Verhey 1983; Hamill 1988; Drewes et al. 1995; Duecker and Miller 1996; Hamill et al. 2001). However, due to the complexity of propeller jet induced scouring processes these approaches were derived empirically for given boundary conditions and vessel types. Aberle and Söhngen (2008) tested some of these approaches and showed that they are characterized by deviations in forecasted scour depths up to a factor of 10 when being applied to independent data. These deviations were attributed to the significance of the near bed flow velocity associated with barge characteristics.

Moreover, most of the existing approaches do not explicitly consider the temporal evolution of scour depth, i.e., they can only be used to estimate the scour depth for larger load-periods. However, information on the temporal evolution of scour depth is required for the estimation of scour depths due starting and/or stopping processes of vessels. Until today, such approaches were only derived for situations without a ship's body (e.g., Hamill 1988, Hamill et al. 2001).

Consequently, there is a need for a more accurate method for the prediction of propeller jet induced scouring processes taking into account peculiarities of the propulsion system. For this purpose, a scale model study is currently carried out in the hydraulic laboratory of the Leichtweiß-Institut für Wasserbau at the Technische Universität Braunschweig, Germany. The objective of this chapter is the description of the experimental setup and the presentation of preliminary results of the ongoing study.

2 Experimental Setup and Procedure

The scale model study (1:16) is carried out in a 15 m long, 5 m wide and 2 m high basin. An observation window at the side enables visual inspection of the scour processes and the water level within the basin is measured by a water gauge. The walls of the basin are equipped with absorbers to minimize backflow effects as well as waves resulting from the reflection of the propeller jet.

A 3.0 m long, 3.6 m wide and 0.335 m high false bottom containing a 1.24 m wide recess area has been installed in front of the observation window (Fig. 2). For the experiments presented in this chapter, the recess area was filled with two types of sediment with a density of $\rho_s = 2650\ kg/m^3$, quartz sand ($d_{50} = 0.8$ mm; $\sigma_g = (d_{84}/d_{16})^{0.5} = 1.3$) and gravel ($d_{50} = 4$ mm, $\sigma = 1.4$), respectively.

The scour experiments are carried out in a maneuvering situation with a 5 m long, 0.7 m wide and 0.4 m deep scale model of a stern of an inland vessel (Großmotorschiff, GMS) typically found on the river Rhine and on the German canal system (see Fig. 2). The vessel can be equipped with two different four-blade-propellers with a diameter of $D_p = 111$ mm (Wageningen B-series propeller and Kaplan propeller in a Kort nozzle, respectively) and two different 115 mm high rudders (central rudder and double rudder, respectively). Taking into account

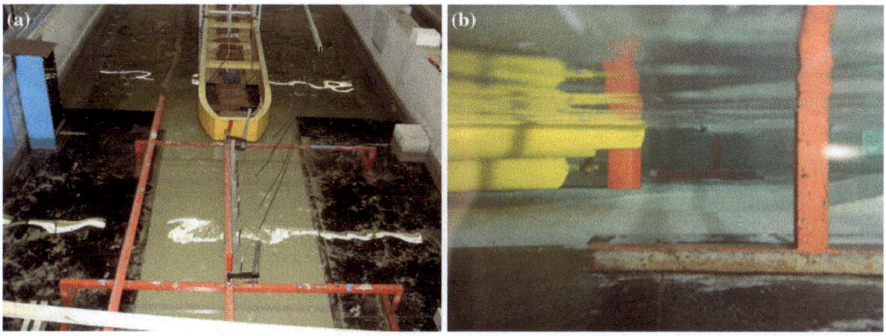

Fig. 2 **a** View of the basin, test area and stern model; **b** side view of the vessel through the observation window during an experiment with the Wageningen propeller, central rudder, and sand bed

relevant setups of propulsion systems, the central rudder is normally used in combination with the Wageningen propeller (Fig. 3a) and the double rudder with the Kaplan propeller (Fig. 3b). The influence of the rudder angle on the scour process is not investigated and, therefore, the rudder angle $\delta = 0°$ in all experiments.

During the scouring experiments, the vessel is firmly attached to a frame in the middle of the basin preventing its movement around the vertical and transverse axis. A wire attached to a drag force sensor holds the ship in its longitudinal position and enables the measurement of the propeller thrust. The propeller torque is measured additionally with a torque-sensor. The revolutions per minute (rpm) of the 440 V DC-motor installed in the ships-body powering the propeller are measured by a photoelectric device and can be continuously adjusted between 0 and 1,660 rpm. The rpm of the motor are linearly related to n_p, i.e., the rpm of the propeller.

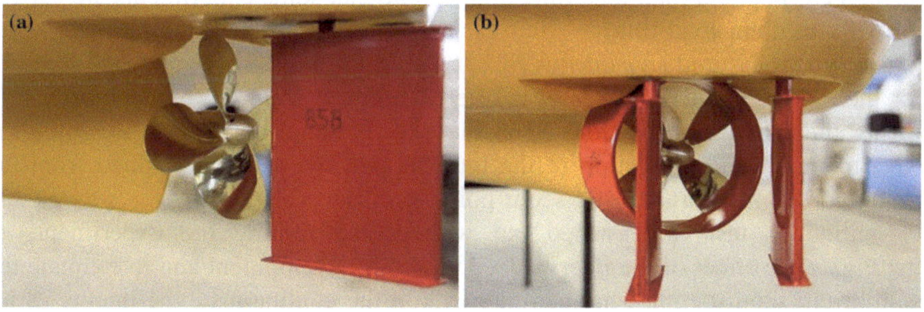

Fig. 3 **a** Wageningen B-series propeller with central rudder (WZ); **b** Kaplan propeller in Kort nozzle with double rudder (KD)

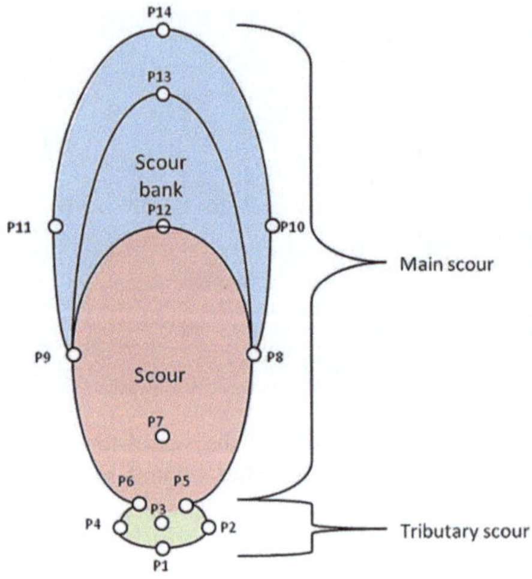

P1: Start of tributary scour
P2: Port side of tributary scour
P3: Deepest point of tributary scour
P4: Starboard side of tributary scour
P5: Port side of transition to main scour
P6: Starboard side of transition to main scour
P7: Scour depth
P8: Beginning of scour ridge (port side)
P9: Beginning of scour ridge (starboard side)
P10: Port side of scour ridge
P11: Starboard side of scour ridge
P12: Beginning of scour ridge
P13: Scour ridge height
P14: End of scour ridge

Fig. 4 Definition of measurement points for determination of scour geometry

Scour depths and geometries are measured manually with a point gauge using a coordinate system that has its origin on the level of the undisturbed sediment bed ($z = 0$) at the port side of the vessel (at the transition of the false bottom to the recess area, $y = 0$) with the propeller position defining $x = 0$. The spatial extent of the scour is recorded at the end of the experiments by surveying a total of 14 characteristic scour points (Fig. 4).

According to Fig. 4, the scour depth ε corresponds to the distance of the deepest point of the scour (P7) to the origin $z = 0$, the scour width b_K to the distance between points P8 and P9, and the scour length l_K to the distance of P5/P6 to P12, which is calculated according to $0.5(x(P5) + x(P6)) - x(P12)$.

The scour experiments are carried out following three different experimental strategies (intermittent, continuous, and long-time tests, respectively). Intermittent tests are carried out to determine the temporal evolution of scour geometry. In these tests, the sediment bed is scoured in reduplicative time intervals ($\Delta t = 10$, 20, 40 s etc.) up to a total load time of 2 h (note that some experiments are carried out with different load times). After each interval, the corresponding main scour geometry is measured subaqueous by surveying P1, P7, P8, P9, and P12 (Fig. 4). For the measurements, the vessel is briefly removed from the test area and, following the measurements, located at the exactly same position to continue the experiment. Upon completion of the interval test, the water is drained from the basin and the 14 characteristic scour points are surveyed.

Continuous tests are carried out without interrupting the experiments to investigate the effect of the interval load on scour development and to estimate the equilibrium scour depth. The load-intervals in the continuous experiments varies mainly between 2,700 and 10,800 s. Long-time experiments cover a load-period of up to 24 h and are approx. interrupted 5 times (e.g., over night and particular load periods). The corresponding scour geometries are measured in dry conditions.

All experiments (intermittent, continuous, and long-time) are carried out with different boundary conditions, i.e., using different propulsion systems, rpm, underkeel clearances, drafts, water depths, and sediment beds. The following section presents preliminary results of some experiments carried out until today.

3 Preliminary Results and Discussion

The first step of the experimental program addressed the reproducibility of the experiments and the validation of the experimental procedure for the intermittent experiments. The reproducibility was investigated by repeating continuous and intermittent experiments using identical boundary conditions and load-periods. Fig. 5 presents scour depths of different repeated continuous experiments which were carried out over the sand bed with identical rpm and water depth but different propulsion systems, vessel drafts and load-periods. The comparison of the scour depths of the individual series (characterized by different colors and markers) shows a reasonable agreement indicating the reproducibility of the experiments. It is worth mentioning that, compared to the differences within the repeated series, larger differences were observed for the experiments carried out with different load-periods, propulsion units and drafts. This issue will be discussed below in more detail.

The reproducibility of the intermittent experiments was also validated by repeating selected experiments. Figure 6a exemplarily presents three scour depth

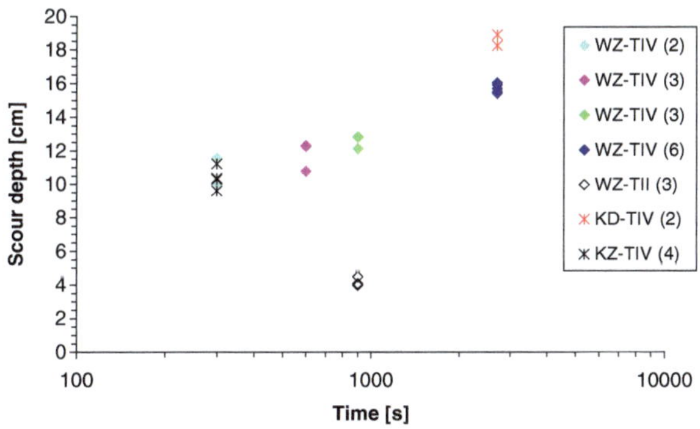

Fig. 5 Scour depths measured in 'repeated' experiments carried out with a water depth of 21.88 cm, $n_p = 1{,}340$ rpm, three different propulsion setups (WZ: Wageningen propeller and central rudder, KD: Kaplan propeller, Kort nozzle, and double rudder, KZ: Kaplan propeller, Kort nozzle and central rudder), two different drafts (TIV: 0.175 m, TII: 0.125 m) and different load-periods. The numbers in brackets in the legend indicate the number of repeated experiments

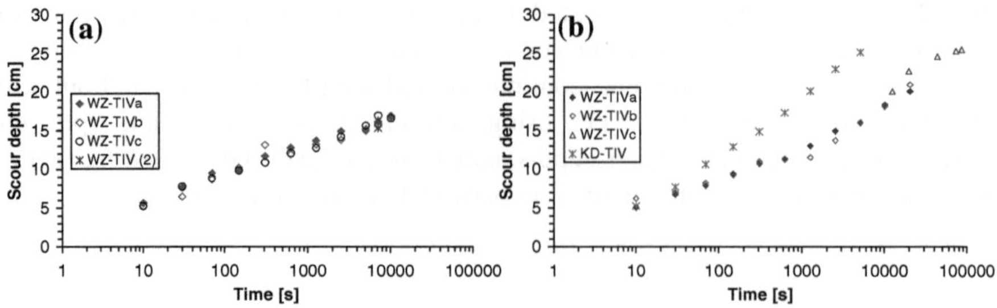

Fig. 6 Scour depth time series of experiments carried out with identical boundary conditions (draft TIV, h = 21.88 cm, n_p = 1,340 rpm) with the Wageningen propeller-central rudder setup (WZ) and Kaplan propeller-double rudder setup (KD); (**a**) three WZ-time series and two WZ-scour depths from continuous experiments for the gravel-bed; (**b**) three WZ-time series (first interruption of series WZ-TIVc after 12,600 s) and one KD-time series for the sand-bed

time series from experiments carried out with the Wageningen propeller-central rudder setup (WZ) over the gravel bed in a semi-logarithmic plot. The repeated time series are characterized by minor scatter which can be considered as negligible due to the complex experimental procedure and boundary conditions. Thus, the good agreement of the three time series is a strong indicator for the reproducibility of the experiments (note that the relatively large difference at t = 320 s was most likely associated with a measurement error).

In order to investigate the effect of the interval load on scour development, scour depths from the intermittent tests were compared to scour depths from comparable continuous experiments at particular time-intervals. This is exemplarily shown in Fig. 6a containing scour depths observed in two continuous experiments after a load-period of t = 7,200 s. The excellent agreement of the scour depths from the intermittent and continuous experiments, which was also confirmed in further experiments, is a strong indicator that the interval load has a negligible effect on the temporal scour development.

The semi-logarithmic plotted time series in Fig. 6a show that, for load periods $t < 10{,}000$ s, scour depth ε is linearly related to the load-period, i.e., $\varepsilon = f(\ln(t))$ which is in agreement with findings reported in the literature (e.g., Hamill 1988). Similar results were observed in our further experiments with both the Kaplan propeller-double rudder setup (KD) and the sand bed (Fig. 6b). Note that Fig. 6b also contains a time series from a 24 h long WZ-experiment which was interrupted at t = 12,600 s for the first time. This time series fits well to the other two WZ-time-series and indicates that near equilibrium conditions were approximately achieved at a load-period of 24 h.

Both Fig. 6a, b reveal relatively large scour depths after short load periods which is in agreement with results reported by Hamill (1988) with a free propeller (i.e., without a ship's body) and Felkel and Steinweller (1972) in a combined laboratory and field study. In fact, the analysis of the data reported in Felkel and Steinweller (1972) showed that the scour depth after a load period of 5 min corresponded to approx. 75 % of the scour depth after 50 min load period. In our experiments, the scour depth after $t = 310$ s, i.e., approx. 5 min, corresponded to 65–80 % of the scour depth observed after 42.5 min (note that the WZ-TIVb series in Fig. 5a yields 95 % due to an outlier).

A closer inspection of Figs. 6a, b reveals distinct features in the time series which are related to the sediment bed and the propulsion system. The semi-logarithmic plotted WZ-time series in Fig. 6b (sand-bed) show a change in the gradient at t = 630s which can not be observed in Fig. 6a (gravel-bed). It is interesting to note that the scour depths were, up to this load-period, reasonably similar although larger scour depths were expected for the sand-bed than for the gravel-bed due to the larger grain diameter of the gravel. The difference between the time-series over the tow sediment beds was also identified in a preliminary analysis of the WZ-scour geometry which revealed that the scour volume in the sand-bed increased, compared to the gravel bed, significantly for load-periods larger than 630s. On the other hand, such a behavior could not be observed for the KD-scours which showed the expected larger scour depths for the sand-bed than for the gravel-bed. As the reproducibility of the experiments was shown above and the repeated WZ-time series shown in Fig. 6b collapsed on almost a single line (including WZ-TIVc), we can currently only speculate that this behavior is associated with near bed jet characteristics determining the bed shear stress.

The influence of the propulsion system and hence propeller jet characteristics is reflected in Figs. 5 and 6b showing that, for comparable boundary conditions (rpm, h_p, draft and sediment), larger scour depths were observed for the KD-setup than for the WZ-setup. The different scour geometry caused by the different propulsion systems could also be observed in time series of scour width and length which showed that the KD-scours were, for t > 20 s, deeper, wider, and longer compared to the WZ-scours (Figs. 6 and 7).

The differences in scour geometry are exemplarily visualized in Fig. 8 for a load period of t = 10,230 s (note that similar scour geometries were observed in our further experiments). Figure 8a shows that the KD-scour was widened towards the port side with a pronounced bow-shaped ridge and that the most distant point of the scour ridge (P14; Fig. 4) was located on the starboard side. This asymmetry

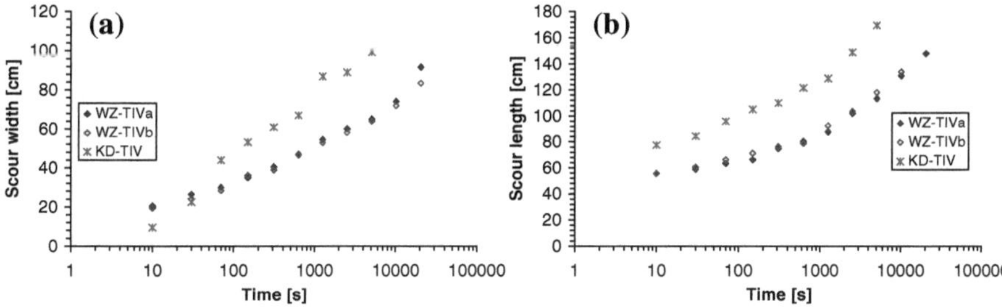

Fig. 7 **a** Scour width development and **b** scour length development for the scour depth time series presented in Fig. 6. Note that scour geometry was not recorded for WZ-TIVc

Fig. 8 Photographs of **a** KD-scour and **b** WZ-scour after a load period of 10,230 s

reflected the rotational sense of the propeller. On the other hand, the WZ-scour was narrower and rather symmetric (Fig. 8b).

The differences in scour geometry and depth were quantified in terms of a dispersion angle α_ε which was calculated using the distance Δh between the level of the propeller axis and scour depth ε (i.e. $\Delta h = h_p + \varepsilon_m$) and the distance Δx of P7 (Fig. 4) to the tip of the propeller axis according to $\alpha_\varepsilon = \arctan(\Delta h/\Delta x)$. Figure 9 presents α_ε (including an additional KZ-experiment with the gravel bed) as a function of the load-period for the experiments presented in Figs. 5–7 and reveals that for short to intermediate load-periods α_ε increases with time for both the WZ and KD-setup reflecting the development of scour geometry. On the other hand, $\alpha_\varepsilon \approx$ const. for $t > 1200$ s indicating that scour geometry has attained a near equilibrium shape (note that the scour is still developing for such load-periods; see Figs. 6 and 7). Figure 9 further revealed no influence of the sediment bed (i.e., gravel or sand) on α_ε.

For short load periods ($t = 10$ s), $\alpha_\varepsilon \approx 14$ and 25° for the KD- and WZ-setup, respectively. These values correspond approximately to the inclination angles of the outer jet boundaries reported in Fig. 1 for SS 1 and SS 2. Hence, the scour geometry clearly reflects the influence of the jet characteristics associated with the corresponding propulsion system.

It is interesting to note that the scour depths for the KD-setup were, for $t > 30$ s, always larger than the scour depths for the WZ-setup for comparable load periods and boundary conditions (Figs. 5 and 6; this difference has been verified in our further experiments to date). Originally, deeper scours were expected for the

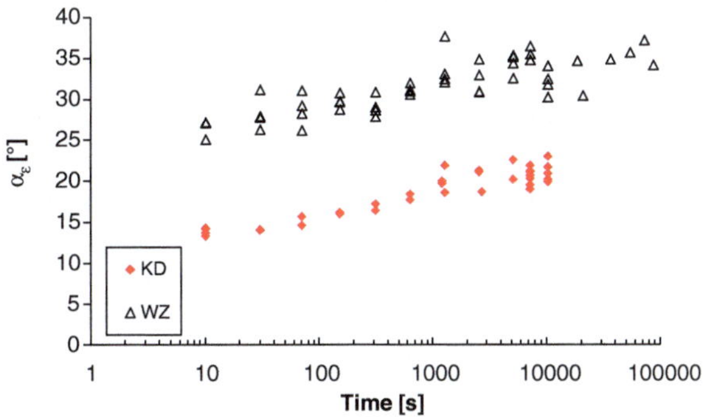

Fig. 9 Development of the propagation angle for the KD- and WZ-setup for scour depths measured to date

WZ-setup than for the KD-setup due to the splitting of the jet at the rudder and associated larger near bed velocities (see E-values in Table 1) and hence bed shear stresses. Currently, we can only speculate that the observed differences for the two propulsion systems are related to differences in the near bed turbulence characteristics of the propeller jet which may have a significant impact on bed shear stress (e.g., Blaauw and van de Kaa 1978). In order to investigate this issue in more detail and to quantify the influence of the propulsion system on near bed velocities and turbulence characteristics, we plan to measure the near bed turbulent flow field with Particle-Image Velocimetry (PIV) in upcoming experiments.

Despite these differences, the results showed for each setup the expected influence of h_p and n_p on scour depth. According to Eq. 1, larger values of h_p reduce, for identical D_p, u_0 and propulsion system, the bottom attack of the jet resulting in lower scour depths. This can bee seen by comparing the WZ-TII scour depths with the WZ-TIV scour depths displayed in Fig. 5. The WZ-TII experiments were carried out with a smaller draught than the WZ-TIV experiments resulting in a larger h_p and hence lower scour depths. We note that the influence of the ship's draught will be investigated in a series of special experiments in which h_p will be held constant but the draught will be varied.

Similarly, lower scour depths were observed in experiments with reduced n_p for the same propulsion system and h_p (e.g., Geisenhainer and Aberle 2012). Due to the reduction of n_p the efflux velocity u_0 is reduced (e.g., BAW 2005; Lam et al. 2011) and hence the near bed velocity.

Finally, it is worth mentioning that small bed forms and tributary-scours were observed in the experiments. Bed forms were mainly observed in the sand-bed experiments at the scour ridge above the initial bed height (Fig. 8). A tributary-scour located directly below the propeller (Figs. 4 and 8) was observed in both the sand- and gravel-bed experiments. Visual inspections during the experiments showed that the tributary-scours were caused by up-flowing water (water sucked towards the propeller) and that their size depended on h_p. The tributary-scours grew in time and, after a certain time, merged with the main-scour.

4 Conclusions

This chapter described the setup and preliminary results of an ongoing experimental study aiming at the investigation of the temporal evolution of propeller jet induced scours taking into account different propulsion systems. The preliminary results showed the reproducibility of the experiments and, in accordance with the literature, that boundary conditions such as propeller characteristics, rudder arrangement, height of the propeller over the sediment bed, propeller-rpm's and sediment characteristics have a significant effect on scour development and scour geometry. In the next steps, the acquired data will be used to test existing approaches for the determination of propeller jet induced scour depths.

The observed differences in scour geometry and the unexpected result that the Kaplan propeller caused, for similar boundary conditions, larger scour depths than the Wageningen-propeller will also be investigated in more detail in our upcoming experiments. In these experiments, the near bed propeller-jet hydraulics will be measured using Particle Image Velocimetry in order to quantify the influence of the propulsion system on the near bed flow field and bed shear stress.

The insights gained from these experiments and the collected data on scour development will finally be used to develop a novel approach for the determination of propeller jet induced scour depths and geometries as a function of time taking into account propeller jet characteristics.

Acknowledgments This study is carried out with support of the Federal Waterways Research and Engineering Institute (BAW), Germany.

References

Aberle J, Söhngen B (2008) Analysis of propeller jet induced scours. In: International Conference on Fluvial Hydraulics River Flow 2008, 3–5 Sept 2008, Türkei, pp 1991-2000

BAW (2005) Principles for the design of bank and bottom protection for inland waterways. Bulletin of the Federal Waterways Engineering and Research Institute, Germany, Bulletin No. 88

Berger W, Felkel K, Hager M, Oebius H, Schäle E (1981) Neue Mittel zum Schutz der Ufer und der Sohle von Wasserstraßen gegen den Angriff von Strömungen und Wellen einschließlich der von Schiffen verursachten Wellen. Deutsche Berichte zum XXv. Internationalen Schiffahrtskongreß Edinburgh, 1–32, in German

Blauuw HG, van der Kaa EJ (1978) Erosion of bottom and sloping banks caused by the screw race of maneuvering ships. Publication No. 202, Delft Hydraulic Laboratory, Delft

Drewes U, Römisch K, Schmidt E (1995) Propellerstrahlbedingte Erosion im Hafenbau und Möglichkeiten zum Kolkschutz für den Ausbau des Burchardkais im Hafen Hamburg. Mitt. des Leichtweiß-Instituts für Wasserbau 134, TU Braunschweig, 285–372 (in German)

Duecker HP, Miller C (1996) Harbour bottom erosion at berths due to propeller jets. In: 11th International harbour congress, Antwerp, pp 411–424

FAS (1972). Wirkung des Propellerstahls auf Sohle und Böschung. Forschungsanstalt für Schiffahrt, Wasser- und Grundbau, Berlin, Abschlussbericht, in German

FAS (1974) Erosionserscheinungen bei Schiffsmanövern. Forschungsanstalt für Schiffbau, Berlin, Abschlussbericht, in German

Felkel K, Steinweller H (1972) Natur- und Modellversuche über die Wirkung der Schiffe auf Flußsohlen aus Grobkies (Breisacher Versuche). Wasserwirtschaft, 62(8): 243–249

Fuehrer M, Römisch K (1977) Effects of modern ship traffic on islands and ocean waterways and their structures. In: Proceedings of 24th PIANC congress, Leningrad, 1977, Sections 1–3

Fuehrer M, Römisch K, Engelke G (1981) Criteria for dimensioning the bottom and slope protections and for applying the new methods of protecting navigation canals. In: XXV PIANC-congress, Edinburgh, pp 29–50

Flamm O (1914) Die Einwirkungen der Schiffsschraube auf die Kanalsohle. Zeitschrift des Vereins Deutscher Ingenieure, Band 58, Nr. 10, 1914, S. 391 ff

Gebers F (1911) Die Entwicklung einer neuen Schleppdampferbauart für Schiffahrtskanäle durch Modellversuche an der königlichen Versuchsanstalt für Wasserbau und Schiffbau. Jahrbuch der schiffbautechnischen Gesellschaft (STG), Band 12, 1911, S. 420–450. in German

Geisenhainer P, Aberle J (2012) Experimental investigation of propeller jet induced scours. Paper submitted to 2nd IAHR-european congress, Munich 2012

Hamill GA (1988) The scouring action of the propeller jet produced by a slowly manoeuvring ship. PIANC Bull 62:85–110

Hamill GA, Johnston HT, Stewart DPJ (1995) Estimating the velocities in a ship's propeller wash. PIANC-Bull 89:46–54

Hamill GA, McGarvey JA, Hughes DAB (2001) The effect of rudder angle on the scouring action produced by the propeller wash of a manoeuvring ship. PIANC-Bull 106:49–62

Kempf G (1939) Die Wirkung des Propellerstrahls auf die Kanalsohle, Congrès Interna-tional des Ingénieurs Navals (CIIN),Liege, 1939, S. 110–116

Lam W, Hamill GA, Song YC, Robinson DJ, Raghunathan S (2011) A review of the equations used to predict the velocity distribution within a ship's propeller jet. Ocean Eng 38:1–10

LWI (2007). Zusammenstellung und Auswertung experimenteller Daten zum Thema Kolkbildung durch Schraubenstrahlen. Leichtweiß-Institut für Wasserbau, Bericht 946, TU Braunschweig, in German, unpublished

Maynord ST (1998) Bottom shear stress from propeller jets. In: Proceedings of ports '98 conference, Long Beach, USA, pp 1074–1093

Oebius H (1984) Loads on beds and banks caused by ship propulsion systems. In: Flexible armoured revetments, T. T. Ltd., London, pp 13–23

Oebius H (2000) Charakterisierung der Einflussgrößen Schiffsumströmung und Propellerstrahl auf die Wasserstraßen. Mitteilungsblatt der Bundesanstalt für Wasserbau, Nr. 82, pp 7–22

PIANC (2003) Guidelines for sustainable inland waterways and navigation. International Navigation Association, EnviCom Report of WG 6

Spitzer D, Aberle J, Geisenhainer P, Söhngen B (2012a) Belastung der Gewässersohle durch Propellerstrahlen – Teil 1: Untersuchungen bis zum Zweiten Weltkrieg. Korrespondenz Wasserwirtschaft, in press (in German)

Spitzer D, Aberle J, Geisenhainer P, Söhngen B (2012b) Belastung der Gewässersohle durch Propellerstrahlen—Teil 2: Untersuchungen nach dem Zweiten Weltkrieg. Korrespondenz Wasserwirtschaft, in press (in German)

Verhey HJ (1983) The stability of bottom and banks subjected to the velocities in the propeller jet behind ships. In: 8th International harbour congress, Antwerp, Belgium.VBD (1999)

Longitudinal Dispersion Coefficients Within Turbulent and Transitional Pipe Flow

James Hart, Ian Guymer, Amy Jones and Virginia Stovin

Abstract The longitudinal dispersion coefficient is used to describe the change in characteristics of a solute cloud, as it travels along the longitudinal axis of a pipe. Taylor (1954) proposed a now classical expression to predict the longitudinal dispersion coefficient within turbulent pipe flow. However, experimental work has shown significant deviation from his prediction for $Re < 20{,}000$. This chapter presents experimental results from tracer studies conducted within the range $2{,}000 < Re < 50{,}000$, from which longitudinal dispersion coefficients have been determined. Initial results are also presented for a numerical model that aims to predict the longitudinal dispersion coefficient over the same range of Reynolds numbers.

1 Introduction

Longitudinal dispersion can be defined as the spreading of a solute along the flow's longitudinal axis. This process leads to a change in characteristics of a contamination cloud from an initial state of high concentration and low spatial variance, to a downstream state of lower concentration and higher spatial variance.

J. Hart (✉) · I. Guymer
School of Engineering, University of Warwick, Coventry, UK
e-mail: James.Hart@Warwick.ac.uk

I. Guymer
e-mail: I.Guymer@Warwick.ac.uk

A. Jones
University of Warwick, Coventry, UK

V. Stovin
Department of Civil and Structural Engineering, The University of Sheffield, Sheffield, UK

P. Rowiński (ed.), *Experimental and Computational Solutions of Hydraulic Problems*, GeoPlanet: Earth and Planetary Sciences, DOI: 10.1007/978-3-642-30209-1_8,

Within potable water networks it is important to quantify the changing characteristics of solutes as they travel through the network. Current water quality models for distribution networks assume steady, highly turbulent flow (Tzatchkov et al. 2009). These assumptions are valid for the majority of the flow conditions experienced in the main network. However, one part of the network for which these assumptions are not valid is the network's periphery, where water leaves the main network and travels to the point of consumption. Here, in the so-called 'dead end' regions of the network, discharge is contingent upon the intermittent demand of the consumer, hence the flow is unsteady, and can assume any flow rate from the relatively high main network rate, through to zero in times of no demand (Lee 2004). In addition, pressure transients caused by any change in the flow conditions (closed valve, leaking pipe, network maintenance etc) can result in periods of both unsteady and low flow in the main network. This scenario is of particular interest in water quality modeling, as negative pressure created in such pressure transients can lead to contaminant intrusion into the network through any leaks in the system (LeChevallier et al. 2003). Thus, contaminant can be released into a low and unsteady flow.

Taylor (1953, 1954) developed two equations to predict the longitudinal dispersion coefficient within steady laminar and turbulent pipe flow, respectively. His equations are still widely used. However, experimental data has shown a significant divergence between predictions made by Taylor's equation for turbulent flow, and experimentally determined longitudinal dispersion coefficients within turbulent flow for $Re < 20{,}000$.

This chapter presents experimentally determined longitudinal dispersion coefficients for steady pipe flow in the range of Reynolds numbers $2{,}000 < Re < 50{,}000$, a range set to highlight conditions under which Taylor's model does not describe experimental data. Furthermore, a simple numerical model is developed for the same range of Reynolds numbers.

2 Background and Previous Work

Longitudinal dispersion is primarily caused by differential advection associated with the flow's longitudinal velocity profile. When fluid flows through a pipe, the velocity varies with radial position from the maximum velocity obtained at the pipe's centreline, to zero at the pipe's boundary. When a cross-sectionally well-mixed tracer is introduced across a pipe, tracer will be advected in accordance with the velocity at its corresponding radial position. Hence, tracer at the centre of the pipe will travel further in a given period of time than tracer at the boundary of the pipe, and thus the tracer disperses. The tracer is further spread in all directions by the effects of molecular and turbulent diffusion. The degree to which these diffusion mechanisms act to spread the tracer directly in the longitudinal direction is negligible when compared to the effects of the differential advection. However, the two diffusion mechanisms are significant with regard to longitudinal dispersion because of their ability to spread the tracer radially. As radial diffusion increases, each particle of tracer experiences a larger number of radial positions and corresponding velocities, thus reducing the effects of the differential advection. Hence, there is an inverse relationship between molecular and turbulent diffusion and longitudinal dispersion.

Taylor (1953, 1954) showed that, after some initial development period, the spatial distribution of the cross sectional mean concentration of a solute is Gaussian, with a variance that increases linearly with distance. Through this, Taylor showed that the cross-sectional average concentration distribution can be described by Fick's second law of diffusion, such that:

$$\frac{\partial c}{\partial t} = D_{xx}\frac{\partial^2 c}{\partial x^2} - \bar{u}\frac{\partial c}{\partial x} \tag{1}$$

where r is the radial position from the centreline, $c(x, r)$ is the concentration at position (x, r), and $D_r(r)$ and $u(r)$ are the radial diffusion coefficient and velocity at position r respectively.

For turbulent flow, Taylor (1954) considered the radial distribution of a concentration in terms of the following partial differential equation:

$$\frac{\partial}{\partial r}\left(D_r(r)r\frac{\partial c(x,r)}{\partial r}\right) = r\left(u(r)\frac{\partial c(x,r)}{\partial x} + \frac{\partial c(x,r)}{\partial t}\right) \tag{2}$$

where r is the radial position from the centreline, $c(x, r)$ is the concentration at position (x, r), and $D_r(r)$ and $u(r)$ are the radial diffusion coefficient and velocity at position r, respectively.

Taylor assumed the radial diffusion coefficient was equivalent to the turbulent diffusion coefficient D_t, such that $D_r = D_t$, and defined the turbulent diffusion coefficient by considering it in terms of Reynolds analogy, i.e. the assumption that the transfer of matter, heat and momentum are analogous, such that:

$$D_r(r) = D_t(r) = \frac{\tau_t(r)}{\rho(\partial u(r)/\partial r)} \tag{3}$$

where $\tau_t(r)$ is the turbulent stress at position r, $\tau_t(r) = \tau \cdot p$, where τ is the wall shear stress and p is dimensionless position $p = r/a$, where a is the pipe's radius.

Taylor assumed a 'universal' velocity distribution, of the form:

$$\frac{u_c - u(r)}{u_*} = f(p) \tag{4}$$

where u_c is the maximum velocity, u_* is the frictional velocity $u_* = \bar{u}\sqrt{f/8}$, where f is the friction factor. $f(p)$ is a geometric relationship for the velocity distribution as a function of dimensionless position p. Taylor derived the values of $f(p)$ as the mean value of the data of Stanton and Pannell (1914) and Nikuradse (1932). In addition, Taylor proposed an empirical relationship for the maximum velocity, $u_c = \bar{u} + (4.25u_*)$.

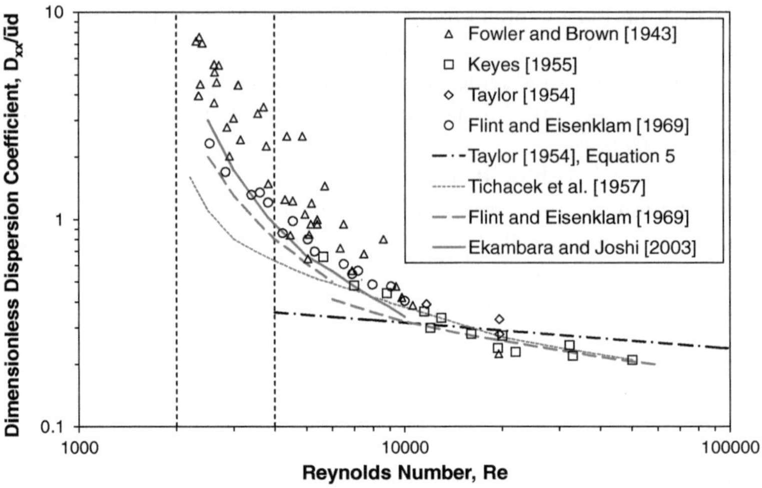

Fig. 1 Comparison between the experimental data of Fowler and Brown (1943), as presented by Levenspiel (1958), Taylor (1954), Keyes (1955) and Flint and Eisenklam (1969) and the theories of Taylor (1954) (Eq. 5), Tichacek et al. (1957), Flint and Eisenklam (1969) and Ekambara and Joshi (2003)

Taylor used this definition of the radial diffusion coefficient (Eq. 3) and velocity profile (Eq. 4) to solve Eq. 2, which gave the following expression for the longitudinal dispersion coefficient within turbulent pipe flow:

$$D_{xx} = 10.1au_* \tag{5}$$

The Reynolds number, $Re = \bar{u}d/v$, where d is the pipe diameter and v is kinematic viscosity, effectively quantifies how turbulent a flow is Fig. 1 shows the results of previous experimental investigations into the relationship between the longitudinal dispersion coefficient and Reynolds number for the range 2,000 $< Re <$ 50,000, compared to Taylor's theory (Eq. 5).

From Fig. 1 it can be seen that for $Re >$ 20,000, the longitudinal dispersion coefficient is relatively small and independent of Reynolds number. For $Re <$ 20,000, it increases significantly to a point that at $Re \approx$ 2,000, its value is approximately 25 times larger at than the value for $Re >$ 20,000. Furthermore, at approximately the same point that the longitudinal dispersion coefficient begins to increase, it also begins to diverge from Taylor's prediction. Some insight into this phenomenon can be gained by considering the relationship between the Reynolds number and the velocity profile throughout this range.

When the flow is laminar, at $Re \approx$ 2,000, the velocity profile is parabolic and thus there is a high degree of spreading due to differential advection. When the flow is turbulent, at $Re >$ 4,000, the velocity is more uniform than the laminar profile leading to a decrease in differential advection. From 2,000 $< Re <$ 4,000 the flow is transitional, and thus the profile transitions from the parabolic profile at $Re \approx$ 2,000,

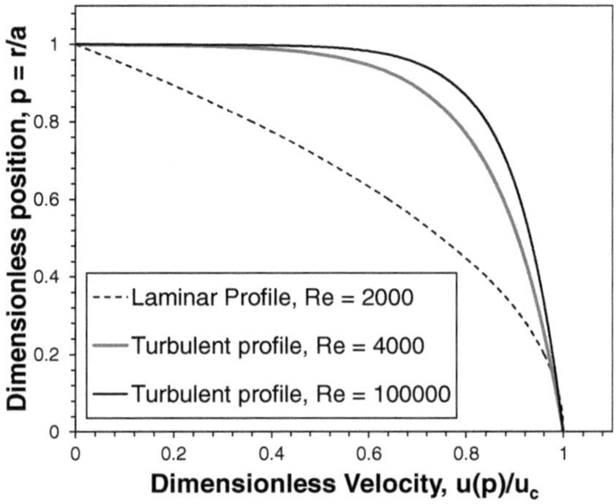

Fig. 2 Comparison between a laminar velocity profile at $Re = 2,000$, (Eq. 6), and the theoretical turbulent velocity profile of Nikuradse (1932), at $Re = 4,000$ and $Re = 100,000$

to the more uniform profile at $Re \approx 4{,}000$. Figure 2 shows a comparison between an analytically predicted laminar velocity profile and two theoretical turbulent velocity profiles (Nikuradse 1932). From Fig. 2 it can also be seen that the is a significant difference in the velocity distribution between 2,000 $< Re <$ 4,000, whereas there is only slight difference between 4,000 $< Re <$ 100,000.

For laminar flow the velocity profile can be determined analytically as White (2008):

$$\frac{u(p)}{u_c} = 1 - p^2 \tag{6}$$

where $u(p)$ is the velocity at position p.

For turbulent flow, the velocity profile cannot be determined analytically, and thus turbulent velocity profiles are generally proposed as empirical expressions. Conventionally turbulent velocity profiles are defined in terms of the dimensionless velocity and distance terms $u^+ = u(p)/u_*$ and $y^+ = u_* y/v$, where y is the actual distance from the wall.

Figure 3 shows the experimentally obtained velocity profile of Durst et al. (1995) for a turbulent flow at $Re = 7,442$. From Fig. 3 it can be seen that there are three parts to a turbulent velocity profile. The main part of the profile is the 'turbulent core', which occurs at $y^+ > 30$. In this part of the flow, the profile is fully turbulent. Here, the profile is logarithmic besides a small portion near the centreline where the profile plateaus, deemed the 'wake' region. For $y^+ < 5$, even in fully developed turbulent flow, the flow remains laminar. This portion of the flow is deemed the 'laminar sub-layer'. From $5 < y^+ < 30$, the flow transitions from being laminar to fully turbulent, a portion of the flow deemed the 'buffer zone'. The region of the flow corresponding to the non-turbulent portion, i.e. the length corresponding to $y^+ < 30$, grows as Reynolds number decreases. At $Re >$ 20,000, the laminar sub-layer and buffer zone constitute less that 5 % of the flow whereas at $Re =$ 4,000, the laminar sub-layer and buffer zone constitute around 20 % of the

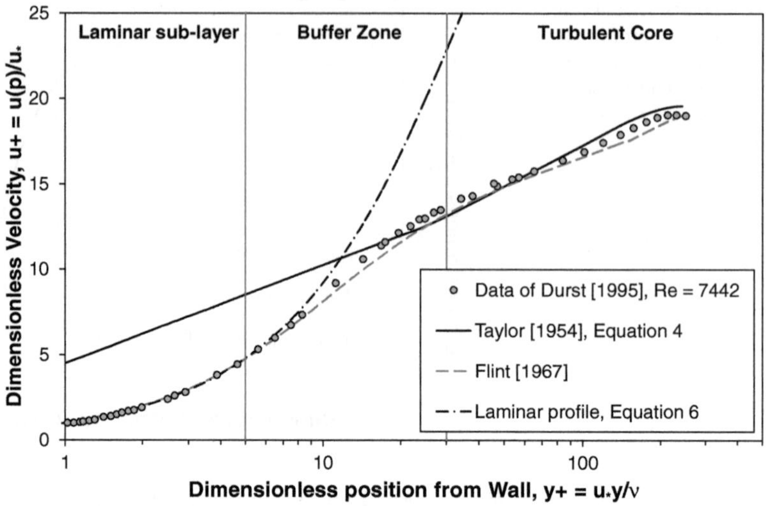

Fig. 3 Comparison between the experimental data Durst et al. (1995), a laminar velocity profile (Eq. 6), Taylor's turbulent velocity profile (Eq. 4) and turbulent velocity profile of Flint (1967)

flow. As Reynolds number reduces from $Re = 4{,}000$, the laminar sub-layer grows further to the point that the entire flow is laminar, at around $Re = 2{,}000$.

Figure 3 shows a comparison between Taylor's velocity profile (Eq. 4), and the data of Durst et al. (1995). Taylor's profile was derived from highly turbulent data, and thus neglects a laminar sub-layer and buffer zone. Due to this, Taylor's equation for the longitudinal dispersion coefficient is only valid for $Re > 20{,}000$, the portion of the flow where the size of the laminar sub-layer and buffer zone are small enough to be considered negligible with regards to longitudinal dispersion.

Several authors have used improved velocity profiles to build upon Taylors original analysis for the longitudinal dispersion coefficient.

Tichacek et al. (1957) solved Eq. 2 using experimental velocity profiles to produce a model for the longitudinal dispersion coefficient for $2{,}200 < Re < 50{,}000$.

Flint (1967) proposed an expression for the velocity profile for turbulent flow which included a laminar sub-layer and buffer zone. Figure 3 shows a comparison between Flints turbulent velocity profile and the data of Durst et al. (1995). In addition, Flint (1967) proposed a further expression for the velocity profile within transitional and low turbulent flow, covering $2{,}500 < Re < 6{,}000$. Flint and Eisenklam (1969) solved Eq. 2 using the theoretical velocity profile of Flint (1967) to produce a model for the longitudinal dispersion coefficient for $2{,}500 < Re < 100{,}000$.

Ekambara and Joshi (2003) solved Eq. 2 using a low Reynolds number $\kappa - \varepsilon$ CFD code to provide a prediction for the longitudinal dispersion coefficient for $2{,}500 < Re < 10{,}000$.

Figure 1 shows a comparison between these models for the longitudinal dispersion coefficient and experimental data.

None of the models discussed cover the whole range 2,000 $< Re <$ 20,000, the range over which Taylors equation do not describe experimental data. Furthermore, all of the models discussed involve solving the problems governing differential equation, and thus are mathematically complex.

The aim of the present work is to experimentally determine the longitudinal dispersion coefficient over the range 2,000 $< Re <$ 50,000, and proposes a relatively simple numerical model for the same range, with particular emphasis on 2,000 $< Re <$ 3,000, the range not previously fully described.

A suitable model for this objective is the 'zonal' model of Chikwendu (1986), which provides an analytical solution for the longitudinal dispersion coefficient for a given velocity profile and radial diffusion coefficient by dividing the flow into N number of zones (See Sect. 5 for more detailed explanation of model).

Thus, the numerical model proposed in this chapter will consider the most suitable definition of the velocity profile for 2,000 $< Re <$ 50,000, in conjunction with the model of Chikwendu (1986).

3 Experimental Setup and Method

A series of experiments was conducted to determine the longitudinal dispersion coefficient for a range of flow rates corresponding to the range of Reynolds numbers 2,000 $< Re <$ 50,000. The tests were conducted using a re-circulating system where the main test pipe was 16.6 m long, with an internal diameter of 24 mm.

The flow rate was obtained by measuring a volume of water collected over a set period of time. For each set of tests at a fixed flow rate, the flow rate was measured three times both before and after each set of injections. Thus, the flow rate for each run was the mean value of six repeats.

Dye injections of Rhodamine WT were made using a computer controlled peristaltic pump. For each injection, Rhodamine WT at a concentration ranging from 700 to 1,500 ppb was injected for a one second period. The relatively large sump volume (around 2,500 l) and the small injection volume (around 18 ml) meant that a stable background concentration was maintained (background rose by approximately 0.5 ppb over the entire series of 33 tests). Injections were made at a distance 3.5 m downstream from the start of the test section, a length sufficient to allow for the flow to become fully developed (White 2008).

The response of the dye to the flow was recorded as cross-sectional average concentration vs. time profiles using two Turner Designs series 10 fluorometers, which were 6 m apart and 7.1 and 13.1 m downstream of the injection point, respectively. The instruments were calibrated before and after the full series of dye injections to confirm the calibration relationship held throughout the tests.

For each flow rate, three injections were made. Thus the longitudinal dispersion coefficients discussed in the Sect. 4 represent the mean value of 3 repeats.

4 Experimental Results and Analysis

An initial estimate of the longitudinal dispersion coefficient was made through the 'method of moments' (Rutherford 1994).

These estimates were optimised through the following routing procedure (Rutherford 1994):

$$c(x_2, t) = \int_{-\infty}^{\infty} \frac{c(x_1, \gamma)\bar{u}}{\sqrt{4\pi D_{xx}\bar{T}}} \exp\left[-\frac{\bar{u}^2(\bar{T} - t + \gamma)^2}{4D_{xx}\bar{T}} \right] d\gamma \quad (7)$$

where $\bar{T}$ is the travel time, the difference between the centroid of two profiles, and γ is an integration variable, or pseudo time.

A routing procedure, such as Eq. 7, takes the experimentally obtained upstream profile, routes it to a downstream position on the basis of the travel time, and spreads it on the basis of the longitudinal dispersion coefficient. Thus, initially the upstream data was routed onto the downstream data using the travel time and longitudinal dispersion coefficient obtained through the method of moments. The routed downstream profile could then be compared to the downstream profile through some criteria of fit, namely R_t^2 (Young et al. 2008). The longitudinal dispersion coefficient and travel time were then optimised to give the best fit to the downstream data on the basis of the value of R_t^2. Figure 4 shows a sample of the results for the downstream concentration profiles, compared to the optimised profiles from Eq. 7 for several representative flow rates covering approximately $2{,}000 < Re < 50{,}000$.

Table 1 summaries the experimental results for each flow rate considered. From Fig. 4 and Table 1 it can be seen that for fully turbulent flow, for $4{,}000 < Re < 50{,}000$, the routing procedure fits the data well. For transitional flow, for $2{,}000 < Re < 4{,}000$, as the profiles begin to deviate from the Gaussian assumption in a non-trivial manner, the goodness of the fit of the model decreases. From Table 1 it can also be seen that for the majority of the tests for fully turbulent flow, for $4{,}000 < Re < 50{,}000$, mass balance is around 100 %. However, for transitional flow, for $2{,}000 < Re < 4{,}000$, mass balance drops to around 90 %. These results could be either a due to tracer being caught in the laminar sub-layer, or experimental error at low flow rates due to the tracer not being cross-sectionally well mixed. Figure 5 shows a comparison between the longitudinal dispersion coefficient determined experimentally in the present work and previous experimental data, from which it can be seen that the results of the present work fall favorably within previous experimental investigation.

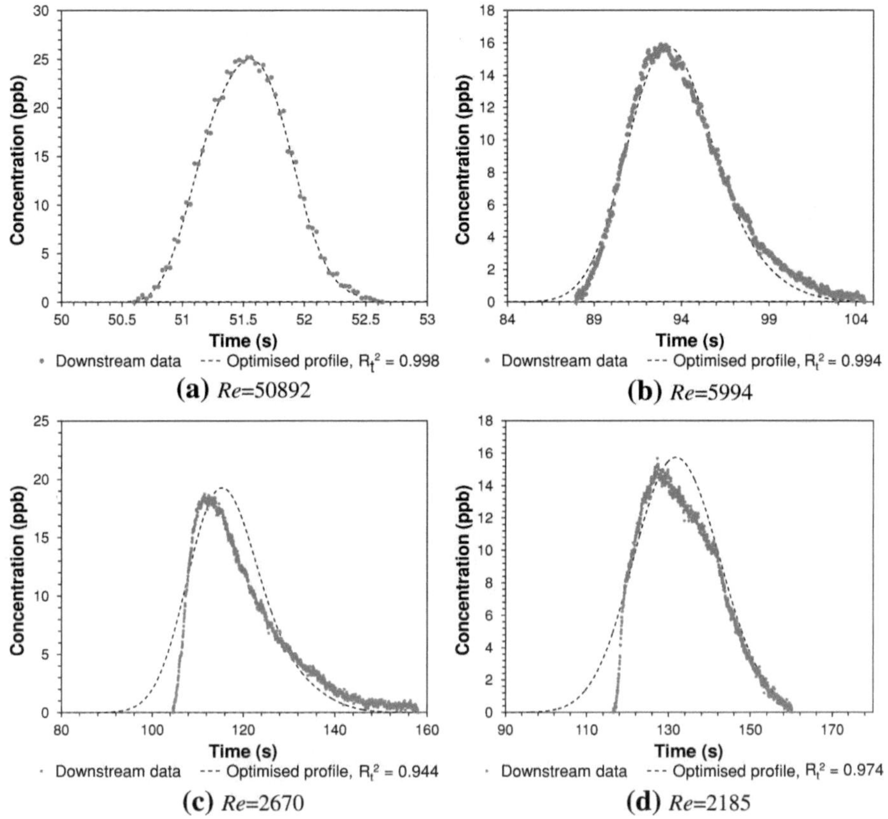

Fig. 4 Comparison between experimental downstream concentration profiles, and concentration profiles routed from upstream data and optimised to best fit downstream data through Eq. 7

Table 1 Summary of experimental results

Re	Mass balance [%]	$D_{xx}/\bar{u}d$	R_t^2
50,892 ± 814	102.3 ± 0.7	0.326 ± 0.042	0.998 ± 0.000
32,363 ± 359	101.0 ± 1.2	0.351 ± 0.020	0.998 ± 0.000
20,380 ± 168	101.0 ± 1.2	0.347 ± 0.020	0.999 ± 0.000
14,815 ± 96	101.6 ± 1.7	0.396 ± 0.003	0.998 ± 0.000
10,365 ± 57	102.6 ± 0.4	0.494 ± 0.005	0.996 ± 0.000
5,994 ± 66	99.037 ± 3.4	0.626 ± 0.009	0.994 ± 0.001
5,148 ± 291	91.4 ± 1.2	0.828 ± 0.033	0.993 ± 0.001
3,784 ± 29	104.7 ± 8.1	1.525 ± 0.076	0.989 ± 0.002
2,670 ± 13	89.3 ± 1.9	2.682 ± 0.596	0.949 ± 0.008
2,185 ± 21	88.4 ± 5.1	4.603 ± 0.383	0.946 ± 0.029

Each value represent the mean value of three repeats
± Represents 1 stand deviation

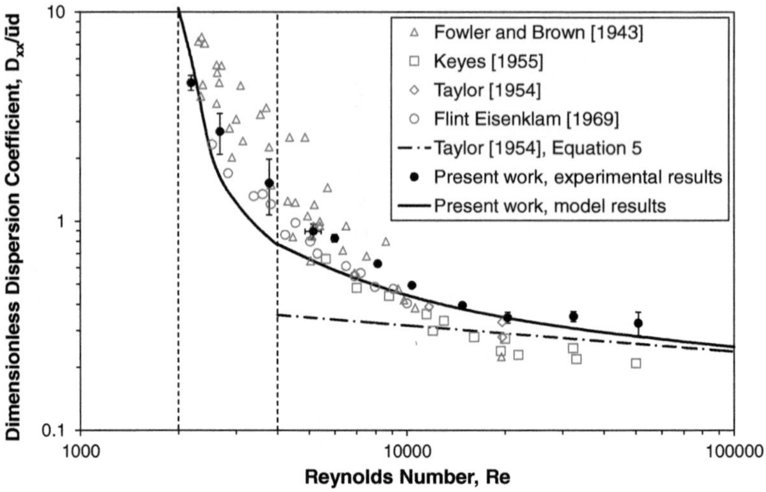

Fig. 5 Comparison between the optimised experimental data of the present work, the experimental data of Fowler and Brown (1943), Taylor (1954), Keyes (1955) and Flint and Eisenklam (1969), Taylor's theory (Eq. 5), and the model of the present work

5 Proposed Numerical Model

The model of Chikwendu (1986) provides a longitudinal dispersion coefficient on the basis of a velocity profile and a radial diffusion coefficient, such that:

$$D(N)_{xx} = \sum_{j=1}^{N-1} \frac{a^2 p_j^4 (1 - p_j^2)^2 [u_{f,1\to j} - u_{s,j\to N}]^2}{4D_{r,j,(j+1)}} (W_j + W_{j+1}) + \sum_{j=1}^{N} q_j D_{x,j} \tag{8}$$

where D_x is diffusion in the longitudinal direction, $W_j = p_j - p_{j-1}$ and $u_{f,1\to j}$ and $u_{s,j\to N}$ are the fast and slow zone velocities:

$$u_{f,1\to j} = \frac{1}{p_j^2} \sum_{k=1}^{j} q_k u_k \tag{9}$$

$$u_{s,j\to N} = \frac{1}{1 - p_j^2} \sum_{k=j+1}^{N} q_k u_k \tag{10}$$

where $q_j = (r_j^2 - r_{j-1}^2)/a^2$.

Chikwendu's model provides an analytical solution for the longitudinal dispersion coefficient by dividing the flow into a discrete number of zones N. Each zone has its own mean velocity and radial exchange between adjacent zones.

Differential advection is accounted for by considering the mean velocity of every zone above each point *j*, the 'fast zone', less the mean velocity of every zone below each point *j*, the 'slow zone'. Through this, a dispersion term is calculated for each zone, which is summed to give the longitudinal dispersion coefficient.

Initially, Chikwendu's model was used to reproduce the Taylor's result for turbulent flow, to demonstrate the applicability of the model and to determine an approximate value for the number of zones required. Taylor's velocity profile (Eq. 4), proposes the velocity profiles as a function of the geometric function $f(p)$. Taylor gives an expression for the function $f(p)$ for $0.9 < p < 1$, but between $0 < p < 0.9$ only gives 14 experimentally derived values for the function. Thus, in order to use his profile at a higher resolution, an expression was fit to Taylor's values for $0 < p < 0.9$. The radial diffusion coefficient, friction factor and the maximum velocity were used as defined by Taylor (1954). In addition, the diffusion term in the longitudinal direction, D_x, was also neglected in accordance with Taylor's analysis. At $N = 3,000$, the model reproduced Taylor's results to within $\sim 0.5\,\%$ over the whole range.

The laminar sub-layer and buffer zone of the turbulent profile of Flint (1967) was added to Taylor's velocity profile for $y^+ < 30$, to better describe experimental data for 4,000 < *Re* < 50,000.

For transitional flow, the transitional velocity profile of Flint (1967) was used for 3,000 < *Re* < 4,000, a range over which it conforms to the experimental data of Senecal and Rothfus (1953). For 2,000 < *Re* < 3,000, Flint's expression diverges from the experimental data of Senecal and Rothfus (1953), and fails to converge upon the analytical prediction of the velocity profile at $Re = 2,000$. Within this range, Flint's expression predicts a distribution that appears more turbulent than the majority of the data, whereas the analytical laminar profile predicts a distribution which appears more laminar than the majority of the data. Therefore, to predict the velocity profile within this range, an expression was suggested that postulates a velocity profile as a combination of the distribution of Flint's profile at $Re = 3,000$, and the analytical laminar profile at $Re = 2,000$. The relative proportion of each distribution used is governed through a transition factor α, such that:

$$f(p)_T = \alpha f(p)_{L(2000)} + (1 - \alpha) f(p)_{F(3000)} \tag{11}$$

where $f(p)$ is a dimensionless velocity distribution, $f(p) = u(p)/u_c$, $f(p)_T$ is the transitional dimensionless velocity distribution between 2,000 < *Re* < 3,000, $f(p)_{L(2,000)}$ is the dimensionless velocity distribution from the analytically laminar profile at $Re = 2,000$ (Eq. 6), and $f(p)_{F(3000)}$ is the dimensionless velocity distribution from Flint's profile at $Re = 3,000$.

Values for α were obtained by fitting Eq. 11 to the 8 profiles of Senecal and Rothfus (1953) between 2,000 < *Re* < 3,000. The trend for α was found to conform to an 'S' trend, which could be described through a sigmoidal function of the form $\alpha = 0.306/(0.2981 + e^{9.747\gamma})$, where $\gamma = (Re - 2000/1000) - 0.5$.

For transitional flow, values for the maximum velocity were obtained by fitting to the data of Senecal and Rothfus (1953), and the friction factor was assumed to vary linearly from the turbulent value at $Re = 3,000$, to the laminar value at $Re = 2,000$.

Figure 5 shows the results using Chikwendus model, with the velocity profiles described in the present work for the range 2,000 $< Re <$ 100,000.

6 Conclusions

Experimental data has been presented by optimising the longitudinal dispersion coefficient through a Fickian type model. Fits to downstream data are good for turbulent flow for 4,000 $< Re <$ 50,000, but the model fails to fully predict the downstream distributions for transitional flow for 2,000 $< Re <$ 4,000. The results for the longitudinal dispersion coefficient compare favorably with previous experimental work and show a deviation from Taylor's result at $Re < 20,000$.

A simple numerical model is proposed that builds upon Taylor's prediction for turbulent flow by adding a laminar sub-layer and buffer zone to Taylor's velocity profile. The use of this velocity profile within Chikwendu's model predicts the general trend in the data for the longitudinal dispersion coefficient of increasing from Taylor's prediction for 4,000 $< Re <$ 20,000. The model presented here was extended to transitional flow by considering two further velocity profiles, that of Flint (1967) for 3,000 $< Re <$ 4,000, and the profile derived in the present work on the basis of the data of Senecal and Rothfus (1953) for 2,000 $< Re <$ 3,000. The use of these expressions within Chikwendu's model predicts the large increases in the longitudinal dispersion coefficient within the transitional region in a manner reasonably consistent with experimental data.

References

Chikwendu SC (1986) Caculation of longitudinal shear dispersivity using an N-zone model as $N \to \infty$. J Fluid Mech 167:19–30

Durst F, Jovanovic J, Sender J (1995) LDA measurments in the near-wall region of a turbulent pipe flow. J Fluid Mech 295:305–335

Ekambara K, Joshi JB (2003) Axial mixing in pipe fows: turbulent and transitional regions. Chem Eng Sci 58:2715–2724

Flint LF (1967) On the velocity profile for turbulent flow in straight a pipe. Chem Eng Sci 22:1127–1131

Flint LF, Eisenklam P (1969) Longitudinal gas dispersion in transitional and turbulent flow through a straight tube. Can J Chem Eng 47:101–106

Fowler FC, Brown GG (1943) Contamination by successice flow in pipe lines. Am Inst Chem Eng 39:491–516

Keyes JJ (1955) Diffusion film characteristics in turbulent flow: dynamic response method. Am Inst Chem Eng 1:305–311

LeChevallier MW, Gullick RW, Mohammad RK, Friedman M, Funk JE (2003) The potential for health risks from intrusion of contaminats into the disstribution system from pressure transients. J Water Health 1:3–14
Lee Y (2004) Mass dispersion in intermittent laminar flow. PhD thesis. University of Cincinnati, Cincinnati
Levenspiel O (1958) Longitudinal mixing of fuids flowing in circular pipes. Ind Eng Chem 50(3):343–346
Nikuradse J (1932) Laws of turbulent flow in smooth pipes. NACA Tech Memorandum 359
Rutherford J (1994) River mixing. Wiley, New York
Senecal VE, Rothfus RR (1953) Transitional flow of fluids in smooth tubes. Chem Eng Prog 49:533–538
Stanton TE, Pannell J (1914) Similarity of motion in relation to the surface friction of fluids. Philos Trans Royal Soc 214:199–224
Taylor GI (1953) Dispersion of soluble matter in solvent flowing slowly through a tube. Proc Royal Soc 219(1137):186–203
Taylor GI (1954) The dispersion of matter in turblent flow through a pipe. Proc Royal Soc 223(1155):446–468
Tichacek LJ, Barkelew CH, Baron T (1957) Axial mixing in pipes. Am Inst Chem Eng 3(4): 439–442
Tzatchkov VG, Buchberger SG, Li Z, Romero-Gomez P, Choi C (2009) Axial dispersion in pressurized water distribution networks-A review. Int Symp Water Manag Hydraul Eng 581–592
White FM (2008) Fluid mechanics. McGraw-Hill, New York
Young P, Jakeman A, McMurtrie R (1980) An instrumental variable method for model order identification. Automatica 16:281–294

Statistical Analysis of Seepage's Influence on Open-Channel Turbulence

Oscar Herrera-Granados

Abstract This chapter presents a statistical analysis concerning the seepage's influence on river dynamics. This analysis is based on the results of several experiments that were carried out at the open-air laboratory of the Wrocław University of Technology. Additionally, a brief review of previous laboratorial works concerning this topic is presented as part of the introduction of this chapter. The output of the laboratorial studies demonstrates that the artificially induced seepage affects the turbulence dynamics of the open-channel flow. Even though the magnitude of the groundwater flow was significantly smaller in comparison with the magnitude of the open-channel flow (not bigger than 0.03 % of the experimental shallow flows), seepage affects the instantaneous velocity field and turbulence dynamics of the open-channel flow. It was observable that seepage modifies the shape of the profiles of the time-averaged parameters of turbulence and in many cases; the open channel flows are not following the laws of isotropic turbulence. As a consequence, seepage modifies the interaction between the flowing water along the channel and the sediments from the bottom.

1 Introduction

It is a matter of fact that in the past, our ancients had been adequating the water bodies and waterways according to the human necessities but not necessarily thinking about the consequences that could arise after modifying the natural water

O. Herrera-Granados (✉)
Institute of Geotechnics and Hydrotechnics, Wrocław University of Technology,
Plac Grunwaldzki 9 Bud. D-2 112 PL-50-377 Wrocław, Poland
e-mail: Oscar.Herrera-Granados@pwr.wroc.pl

P. Rowiński (ed.), *Experimental and Computational Solutions of Hydraulic Problems*, GeoPlanet: Earth and Planetary Sciences, DOI: 10.1007/978-3-642-30209-1_9,

conditions. Higher seepage intensities through the soils that surround man-made infrastructure can be considered as one of these consequences. Also, these constructions not only affect the flow regime but also change the sediment balance of river systems. Seepage exists if the material that constitutes any hydraulic structure or river bed is permeable enough and if there is a difference between the upstream water level and the tail water level downstream. Due to the fact that groundwater flow rates are much smaller than open-channel flow rates, seepage is commonly neglected by river engineers. Nonetheless, this small groundwater flow can represent an important factor in many environmental issues.

The aim of this contribution is to analyze the effect of the groundwater flow in open-channel hydrodynamics based on the results of several series of turbulent flow measurements, which were carried out in a re-circulating flume where upward seepage was induced by external hydrostatic pressures acting in the lowest part of the bed. Some turbulence time-averaged parameters [*Reynolds Shear Stresses* and *Turbulent Kinetic Energy* (*TKE*)] were obtained using one-point statistics. Turbulence spectra were estimated for several time series as well. Even though the magnitude of the induced seepage was significantly smaller in comparison with the magnitude of the shallow flows (not bigger than 0.03 % of the experimental open channel flows), the output of the experiments demonstrates that upward seepage is affecting the instantaneous velocity field and turbulence dynamics of the open-channel flow.

1.1 Seepage's Influence in Open-Channel Dynamics: Previous Laboratorial Works

Experimental works on this topic were previously carried out by Cheng and Chiew (1998), Ali et al (2003), Chen and Chiew (2004), Dey and Sarkar (2007), Lu and Chiew (2007), Fontana (2008), Herrera Granados (2008), Sreenivasulu et al (2010) and Herrera Granados and Kostecki (2011).

Ali et al. (2003) carried out an investigation into the effect of seepage on sediment transport and the resulting bed forms without modifying the velocity distributions and bed shear stresses. Ali et al found that seepage caused a reduction in bed load transport; seepage also caused longer and higher ripples than in the cases without seepage. Chen and Chiew (2004) investigated theoretically and experimentally the velocity distributions of turbulent open channel flow with bed suction. A velocity profile with a slip velocity at the bed surface and an origin displacement under the bed surface was proposed. Their data showed a significant increase in the near bed velocity and a velocity reduction near the water surface, resulting in the formation of a more uniform velocity distribution. Dey and Sarkar (2007) carried out experiments on scour downstream of an apron due to submerged jets under upward seepage. Dey and Sarkar observed that many of the scour hole characteristics increased with an increase in seepage velocity. Their experiments

Fig. 1 The experimental facilities at the open air laboratory—Wrocław University of Technology

showed that seepage influenced the decay of the velocity and turbulent intensities and the growth of the boundary layer. Another interesting output of these experiments is that the decay rate of the local maximum Reynolds stress and the turbulence intensities within the scour holes are slowed down by the upward seepage velocity. Lu and Chiew (2007) developed investigations on seepage's influence on dune dimensions. Their studies demonstrate that suction (downward seepage) increased the equilibrium height of dunes and injection (upward seepage) reduced it. Lu and Chiew showed that seepage effects on the lee-side slope of dunes are the same as that on the critical slope of sediments. Fontana (2008) demonstrated that any hydraulic structure subjected to seepage must be evaluated for stability with reference to seepage forces. Sreenivasulu et al. (2010) reaffirmed that velocity profiles obtained by solving the gradually varied flow should be corrected by including the effect of seepage. This required an adjustment to Manning's n or Darcy's friction coefficient.

Notwithstanding, it is necessary to highlight that in the majority of the previously mentioned laboratorial works, the experimental seepage flow rates were considerably high (due to the way of inducing groundwater flow at the laboratory). But in natural conditions, groundwater flow rates are much smaller in comparison with stream and river flows. For the experimental research presented in this contribution, the maximum registered experimental seepage's rate was less than 0.03 % of the open-channel flow value.

2 Experimental Research

A rectangular flume (see Fig. 1) was modified in a way that free-surface water could flow over a movable sandy bed where upward seepage existed due to artificially induced hydrostatic pressures acting on the lowest part of the flume's bed.

Several time velocity series were recorded with a programmable *Electro-Magnetic Liquid Velocitymeter* (P-EMS) and its accessory E-30. This apparatus (with its different probes) is able to measure simultaneously the flow velocity components in two (mutually perpendicular) different directions (*x*, *y*) or (*x*, *z*) with an accuracy of 0.001 m s^{-1} and a frequency of 10 Hz in the point where the probe is fixed.

2.1 Experimental Setup

The total length of the flume is sixteen meters of which an 8.00 m length zone has a window that allows the visualization of the experiments. Thus, six meters of this section were chosen to perform the experiments (6.0 m-L × 0.5 m-W × 1.0 m-H).

The porous medium was constituted by a 0.20 m movable soil layer (a mixture of sand with gravel). The porous layer was allocated 0.30 m over the bottom of the channel in order to avoid damaging the pipes connected to piezometers that previously existed as shown in Fig. 2. This figure depicts as well the location where the probe E-30 was fixed to record the velocity time-series. These pipes (see Figs. 2 and 3—left side) played a crucial role because water from an external tank had been flowing through them to provoke hydraulic pressures acting on the lowest part of the channel bed. The external tank was able to vary its position in order to have different water levels. This was the mechanism that controlled the seepage intensity. The metal base that supports the bed was symmetrically drilled in order to allow water flow though the porous medium. Seepage flow rates were calculated with the registered water volumes flowing from the tank (using the gauge depicted in Fig. 3—left side). A thin layer of geotextile was situated in between the sandy bed and the drilled base in order to distribute the induced hydraulic pressures uniformly.

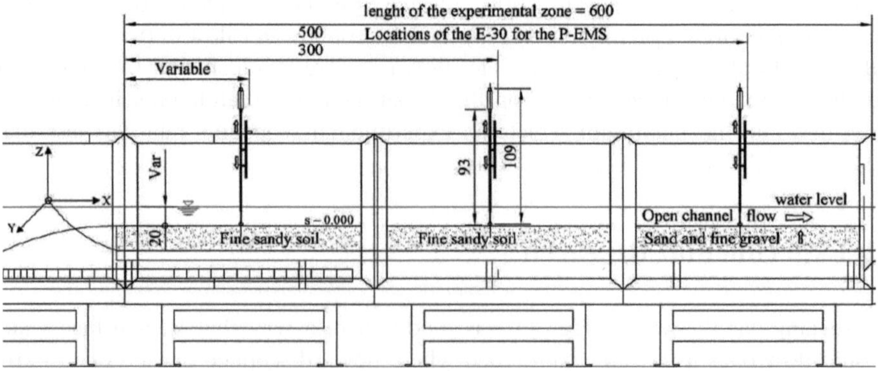

Fig. 2 General scheme of the experimental setup and the location of the probe E-30

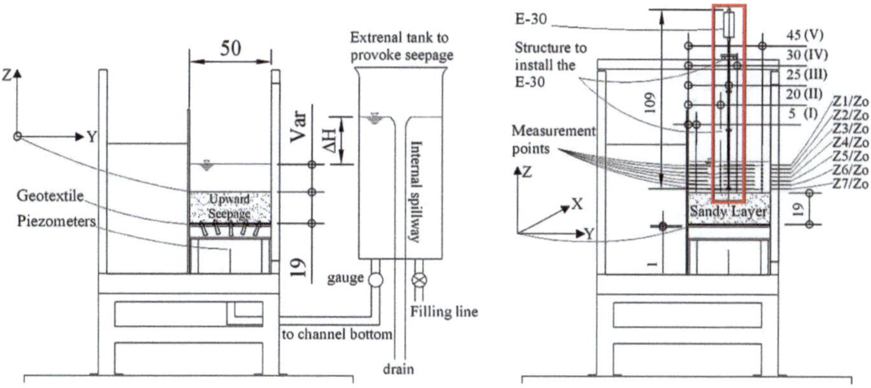

Fig. 3 *Left side* Connection between the external tank and the flume's *bottom. Right side* Location of the points where velocity time-series were recorded for the statistical analysis

The left side of Fig. 3 depicts the general scheme of the external tank and its connection to the bottom of the flume. In the same figure the small vertical line represents the aligned piezometers. The induced hydraulic pressures (depicted as ΔH) provoke upward groundwater flow through the porous medium depending on the water levels inside the tank and on the free surface water flowing along the flume. The right side of Fig. 3 depicts the points where the velocity time series were recorded as a function of the normal depth of each free surface flow. The reference frame that was used to identify the direction of the shallow flow components (x—streamwise, y—spanwise and z—vertical direction) is depicted in both Figs. 2 and 3.

Three different open channel flow rates were used in the series of experiments under steady state flow, as shown in Table 1. The intensity of the induced seepage varied according the type of analysis that was carried out.

2.2 Descriptive Statistics

For small seepage flow rates, there is not enough evidence of the seepage's influence on the time-averaged statistical parameters (Herrera Granados and Kostecki 2011). It was demonstrated as well that seepage is influencing the behavior of the velocity fluctuations rather than the time-averaged open-channel velocity.

Table 1 Open channel flow rates that were used during the experiments

ID	Flow rate $\left(dm^3\ s^{-1}\right)$
Q_1	11.4
Q_2	20.0
Q_3	30.0

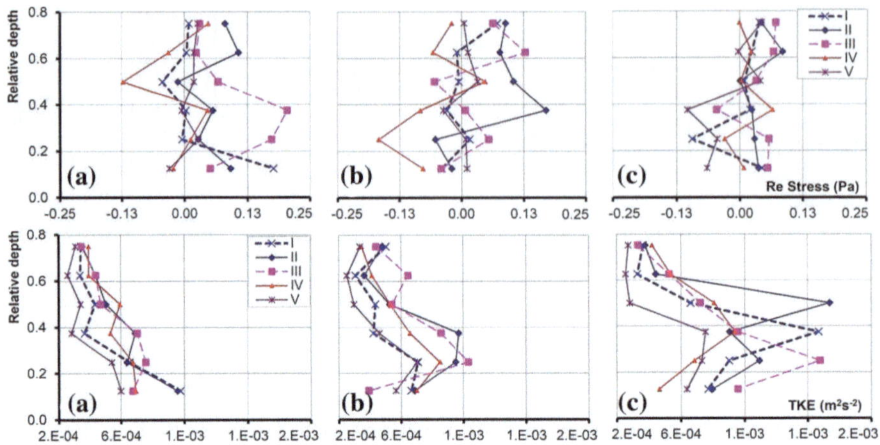

Fig. 4 Time-averaged turbulence parameters profiles at X = 300 cm for Q_2

For this contribution, seepage's influence on open-channel dynamics was analyzed by estimating two turbulence parameters, namely the Reynolds (*Re*) Shear Stresses in the *xy*-plane $\left(-\rho\overline{V_x V_y}\right)$ and Turbulent Kinetic Energy *TKE*. These parameters were analyzed for two experimental open channel flows Q_2 (Fig. 4) and Q_3 (Fig. 5). Five different profiles were plotted that correspond to the open-channel flow subjected to three different seepage intensities: (**A**) ΔH = 00 cm—no seepage; (**B**) ΔH = 10 cm and (**C**) ΔH = 30 cm. The plotted lines correspond to the time-averaged profiles at a distance of 5 (I, blue dashed line), 20 (II, blue continuous line), 25 (III, magenta dashed line), 30 (IV, red continuous line) and 45 (V, purple line) cm away from the left wall of the flume (see Fig. 3—right side) at the cross section X = 300 cm.

Regardless the small magnitude of the groundwater flow, the behavior of the open channel flow is considerably affected by seepage. There is a relation between the hydraulic head (ΔH) that induced seepage and the shape of the turbulent stresses profiles. In the case of no seepage flowing through the porous medium, the *Re* stresses are more irregular while for the highest seepage intensity, the turbulent stresses became similar between them along the cross section and closer to the value of zero. The time averaged *TKE* behaves in the other way. The turbulent energy profiles are more irregular when seepage is more intense.

Descriptive statistics are limited for a better analysis of turbulence dynamics. Hence, spatial correlations and structure functions were obtained in order to check the seepage's impact on the local isotropic behavior of the open-channel flow, which is presented in the next part of this contribution.

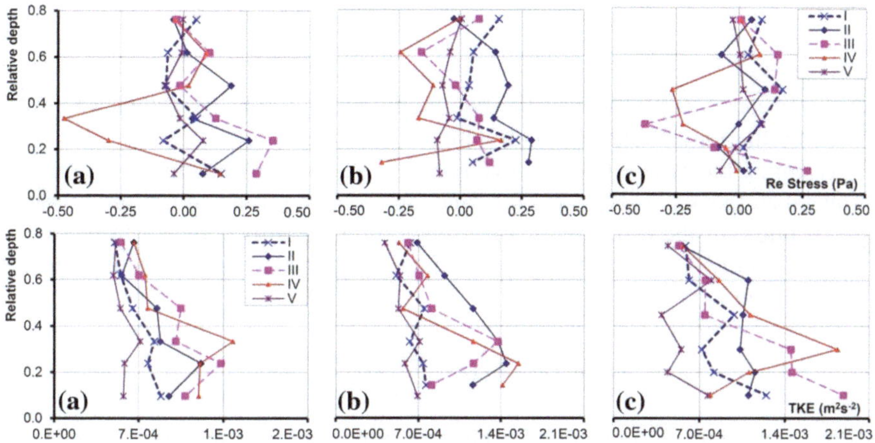

Fig. 5 Time-averaged turbulence parameters profiles at X = 300 cm for Q_3

2.3 Concerning the Statistical Theory of Turbulence

The Kolmogorov hypotheses have no direct connection with the Navier Stokes Equations (NSE). Taylor (1935), as well as Von Kármán and Howarth (1938), were the first to attempt it based on the two-point correlation R_{ij}. In homogeneous turbulence, the two-point correlation (Eq. 1) can be defined as

$$R_{ij}(\mathbf{r},t) \equiv \langle v_i(\mathbf{x}+\mathbf{r},t)\ v_j(\mathbf{x},t)\rangle \tag{1}$$

where the second term of (1) represents the velocity difference between two points separated by the distance $\mathbf{r}$. At the origin, the two-point correlation is equal to $v_{ij}'^2$; which are the root mean square value of the velocity difference and the Kronecker delta respectively. In addition, a consequence of isotropic turbulence is that R_{ij} can be determined by two scalar functions as follows

$$R_{ij}(\mathbf{r},t) = v'^2\left(g(r,t)\delta_{ij} + [f(r,t) - g(r,t)]\frac{r_i r_j}{r^2}\right) \tag{2}$$

with $\mathbf{r} = \mathbf{e_1 r}$, Eq. (2) becomes

$$\begin{aligned} R_{11}/v'^2 &= f(r,t) = \langle v_1(\mathbf{x}+\mathbf{e}_1 r,t)v_1(\mathbf{x},\mathbf{t})\rangle/\langle v_1^2\rangle, \\ R_{22}/v'^2 &= g(r,t) = \langle v_2(\mathbf{x}+\mathbf{e}_1 r,t)v_2(\mathbf{x},\mathbf{t})\rangle/\langle v_2^2\rangle, \\ R_{33} &= R_{22}, \qquad R_{ij} = 0, \quad \text{for} \quad i \neq j, \end{aligned} \tag{3}$$

where f and g are identified as the longitudinal and transverse autocorrelation functions. Using the properties of the structure functions and in combination with Eq. (2); $g(r,t)$ yields

$$g(r,t) = f(r,t) + \frac{1}{2} r \frac{\partial}{\partial r} f(r,t) \quad (4)$$

Hence, in isotropic turbulence the two-point correlation R_{ij} is completely determined by the longitudinal autocorrelation function $f(r,t)$.

Assuming the Taylor frozen's hypothesis as valid, time velocity time series can be switched to spatial series and the function f can be estimated as time-invariant. One example of the estimated autocorrelation functions is depicted in Fig. 6.

Based on the statistical theory of turbulence, the second order structure function D_{ij} is defined as follows (5)

$$D_{ij}(\mathbf{r},t) = C_2(\varepsilon r)^{2/3}\left(\frac{4}{3}\delta_{ij} - \frac{1}{3}\frac{r_i r_j}{r^2}\right) \quad (5)$$

In the *inertial subrange*, the Kolmogorov hypotheses are sufficient to determine the second order structure function as a function of ε, r and the universal constant C_2 (Pope 2000). In addition, another important function that allows to visualize the energy cascade process is the energy spectrum E(κ), defined as Eq. (6):

$$E(\kappa) = \varepsilon^{2/3}\kappa^{-5/3}\Psi(\kappa\eta) \quad (6)$$

where ε is the energy dissipation rate; κ is the wavenumber and Ψ is the compensated Kolmogorov Spectrum Function, which is function of the wavenumber; and the Kolmogorov's microscale η. In literature, this 1D energy spectrum is commonly referred as *Power Spectrum.*

Figures 7 and 8 depict the second order structure functions and power spectra of the same velocity series that were analyzed for the autocorrelation functions of Fig. 6. The analyzed spatial series are the transformed velocity time series for the experimental flow rate Q_1 recorded at the lowest point, where the probe E-30 was fixed at the location X = 100 cm and Y = 25 cm. This point was selected because at the beginning of the channel, seepage influence should be higher than in further

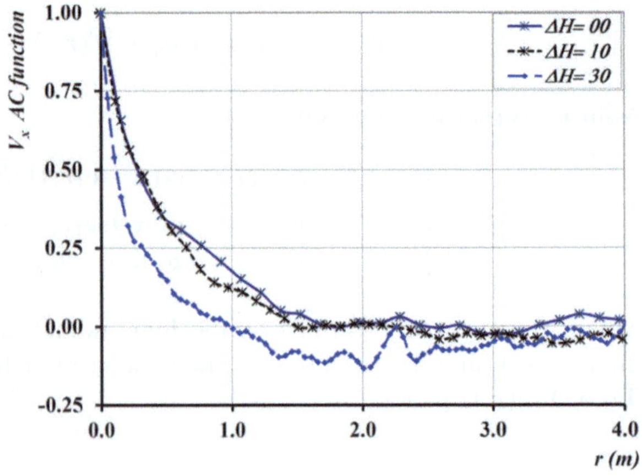

Fig. 6 V_x longitudinal autocorrelation functions estimated from three time series recorded close to the bottom subjected to different induced ΔH. Velocity series without induced seepage (continuous line, dark blue); Velocity series affected by a small upward seepage ($\Delta H = 10$ cm, short-dashed black line) & Velocity series affected by a larger upward seepage ($\Delta H = 30$ cm, long-dashed blue line).

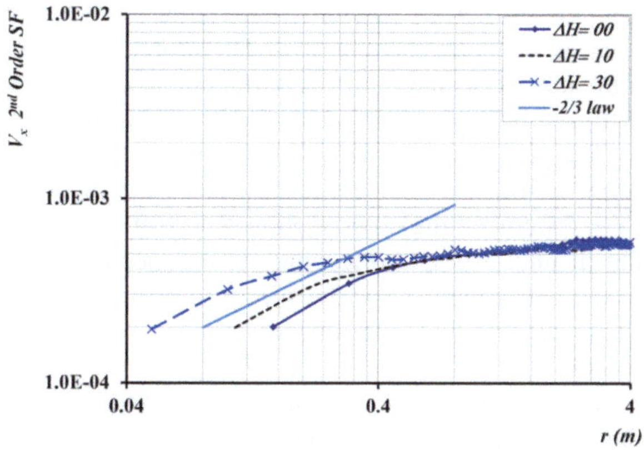

Fig. 7 Second order structure functions estimated from three time series subjected to different induced ΔH. The same legends of Fig. 6 are used. The straight line represents the 2/3 power law, which is representative in isotropic turbulence.

cross sections of the flume. The experimental flow rate Q_1 was analyzed due to the fact that for higher open-channel flow rates, the movable bed was eroded very fast and the output of the experiments would lose confidence. It is necessary to mention that the presented results are representative of more than 100 runs of experiments carried out under similar hydrodynamic conditions for small seepage intensities.

In any turbulent flow with high Reynolds number, the high wavenumber portion of the velocity adopts universal forms Pope (2000). Therefore, the energy cascade process can be pictured by the well-known *Kolmogorov Spectrum* according to his theory of isotropic turbulence. Recalling the Kolmogorov's hypotheses, there are two energy ranges: The *Energy containing* range and the *Universal equilibrium range*. Additionally, the Universal equilibrium range is subdivided in two sub-ranges: The *Inertial* subrange and the *Dissipation* subrange. The inertial subrange is this one that is classically represented by the $-5/3$ and $2/3$ power laws.

Based on Figs. 7 and 8, the initiation of the inertial subrange is dependent on the induced seepage's intensity. In addition, the time series that is subjected to the highest seepage intensity is better autocorrelated. It seems to be that the flow

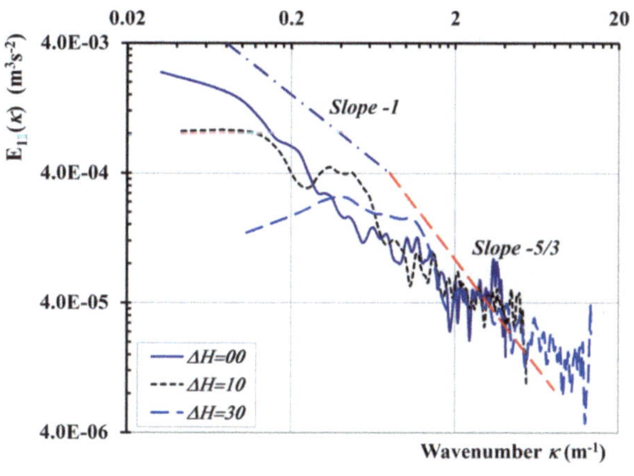

Fig. 8 V_x longitudinal Power Spectra estimated from three time series subjected to different induced ΔH. The same legends of Fig. 6 are used. The straight red line represents the classical $-5/3$ Kolmogorov Power Law and the blue straight line represents the -1 Power Law representative in shallow flows.

through the porous medium is affecting the homogeneity of the velocity fluctuations of the open channel flow. Therefore, in the last part of this contribution, the influence of seepage on the turbulent isotropic behavior of the shallow flow is analyzed by the spectral analysis of one profile for the experimental flow rate Q_1.

The power spectra (E_{11}) of Fig. 9 are those estimated from the velocity time series in the streamwise direction that were recorded in the middle of the channel (Y = 25 cm) at the cross section X = 100 cm and at five different distances from the channel bed: *Z*1 corresponds to the spectrum of the time series recorded closest to the bottom and *Z*5 is the spectrum of the furthest time series from the bottom. There was no induced upward seepage while recording the time series of this profile.

The power spectra (E_{11}) of Fig. 10 correspond to the velocity series analyzed in the same location for the same distances from the flume's bed but that correspond to the open channel flow subjected to an induced hydraulic head $\Delta H = 20$ cm. The spectra of open channel flow affected by seepage behave more isotropically.

It is observable as well that the transition between the productive and energy dissipation range occurs for lower wavenumbers in the spectra that are influenced by seepage (Fig. 10). Spectra of Fig. 9 even do not show clearly initiation of this energy cascade; it seems to be that these spectra follows more the -1 power law that is representative for shallow flows (Nikora, 1999). Hence, there is a relation between the seepage intensity and the initiation of the inertial subrange. As consequence, the value of the turbulent scales should be affected.

Figure 11 depicts the power spectra (E_{22}) in the spanwise direction for the shallow flow that was not influenced by seepage. These spectra show that for these velocity series, the classical −5/3 power law for isotropic turbulence is not visible at all. Above all for the velocity series that was recorded closest to the bed.

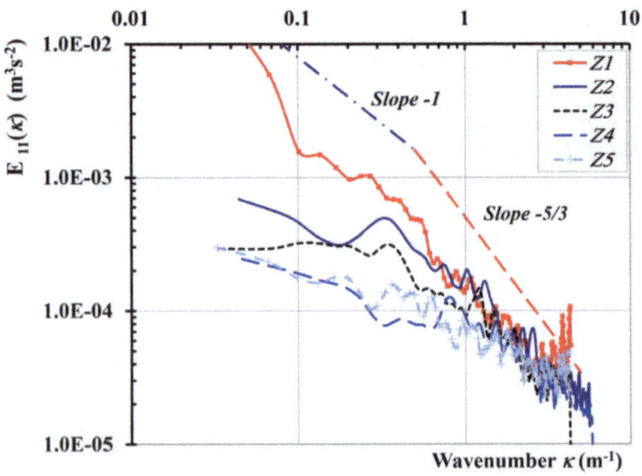

Fig. 9 1D longitudinal spectra in the streamwise direction without induced groundwater flow through the flume's bed. The plotted spectra correspond to the same profile at different depths. The straight red line represents the classical $-5/3$ Kolmogorov Power Law. The *blue straight line* represents the -1 Power Law.

Fig. 10 1D longitudinal spectra in the streamwise direction with induced groundwater flow through the flume's bed. The plotted spectra correspond to the same profile at different depths. The straight red line represents the classical −5/3 Kolmogorov Power Law. The *blue straight line* represents the −1 Power Law.

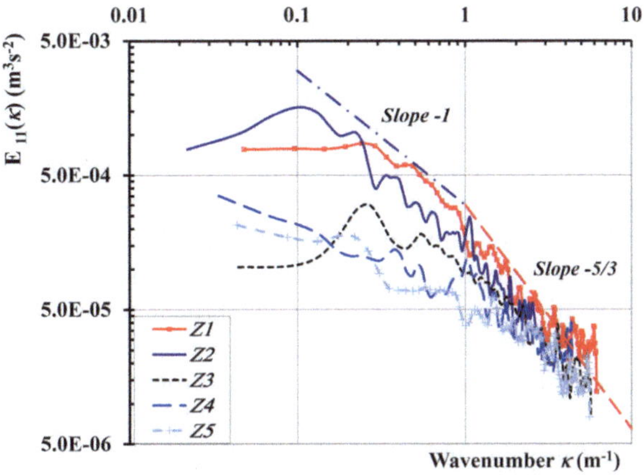

Fig. 11 1D spectra in the spanwise direction without induced groundwater flow through the flume's bed. The plotted spectra correspond to the same profile at different depths. The *straight red line* represents the classical −5/3 Kolmogorov Power Law and the *blue straight line* represents the −1 Power Law.

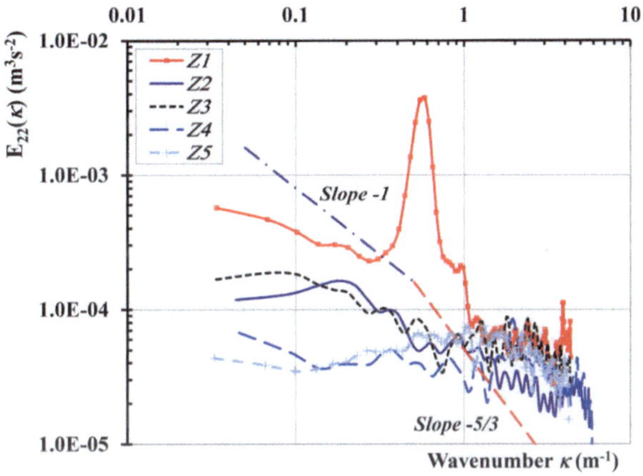

3 Conclusions and Final Remarks

The output of this experimental research demonstrates that the induced flow through the porous medium is considerably affecting the turbulence dynamics of the open-channel flow (regardless the small seepage's magnitude).

The presented Figures (descriptive statistics) showed how the instantaneous velocity field of open channel flows changed when flow through the hyporheic zone existed. Thus, the time averaged *TKE* and *Re* stresses profiles are affected. Seepage is changing the statistical homogeneity of the velocity fluctuations. Therefore, seepage seemed to affect the local isotropic behavior of the open channel instantaneous velocity field. One could observe a direct correlation between the intensity of the induced seepage and the boundaries of the energy ranges of the Kolmogorov's spectrum. In the case of the presented E_{11} spectra without flow through the porous medium, the initiation of the internal subrange is even not clearly observable.

The impact of seepage in many environmental issues could be important because seepage is modifying the flow velocity distribution, bed shear stresses and turbulence intensities of the open channel flow. This small groundwater flow can exert an additional hydrodynamic force on the bed sediment (Lu et al. 2008). Hence, the onset of sediment motion changes and other sediment transport mechanics are affected (Herrera Granados 2008). Results from recent research efforts in this area are often inconclusive and sometimes conflicting. In addition, there is the necessity of accurate measurements at small values of seepages velocities to verify existing theoretical analyzes (Lu et al. 2008). The seepage intensities that were used for the experiments, described in this chapter, are in the range that Lu et al mentioned. Therefore, a further step of this research is to adapt and compare the output of these laboratorial works to previously developed experimental and theoretical analyses.

References

Ali KHM, Achterberg J, Li M, Zhu Y (2003) Effect of seepage on sediment transport in channels. In: International conference on estuaries and coasts, pp 461–466

Chen X, Chiew Y (2004) Velocity distribution of turbulent open-channel flow with bed suction. J Hydraul Eng 130(2):140–148

Cheng NS, Chiew YM (1998) Turbulent open channel flow with upward seepage. J Hydraul Res 36(3):415–431

Dey S, Sarkar A (2007) Effect of upward seepage on scour and flow downstream of an apron due to submerged jets. J Hydraul Eng 133(1):59–69

Fontana N (2008) Experimental analysis of heaving phenomena in sandy soils. J Hydraul Eng 134(6):794–799

Herrera Granados O (2008) Effect of seepage flow on channel bed evolution at the laboratory scale. In: Altinakar M et al. (eds) River flow 2008, Kubaba Congress Department, pp 1117–1126.

Herrera Granados O, Kostecki S (2011) Experimental methods in hydraulic research, chap analysis of turbulent flow measurements in a flume with induced upward seepage, Springer, Berlin, pp 163–173.

Lu Y, Chiew YM (2007) Seepage effects on dune dimensions. J Hydraul Eng 133(5):560–563

Lu Y, Chiew YM, Cheng NS (2008) Review of seepage effects on turbulent open-channel flow and sediment entrainment. J Hydraul Res 46(4):476–488

Nikora V (1999) Origin of the "−1" spectral law in wall-bounded turbulence. Phys Rev Lett 83(4):734–736. doi: 10.1103/PhysRevLett.83.734

Pope S (2000) Turbulent flows. Cambridge University Press, Cambridge

Sreenivasulu G, Ramakrishna A, Kumar B, Tripathi S (2010) Analysis of gradually and spatially varied flow in sand-bed channels. J Hydraul Res 48(2):274–279

Taylor G (1935) Statistical theory of turbulence: parts I-III. Proc R Soc London Ser A 151:421–464

Von Kármán T, Howarth L (1938) On the statistical theory of isotropic turbulence. Proc R Soc London Ser A 164:192–215

Flow Capacity Coefficient of Strainers

Tomasz Kałuża and Paweł Zawadzki

Abstract The values of hydraulic specifics of the prototypes of new DN 50, DN 80, DN 100, DN 150 and DN 200 strainers have proven their enhanced hydraulic properties. These are a result of both increased screening surface of the filter insert, as well as the improved shape of the filter bodies. These conclusions were confirmed in the course of examinations of the DN 80 and DN 200 filter bodies. The performed measurements of new filters with screen inserts have revealed an average growth of 43 % of the K_v flow coefficient in case of all examined products. The maximum growth of the flow coefficient, amounting to 55 %, was recorded with regard to filter DN 200. During the tests also the possibility of estimating the K_v coefficient value was checked with regard to the entire filter series. By means of dimensional analysis and the similitude concept, on the basis of a given product it is possible to obtain K_v values of filters at a maximum tolerance of 10 %.

1 Introduction

Nowadays, the designing of water, sewage or gas network fittings necessitates the cooperation of various specialists, only to mention constructors, casters, technicians and hydromechanics. Thanks to their mutual cooperation, supported by modern designing, modelling and verification tools for prototype devices, it is

T. Kałuża (✉) · P. Zawadzki
Department of Hydraulic and Sanitary Engineering, Poznan University of Life Sciences, ul. Piątkowska 94A, 60-649 Poznań, Poland
e-mail: kaltom@gmx.net

P. Zawadzki
e-mail: pzaw@up.poznan.pl

P. Rowiński (ed.), *Experimental and Computational Solutions of Hydraulic Problems*, GeoPlanet: Earth and Planetary Sciences, DOI: 10.1007/978-3-642-30209-1_10,

possible to develop new products characterised by better operating parameters. In the course of preparing a device concept it is necessary to analyse numerous alternatives, whereas their verification is often possible only at the stage of prototype development. This study presents an example of research and verification of hydraulic parameters of diagonal strainers manufactured by the Hawle (Hawle Armaturenwerke).

2 Purpose and Scope of Research

In order to ensure the protection of the waterworks system, as well as measurement and adjustment instruments fitted on water pipelines, additional mechanical filters are installed. Their filtering function is intended to separate solids from water by one or several porous compartments. Rapid colmatage stands as an important problem of the filtering function. It necessitates frequent cleaning or replacement of filter inserts. In case of most of diagonal strainers, the screen insert is accessed and installed from the bottom which hinders their replacement. On top of that, significant hydraulic losses have occurred in such filters in the course of water flow and accumulation of foreign bodies.

The relevant feedback received from the users has inspired the Hawle to develop a new filter structure with improved hydraulic parameters and a more easily accessible screen insert. The first version of the filter body was prepared by means of the (Computer Aided Three-dimensional Interactive Application) CATIA computer software intended for supporting the mechanical designing process. This programme is a modern, integrated CAD/CAM/CAE system which offers an abundant set of computer tools enhancing the entire cycle of operations within the construction and production process.

While developing the design, the previously defined strength parameters of the specific device elements were considered. On this basis, construction drawings were generated. In the process of designing the new filter, it was assumed that the new screen insert should have a larger diameter than the former one and that it should be fitted laterally into the filter body. The entire prototype development process lasted for approx. 1 year. In this time, during the first 2 months, a concept was prepared and the filter was designed, whereas during the next 9 months the forms and casts were developed (Fig. 1).

In the first place, prototypes of new strainers were created, featuring the DN 80 and DN 200 diameters. On this basis, still before starting the production of the entire series (DN 50, DN 80, DN 100, DN 150, DN 200), the hydraulic parameters of prototypes were assessed in order to establish the suitability of the new device for the operating conditions of a water network. For the purpose of estimating the hydraulic parameters of other than the two above-mentioned diameters, scaling calculations were used on the basis of a dimensioning analysis of the complete prototypes. The prototypes were subjected to pressure tests and laboratory examinations in order to determine the value of K_v flow coefficient. The scope of

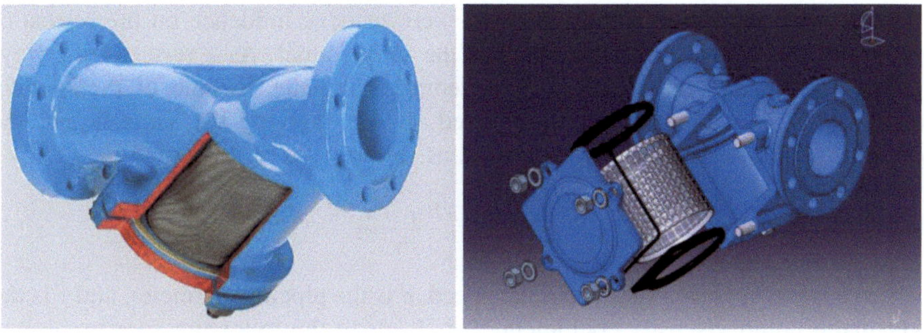

Fig. 1 Former strainer and the designing process of the new strainer by means of the CATIA software

survey was also extended by the measurements of hydraulic specifics of old and new DN 80 and DN 200 filter bodies (without screen inserts). The results of these tests have enabled the assessment of the influence of new profile of the filter body on the occurrence of hydraulic losses.

3 Dimension Analysis and Similitude

Hydraulic modelling of water pipeline fittings is based on the concept of similitude (Czetwertyński 1958; Franzini and Finnemore 1997). This theory enables qualitative and quantitative recording of a real phenomenon on basis of measurements performed by examining a similar model phenomenon. While speaking of dynamic similitude it is assumed that liquid movement is analysed in two different systems, resembling each other in terms of their geometry which in practice means the comparison of movement in nature and in a model. The most general law of dynamic similitude of two systems is known as the Newton's Law, however the fulfilment of the similitude principle therein in all its scope is virtually extremely difficult, as it requires a selection of liquids of different densities. Therefore, in specific cases the Froude's number (for flows with a free water surface) or Reynolds' theorem is employed, which is applicable to pressurised pipeline flows.

Dynamic similitude necessitates the preserving of geometric similarity. While considering two geometrically similar systems: real N with natural dimensions and model M with miniaturised dimensions, we state that the geometrical similitude shall be maintained, as for each of the two sections which are L_N and L_M long, the following relation shall apply:

$$L_N = \lambda \cdot L_M \tag{1}$$

where λ is the geometric scale.

For the purpose of modelling viscous liquids in pipelines, the Reynolds theorem stands as the essential principle, worded as follows: 'In two similar flows of real liquids, internal friction forces whereof exert decisive influence on the course of phenomenon, the Reynolds number has the same value'. As a result of the Reynolds theorem and the adoption of a geometrical scale, a scale of the specific elements of liquid movement is obtained. While applying the same liquid, the criterion of similitude may be expressed as follows:

$$\frac{u_N \cdot d_N}{\nu} = \frac{u_M \cdot d_M}{\nu} \tag{2}$$

where u is the average cross-sectional speed, d is the pipeline diameter, and ν is the kinematic liquid viscosity coefficient. This leads to the following speed scale:

$$\frac{u_N}{u_M} = \frac{1}{\lambda} = \lambda^{-1} \tag{3}$$

and water flow rate scale:

$$\frac{Q_N}{Q_M} = \frac{u_N A_N}{u_M A_M} = \lambda^{-1} \cdot \lambda^2 = \lambda \tag{4}$$

The power scale may be expressed by means of the relation:

$$\frac{F_N}{F_M} = \frac{\mu \cdot u_N \cdot D_N}{\mu \cdot u_M \cdot D_M} = \lambda^{-1} \cdot \lambda = 1 \tag{5}$$

and finally, the pressure scale:

$$\frac{p_N}{p_M} = \frac{\frac{F_N}{A_N}}{\frac{F_M}{A_M}} = \frac{F_N \cdot A_M}{F_M \cdot A_N} = 1 \cdot \lambda^{-2} = \frac{1}{\lambda^2} \tag{6}$$

By employing the relation provided in (6), pressure level scale or pressure loss scale may be determined:

$$\frac{H_N}{H_M} = \frac{\frac{p_N}{\rho \cdot g}}{\frac{p_M}{\rho \cdot g}} = \frac{p_N}{p_M} = \frac{1}{\lambda^2} \tag{7}$$

whereas by means of the relation between (3) and (7) the scale of the K_v flow coefficient is obtained:

$$\frac{K_{vN}}{K_{vM}} = \frac{Q_N \cdot \sqrt{\frac{\rho}{H_N \cdot \rho_0}}}{Q_M \cdot \sqrt{\frac{\rho}{H_M \cdot \rho_0}}} = \frac{Q_N}{Q_M} \cdot \sqrt{\frac{H_M}{H_N}} = \lambda \cdot \sqrt{\lambda^2} = \lambda \cdot \lambda = \lambda^2 \tag{8}$$

On the basis of the modelled flow and resistance rates, hydraulic parameters of devices with different nominal diameters may be thus estimated.

4 Research Method

The research environment at the Water Laboratory of the Department of Hydraulic and Sanitary Engineering of the Poznan University of Life Sciences was prepared on the basis of fittings provided by the Hawle. The scope of research included:

- volumetric flow rate, Q ($m^3\ s^{-1}$)
- pressure difference measurement, H (m),
- water temperature measurement, t (°C).

On this basis, the calculation of the necessary physical values was performed, including the determination of K_v flow coefficients.

The study was performed under steady flow conditions and closed water circulation. Water flow rate was measured with an electromagnetic flow meter, model SITRANS FM MAGFLO, type MAG 5000, manufactured by SIEMENS. Measurement accuracy amounted to ±0.5 % of the measured flow value. The pressure rate was adjusted by means of electronically set pump frequency via an inverter, model HYDROVAR HV 4.075, manufactured by Vogel Lowara. The range of flow rates during the research amounted to 1–120 $l\ s^{-1}$.

The water pressure loss was measured with a pressure difference transmitter manufactured by Siemens, model SITRANS P, DS III series, at an accuracy of ±0.02 % of the set measurement scope. The pressure measurement hole diameters and location of piezometers was adopted in accordance with standard no. PN-EN 1267. Measurement points were placed before the filter (inlet pressure) at a distance of two pipeline diameters and behind the filter (outlet pressure) at a distance of ten pipeline diameters. In each measurement cross-section, two piezometers with a 4 mm hole diameter were installed (for DN > 50, the diameter should amount to 3–5 mm according to PN-EN 1267). The measurement time was determined by means of a test (a period of 2–3 mins was sufficient to stabilise the pressure difference and the indications of the pressure transmitter).

The K_v flow coefficients and the necessary calculations were performed in accordance with the relations included in standard no. PN-EN 1267:

$$K_V = Q \cdot \sqrt{\frac{\rho}{p \cdot \rho_0}} \left(m^3 \cdot h^{-1}\right) \quad (9)$$

where Q is the volumetric flow rate ($m^3\ h^{-1}$), ρ is the water density ($kg\ m^{-3}$), ρ_o is the water density at 15 °C ($kg\ m^{-3}$), and p is the pressure loss (bar).

The water temperature was measured with a technical thermometer at an accuracy of ±0.5 °C. Due to the fact that water temperature amounted to 17 °C during all measurements, the influence of density fluctuation was ignored.

Each test was initiated at a low flow rate which was then gradually increased at 3–5 $l\ s^{-1}$ up to the maximum flow rate. The flow rate was adjusted by the extent of gate opening and pump settings. Pressure loss was measured for each flow rate value. For calculation purposes, average values from approx. 20 measurements were adopted. In case of all tests, the minimum value of Reynolds number amounted to approx. 1.1×10^5, which was higher than the acceptable value 4.0×10^4 required by standard no. PN-EN 1267. The acceptable value is justified by the necessity to obtain high pressure differences which influences the accuracy of measurements.

5 DN 80 and DN 200 Filter Body Test Results

During the first stage of the research, hydraulic specifics of the old and new DN 80 filter body were determined (after the removal of the screen insert). In each research, 8 measurement series were carried out at flow rates ranging from approx. 8 l s^{-1} to above 16 l s^{-1}. For each measurement series, an average flow rate and average pressure loss was determined.

On the basis of Eq. (9), K_v coefficient was calculated for the new DN 80 filter body. The results of all measurement during this and subsequent survey stages (also for DN 200) referred to water temperature of 15 °C. The average value of the flow coefficient for the new DN 80 filter body determined on the basis of all measurement series amounted to $K_v = 347$ m^3 h^{-1}. This value referred to the standard divergence value enabled the determination of the relative measurement accuracy which amounted to 2 %.

In order to compare the values obtained for the new DN 80 filter body, they were confronted with the research results of the old DN 80 filter body provided by the Hawle. Based on the existing research environment, an old filter was installed in the position of the new filter body. Measurements were undertaken with unchanged location of piezometers and within the previously developed research method. Similarly to the previous session, the survey featured 8 measurement series performed at flow rates, ranging from approx. 4 l s^{-1} to above 16 l s^{-1}. The values of flow rates in m^3 h^{-1} and losses converted into Pa both for the new and old filter body are provided in Fig. 2. The measurement points, according to examples provided in literature (Kurganow and Fedorow 1986; Mitosek 2001) were equalled by means of an exponential relation. The obtained results indicated a significant improvement of throughput in the new DN 80 filter body. The new filter body generated much smaller losses at identical flow rates. The average flow coefficient value of the old filter body was determined on basis of all measurement series and it amounted to: $K_v = 199$ m^3 h^{-1}, whereas relative measurement tolerance amounted to ±3 %. It is thus evident that the new filter body indicates a growth of the K_v flow coefficient value by 74.4 %.

The research environment for the DN 200 filter corresponded with the one used for the DN 80 filter. During the first stage of the research, hydraulic specifics of the new DN 200 filter body were determined (after the removal of the screen insert). The survey featured 7 measurement series featuring flow rates from approx. 15 l s^{-1} to above 26 l s^{-1}. For each measurement series, an average flow rate and average pressure loss rate was determined. By means of Eq. (9) K_v coefficient was calculated for the new DN 200 filter body. The average value of the flow

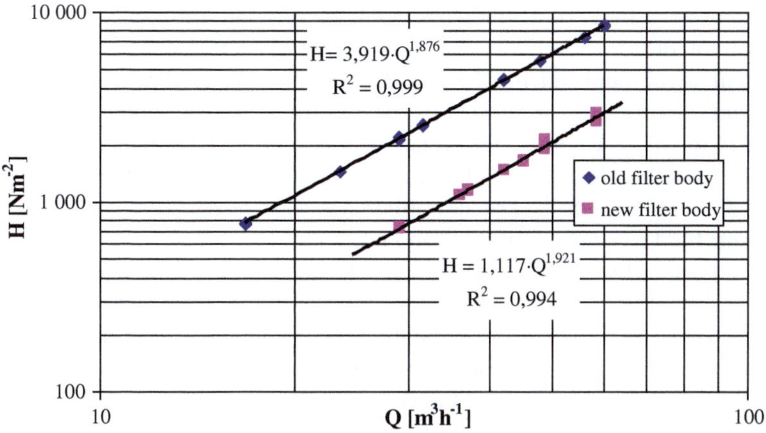

Fig. 2 Relation between pressure loss and flow rate of new and old DN 80 filter body

coefficient determined on the basis of all measurement series amounted to $K_v = 199\ m^3\ h^{-1}$, whereas relative measurement tolerance amounted to ±5 %. The average flow coefficient value of the old DN 200 filter body amounted to $K_v = 946\ m^3\ h^{-1}$, whereas the relative measurement tolerance amounted to ±4 %. The new filter body thus indicated a growth of the K_v coefficient by 106.1 % in comparison to the old body.

By means of similitude principles, the possibility of determining the hydraulic parameters of the bodies was checked. It was assumed that the DN 80 and 200 bodies are identically profiles and they differ only in terms of their geometrical scale, that is $\lambda = 2.5$. On the basis of the experimentally identified value the flow coefficient for the new DN 80 filter body, the coefficient of DN 200 was calculated, which amounted to $K_v = 2{,}168\ m^3\ h^{-1}$. The relative estimation error referred to the measured DN 200 value amounted to approx. 11 % in this respect.

6 DN 80 and DN 200 Strainer Test Results

In the second part of the research, the values of flow coefficients of old and new DN 80 and 200 filters were determined. For the research purposes, the existing measurement installation was used. Measurements were undertaken with unchanged location of piezometers and within the previously developed research method.

The examination of the DN 80 filter (together with the screen insert) involved the measurement of flow specifics of both the old and new filter. The survey featured 21 measurement series performed at flow rates ranging from approx. $9\ l\ s^{-1}$ to above $21\ l\ s^{-1}$. For each measurement series, an average flow rate and average pressure loss rate (on the basis of 15 flow-meter indications) and average loss rate (on the basis of 15 indications of the pressure difference transmitter for each pair of piezometers) was determined (Fig. 3). On the basis of Eq. (9), K_v coefficient was calculated for the new DN 80 filter body. The value of the flow coefficient determined on the basis of all measurement series amounted to: $K_v = 207\ m^3\ h^{-1}$, whereas the relative measurement tolerance amounted to ± 1.5 %. Also the old DN 80 filter was examined. Its flow coefficient was lower, that is, $K_v = 146\ m^3\ h^{-1}$. The new filter has thus provided a growth of the K_v coefficient by 42 % against the old model.

The survey of the new DN 200 filter (together with a screen insert) featured 16 measurement series performed at flow rates ranging from approx. 17 $l\ s^{-1}$ to above 90 $l\ s^{-1}$. For each measurement series, an average flow rate and average pressure loss rate was determined (Fig. 4). By using the relation provided in (9), K_v coefficient was calculated for the new DN 200 filter. The average value of the flow coefficient determined on the basis of all measurement series amounted to $K_v = 1{,}069\ m^3\ h^{-1}$, whereas the average measurement tolerance amounted to ± 0.30 %. By referring the value of the K_v coefficient to the standard divergence value it was possible to determine the relative measurement accuracy which amounted to ± 0.82 %. Also the old DN 200 filter was examined. Its flow coefficient was lower, that is, $K_v = 691\ m^3\ h^{-1}$. The new filter has thus provided a growth of the K_v coefficient by 55 % against the old model.

Similarly to the filter body analysis, also in this case the hydraulic parameters of the filters were defined with the principles of similitude. It was assumed that the DN 80 and 200 bodies differ only in terms of their geometrical scale, that is, $\lambda = 2.5$. On the basis of the experimentally identified value of the flow coefficient of the DN 80 filter, first the coefficient of DN 200 was calculated, which amounted to $K_v = 1{,}293.6\ m^3\ h^{-1}$. The relative estimation error referred to the measured DN 200 value amounted to approx. 21 %.

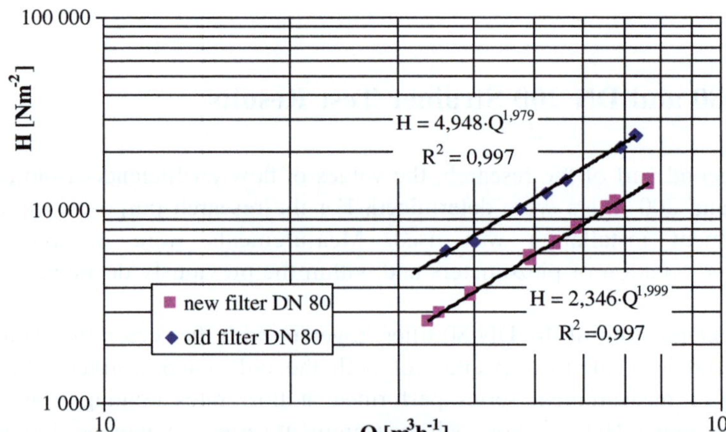

Fig. 3 Relation between pressure loss and flow rate of the new and old DN 80 filter

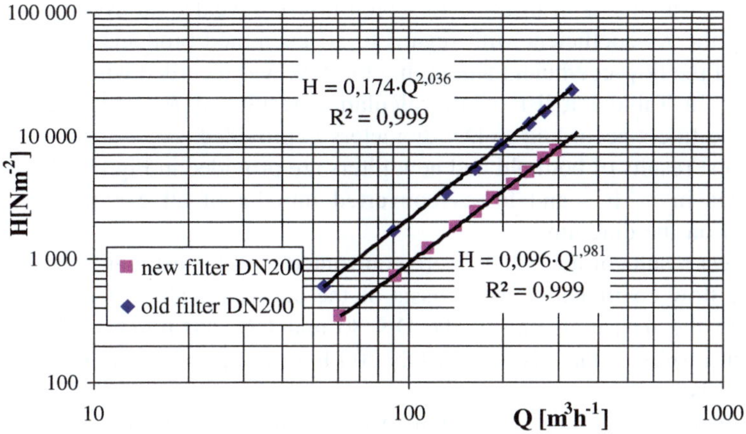

Fig. 4 Relation between pressure loss and flow rate of the new and old DN 200 filter

By means of a corresponding method, the flow coefficient values were determined also for other, not yet manufactured filters of the entire designed series. The calculations were performed initially for the product diameter of DN 80 and repeated for DN 200. The results are presented in Table 1.

The results of estimations presented in Table 1 indicate a significant span of the acquired results. The K_v coefficient values determined on the basis of the DN 80 filter measurement against the values estimated on the basis of the DN 200 filter indicate a growth of the K_v coefficient by 21 %.

7 Test Results of the Entire Filter Series

On the basis of the performed research of the DN 80 and DN 200 filters which have proven their enhanced hydraulic parameters, it was decided to prepare the prototypes of the entire series. Also new filter prototypes were subjected to laboratory examinations. The examination of the values of flow coefficients of the new DN 50, DN 100 and DN 150 filters were supplemented with the tests of old filters. For the research purposes, the existing measurement installation was used. Measurements were performed with unchanged location of piezometers and within the previously developed research method.

Table 1 Estimation of filter flow coefficients on the basis of the adopted scale of similitude in reference to the measurements of the DN 80 and DN 200 filters

	DN 50	DN 80	DN 100	DN 150	DN 200
K_v (m^3 h^{-1})	67	171	267	601	1,069
K_v (m^3 h^{-1})	81	207	323	728	1,294

The measurements of filters with screen inserts revealed an average growth by 43 % of the K_v coefficient with regard to all examined products. The maximum growth of the flow coefficient amounting to 55 % was recorded with regard to filter DN 200. In Table 2 a list of all calculation results of K_v flow coefficients of respective filters according to DN diameters is provided.

The test results of both old and new filters are provided in Fig. 5. The values of flow rates in $m^3\ h^{-1}$ and losses converted into Pa of all examined filters are provided on the diagram.

The test results of new filters were confronted with the estimation results of the K_v filter flow coefficients on the basis of the adopted scale of similitude in reference to the measurements of the DN 80 and DN 200 filters (Table 1). Measurement results of the DN 80 and DN 200 filter are provided in Tables 3 and 4, respectively.

The measurement and estimation results compiled in Tables 3 and 4 on the basis of the scale of similitude indicate that with regard to the measurements of the DN 80 filter, the error factor against DN 50 and DN 100 is marginal and amounts to ±5 %, whereas in the case of higher diameters, DN 150 and 200, it exceeds 20 %. An opposite trend may be observed on the list of measurement results of the DN 200 filter. For the DN 50, 80 and 100 diameters, the error exceeds 10 % (for

Table 2 Calculated values of the K_v ($m^3\ h^{-1}$) flow coefficients of respective filters according to DN diameters

	DN 50	DN 80	DN 100	DN 150	DN 200
Old filter	–	146	202	475	691
New filter	85	207	307	585	1,069
Coefficient growth K_v (%)	–	42	52	23	55

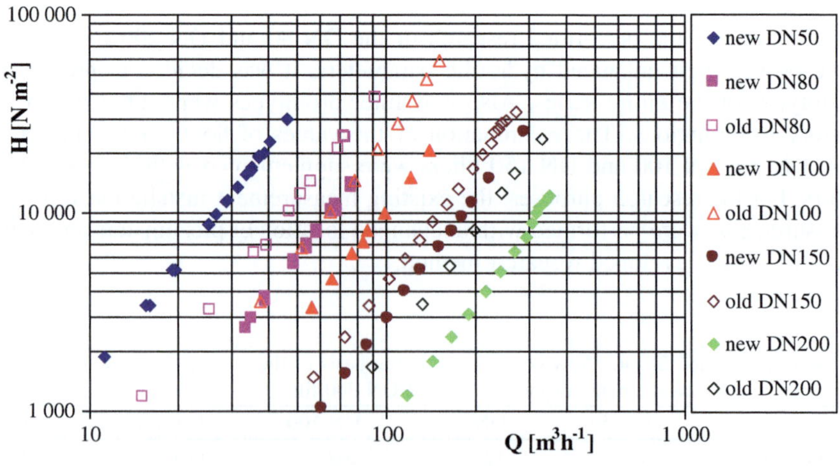

Fig. 5 Relation between H pressure loss and Q flow of old and new filters

Table 3 K_v coefficient estimation error on the basis of the adopted scale of similitude in reference to the measurements of the DN 80 filter

	DN 50	DN 80	DN 100	DN 150	DN 200
Estimation of K_v ($m^3 h^{-1}$)	81	207	323	728	1,294
Measurement of K_v ($m^3 h^{-1}$)	85	207	307	585	1,069
Estimation error %	5	0	−5	−24	−21

Table 4 Estimation error of filter coefficient flow rates on the basis of the adopted scale of similitude in reference to the measurements of the DN 200 filter

	DN 50	DN 80	DN 100	DN 150	DN 200
Estimation of K_v ($m^3 h^{-1}$)	67	171	267	601	1,069
Measurement of K_v ($m^3 h^{-1}$)	85	207	307	585	1,069
Estimation error %	21	17	13	−3	0

DN 50 it even reaches 21 %), whereas for DN 150 it is only 3 %. The grouping to these results performed on the basis of Tables 3 and 4 is caused by the different types of screen inserts used in these product groups. The filters were equipped with fine stainless steel screens with a mesh size of 0.5 mm (DN 50–DN 100 mm) or 0.6 mm (DN 150–DN 300). The mesh size affected directly the flow resistance and thus also the value of flow coefficient. Nevertheless, having analysed the possibilities of estimation of the hydraulic parameters on the basis of the scale of similitude, the applied scale was retained with regard to filter bodies and the size of screen inserts, which has not been possible for the mesh size yet.

8 Conclusion

The research results of strainer prototypes produced by the Hawle provide an example of effective cooperation between the specialists of various fields. Thanks to the performed hydraulic verification of the prototypes, the improvement of the operating parameters of the new filters could be confirmed. The assumptions adopted at the preparation stage of the construction concept were supported by the test results.

The values of hydraulic specifics of the prototypes of the new DN 50, DN 80, DN 100, DN 150 and DN 200 filters have proven their enhanced hydraulic properties. These are a result of both increased screening surface of the filter insert, as well as the improved shape of the filter bodies. These conclusions were confirmed in the course of examinations of the DN 80 and DN 200 filter bodies. It is thus evident that the new DN 80 filter body has indicated a growth of the K_v flow coefficient by 74.4 % against its former version, whereas in the case of DN 200 the recorded growth accounted for as much as 106.1 %. The performed measurements of new filters with screen inserts have revealed an average growth of 43 % of the

K_v coefficient in case of all examined products. The maximum growth of the flow coefficient, amounting to 55 %, was recorded with regard to filter DN 200.

During the tests also the possibility of estimating the K_v coefficient value was checked with regard to the entire filter series. By means of dimensional analysis and the similitude concept, on the basis of a given product it is possible to obtain K_v values of filters at a maximum tolerance of 10 %. Having considered the impact of the scale effects, it is also essential to consider the mesh size of screen inserts apart from reflecting the scale of the filter bodies and inserts.

References

Czetwertyński E (1958) Hydraulika i hydromechanika (Hydraulics and hydromechanics). PWN, Warsaw. (in Polish)

Franzini JB, Finnemore EJ (1997) Fluid mechanics with engineering applications. WCB/ McGraw-Hill, Boston

Kurganow AM, Fedorow NF (1986): Gidrawlicheskie raschety system wodosnabzhenija i wodootwenija. Wyd. Strojzdat St. Petersburg (in Russian)

Mitosek M (2001) Mechanika płynów w inżynierii i ochronie środowiska (Fluid mechanics in engineering and environment protection). Wydawnictwo Naukowe PWN, Warsaw (in Polish)

Experimental Analysis of Using Cavitation to Treat Ballast Water

Apoloniusz Kodura

Abstract Water is the most commonly used substance to obtain proper weight of vessels. The ballast tanks can be easily filled with a cheap material—water. However, ballast water may contain additional unwanted species—a variety of sea microorganisms. They can be dangerous for the environment of destination harbor area and can have strong influence on human activity. A problem of ballast water treatment is becoming more and more important nowadays. This is the main reason of investigation on methods of treatment. In the presented paper, an experimental analysis of using cavitation to treat ballast water is described. During experiments a physical model was designed and made. Laboratory tests were made to confirm a concept of obtaining the cavitation phenomenon. After that, the model was moved to a lake shore to make field tests in real conditions. The field tests are promising—the efficiency of elimination of microorganisms from water equals 80 %.

1 Ballast Water

Shipping is the most popular mean of cargo transportation. One of the main problems of marine transport is to obtain similar nautical conditions of each vessel during its voyage. This target is difficult to achieve due to different weights and volumes of cargo. As an answer, a ballast material is used to obtain optimally constant conditions. The ballast material has to be cheap, easy to load and unload and the weight of ballast has to be properly designed. Additional boundary conditions refer to small volume of ballast and easy method of changing the center of weight of a ship. The best material is water. In addition, water can be taken directly from harbor.

A. Kodura (✉)
Warsaw University of Technology, Environmental Engineering Faculty, Warsaw, Poland
e-mail: apoloniusz.kodura@is.pw.edu.pl

P. Rowiński (ed.), *Experimental and Computational Solutions of Hydraulic Problems*, GeoPlanet: Earth and Planetary Sciences, DOI: 10.1007/978-3-642-30209-1_11,

During the process of ballast water loading, numerous marine species are contained in the ballast water, and are discharged with ballast water at the destination. Most of them do not survive the voyage, but some do and can become invasive in area of destination port (Chase et al. 2008). Most of these organisms include bacteria and other microbes, like small invertebrates and the eggs, cysts and larvae of various species. They can be dangerous for the ecosystem of the destination. Some of them have their life cycles during transfer in ballast water (Disinfecting ballast water http://www.lenntech.com/ballast_water_eng.htm). Some marine organisms, that are unlikely to be taken on in water ballast as adults, can be transported in ballast during their infancy.

Modern, larger and faster vessels reduce natural barriers for marine species. In a very short time, water from tropical zones can be transported to northern ocean areas and vice versa. As a result, whole ecosystems can be influenced and—which is very negative—these changes can happen very quickly. It means that the local ecosystem has no time to prepare for invasive organisms from different parts of the world. Quick changes can lead to disturbance of the ecosystem balance. Their final effects are more complex and have impact on human health and economy (Liang et al. 2007; Matheickal et al. 2004). Each year, vessels transport over 80 % of the world commodities, and carry up to 10 billion tones of ballast water. Invasive aquatic species carried in the ballast water, which can cause extremely severe environmental, economic and public health impacts, are one of the four greatest threats for the world's oceans. Another three are land-based sources of marine pollution, overexploitation of living marine resources and physical alteration/ destruction of marine habitats.

In the whole world, additional examples can confirm the range of this problem. As the answer, many countries introduce law solutions to protect their water before unwanted species (International Maritime Organization 1997, 2002, 2003). According to IMO regulations, each vessel constructed after 2012 has to be equipped in ballast water treatment system. Next step will be to apply these systems for existing vessels.

With reference to the regulations, new solutions were introduced. Each has advantages and disadvantages. The most commonly used are (Perakis and Yang 2002; Waite 2001; Brown 2007):

- Changing ballast water during voyage—the idea is very easy, however, it costs a lot of money and time;
- Heating ballast water—this method needs a special system of heating ballast water and has significant consumption of energy;
- Screens and filter—sets of screens and filters are used to treat ballast water; however, this method is limited by size of a screen and has significant consumption of time during loading ballast water;
- Electromagnetic radiation—a method needs special equipment;
- Chemicals—adding chemicals aids is complicated during voyage—dosing of chemicals has to be directly calculated; in addition, a proper place for storage of chemicals has to be prepared.

Otherwise there are additional solutions—one of them is to use ballast-free vessels (Pearsons and Kotinis 2007)—there is a new project of using flow reservoirs of ballast water, so far that solution is under investigation.

2 Cavitation

Cavitation is a definition of a process which occurs when a liquid flows into a region where its pressure is reduced to vapor pressure. Liquid boils and vapor pockets develop on it. The vapor bubbles are carried along with the liquid until a region of higher pressure is reached, where they suddenly collapse (Streeter and Wylie 1979). If the vapor bubbles are near to a solid boundary when they collapse, the forces exerted by liquid rushing into the cavities create very high localized pressures that can cause pitting of the solid surface (Tropea et al. 2007). The phenomenon is accompanied by noise and vibrations.

The cavitation phenomenon is mostly an unwanted process in pressure systems, due to very high amplitude of pressure. Cavitation bubbles may form on nuclei, grow, and then move into area of higher pressure and collapse, all in a few thousandths of a second (Tropea et al. 2007). Formation and collapse of great numbers of bubbles on a surface subject that surface to intense local stressing, which appears to damage the surface by fatigue (Streeter and Wylie 1979). This phenomenon, which is normally unwanted in pressure systems and turbomachinery, can be useful for ballast water treatment. The background of the concept was to eliminate microorganism in ballast water by applying processes that would be destructive for solid materials. To find a proof of this hypothesis, physical tests were made.

So far, cavitation is used in two methods of ballast water treatment. Both of them are under investigation. In both cases, cavitation is used as a way to obtain better results. The first method assumes decrease of oxygen contents in ballast water (Tamburri and Ruiz 2005). This target is achieved by using Venturi nozzle. Ballast water flows through nozzle and due to very low pressure, a neutral gas, with lower volume of oxygen is delivered. Cavitation process in Venturi nozzle leads to obtain better mixing of water and gas.

In the second method, cavitation set is used to obtain homogenous mixture of water and chemical components that are destructive for microorganisms (http://www.oceansaver.com).

3 Laboratory Tests

To find an answer of a possibility of using cavitation to eliminate microorganism in ballast water, physical tests were performed. The main target was to make field tests and check an influence of collapsing vapor pressure on biological life. This aim can be achieved only by designing cavitation set which can be tested in real conditions.

Before the field tests, a unique cavitation set (Fig. 1) has to be analyzed in laboratory. There are many ways to obtain vapor pressure in pipelines. The main idea of using cavitation to treat ballast water was to obtain cavitation in some volume of water and then return to higher pressure. Due to destructive influence of this phenomenon on pipelines, the area of cavitation should be limited. If the tests are positive, the analyzed cavitation set can be used as a prototype of a treatment system. Boundary conditions led to design a large cavitation set, which can be used in a real system.

The cavitation set consists of two cones and a cylinder. It was made from aluminum and connected in an invisible acrylic pipeline. By using this part, cavitation bubbles can be observed and stored by taking pictures. The cavitation set was connected to a pipeline of 150 mm diameter. This element was located horizontally. Water was delivered to the model by a pump type Goulds i9136. The characteristics of the pump and the pipeline (Fig. 2) were calculated to find a theoretical point of cooperation of the pump and the pipeline.

The model was equipped with a flow sensor (Signet 515 Rotor X) to collect data of discharge. The sensor was located directly upstream of the cavitation set. Pressure was measured by four pressure transducers (PX219-100G5 V). The first transducer was located at the beginning of the cavitation set, the second and the third were connected directly in a cross-section of a cavitation cylinder. The fourth transducer was located at the end of the cavitation set. The whole measurement equipment was connected to a computer. Whole data were read and stored during experiments. The example of the stored data is presented in Fig. 3. During experiments, due to very high local velocity (violet line in Fig. 3) in crack between the cavitation cylinder and a pipe wall, pressure (blue and green lines in Fig. 3) decreased significantly. Minimum values of pressure were measured downstream

Fig. 1 Cavitation set

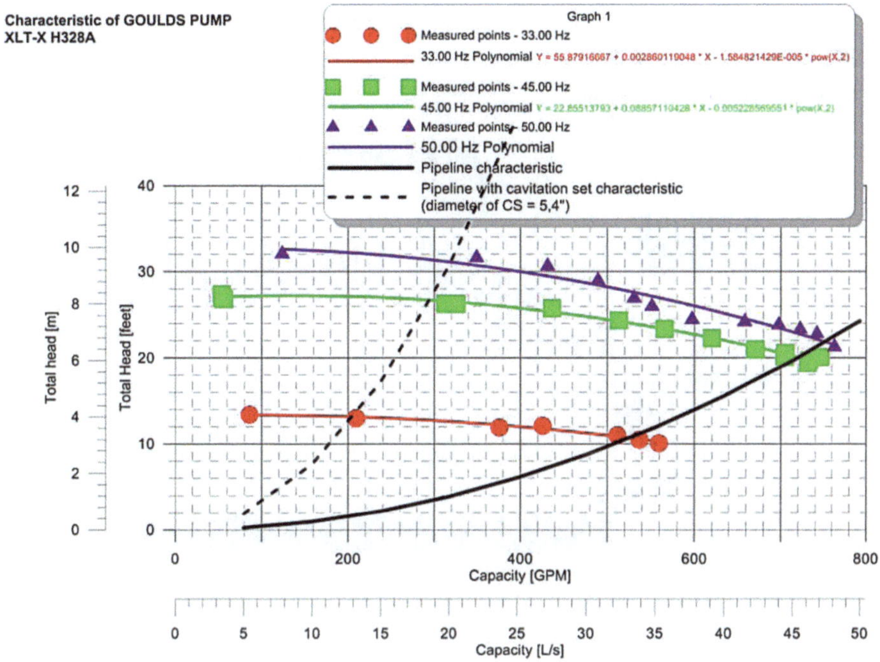

Fig. 2 Characteristic of pump and pipeline

of the cavitation cylinder. In this zone, the maximum amplitude and frequency of pressure changes was also observed.

The target of these measurements was to determine the value of flow, which is needed to obtain the cavitation phenomenon.

Numerous pictures were taken during experiments. At the downstream end of aluminum cylinder a zone of bubbles was observed (Fig. 4). This process was very intensive. Cavitation bubbles were moved downstream and observed at the end of the acrylic pipeline (Fig. 5) like a very homogeneous mixture of water and small white points.

4 Field Tests

The next step of the experiment was to move the model to a lake shore and to repeat tests by using water from the lake. All elements of model were transported to Muskegon channel which connects Lake Muskegon and Lake Michigan. One of harbors in the Great Lakes is located in Muskegon. The experimental model was set up on board of a small ship (Fig. 6). Water to the experiments was taken directly from the channel. It flowed through the pump, the pipeline system,

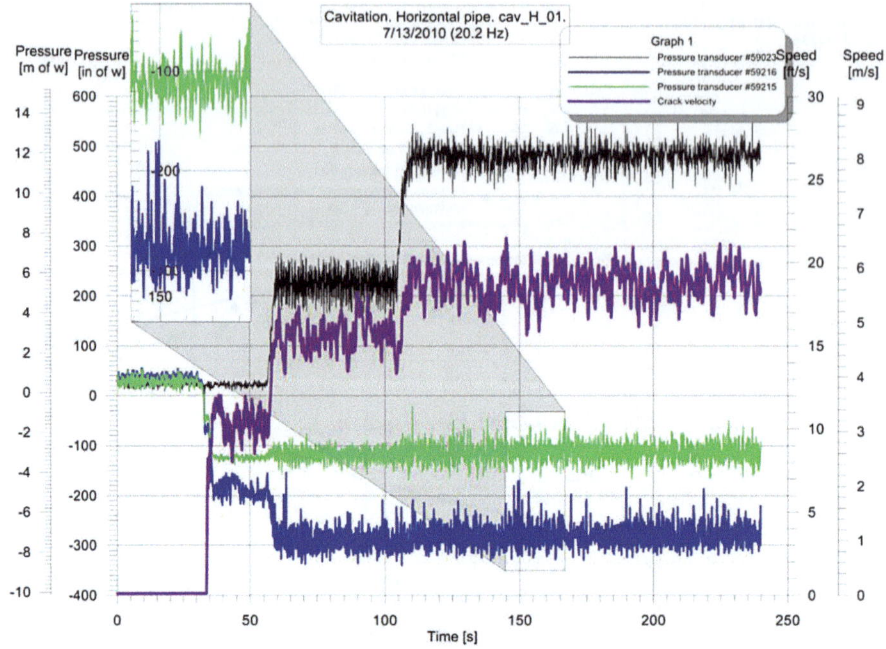

Fig. 3 Example of measured values during laboratory tests

Fig. 4 Cavitation process at the end of crack

cavitation set and outlet of model. It means that each portion of water flowed though the cavitation set only once.

This part of experiments consists of obtaining cavitation (Fig. 7) and collecting samples of water from the channel and from the model downstream of the cavitation set. These samples were analyzed by using biological tests. Results were

Fig. 5 Cavitation bubbles at the downstream end of acrylic pipeline

promising—approximately 80 % of microorganisms were killed by the cavitation process. The efficiency could be increased by using few steps of cavitation. However, the main target was achieved—water microorganisms can be eliminated by using cavitation.

Outside of taking analyzing samples, main physical and chemical parameters of water were measured by using YSI sensor (Fig. 8) Solid lines were used to mark parameters in water before cavitation, dashed—after this process. The sensor is used to measure: temperature, specific conductivity, pH, oxidation reduction potential (ORP) and oxygen saturation (ODO).

A small increase of temperature was observed. Initially, the temperature after cavitation was higher by 1, 2 °C than before the process. However, after some time the difference was much smaller. A method of taking measurements by the YSI

Fig. 6 Experimental model on the board of ship

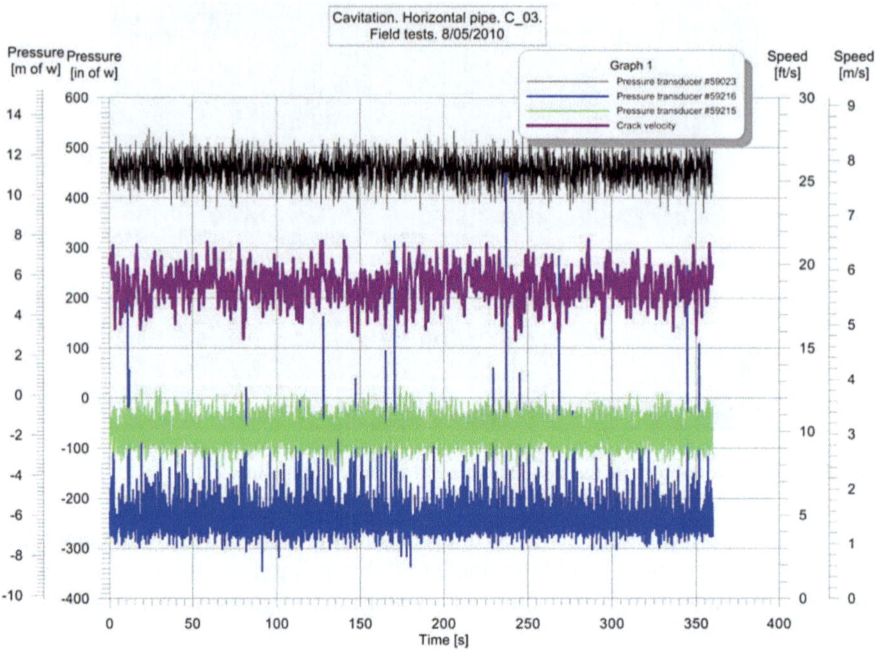

Fig. 7 Example of measured values during field tests

sensor can be one of reasons of temperature changes. The sensor has a diameter of approx. 15 cm and a height of approx. 40 cm. Before the cavitation, the sensor was located in the channel. After the process, a bucket of water was taken and the sensor was located in this vessel. Due to wind, the temperature of water can decrease during the collecting data.

No influence of the cavitation phenomenon on specific conductivity was observed.

A small difference of pH was obtained. In the first phase there is an increase of pH; however, there is a time of sensor stabilization, after which the value of pH is similar like before the cavitation phenomenon.

Unexpected differences of ORP and ODO were collected. The oxidation reduction potential is a measure of tendency of chemical species to acquire electrons and thereby be reduced. This parameter was significantly lower after cavitation. The initial difference is higher (approx. 50 mV); however, this can be a result of the time of sensor stabilization. The ending values of ORP were still lower (approx. 20 mV) than before the cavitation.

Oxygen saturation is a relative measure of the amount of oxygen dissolved in water. This parameter was significantly higher after the cavitation. The initial increase can be a result of the time of sensor stabilization. However, ODO values stabilized at the end and achieved approx. 10 % increase in comparison to the results before cavitation.

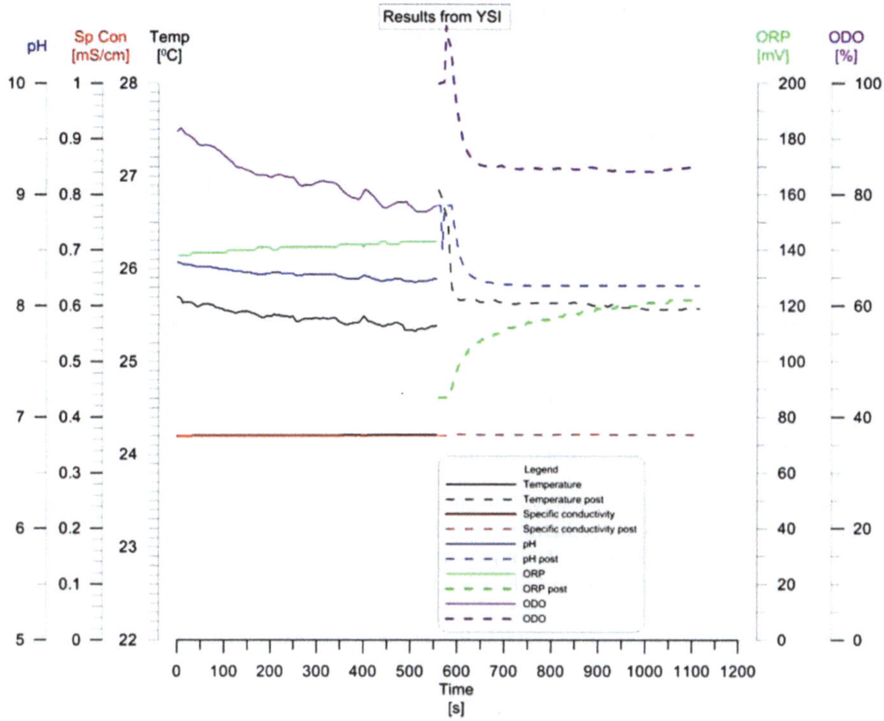

Fig. 8 The main parameters of water before and after cavitation

Differences of ORP and ODO parameters are symptoms of effective mixing of water during cavitation phenomenon, which can be useful during a process of designing a cavitation set to treat ballast water.

5 Conclusions

The results collected during the experimental analysis of using cavitation to treat ballast water, were interesting. The first was the designing of a unique cavitation set to obtain cavitation process. The process was observed by measuring pressure and taking pictures. After applying small improvements, the cavitation set was reported for a patent application.

After analyzing samples collected during the field tests, high efficiency of using cavitation to eliminate microorganism was confirmed. The result of 80 % destroyed species after one stage of cavitation is very promising. Next experiments and analysis are necessary to improve the method and to design a prototype of ballast water treatment system by using cavitation.

In addition, a very interesting phenomenon of changing values of ORP and ODO was observed, which is a proof of efficient mixing water during cavitation process and should be tested in next steps.

Acknowledgments Laboratory and field tests were made in cooperation with the Marine Hydrodynamics Laboratories of the Naval Architecture and Marine Engineering Department in the University of Michigan. Special thanks to Professor Guy Meadows, PhD Director of Marine Hydrodynamics Laboratories, and his staff.

References

Brown NAK, Reynolds G (2007) Lloyd's register releases guide to ballast water treatment technology. www.lr.org

Chase C, Reilly C, Pederson J (2008) Marine bioinvasions fact sheet: ballast water treatment options, sea grant. massbay.mit.edu/resources/pdf/ballast-treat.pdf

International Maritime Organization (1997) Guidelines for the control and management of ship's ballast water to minimize the harmful aquatic organisms and pathogens RESOLUTION A. 868 (20) adopted on 27 Nov 1997. http://www.imo.org/

International Maritime Organization (2002) Stopping the ballast water stowaways!, programme coordination unit global ballast water management programme. http://www.imo.org/

International Maritime Organization (2003) In: 2nd International ballast water treatment R&D symposium abstracts, London, http://www.imo.org/

Liang SY, Oon SL, Tan LH, Hor BH, Liang DT, Tay JH (2007) Ballast water treatment technologies and shipboard demonstrations, OCEANS 2006—Asia Pacific, 16–19 May 2007

Matheickal J, Raaymakers S, Tandon R (2004) Ballast water treatment. R&D directory 2nd edn. International Maritime Organization, London

Pearsons MG, Kotinis M (2007) Hydrodynamic optimization testing of ballast-free ship design grant report. Great Lakes Maritime Research Institute, London

Perakis AN, Yang Z (2002) Economic impact of measures to limit the introduction of non-indigenous species on st Lawrence Seaway Shipping. Final Report Michigan Sea Grant Program

Streeter VL, Wylie BE (1979) Fluid mechanics. McGraw-Hill, New York

Tamburri MN, Ruiz GM (2005) Evaluations of a ballast water treatment to stop invasive species and tank corrosion. In: SNAME maritime technology conference and expo

Tropea C, Yarin AL, Foss JF (eds) (2007) Springer handbook of experimental fluid mechanics. Springer-Verlag Berlin Heidelberg. ISBN: 978-3-540-25141-5, e-ISBN: 978-3-540-30299-5

Waite TD, Kazumi J (2001) Ballast water treatment standards: concepts and issues, University of Miami. In: Proceedings of the 2nd international conference on marine bioinvasions, New Orleans, pp 143–144

Design, Studies, Construction and Operation Problems During 40 Years of the Exploitation of Hydraulic Project Włocławek

Wojciech Majewski

Abstract Hydraulic Project Włocławek was commissioned in 1970 as the first project of the Lower Vistula Cascade (LVC). It was planned to construct eight similar hydraulic projects along the Lower Vistula. The aim of the LVC was to create the source of hydro-energy and to form the route for inland navigation connecting the center of Poland with the harbour Gdańsk. Włocławek project is now one of the most controversial hydraulic projects in Poland. During past 40 years of Włocławek project operation, ecological problems and sustainable development became very important. The chapter presents the basic hydrologic and hydraulic data about the Lower Vistula, description of the hydraulic project Włocławek, hydraulic model investigations completed in connection with the design of the project, run of the project construction and main problems, which appeared during 40 years of project exploitation. Proposals for the construction of the next hydraulic project downstream from Włocławek are also included.

1 Introduction

Hydraulic Project Włocławek was commissioned in 1970 on the Lower Vistula. Vistula is the largest Polish river flowing from the south of Poland in the mountains to the Baltic Sea in the north. Length of the Vistula is 1,047 km and its

W. Majewski (✉)
National Research Institute, Institute of Meteorology and Water Management,
61 Podleśna str, Warsaw 01-673, Poland
e-mail: wojciech.majewski@imgw.pl; wmaj@ibwpan.gda.pl

W. Majewski
Committee of Water Resources Management, Polish Academy of Sciences,
61 Podleśna str, Warsaw 01-673, Poland

P. Rowiński (ed.), *Experimental and Computational Solutions of Hydraulic Problems*, GeoPlanet: Earth and Planetary Sciences, DOI: 10.1007/978-3-642-30209-1_12,

catchment amounts to 194,000 km^2. The catchment in Poland constitutes of 169,000 km^2, thus occupying 54 % of Polish territory. Polish water resources in relation to the population or country area are small. Similarly, the volume of retention reservoirs and the hydro-energy potential are meager. At present the total volume of retention reservoirs amounts to about 6 % of the average annual outflow from the Polish territory. Hydraulic power plants in Poland produce about 1.5 % of the total electric energy, which is produced mainly in thermal power plants. The map of Poland with the Vistula River is shown in Fig. 1.

There are, however, several locations where new retention reservoirs can be built and new hydraulic power plants installed. The area which includes such locations is the catchment of the upper Vistula including mainly mountain tributaries. Similar situation exists along the section of the Lower Vistula, where estimated hydro-energy potential amounts to about 30 % of the total hydro potential of the country.

Ecologists state that Vistula is the only one wild, natural river (not trained) in Europe and should stay in the same state, keeping to the maximum degree its ecological values. On the other hand, water resources specialists consider the Vistula as the only one river of this size in Europe, which does not bring economical and social benefits to the country. In fact, Vistula is a highly deteriorated and polluted river, without any long-term plan for management.

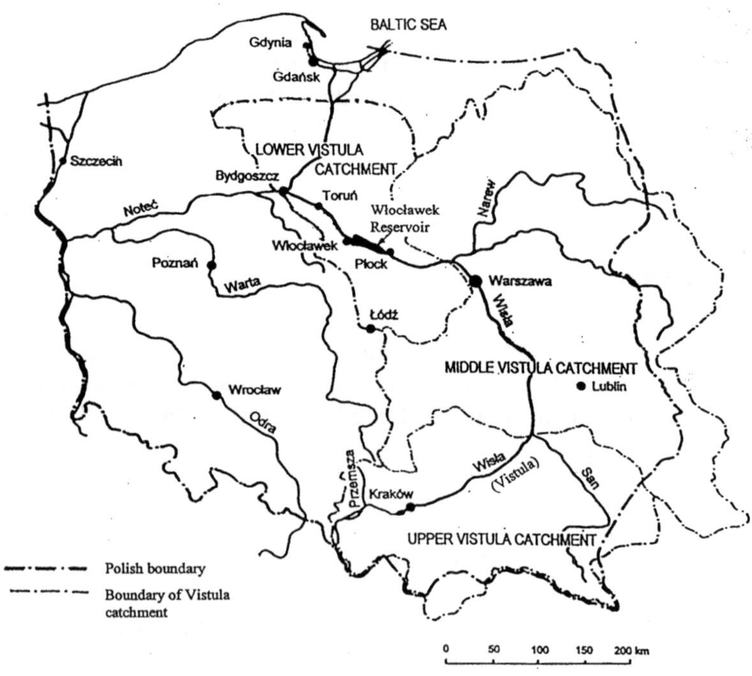

Fig. 1 Map of Poland with Vistula River catchment

In the seventeenth century, Vistula was the most navigable river in Europe. Despite very primitive means of river transport, more than quarter million tons of goods and resources were transferred annually between Poland and the harbor Gdańsk, which at this time was Poland's window to the world. After partitioning of Poland (eighteenth century) the Vistula lost its significance and gradually deteriorated, while other European rivers developed very dynamically with respect to inland navigation, production of hydro-energy and forming important source of water supply for people, industry and agriculture.

From hydrographic point of view, Vistula can be divided into three distinctly different sections and sub-catchments. These are: Upper Vistula, Middle Vistula, and the Lower Vistula (Fig. 1). Upper Vistula is from the source on Barania Mountain (1,116 m a.s.l.) up till the tributary San. Middle Vistula is from tributary San to the tributary Narew and the Lower Vistula is from tributary Narew to the outlet into Gdańsk Bay. These sections of the Vistula River represent three parts of the river, differing significantly from hydrological point of view.

After the Second World War, the project of establishing Lower Vistula Cascade (LVC) was put forward. It consisted of eight run-of-river hydraulic power plants of the total capacity of 1,300 MW and annual energy production of 4,200 GWh. The main aim of the LVC was energy production and formation of inland navigable route connecting the center of Poland with the harbor Gdańsk.

In 1970 the first hydraulic project Włocławek was put into operation with power 160 MW and electric energy production of 750 GWh in the average hydrological year. Till the present time, Włocławek operates as a single project bringing a lot of advantages, but also a lot of drawbacks, thus creating a lot of controversies. The most important problem of the operation of Włocławek project was the erosion downstream from the project and the deposition of sediment in the upper part of the reservoir. Other problems were created by ice phenomena and possibilities of ice jams. In 1982 a very important winter flood occurred on Włocławek Reservoir. Now enterprise ENERGA wants to develop next hydraulic project downstream from Włocławek project.

The chapter presents the description of the Lower Vistula, concept of Lower Vistula Cascade, description of the project, hydraulic investigations, which were carried out in connection with this project, run of the project construction, and encountered problems during 40 years of project operation. These include, in particular, the erosion downstream from the project, sedimentation in the upper part of the reservoir ice phenomena along this river section, ice jam flood in 1982, ecological aspects and water quality problems. The idea of next hydraulic project downstream from Włocławek is outlined.

2 Lower Vistula

Lower Vistula is the longest river section (391 km), but has the smallest catchment, without bigger tributaries which could influence the change of river discharge. The average multiannual discharge of the Vistula at the mouth is 1,080 m^3/s. The maximum recorded discharge at the mouth of the Vistula is 7,840 m^3/s, while the minimum discharge is 253 m^3/s. The average annual outflow of the Vistula to the sea is 34 km^3, while the minimum and maximum outflow are 20.5 and 50.8 km^3 respectively.

Along the Lower Vistula there are several important cities and industrial centers, which use the Vistula as the source of water. These are: Płock, Włocławek, Toruń, Bydgoszcz, Grudziądz, Tczew, Elbląg, and Gdańsk. Due to high discharge, Lower Vistula has a high hydro energy potential, which is estimated as 30 % of the total hydro energy potential of Poland.

Besides, this river section creates good navigation connection to European navigable network. The map of the Lower Vistula is shown in Fig. 2.

3 Lower Vistula Cascade

The idea of the LVC was proposed already in the 1960s of the past century (PROEKO 1993). The main purpose of the LVC was production of electric energy and formation of the navigable route of international class. The LVC is a very complex investment project which encompasses the economic, social and

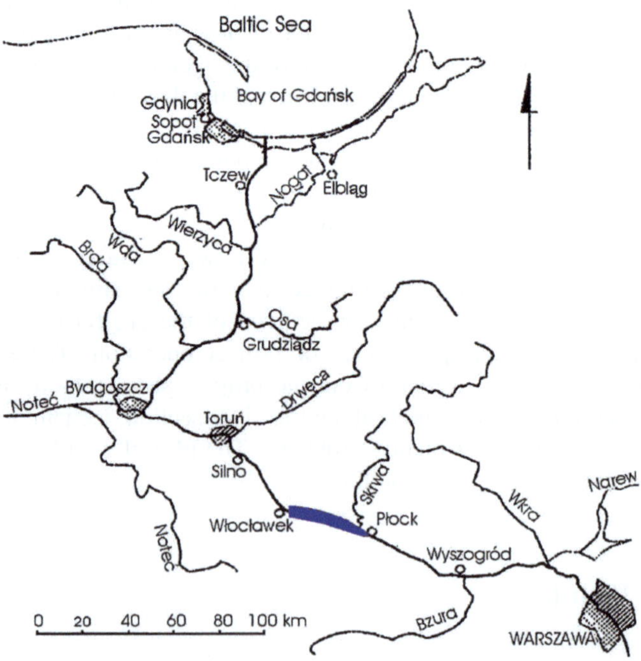

Fig. 2 Map of the Lower Vistula

environmental aspects of previous six and present three vojvodeships. It will affect 12 % of the Poland area and 14 % of Poland population. The LVC will include eight hydraulic impounding projects, including the present project Włocławek, which is in operation since 1970. The principal effect of the LVC is the production of environmentally sound and renewable electricity. Total annual production of electricity of LVC was estimated as 4,200 GWh with total power output of 1,300 MW, which can perform important regulation and peak-load functions in the energy supply system. The energy production is closely linked with the basic objectives of water management, i.e., water supply, stabilizing of free surface and ground water levels, partial flood control, development of navigation, sport and recreation facilities.

The project will stimulate construction of additional road network associated with new crossings over the Vistula River, and will support significant economic growth, creating jobs for thousands of people and reducing unemployment in the area.

The LVC should be closely linked to a comprehensive program of the water quality improvement throughout the Vistula catchment, thus reducing the pollution load carried to the Baltic Sea. The LVC will bring about many ecological changes; however, they will affect only the area between present flood dykes, which will be changed into side dams. The need to limit negative effects on nature to a minimum will require numerous interdisciplinary studies and close cooperation between experts from various fields. Similar problems have been solved in many places in Europe. The plan of the LVC is shown in Fig. 3 and longitudinal cross-section in Fig. 4.

4 Hydraulic Project Włocławek

In 1970 the first hydraulic project Włocławek of the LVC has been completed and put into operation. The project consists of earth dam, ten bay weir with spillway for discharging water and ice, hydraulic power plant consisting of six units (Vertical Kaplan turbines), navigation lock and fish pass. The layout of the project is presented in Fig. 5. Włocławek project forms the run-of-river reservoir of the initial volume 400 hm^3. In several places flood dykes were transformed to side dams. Over the crest of the project runs an important road connecting both banks of Vistula. The design of the project was done by Hydroproject and Energoproject. Construction and assembly work was completed by hydro-engineering enterprise Hydrobudowa.

The project was located in 647.75 km of the Vistula River. Normal water level is 57.30 m, while maximum and minimum water levels are 58.50 and 56.50 m. Tail water level in case of the construction of next hydraulic project was assumed to be 46.00 m. This level guaranteed good performance of the weir and hydraulic power plant as well as stability of the whole project. The average discharge in project cross-section is 890 m^3/s, the flow of 1 % probability was calculated as

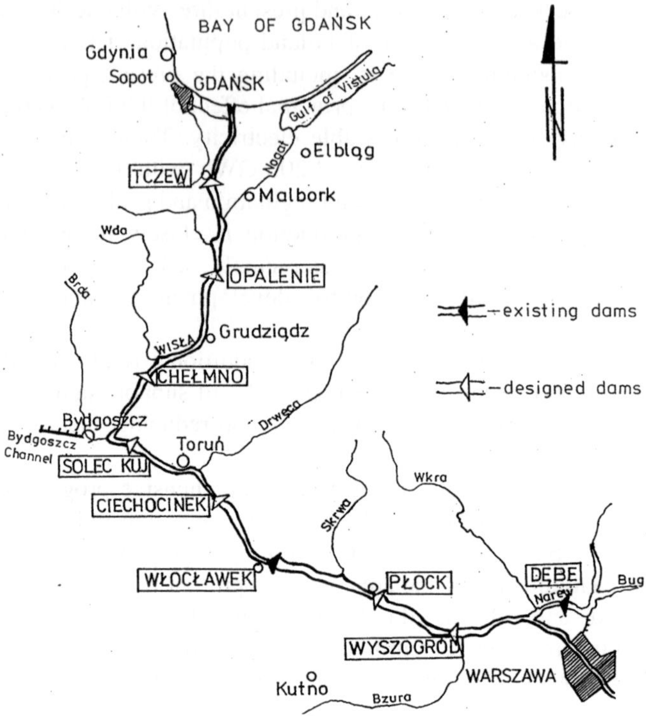

Fig. 3 Plan of the Lower Vistula Cascade

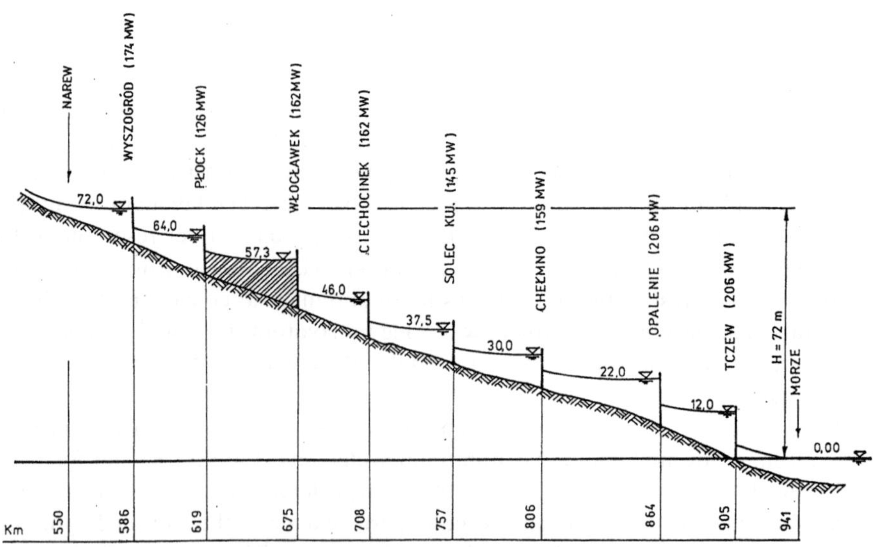

Fig. 4 Longitudinal section of the Lower Vistula Cascade

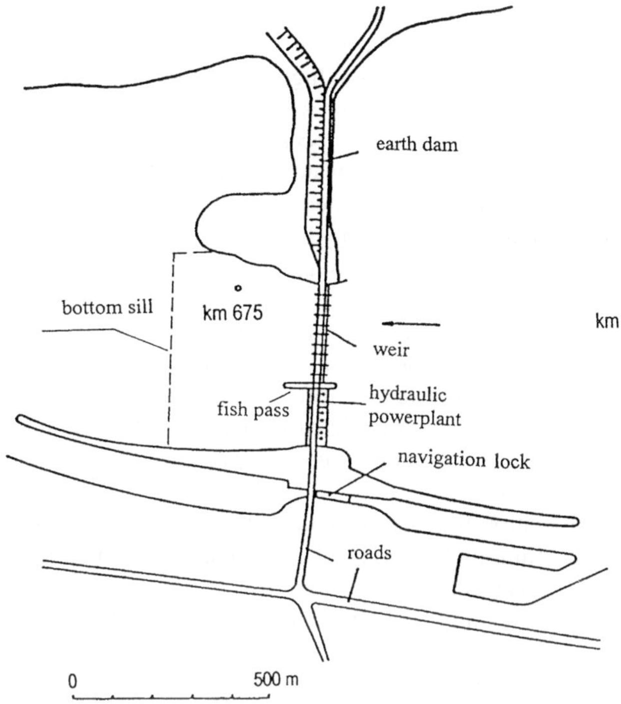

Fig. 5 The layout of the hydraulic project Włocławek

8,700 m^3/s, and the control flow of the probability 0.3 % was 10,280 m^3/s. The biological flow was initially assumed as 350 m^3/s and later increased to 450 m^3/s. The maximum discharge at maximum water level (58.50 m) was estimated as 11,150 m^3/s. According to corrected later conveyance calculations this discharge is 9,590 m^3/s, which is less than initially designed. Navigation lock has the dimensions 12 × 115 m and was designed for the annual conveyance of six million tones. The bottom of the lock chamber is at the level of 41.80 m. The upper gate is of segment type, while the lower gate is of wicket type.

Hydraulic power plant is equipped with six Kaplan turbines of the total power of 162 MW and energy production in the average hydrological year 750 GWh. Installed discharge of the power plant is 2,190 m^3/s. The turbines can operate in the range of the head from 5.2 to 12.7 m.

5 Hydraulic Model Investigations

Important hydraulic project required detailed hydraulic model investigations. There were several problems which should be investigated on separate models. These were.

- Spatial hydraulic model of the inflow to the weir from the reservoir. The model was constructed in the scale 1:100 without vertical scale distortion. The aim was to determine flow pattern especially at high discharges with all bays of the weir fully opened. Special guiding structures were proposed to guarantee uniform inflow to all bays of the weir. It has been checked whether, during high discharges, the maximum permissible water levels in the reservoir will not be exceeded. This requirement was fulfilled.
- Spatial hydraulic model concerning hydraulic conditions in river cross-section of the project narrowed by the cofferdam in which navigation lock, hydraulic power-plant and weir were constructed. Scale of the model 1:80.
- Sectional model of the weir including two bays with piers constructed in the scale 1:50. The aim of the model was determination of flow conditions for various configuration of the gate and different discharges of water and ice floes. The efficiency of the stilling basin was also investigated. Several corrections to the proposed shape of the structure were introduced. Discharge coefficients of the weir were determined.
- Model of the navigation lock in the scale 1:20. The aim of the model was the determination of filling and empting of lock chamber with simultaneous measurements of the forces acting on the barge.
- Model study of the closing of the remaining river channel after the completion of the weir, navigation lock and hydraulic power-plant. This study was very important because closing operation cannot be repeated shortly in case of failure. This operation can be performed in two ways: from a floating bridge or from two banks simultaneously.

6 Construction of the Project

Construction of the project was performed by Polish firms Hydrobudowa and Energobudowa. One of the most important and crucial stages was the closing of the remaining river channel. This operation was carried out exactly to the indications of hydraulic model study and performed simultaneously from the two banks. This operation was carried out during low water discharge and was successful. This way, the construction of hydraulic project was completed.

7 Operation of Hydraulic Project

Hydraulic project was commissioned in 1970 and since that time, for more than 40 years, has been operated not in conformity with the initial project. It has been one of the most controversial hydraulic projects in Poland. The fundamental drawback was the water level downstream from the project. This caused accelerated erosion, which resulted in decreasing tail water level and in consequence deteriorating the exploitation conditions of the whole project. The main problem was the deterioration of operation conditions of the power plant and performance of the stilling basin of the weir. In order to improve this situation, underwater bottom sill was constructed downstream from the power plant and the spillway section (Fig. 4). This, however, did not improve the stability of the dam and operation of the navigation lock. Now, especially during low discharges, the passage of the ships through the navigation lock is practically impossible.

Operation of Włocławek project caused also a change of ice regime on the reservoir but also downstream from the project. Important problem creates the river section upstream from the reservoir, which is close to natural conditions. Here large amounts of frazil ice form. This ice is transferred to the reservoir, where solid ice cover already formed. In consequence frazil ice forms hanging dams, which deteriorates flow conditions and increases water levels. In order to improve this situation, ice booms were installed every year in the upper part of the reservoir.

The Vistula River upstream from the reservoir was severely polluted and considerable amount of pollutants settled in the reservoir, thus forming dangerous deposits.

The operation of fish pass was not very successful, which deteriorated the movement of migrating fish.

Advantages of the hydraulic project Włocławek included:

- Production of electric energy on the average 750 GWh per year, which nearly covered full demand of electricity for the two cities Włocławek and Płock situated on the ends of reservoir. Production of this amount of electric energy in a thermal power plant would require supply and burning of nearly 900 tons of coal every day, with all consequences of air pollution.
- Formation of additional passage over Vistula for the city Włocławek.
- Possibilities for the abstraction of water for industrial, domestic and agricultural purposes.
- New possibilities for recreation, and water sports.

Hydraulic project and the reservoir were always very controversial and blamed, especially by the ecologists for all inconveniencies, which appeared during the operation of the project.

There were numerous conferences, meetings and expertise concerning future operation of hydraulic project. The first solution, which was indicated by the specialists of water resources management, was the construction of next hydraulic project downstream, which would increase tail water level and thus solve undesirable consequences of downstream erosion, which endangered the stability of all parts of the project, improve the operation of hydraulic power plant, improve the operation of stilling basin of the weir, and facilitate the use of navigation lock. This solution was highly supported by local society and local authorities.

Ecologists (WWF) on the other hand put forward the idea of dismantling the whole project and return the Vistula to the previous natural state. The first and the second solution was very costly and therefore government authorities were avoiding any decision putting forward arguments of ecologists that next project will bring only additional problems mentioned before in connection with Włocławek Project.

Włocławek project was not designed for flood protection, because of small volume of the reservoir for flood wave. However, first attempts were recently performed. These were based on emptying certain volume of the reservoir according to flood wave forecast. This allows certain mitigation of flood wave when it proceeds downstream. This was proved during flood on the Vistula in 2010.

8 Physical, Chemical and Ecological Processes Encountered During Operation of Hydraulic Project Włocławek and their Solutions

8.1 Erosion and Sedimentation

The process of erosion downstream from the hydraulic project and sedimentation in the upper part of the reservoir started immediately after the commissioning of the project. Sedimentation of the bed load is the natural phenomenon caused by the higher depth and thus smaller velocities. Predominant part of suspended sediment passed through the reservoir. Downstream erosion was caused predominantly by peak operation of the power plant, which resulted in the highly unsteady flow and thus accelerated erosion. High discharges were also responsible for accelerated erosion. This caused lowering of tail water level, which could not been prevented by the controlled water level from the next hydraulic project. It was designed in the location of Ciechocinek, about 30 km downstream. Unfortunately, financial crisis did not allow to start the construction of next hydraulic project. Downstream erosion was increasing in depth and extending downstream, which caused ecological effects.

Lowering tail water level caused problems of the operation of the power plant due to turbine cavitation and also not efficient operation of the stilling basin of the weir. Therefore, an underwater sill about 500 m downstream from the power plant, and the weir was constructed. This improved the operation of hydraulic power plant and the stilling basin of the weir. It was also decided that hydraulic power plant will not operate in the peak way.

8.2 Ice Phenomena

The Vistula River section which was transferred into the reservoir was the most ice-jam prone. Formation of the reservoir changed also the ice regime of this river section. Ice run and frazil ice run were much shortened and the solid ice cover period was significantly extended. The main problem for the reservoir was the formation of frazil ice upstream from the reservoir along the Vistula River, which is considered as natural. In fact, this situation was one of the reasons of the ice-jam flood on the upstream part of the reservoir.

The solution to ice problems on the Włocławek Reservoir is the correct operation of the weir to discharge floating ice floes downstream over the lowered weir gates. Next solution is the operation of the ice-breakers fleet, which has to break ice on the reservoir and form the ice-free channel for discharging ice floes downstream. The third solution, which was introduced after the flood in 1982, were the floating ice booms in the upper part of the reservoir. They prevented the inflow of frazil ice from the upstream part of the Vistula into the reservoir and also accelerated the formation of ice cover upstream from the reservoir, which prevented the formation of frazil ice. The ice jam flood which occurred in 1982 was caused by the coincidence of extreme hydro-meteorological conditions.

8.3 Water Quality

The Vistula River was in the past, and is still very polluted. These pollutions entered the reservoir and part of them settled in the form of toxic sediments. Very often, Włocławek Reservoir was blamed for the polluted water and sediments in the reservoir and not the river which transported these pollutants. Studies concerning water quality in the reservoir revealed that the quality of water leaving the reservoir is much better than that of water which entered the reservoir. It may be easier to remove toxic sediments from the bottom of the reservoir than to find them at the beaches near the mouth of Vistula. Arguments about polluted water and toxic sediments in the reservoir were used by ecologists as an argument against the existence of this reservoir and proposal of the next one. Such arguments are a complete nonsense.

8.4 Ecological Aspects

The Vistula River as a whole is known as a large ecological corridor which is used by various species. Quite a significant part of the river is now occupied by the program NATURA 2000. This especially concerns the whole section of the Lower Vistula. Only the existing Włocławek Reservoir is excluded from this program. Along Vistula River there are several National Parks and Parks of the Landscape. All ecologists regard Vistula River as the only wild river in Europe. On the other hand, water resources specialists consider Vistula as the only river which does not bring any economical advantages to the country. It is difficult to understand how numerous European rivers serve ecological purposes but also bring important benefits in the form of inland navigation, recreation, water supply, hydro-energy etc.

Forty years of the operation of Włocławek Reservoir revealed that biodiversity in the reservoir is much better than in the river upstream. There is abundance of new species of fish, birds and other living species. Water quality in the reservoir is at present satisfactory, and water quality downstream from reservoir improved significantly. Włocławek Reservoir formed a new large ecological system, which has a very important influence on ground waters surrounding the reservoir.

9 Next Hydraulic Project Downstream from Włocławek

After the completion of hydraulic project Włocławek,first preparations were carried out to develop next projects, downstream (Ciechocinek) and upstream (Wyszogród). As concerns the project in the localization Ciechocinek, technical design was completed and the preparation of grounds for the execution of the project was in progress. Unfortunately, very bad economic conditions of the state resulted in the suspension of the project.

In 2005 Design Office Hydro project Warsaw (Hydraulic project Nieszawa, concept design 2005) completed the concept of the new hydraulic project in the region Nieszawa-Ciechocinek. This concept was prepared on the order of the Regional Board for Water Management in Warsaw.

This project was designed as a weir with 16 bays, 20 m long. The weir had low stable sill and segment gates with ice flaps. The normal water level in the reservoir was assumed at the elevation of 46 m. Hydraulic power plant will consist of six units with bulb turbines of the total power of 46.4 MW, design discharge of 1,150 m^3/s and the heads in the range 4.80–7.15 m. Production of electric energy in the average hydrological year was estimated as 276 GWh. The project will include also the navigation lock of the length of 120 m and the width of loch chamber of 12 m. Two fish passages were also designed. Technical fish pass was located in the pier between the weir and hydraulic power plant. The ecological fish pass is on the left hand bank and is very close to natural streams or mountain torrents.

10 Conclusions

- Lower Vistula represents important hydro-energy potential, is suitable for the development of inland navigation and can also bring numerous ecological and social benefits.
- In 1970, hydraulic project Włocławek was commissioned as the first project of the LVC. It operates till today as a single project bringing a lot of benefits but also several drawbacks.
- Several times the construction of the new hydraulic project downstream from Włocławek was proposed, but unfortunately its realization always brings opposition from various ecological organizations.

- At present ENERGA SA is willing to build a next hydraulic project on the Lower Vistula River and only an acceptance from the authorities is necessary.

References

Lower Vistula Cascade (1993) PROEKO
Hydraulic project Nieszawa (2005) concept design, Hydroprojekt Warsaw

The Uncertainty of Measurements in River Hydraulics: Evaluation of Friction Velocity Based on an Unrepeatable Experiment

Magdalena M. Mrokowska, Paweł M. Rowiński and Monika B. Kalinowska

Abstract The chapter addresses the issue of evaluating uncertainty of measurements in river hydraulics. As replication of experiments in natural settings is not possible, the chapter highlights the case of unrepeatable experiments. Evaluation of friction velocity under flood conditions basing on the Saint-Venant set of equations serves as an example of indirect measurement of physical quantity.

1 Introduction

River hydraulics draws from the principles of fluid mechanics, and as such deals with physical quantities; hence it is absolutely critical that it should follow a physical experimental methodology. Good practice in experimental work requires complementing a measured value by a quantified level of uncertainty. Nonetheless, the uncertainty appears to be disregarded in river hydraulics, as an uncertainty analysis is ignored in many studies. Due to this shortcoming the validity of results as well as proposed theories could be challenged, if data which they are based on are of questionable quality. Various aspects of uncertainty analysis in hydraulics/hydrology studies are discussed from time to time in the literature see, e.g., (Romanowicz and Beven 2006; Baldassarre and Montanari 2009; Kalinowska and Rowiński 2012; Domeneghetti et al. 2012).

M. M. Mrokowska (✉) · P. M. Rowiński · M. B. Kalinowska
Institute of Geophysics PAS, ul. Ks. Janusza 64 01-452 Warsaw,
e-mail: m.mrokowska@igf.edu.pl

P. M. Rowiński
e-mail: p.rowinski@igf.edu.pl

M. B. Kalinowska
e-mail: monika.kalinowska@igf.edu.pl

P. Rowiński (ed.), *Experimental and Computational Solutions of Hydraulic Problems*,
GeoPlanet: Earth and Planetary Sciences, DOI: 10.1007/978-3-642-30209-1_13,

The aim of the chapter is to present the basis of uncertainty analysis and its application to practice. As there seems to be a gap in the evaluation of quality of measurement data in river hydraulics, the chapter addresses the issue of hydraulic measurements and experiments carried out in natural settings. Since measurements cannot be replicated due to the impossibility of keeping the same conditions in a river, the chapter presents a methodology of the uncertainty analysis in the case of unrepeatable experiment.

In the chapter, the evaluation of the friction velocity under flood conditions derived from the St Venant set of equations serves as an example of an indirect measurement of a physical quantity which is the subject of uncertainty.

2 Friction Velocity

The friction velocity is a crucial variable occurring in a wide range of hydraulic problems, such as the resistance to flow, pollutants transport, bed load and suspended load transport. The friction velocity expresses the bed shear stress in velocity units, and is defined as:

$$U_* = \sqrt{\frac{\tau_0}{\rho}}, \tag{1}$$

where τ_0—bed shear stress [N/m^2], ρ—water density [kg/m^3].

Many methods for evaluating the friction velocity have been invented so far; an in-depth review of these methods could be found in (Graf and Song 1995; Nezu et al. 1997; Afzalimehr and Anctil 2000; Yen 2002; Rowiński et al. 2005; Pokrajac et al. 2006; Sime et al. 2007; Kim et al. 2000).

In the chapter, the application of a method dedicated to flood waves is scrutinized, namely, the evaluation of the friction velocity basing on the St Venant set of equations Rowiński et al. (2000). The St Venant model is widely used in modeling of shallow water flow. Nonetheless, it could be treated as a set of equations, which enables to derive one of its variables. This approach has been recommended to evaluate the friction velocity in the following articles on the subject of resistance to flow: (Graf and Song 1995; Rowiński et al. 2000; Shen and Diplas 2010; Ghimere and Deng 2011).

In the chapter, the case of trapezoidal channel is presented. In this case the friction velocity formula is expressed as:

$$U_* = \left[gR\left(I + \frac{U}{g}\frac{b+mh}{bh+m\frac{h^2}{2}}\eta + \left(\frac{U^2}{g}\frac{b+mh}{bh+m\frac{h^2}{2}} - 1 \right)\vartheta - \frac{1}{g}\zeta \right) \right]^{\frac{1}{2}}, \tag{2}$$

where: $\zeta = \frac{\partial U}{\partial t}$—temporal derivative [$m/s^2$], $\eta = \frac{\partial h}{\partial t}$—temporal derivative [$m/s$], $\vartheta = \frac{\partial h}{\partial x}$—spatial derivative [-], I—bed slope [-], U—mean cross-sectional velocity

$[m/s]$, R—hydraulic radius [m], b—river bed width [m], h—the maximum depth of channel section (trapezoidal height) [m], $m = m_1 + m_2$, m_1 and m_2—side slopes [-], x—length [m].

3 Field Measurements

In the study, field data collected during experiment on the Olszanka River (Central Poland) were used. The experiment was carried out by the Institute of Meteorology and Water Management in the 1990s of 20th century. A set of data that is analyzed in the chapter originates from a dam-break like experiments. Description of this kind of experiments performed among others in Olszanka River could be found in Szkutnicki (1996), Rowiński and Czernuszenko (1998), Rowiński et al. (2000). In this experiment, the variation of a water depth and a mean velocity during flood wave propagation were measured. Temporal variability of mean velocity and water level in one cross-section of Olszanka River are shown in Fig. 1. The bed slope of a channel at the cross-section was 0.0012.

4 The Theory of Measurement Uncertainty

The uncertainty analysis is the subject of metrology—a measurement science (ISO 1999). As metrology aims at continuous refinement of measurements and presentation of results, the metrology concepts are evolving. One example of this evolution is the concept of error and uncertainty. Although an error analysis and an uncertainty analysis are similar in underlying principles, ISO (International Organization for Standardization) recommends using the uncertainty analysis instead of an error analysis, emphasizing the fact that evaluating all components of an error is not enough to assess the real value of quantity, as there is still some remaining uncertainty about the correct value (JCGM 2008). Accordingly, in the chapter a term uncertainty is used.

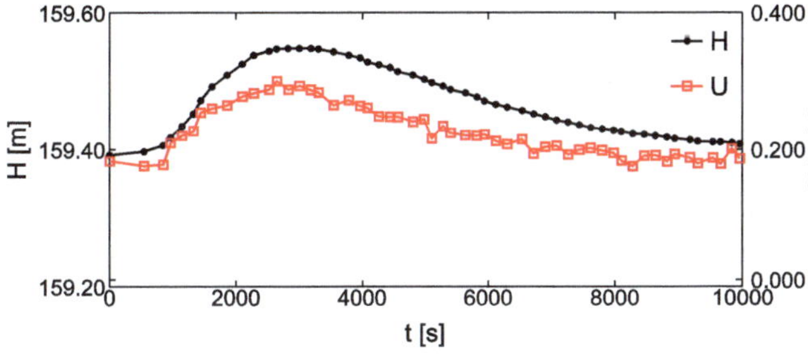

Fig. 1 Results of water level (in Kronstadt reference level) and mean velocity measurements

Methods of uncertainty analysis may seem to be confusing because of ambiguous and incoherent concepts (Zieba 2001); to exemplify, uncertainty is quantified statistically by standard deviation or simply by bounds of uncertainty interval, which is a deterministic approach. The question is how to estimate uncertainty effectively in the case of hydraulic measurements, which are frequently unrepeatable in nature.

4.1 The Uncertainty of Direct Measurements

Two approaches are presented in this section, namely statistical [which is put forward in a guide (ISO 1999)] and deterministic [which is widely used in measurement practice (Zieba 2001)]. In the chapter, a case of interrelation of these two approaches is presented.

4.1.1 Statistical Approach

This section presents some knowledge bases on evaluating uncertainty recommended by the ISO. In order to keep the theory transparent and consistent, the same terms and symbols are used in the chapter as in the ISO (1999).

First of all, let X_i be a physical quantity (measurand), and x_i an estimate of the measurand X_i. x_i is a result of a measurement, and it is only the approximation of a real physical value X_i, hence it is a subject of uncertainty (ISO 1999; JCGM 2008). The uncertainty is quantified either by variance u^2 or by standard deviation u, which is called standard uncertainty.

ISO (1999) recommends two means of evaluating uncertainty, namely the *Type A evaluation* and the *Type B evaluation*. Both of them are based on probability distributions; notwithstanding, while in the case of the Type A evaluation a series of repeated observations is used to compute the probability density of a measurand, the Type B evaluation is acceptable when there are no repeated measurements, and assumed probability density (*subjective probability*) based on scientific judgment (information available on variability of X_i) is applied. Since in the case presented in the chapter the Type B evaluation of uncertainty applies, further theoretical descriptions are focused only on this approach.

In the case of an unrepeatable experiment which is described in the chapter, it is assumed that each directly measured quantity X_i has a uniform (rectangular) distribution, as not enough information was available to choose another distribution. Figure 2a shows a uniform distribution within an interval $[x_i - \Delta x_i, x_i + \Delta x_i]$, where Δx_i is a maximum uncertainty; variance of the distribution equals

$$u^2(x_i) = \frac{\Delta x_i^2}{3}, \tag{3}$$

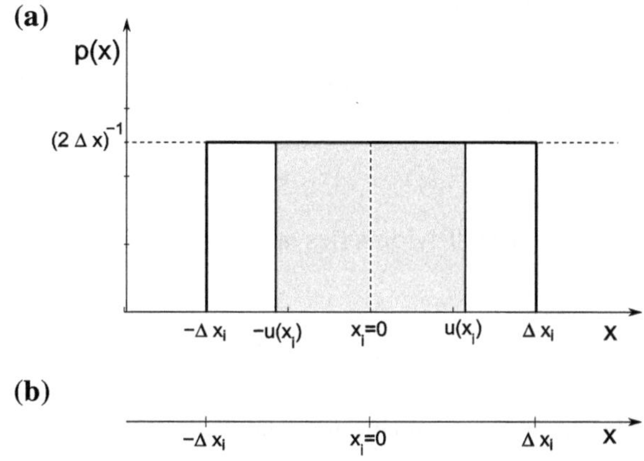

Fig. 2 Relation between standard uncertainty $u(x_i)$ of uniform distribution (**a**) and the maximum deterministic uncertainty (Δx_i) (**b**) for measured variable $x_i = 0$

and standard uncertainty is

$$u(x_i) = \frac{\Delta x_i}{\sqrt{3}}. \tag{4}$$

4.1.2 Deterministic Approach: Maximum Uncertainty Theory

In this approach a measured quality x_i is given with the maximum uncertainty denoted as Δx_i, which is shown in Fig. 2 (b). Evaluation of Δx_i is parallel to the evaluation of a uniform distribution bounds in the case of the Type B evaluation of uncertainty. In that, a standard uncertainty $u(x_i)$ and a deterministic maximum uncertainty Δx_i are interrelated as follows according to the formula (4), which is also shown in Fig. 2.

4.2 The Uncertainty of Indirect Measurements

In the case of an indirect measurement, the measured value is determined through a functional relationship (ISO 1999; JCGM 2008; Fornasini 2008): $y = f(x_1, x_2, \ldots, x_n)$, where y—the estimate of the measurand (dependent variable), x_i—input estimates. Generally, y and x_i are called an output quantity and input quantities, respectively. The output quantity y depends on input quantities which are directly measured or depend on some other quantities.

The uncertainty of indirectly measured quantity is expressed by *the law of propagation of uncertainty*. In the following subsections the law is presented for both a deterministic approach and a statistical approach.

4.2.1 Deterministic Approach

The change of an output quantity y caused by the change of input quantities x_i is evaluated as:

$$\Delta y = f(x_1 + \Delta x_1, x_2 + \Delta x_2, \ldots, x_n + \Delta x_n) - f(x_1, x_2, \ldots, x_n). \tag{5}$$

The first-order Taylor series approximation applied to formula (5) leads to the total differential:

$$\Delta y \simeq \frac{\partial f}{\partial x_1}\Delta x_1 + \frac{\partial f}{\partial x_2}\Delta x_2 + \cdots + \frac{\partial f}{\partial x_n}\Delta x_n. \tag{6}$$

Maximum deterministic uncertainty is calculated as the sum of absolute values of the component uncertainties basing on the total differential according to the following equation (Zieba 2001):

$$\Delta y_{max} \simeq \sum_{k=1}^{n}\left|\frac{\partial f}{\partial x_i}\right|\Delta x_i. \tag{7}$$

4.2.2 Statistical Approach

The standard uncertainty of dependent variable y evaluated on the basis of standard uncertainties of input estimates x_i is called the combined standard uncertainty and is denoted by $u_c(y)$.

Application of Taylor series expansion around $(x_1, \ldots, x_n)$ leads to the following estimation of the uncertainty (the law of propagation of uncertainty):

$$u_c(y) \simeq \sqrt{\sum_{k=1}^{n}\left(\frac{\partial f}{\partial x_i}\right)^2 u^2(x_i)}. \tag{8}$$

Evaluation of the maximum combined standard uncertainty which is always true (for both statistically dependent and statistically independent measurements) is as follows (Fornasini 2008):

$$u_c(y)_{max} \simeq \sum_{k=1}^{n}\left|\frac{\partial f}{\partial x_i}\right| u(x_i). \tag{9}$$

Considering a statistical and a deterministic approach, the maximum combined standard uncertainty and the maximum deterministic uncertainty are related as follows:

$$u_c(y)_{max} \simeq \sum_{k=1}^{n}\left|\frac{\partial f}{\partial x_i}\right| u(x_i) \simeq \sum_{k=1}^{n}\left|\frac{\partial f}{\partial x_i}\right|\frac{\Delta x_i}{\sqrt{3}} = \frac{1}{\sqrt{3}}\sum_{k=1}^{n}\left|\frac{\partial f}{\partial x_i}\right|\Delta x_i = \frac{1}{\sqrt{3}}\Delta y_{max}. \tag{10}$$

It should be stressed that the law of propagation of uncertainty presented here was derived under the assumption that the function is locally linear in a surrounding of $(x_1,\ldots,x_n)$. As long as the uncertainty of input variables is small enough, linear approximation results in good assessment of dependent variable uncertainty (Fornasini 2008).

In the case of more complicated nonlinear relations, uncertainty can be evaluated by incorporating terms of higher order in Taylor expansion series as is presented in ISO (1999), Mekid and Vaja (2008), Barwick et al. (2001).

4.2.3 Sensitivity Analysis

The sensitivity analysis is a valuable addition to the uncertainty analysis, as it provides the information which input variables x_i have the biggest impact on the uncertainty of an output variable y.

The effect of a small perturbation of input variable x_i on output variable y is assessed by a sensitivity coefficient (Gutenbaum 1992; ISO 1999), which is defined as a partial derivative:

$$WB_{x_i}^{y} = \frac{\partial y}{\partial x_i}. \tag{11}$$

The sensitivity coefficient could be written in a non-dimensional form as well:

$$WW_{x_i}^{y} = \frac{\partial y}{\partial x_i}\frac{x_i}{y}. \tag{12}$$

The application of sensitivity coefficient could be found for example in hydrological research (Gong et al. 2006; Hupert and Vanclooster 2001; Qiu et al. 1998).

5 The Application of the Uncertainty Analysis to the Friction Velocity Estimation

In this section, the basics of uncertainty analysis is applied to the evaluation of the friction velocity.

The friction velocity calculated from formula 2 is a dependent variable expressed by the functional relationship:

$$U_* = f(I, R, U, h, \zeta, \eta, \vartheta, b, m). \tag{13}$$

As could be seen, in relation (13) there are nine input variables, which friction velocity depends on. In the experiment on the Olszanka River the following quantities were measured directly: U, h, b, while the other variables were measured (determined) indirectly.

First of all, the sensitivity analysis has been performed to assess which input variables have the biggest effect on the friction velocity U_*. Figure 3 presents values of relative sensitivity coefficients of three input variables, namely: bed slope I, hydraulic radius R, and spatial derivative $\frac{\partial h}{\partial x}$. Results of sensitivity analysis for other input variables are not shown, since sensitivity coefficients of these variables are negligibly small. As could be seen in Fig. 3, a bed slope I largely determines the friction velocity U_*. It means that it is the uncertainty of a bed slope evaluation that affects the uncertainty of the friction velocity most. Hence, the bed slope is a variable which should be measured with a particular caution to minimize its uncertainty.

Next, the maximum deterministic uncertainty has been evaluated by applying the law of propagation of uncertainty (formula 7) to the friction velocity (formula 2):

$$\begin{aligned} \Delta U_{*max} \simeq & \left|\frac{\partial U_*}{\partial I}\right| \Delta I + \left|\frac{\partial U_*}{\partial R}\right| \Delta R + \left|\frac{\partial U_*}{\partial U}\right| \Delta U + \left|\frac{\partial U_*}{\partial h}\right| \Delta h + \\ & + \left|\frac{\partial U_*}{\partial \zeta}\right| \Delta \zeta + \left|\frac{\partial U_*}{\partial \eta}\right| \Delta \eta + \left|\frac{\partial U_*}{\partial \vartheta}\right| \Delta \vartheta + \left|\frac{\partial U_*}{\partial b}\right| \Delta b + \left|\frac{\partial U_*}{\partial m}\right| \Delta m. \end{aligned} \tag{14}$$

The maximum deterministic uncertainties of input variables have been evaluated basing on intuitive scientific judgment as follows:

- $\Delta h = 0.01$ m,
- $\Delta U = 10$ % U, according to recommendations of executors of the experiment (measurement was performed by propeller current meter),
- $\Delta R = 0.01$ m,
- partial derivatives: $\zeta = 0.0001$ $[m/s^2]$, $\eta = 0.0001$ $[m/s]$, $\vartheta - 0.00001$ [-],
- $\Delta I = 0.0001$.

Figure 4 presents the evaluated friction velocity (black line) with a maximum deterministic uncertainty (red bars). In addition, values of upper and lower limits of the uncertainty intervals were calculated directly from formula 2 using the relations:

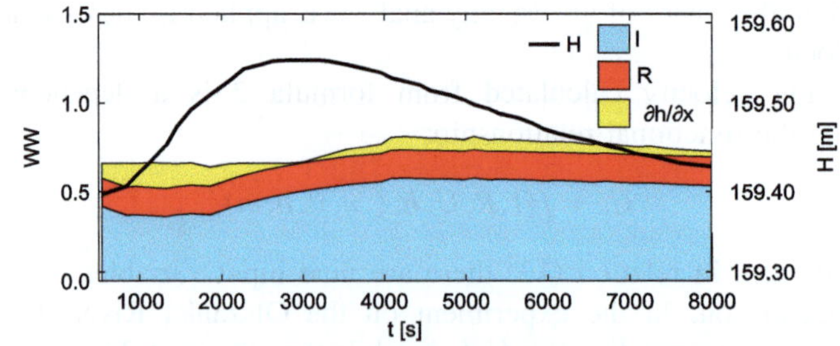

Fig. 3 The sensitivity analysis of friction velocity derived from the St Venant set of equations. Each area corresponds to the sensitivity coefficient of a particular variable

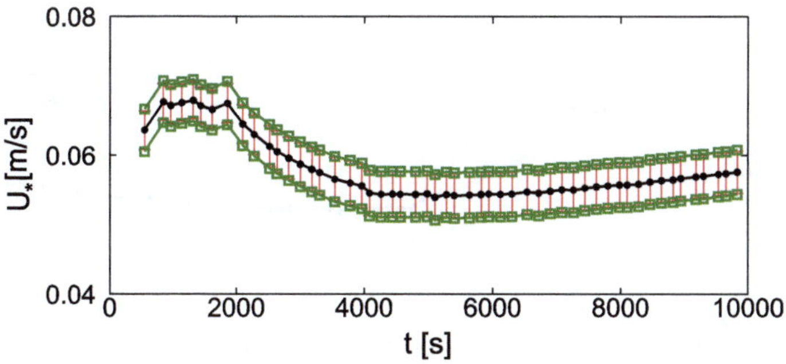

Fig. 4 Results of uncertainty analysis of evaluating the friction velocity. (*black line*—values of the friction velocity obtained from formula (2), *red bars*—the maximum deterministic uncertainty, *green squares*—the maximum uncertainty obtained directly from the formula (2)

$$U_* = f(I + \Delta I, R + \Delta R, U + \Delta U, h + \Delta h, \zeta + \Delta\zeta, \eta + \Delta\eta, \vartheta + \Delta\vartheta, b + \Delta b, m + \Delta m) \tag{15}$$

and

$$U_* = f(I - \Delta I, R - \Delta R, U - \Delta U, h - \Delta h, \zeta - \Delta\zeta, \eta - \Delta\eta, \vartheta - \Delta\vartheta, b - \Delta b, m - \Delta m) \tag{16}$$

for upper limit and lower limit, respectively.

It can be stated that the assumption about linear approximation for assessed level of uncertainty of input variables is correct, and leads to reliable results, as the bounds of the maximum uncertainty obtained directly from the friction velocity formula and from the law of propagation of uncertainty are alike.

In Table 1 a few example results of uncertainty analysis shown in Fig. 4 are presented in more detail. The maximum deterministic uncertainty is about 4–5 % values of the bed shear stress, and a standard uncertainty is about 2–3 %

In Fig. 5 there is shown the comparison of the friction velocity uncertainties for the range of bed slope *I* uncertainties. As could be seen, the maximum uncertainty of friction velocity varies considerably with changes of bed slope uncertainty.

Table 1 Selected results of uncertainty analysis

time [s]	U_*	ΔU_*	$\Delta U_*/U_*$	$u_c(U_*)$	$u_c(U_*)/U_*$
0	0.064	0.003	0.044	0.002	0.03
840	0.068	0.003	0.040	0.002	0.02
2,520	0.061	0.003	0.048	0.002	0.03
3,540	0.057	0.003	0.056	0.002	0.03
9,300	0.057	0.003	0.053	0.002	0.03

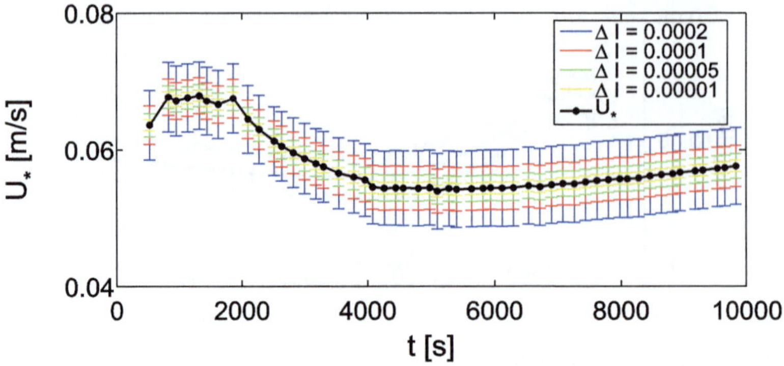

Fig. 5 Results of uncertainty analysis of evaluating the friction velocity for varying uncertainty of bed slope I

6 Discussion

To bring the chapter to a close, the main points are summarized here.

The results of this study refer to a single experiment performed in the Olszanka River in Poland, so the conclusions are valid to the particular type of the flood wave that is close to the one that was studied herein. However, presented method could be applied to any hydraulic measurements.

First of all, the uncertainty analysis is a crucial element of measurements and data processing, which cannot be disregarded in river hydraulics. The methodology of uncertainty analysis proposed by ISO is widely accepted in physics and related sciences (Fornasini 2008; Zieba 2001) therefore it should be also applied in hydraulic research.

Uncertainty analysis provides the information about the quality of data, which improves the reliability of research results and conclusions. For this reason, quantification of uncertainty should be presented to assure about the completeness of an analysis. With reference to an example presented in the chapter, the bed friction velocity is usually used as an input variable in further analyses, e.g. bed load transport, the rate of erosion, pollutants transport, hence the uncertainty of the friction velocity evaluation is crucial for the reliability of further calculations.

Secondly, as experiments carried out in rivers are usually unrepeatable, in that the same measurement conditions in a series of observations are impossible to keep, advanced statistical analysis is not applicable. For this reason, simple but efficient method of uncertainty analysis is recommended in the chapter based either on the maximum deterministic uncertainty concept or on standard uncertainty concept in the case of uniform distribution, which are, in fact, paralell. These two classes of uncertainty in the form presented in the chapter can be easily recalculated from each other depending on needs of further analyses.

Next, the presented study has been focused on evaluation of uncertainty of indirectly measured variable—the friction velocity. In the chapter, the application of the law of propagation of uncertainty which assumes linear approximation of U_* formula has been scrutinized. It has been shown that for presented case the evaluation of the maximum uncertainty from the law is reliable, and incorporation of higher order terms of Taylor series is not necessary to improve the accuracy.

Last but not least, the uncertainty analysis has been complemented by the sensitivity analysis, which helped to identify variables which have the biggest impact on the dependent variable. As has been shown, identification of input variables of significant sensitivity coefficients values helps to find the source of dependent variable uncertainty.

Whether to incorporate the uncertainty analysis into a research or not is a choice of a researcher, however the estimation of uncertainty is always an asset to a study, since it makes it more reliable.

Acknowledgments This study has been financed by National Science Centre. Grant No. DEC-2011/01/N/ST10/07395. The authors would like to express their appreciation to Jerzy Szkutnicki from the Institute of Meteorology and Water Management for his help in obtaining and interpreting field measurement data.

References

Afzalimehr H, Anctil F (2000) Accelerating shear velocity in gravel-bed channels. Hydrol Sci-J-des Sci Hydrol 45(1):113–124

Baldassarre GD, Montanari A (2009) Uncertainty in river discharge observations: a quantitative analysis. Hydrol Earth Syst Sci 13:913–921

Barwick VJ, Ellison SL, Lucking CL, Burn MJ (2001) Experimental studies of uncertainties associated with chromatographic techniques. J Chromatogr A 918:267–276

Domeneghetti A, Castellarin A, Brath A (2012) Assessing rating-curve uncertainty and its effects on hydraulic model calibration. Hydrol Earth Syst Sci 16:1191–1202

Fornasini P (2008) The uncertainty in physical measurements an introduction to data analysis in the physics laboratory. Springer.

Ghimere B, Deng ZQ (2011) Event flow hydrograph-based method for shear velocity estimation. J Hydraul Res 2:272–275

Gong L, Xu C, Chen D, Halldin S, Chen Y (2006) Sensitivity of the penman-monteith reference evapotranspiration to key climatic variables in the changjiang (yangtze river) basin. J Hydrol 329:620–629

Graf W, Song T (1995) Bed-shear stress in non-uniform and unsteady open-channel flows. J Hydraul Res 33(5):699–704

Gutenbaum J (1992) Mathematical modeling of systems. Omnitech Press, Warsaw

Hupert F, Vanclooster M (2001) Effect of the sampling frequency of methodological variables on the estimation of the reference evapotranspiration. J Hydrol 243:192–204

ISO (1999) Guide to expression of uncertainty in measurement, ISO, 1995. Polish translation, Central Office of Measures.

JCGM (2008) Evaluation of measurement data–guide to the expression of uncertainty in measurement. JCGM 100:2008 GUM 1995 with minor corrections. Joint Committee for Guides in Metrology.

Kalinowska MB, Rowiński PM (2012) Uncertainty in computations of the spread of warm water in a river lessons from environmental impact assessment. Hydrol Earth Syst Sci Discuss 9:5871–5904

Kim SC, Friedrichs T, Maa JY, Wright L (2000) Estimating bottom stress in tidal boundary layer from acoustic doppler velocimeter data. J Hydraul Eng 126(6):399–406

Mekid S, Vaja D (2008) Propagation of uncertainty: expressions of second and third order uncertainty with third and fourth moments. Measurement 41:600–609

Nezu I, Kadota A, Nakagawa H (1997) Turbulent structure in unsteady depth-varying open channel flows. J Hydraul Eng 123(9):752–763

Pokrajac D, Finnigan J, Manes C, McEwan I, Nikora V (2006) On the definition of the shear velocity in rough bed open channel flows. River Flow 2006. Taylor and Francis Group, London.

Qiu GY, Yano T, Momii K (1998) An improved methodology to measure evaporation form bare soil based on comparison of surface temperature with a dry soil surface. J Hydrol 210:93–105

Romanowicz R, Beven K (2006) Comments on generalized likelihood uncertainty estimation. reliab eng syst saf. Reliab Eng Syst Saf 91:1215–1321

Rowiński PM, Czernuszenko W (1998) Experimental study of river turbulence under unsteady conditions. Acta Geophys Pol XLVI 4:462–480

Rowiński PM, Czernuszenko W, Pretre JM (2000) Time-dependent shear velocities in channel routing. Hydrol Sci J 45(6):881–895

Rowiński PM, Aberle J, Mazurczyk A (2005) Shear velocity estimation in hydraulic research. Acta Geophys Pol 53(4):567–583

Shen Y, Diplas P (2010) Modeling unsteady flow characteristics for hydropeaking operations and their implications on fish habitat. J Hydraul Eng 136(12):1053–1066

Sime L, Ferguson RI, Church A (2007) Estimating shear stress from moving boat acoustic doppler velocity measurements in a large gravel bed river. Water Resour Res 43:W03418

Szkutnicki J (1996) Assessment of roughness of a small river-channel influenced by vegetation. Materiały badawcze. Seria: Hydrologia i Oceanologia 19. Institute of Meteorology and Water Management, Warszawa.

Yen B (2002) Open channel flow resistance. J Hydraul Eng 128(1):20–39

Zieba A (2001) The nature of the measurement uncertainty theory and its new codification. Adv Phys 52(5):238–247

Hydraulic Operation of a Diversion Structure: Physical Modeling

F. Rivera-Trejo, J. Barajas-Fernández and G. Cortés-Soto

Abstract The City of Villahermosa, capital of Tabasco, México, constantly suffers from the brunt of floods. In order to mitigate them, the regulation of levels and distribution of flows was envisaged in the vicinity of the city through a series of diversions in the main rivers of the state. Here are the basics of physical modeling and the results of experimental work to verify the operation of a major structure which is to be built to limit the flow in the river of Sierra that runs freely to the Villahermosa City, to reduce the risk of flooding. The results helped to test the hydraulic functioning of the structure and determined the expected flow that can be discharged under design conditions.

1 Introduction

The City of Villahermosa is located in southeastern México (Fig. 1) and is surrounded by a network of rivers (Fig. 2), the dynamics of which generate recurring flooding problems (Rivera-Trejo et al. 2010). One of the most important flows that go to the city is the one that carries the river from the mountains with an average flow of 450 m^3/s up to 1,800 m^3/s in its limit scale.

F. Rivera-Trejo (✉) · J. Barajas-Fernández
Juárez Autonomous University of Tabasco, Cunduacán, Tabasco, México
e-mail: jose.rivera@ujat.mx

J. Barajas-Fernández
e-mail: juan.barajas@ujat.mx

G. Cortés-Soto
Metropolitan Autonomous University, Azcapotzalco, México
e-mail: gsc@correo.azc.uam.mx

P. Rowiński (ed.), *Experimental and Computational Solutions of Hydraulic Problems*, GeoPlanet: Earth and Planetary Sciences, DOI: 10.1007/978-3-642-30209-1_14,

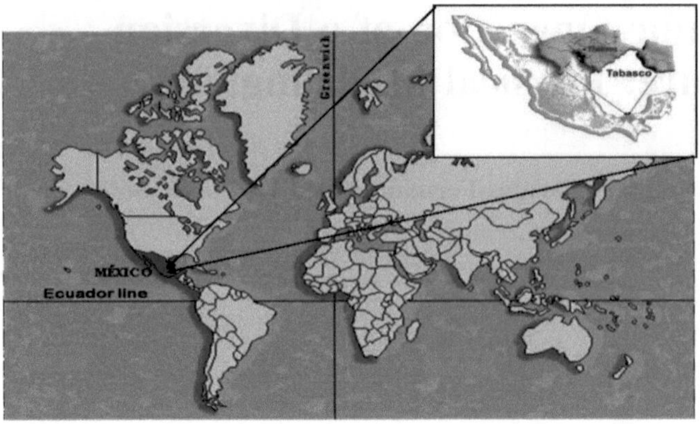

Fig. 1 Tabasco, México

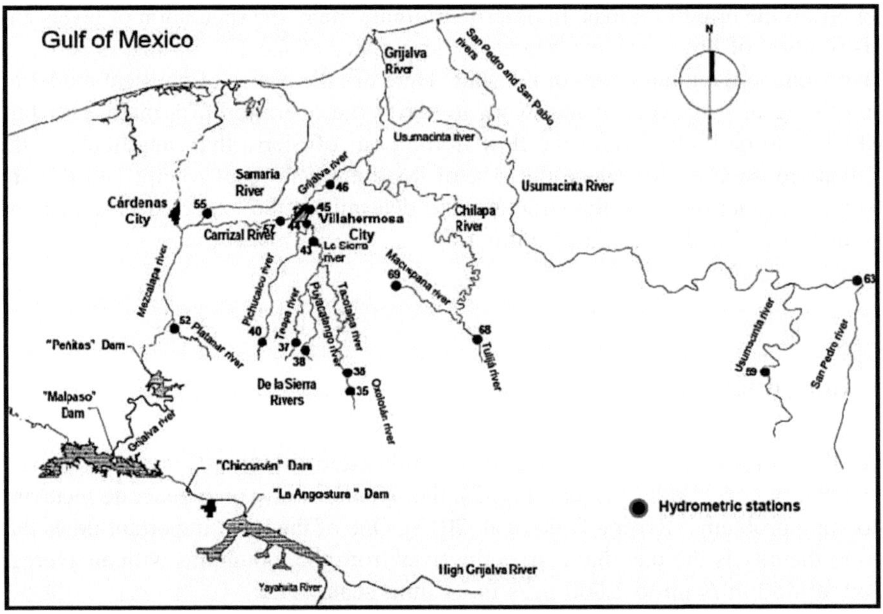

Fig. 2 Fluvial system around Villahermosa City, Tabasco, México

To control the runoff from the river, the National Water Commission of México (CONAGUA) designed a control structure to be located on a bend of the river, 8 km upstream of the City. This work consists of 200 m of lateral spillway, with 800 m of bypass channel to discharge into a lagoon area (Fig. 3). A series of seven groynes has been designed to protect the right bank of the river bed downstream from the work of diversion.

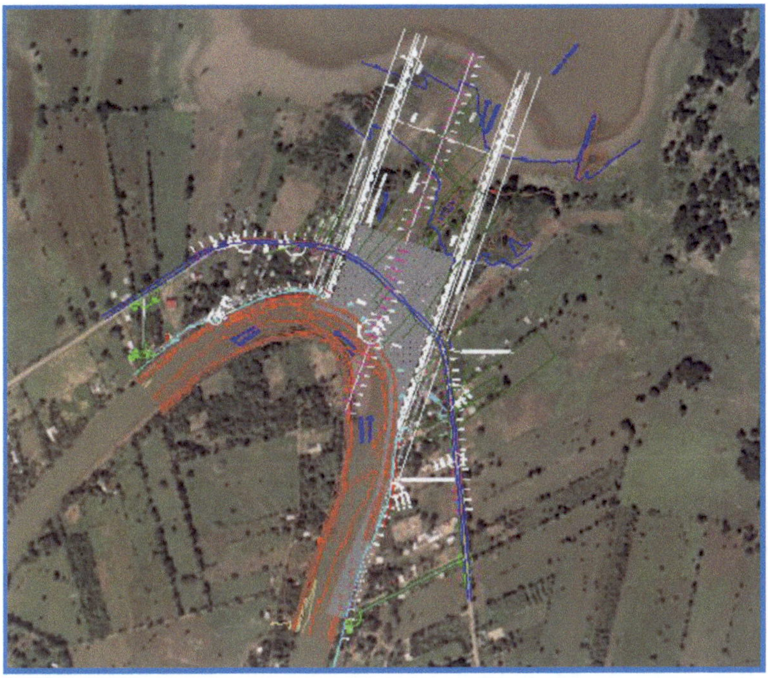

Fig. 3 Diversion channel

2 Physical Modeling

The modeling focused on finding: the flow that can be discharged by the proposed structure under maximum operating conditions; the level of the free surface of water when the spillway begins to work; and the effects of imposing the lagoon border (with or without load) on the functioning of the structure when it began to discharge.

The physical model was designed and built by professors and students from the engineering faculty of the Juárez Autonomous University of Tabasco (UJAT), on an area of 2,500 m^2 and at a geometric scale 1:40, which was the maximum dimension given the size of the laboratory (Fig. 4). It was decided that the model be carried out without distortion to avoid problems of scale (Echávez Aldape 1996), and also to include the equivalent of 500 m upstream and downstream of the structure.

Based on the selected scale and according to the roughness of the proposed construction material in the prototype, the roughness of the model was $n_m = 0.013$. The roughness of the material was imposed artificially (Julien 2002), using a mortar.

The design data were PCN (2009): flow in the river of 1,800 m^3/s; derivate flow of 550 m^3/s; load conditions in the lagoon of 6.00 m.a.s.l. These flows in the

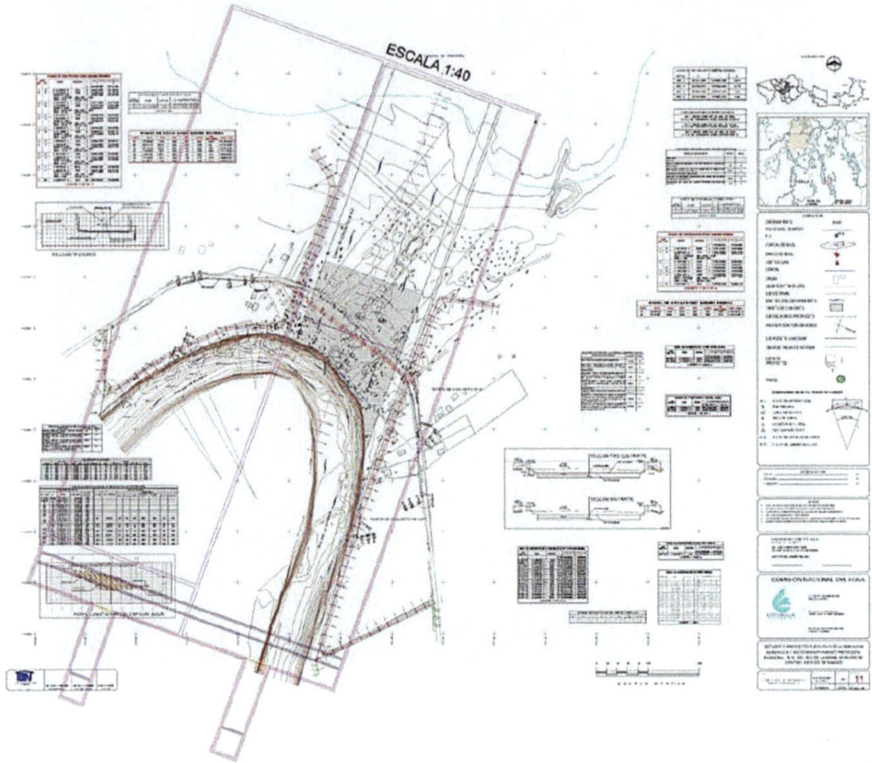

Fig. 4 Physical model proposed in hydraulics laboratory UJAT

model correspond to 0.178 and 0.054 m^3/s, respectively. The discharge flow in the model was provided by a system of five pumps: three primaries and two as backups.

3 Construction

Since the construction of the model was designed to study the hydraulics of the structure and evaluate the stability of the channel, it was decided that the model was of fixed bed. The construction was based on cross-sections of the river every 50 m, manufactured in wood. The spaces between sections were filled with sand and coated with polished mortar (Fig. 5). The details were refined and the channel section and the proposed structure were obtained according to the physical model (Bureau of Reclamation 1980).

Fig. 5 Overview of the physical model

Fig. 6 Wave probe monitor

4 Instrumentation

In the particular study case, the aim was to determine the derivate discharge flow (Q), which was subject to loading conditions (levels-h) in the area, in addition to the measurement of velocities in the diversion channel (v).

Level measurement. It had has a Wallingford digital monitor, equipped with 6 wave gauge probes (Fig. 6). Each probe consists of two stainless steel parallel bars with a plastic head and foot (Fig. 7). The head is attached to the monitor calibration that allows measurement of water level in any position with an accuracy of ± 1 mm.

Discharge measurement. Triangular weirs were used (Fig. 8), equipped with Mitutoyo digital point gauges (Fig. 9) with an accuracy of ± 0.01 mm.

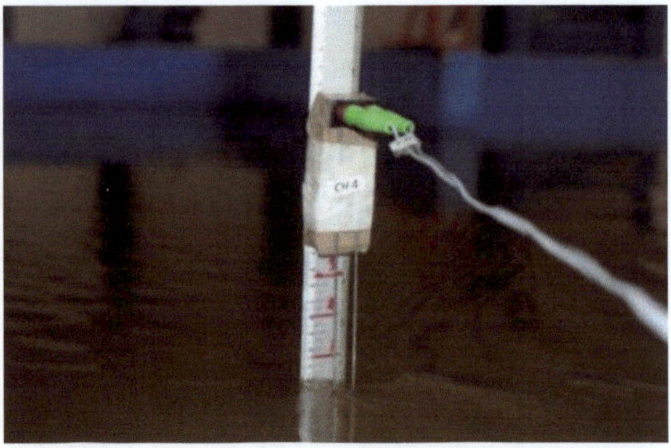

Fig. 7 Wave gauges

Fig. 8 Triangular weirs

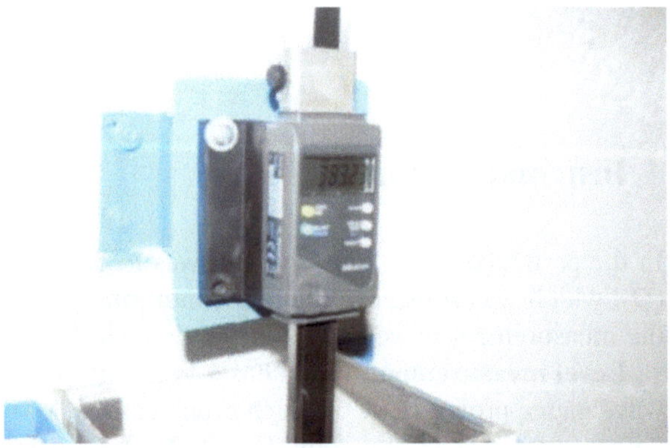

Fig. 9 Mitutoyo digital point gauges

Velocity measurements. Speeds that may occur in the diversion channel were important data, because they predict the strength that the coating has to withstand to avoid erosion, or, as is the case of prefabricated materials used, the problems with lifting. In the velocity measurement on the model, miniature propeller meters

Fig. 10 Digital indicator

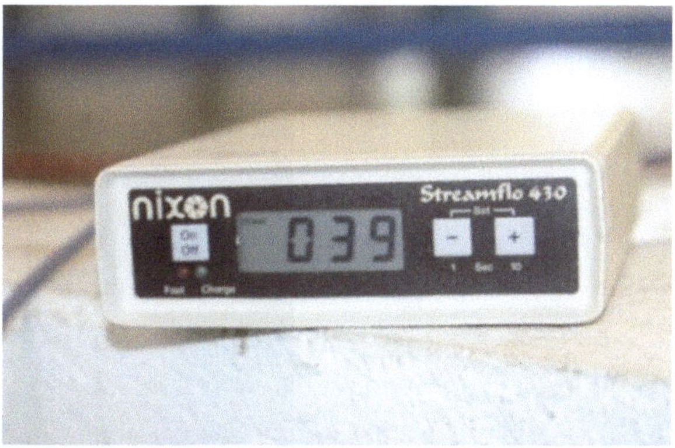

Fig. 11 Nixon miniature propeller meter

produced by Nixon were used, for a range of speeds of 5–150 cm/s, with an error of 2 % (Figs. 10 and 11).

Topographic profile measurement. Good representation of topography is essential to reproduce three-dimensional effects on a physical model. This study used a topographic bed profiler produced by Wallingford, with an accuracy of ±1 mm in the horizontal and vertical directions (Figs. 12 and 13).

5 Calibration

A critical step prior to the experiment is the calibration and tuning of the experimental model, to ensure geometric correspondence between the model and the prototype. This was achieved through a careful process of verification of topography with the bed profiler equipment.

As concerns the experimental equipment, all necessary measures were applied to ensure its proper operation and calibration. For example, in the case of weirs used for gauging, calibration was performed in strict accordance with the regulations

Fig. 12 2D bed profiling system

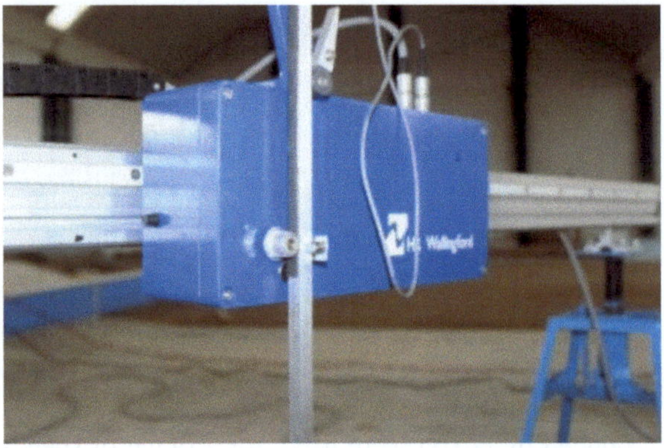

Fig. 13 The profiler carriage

(BS 3680: Part 4A 1981). In the case of digital level gauges, calibration relied on two exit gates to regulate the levels, one for the river channel and one for the diversion.

The experimental modeling needed the river level downstream as a boundary condition; these data are not available for the design condition, so it was decided to infer them from the last reported value in the design data, which was of 1,400 m^3/s (PCN 2009). The graph in Fig. 14 shows the comparison between experimental results and those generated numerically (PCN 2009). On the horizontal axis, the flow through the river (Q_T) is plotted, while the vertical axis shows the levels in the downstream section (H); the numerical results are in black and the experimental ones are blue.

This case shows that the levels found experimentally are higher than those calculated numerically, so the experimental values were selected as a reference to set the downstream boundary conditions.

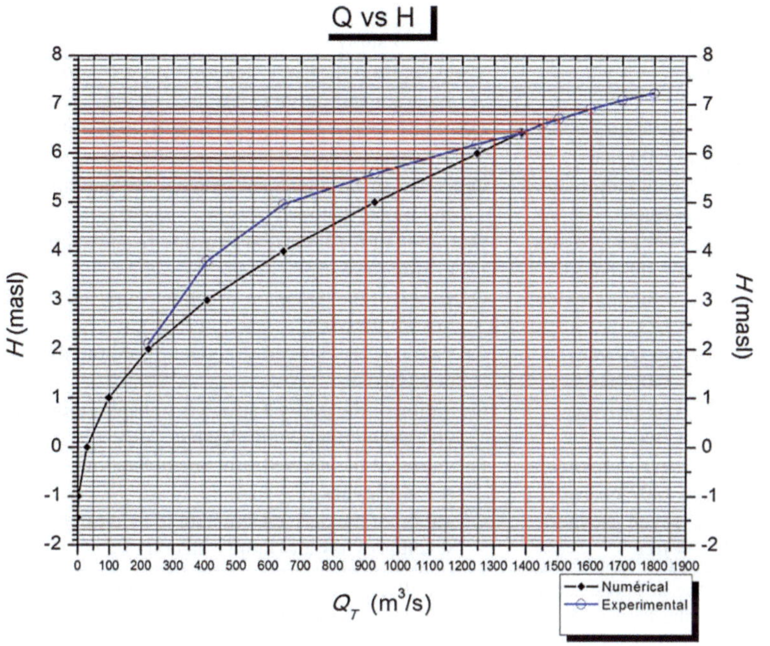

Fig. 14 River discharge curve

6 Tests

The revision was made of the hydraulic operation of the structure for the design condition (1,800 m^3/s). Additionally, the discharge diversion for different operating conditions at the time of floods was analyzed. Also, since the design of the structure provides for the protection of the right bank of the river downstream of the structure through seven groynes, we examined its effect on the discharge side of the spillway. In these terms, variants considered for the experimental performance of the work were: (1) without groynes, (2) with a one groyne, and (3) with the seven groynes. Table 1 shows the testing matrix.

7 Results

Figure 15 shows the discharge curve of the derivation structure, working without groynes; in Fig. 16 we show the same condition but working with one groyne; and in Fig. 17 the system works with seven groynes. On the horizontal axis the discharge in the model (Q_T) is plotted, while the right vertical axis represents the water level of the free surface *(H)* measured at the center of the curve of the river near the beginning of the spillway, while the left axis shows the diversion discharge (Q_{div})

Table 1 Testing matrix

Test	Conditions	Feature to study
I	Natural, Q_D = 1,800 m^3/s	Verify design discharge, bankfull discharge (Q_{bf}) and experimental curve Q–H
II	Q = Q_{bf}, without/groynes, H_{lag} = 6.00 m.a.s.l	Operation of the derived structure to bankfull discharge
III	Q = Q_{bf}, without/groynes, H_{lag} = free	Operation of the diversion structure to bankfull discharge and empty lagoon
IV	Q = Q_{bf}, without/groynes, H_{lag} = 6.00 m.a.s.l	Operation of the diversion structure from bankfull discharge to stop deriving
V	Q = Q_{bf}, without/groynes, H_{lag} = free	Operation of the diversion structure from bankfull discharge to stop deriving
VI	Q = Q_{bf}, one groyne, H_{lag} = 6.00 m.a.s.l	Operation of the derived structure to bankfull discharge with one groyne
VII	Q = Q_{bf}, one groyne, H_{lag} = free	Operation of the derived structure to bankfull discharge with one groyne and empty lagoon
VIII	Q = Q_{bf}, seven groynes, H_{lag} = 6.00 m.a.s.l	Operation of the derived structure to bankfull discharge with seven groynes
IX	Q = Q_{bf}, seven groynes, H_{lag} = free	Operation of the derived structure to bankfull discharge with seven groynes and empty lagoon

where Q_D is the design discharge; Q is the discharge; Q_{bf} is the bankfull discharge; H_{lag} is the water level in the lagoon; H is the water surface elevation

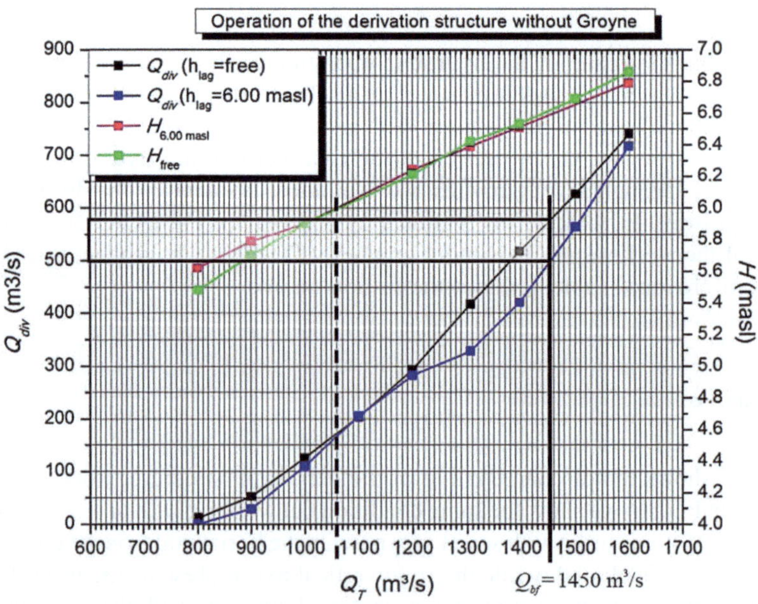

Fig. 15 Discharge curve of the derivation structure, working without groynes

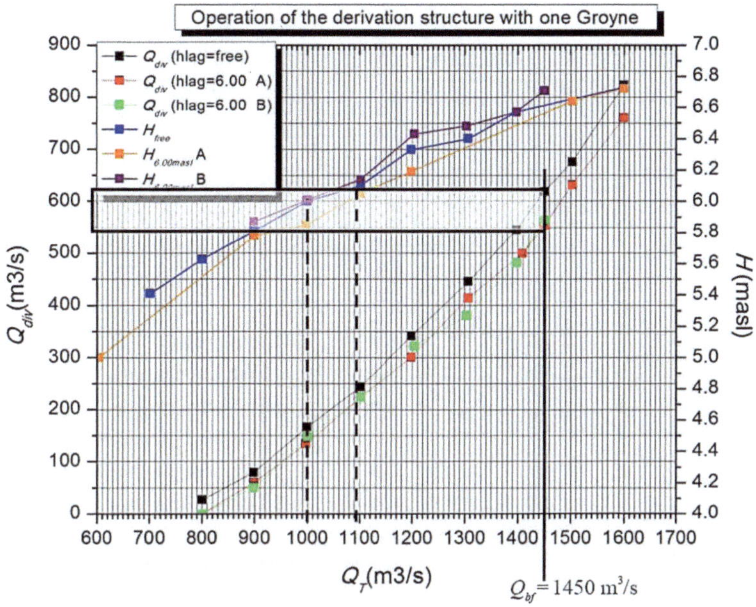

Fig. 16 Discharge curve of the derivation structure, working with one groyne

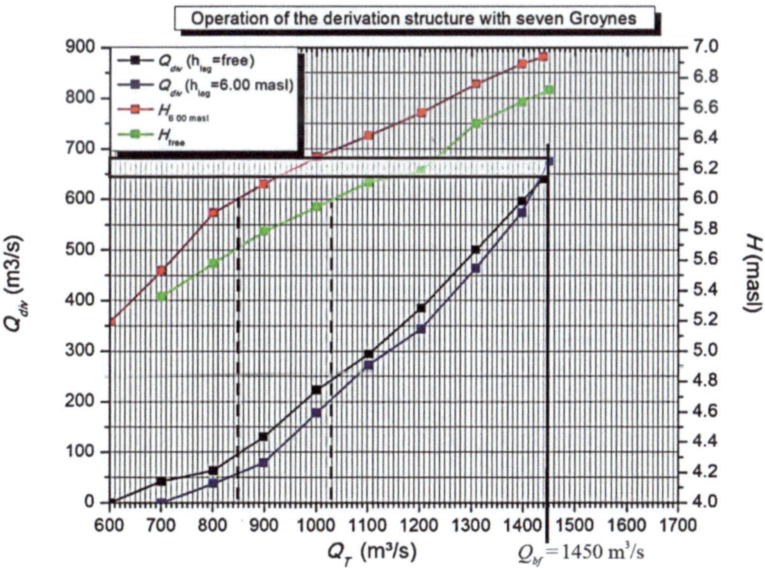

Fig. 17 Discharge curve of the derivation structure, working with seven groynes

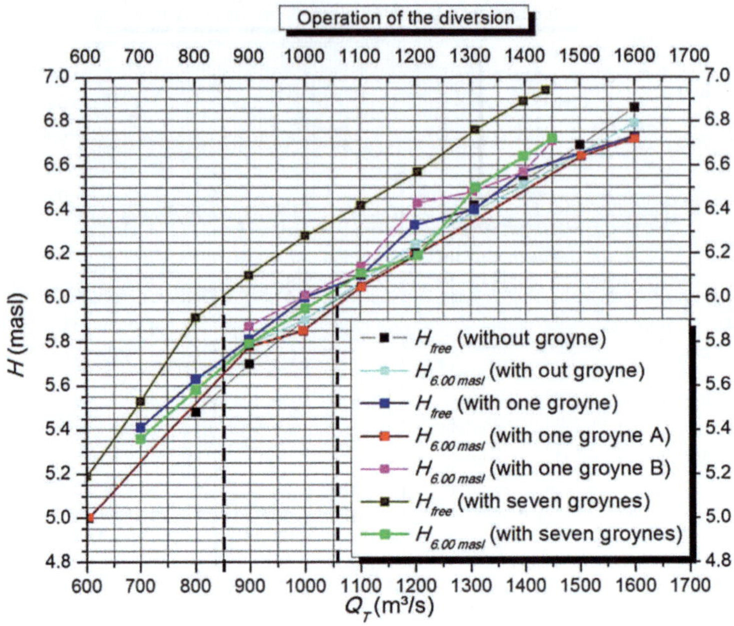

Fig. 18 Level water operation of the diversion channel

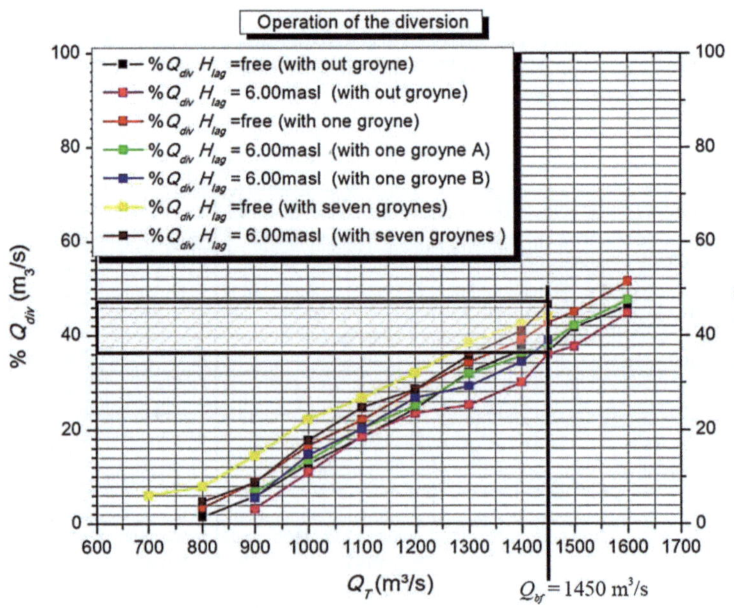

Fig. 19 Discharge operation of the diversion channel

under two output conditions: (1) the discharge when the lagoon is empty and the diversion channel runs free (H_{free}), and (2) the discharge when the lagoon is full and has a level corresponding to a height of 6.00 m above sea level (m.a.s.l.) ($H_{6m.a.s.l}$). All the experiments are grouped in Fig. 18, which shows the variation in the water level over the diversion channel. Figure 19 shows the derived discharge resulted in percentage with respect to the total discharge in the river.

In the figures, the continuous lines show the bankfull discharge (Q_{bf}), while the dashed lines show the water level over the weir.

It is observed that for the water levels of free surface at 6.00 m.a.s.l. the lagoon is full. The diversion discharge is shown in a range (rectangle) that depends on the operating conditions of the structure, i.e., the boundary conditions imposed.

8 Conclusions

The design of the diversion structure proposal is considered to control the discharge of 515 m^3/s by using a lateral weir, for a river design discharge of 1,800 m^3/s and a level of 6.00 m.a.s.l. in the downstream lagoon. However, the experimental study found that the maximum discharge to bankfull (Q_{bf}) that can run without overflowing over the river is 1,450 m^3/s, so it does not reach the 1,800 m^3/s; however, the derived discharges are around 50–30 % (Fig. 19) of the total discharge through the river. Moreover, in case of flow rates less than 1,000 m^3/s in the river and with the lagoon filled (6.00 m.a.s.l.), we have a possibility of water comeback from the lake to the river due the level difference, because at this condition the river level is lower than the lagoon level (Fig. 18).

In this chapter we showed the goodness of the application of physical models to test the operation of hydraulic works, and were found them to be a useful tool for solving engineering problems where numerical models are not sufficiently reliable.

References

BS 3680: Part 4A (1981) Liquid flow in open channels. British Standard methods of measurement, pp 1–21

Bureau of Reclamation (1980) Hydraulics Laboratories Techniques. U.S. Department of the Interior, Denver, Colorado, p 230

Echávez Aldape G (1996) Introduction to hydraulic models of bed fixed and the engineering experiment, UNAM, 1996, p 155 (in Spanish)

Julien P (2002) River Mechanics. Cambridge University Press, New York

PCN (2009) Study and executive project of the Sabanilla diversion channel and its corresponding marginal protection (RB) on the La Sierra river, Tabasco, PCN Constructions. S.A. DE C.V. Sept 2009, p 14 (in Spanish)

Rivera-Trejo F, Soto-Cortés G, Méndez-Antonio B (2010) The 2007 flood in Tabasco, Mexico: an integral analysis of a devastating phenomenon, Int J River Basin Manage, first published on 15 Sept 2010 (iFirst)

Experimental Investigation of Flood Waves from Open-Channel Levee Breach

Ahmed M. A. Sattar

Abstract Levees are commonly encountered in various places of the world on open channels controlling high flows and protecting surrounding areas from inundation. For several reasons, such as soil erosion and catastrophic high floods, levees are subject to being breached at several locations. Unsteady flow resulting from levee breach is inherently three dimensional in character and similar in some aspects to dividing flows in open channel and in other aspects to dam break flow. In the current work, flow characteristics in a rectangular open channel levee breach are experimentally collected while treating flow as 3D. Depth measurements are made using point gauges and velocity measurements are obtained using a Sontec ADV over grids defined throughout the levee breach region. Various discharges and area ratios for levee breach configuration are run experimentally. Hydraulic characteristics of secondary flows developed in levee breach in an open channel breach region and outside open channel are presented, which are vital for breach opening and breach area stability in case of erodible channel bed and sides. One of the fastest and most efficient methods for breach closure is to use sand bags. The current work is extended to testing various sizes and shapes of sand bags for breach closure and flow characteristics at incipient motion for bags are collected and analyzed.

A. M. A. Sattar (✉)
Department of Irrigation & Hydraulics, Faculty of Engineering, Cairo University,
Orman Street, Giza, Egypt
e-mail: ahmoudy77@yahoo.com

P. Rowiński (ed.), *Experimental and Computational Solutions of Hydraulic Problems*, GeoPlanet: Earth and Planetary Sciences, DOI: 10.1007/978-3-642-30209-1_15,

1 Introduction

The main purpose of levee is to prevent flooding and inundation of the adjoining areas; they also confine river flow under normal river operation. Levees were constructed over 3,000 years ago in ancient Egypt along the left bank of River Nile for more than 1,000 km stretching from Aswan to Nile Delta. Recently, levee systems have been built along the Mississippi and Sacramento Rivers in the U.S. and Rhine, Loire, Vistula, Scheldt and Danube Rivers in Europe. In Egypt, smaller levee systems have been constructed around main irrigation canals mostly from earthworks to protect surrounding lands from inundation in case of an increase in release of flow through them. A very dangerous form of levee failure is a breach, which is the formation of a large opening in the levee allowing water to inundate the surrounding lands protected by levees. Lately, levee breaches have often been encountered all round the world, leading to catastrophic events, such as life and property loss and changes in socio-economic conditions of the inundated areas adjacent to levees, e.g., 2012 levee breach on Nasr Noba Canal in Egypt, 2011 levee breach on Black River in Missouri, 2010 levee breach in Vendee in France, etc. Understanding the various hydraulic aspects of open channel levee breaches is thus of major importance in hydraulic and environmental engineering, to help devise mitigation measures and to decrease the adverse impacts on surrounding areas. The levee breach is hydraulically similar to dividing open channel flow and dam break flow. Figure 1 shows some characteristics of breach flow; approaching the breach, flow in the main channel starts to divide and a dividing streamline is shown to indicate the parts of the flow entering the breach and flowing in the main channel. A separation zone appears in the main channel after the breach location and a stagnation zone appears near the downstream corner of the breach. At the breach outfall, a contracting flow region is observed with free overfall conditions of dam-break similar waves. All of these features depend mainly (beside the Froude number of flow in the main channel) on the ratio, $Q_r = Q_b/Q_u$, where Q_b is the breach discharge and Q_u is the main channel flow upstream the breach. Figure 1 shows two cases for breach flow; a general case where part of channel flow goes through the breach and an extreme case where all channel flow goes through the breach. In the second case, it is observed that the stagnation zone increases to extend from the breach to the channel end and the separation zone decreases in size and almost disappears.

Despite the importance of open channel levee breach hydraulics, only few researchers have studied it. Yen (1995) studied the impact of levee breach on water level in the main channel upstream the breach. He discussed various methods for channel capacity determination and identification of bottlenecks suitable for levee improvements. Jaffe and Sanders (2001) studied the hydraulic impact of engineered levee breaches over the peak flood stages in the main channel. Using a shallow water model, they studied the optimal design of engineered levee breaches to reduce flood stages. Apel et al. (2008) studied the effect of levee breaches on hydraulic characteristics of the main channel flood frequency distributions. Sattar

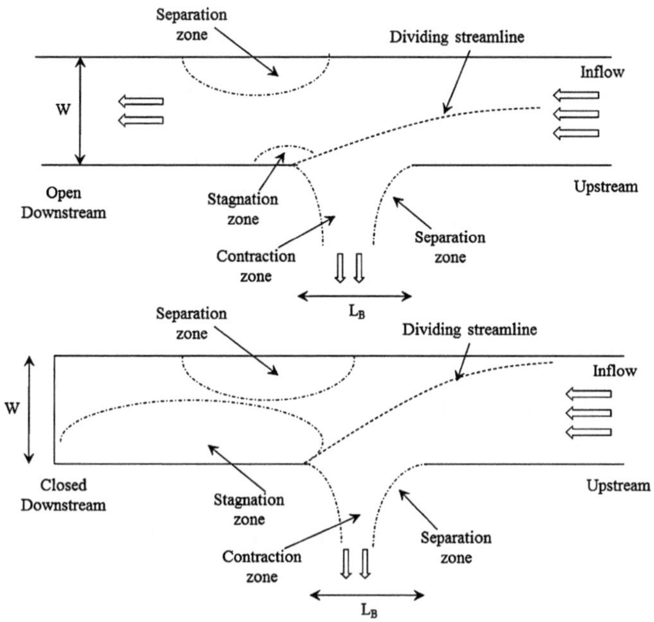

Fig. 1 Flow characteristics of an open channel levee breach flow

et al. (2008) studied the flow hydraulics of the breach that occurred in the 17th Street Canal in New Orleans following Hurricane Katrina and linked the flow velocities to sandbag sizes and investigated various possible ways to close the breach. This was done using a 1:50 physical model including the inundation area downstream the breach. Yu et al. (2009) applied a 2D numerical model to study the hydraulics of levee breach flood propagation under complex boundary conditions and unsteady flood processes in river. Their study dealt with flood depth, inundation area, and flood wave travel time for possible levee flood management. Jia et al. (2010) utilized the experimental setup of Sattar et al. (2008) to validate CCHE2 numerical model and further apply it to investigate the hydraulic characteristics of breach flow during various closure procedures. Zolghad et al. (2010) used MIKE21 to study flood propagation characteristics through levee breach in Helleh River in Iran.

While most of the available research studied the breach flow hydraulics in one or two-dimensional domains, the levee breach flow is inherently three dimensional in nature, especially in regions of secondary currents and during highly turbulent flow events and at sharp water surface gradients. Therefore, this chapter presents experimental results for some of the hydraulic characteristics of rectangular open channel levee breach in 3D. Data presented is composed of water surface mapping and velocity distribution at selected sections at the vicinity of the breach. Afterwards, experiments are conducted on sandbags with different sizes and shapes to determine a relation between breach flow velocity and sand bag size that can be used in breach closure.

2 Experimental Setup

The experiments are performed in a straight flume, which is constructed in 11.6 × 6.7 m basin (Fig. 2), located at the Irrigation and Hydraulics Department at Cairo University. The flume is 6.4 m long, 0.6 m deep and 0.4 m wide. The average manning roughness coefficient of bottom and wall was calculated from velocity and water surface profile measurements and is found to be approximately 0.01. The discharge is measured by an orifice plate and a manometer installed on pipe on pump discharge side. The current study is concerned with a single lateral breach in the channel floodwall. The lateral breach is located 3 m from the channel entrance to ensure developed flow just before its location. The floor of the entire flume is horizontal and the breach discharges into a horizontal platform constructed to have the same level as the channel. The downstream end of the flume is left opened with an adjustable tailgate for controlling the water level in the channel downstream. Two 0.16 m^3/s (2500 GPM) axial pumps are used for flow supply from an underground sump to the flume through a 0.305 m supply pipe. The pumps supply the flow to an overhead tank that supplies the flume with its flow. To ensure properly developed flow at the breach location, perforated screens and 60 mm long honeycomb are places at the flume inlet.

A point gauge is used to measure water surface in channel with an accuracy of 0.1 mm on vernier scale. A 10 MHz 3D Sontec Acoustic Doppler Velocimeter (ADV) is used to measure 3D velocity and turbulent components of flow fields. To ensure repeatability of experiments and determine error in velocity measurements, the experiment was run several times and velocity components at selected locations have been recorded; it is found that experiments are repeatable and error in velocity measurements was less than 1 % at all locations.

A coordinate system is chosen such that the positive x-axis is in the upstream direction of the main channel. The positive y-axis is pointing towards the channel

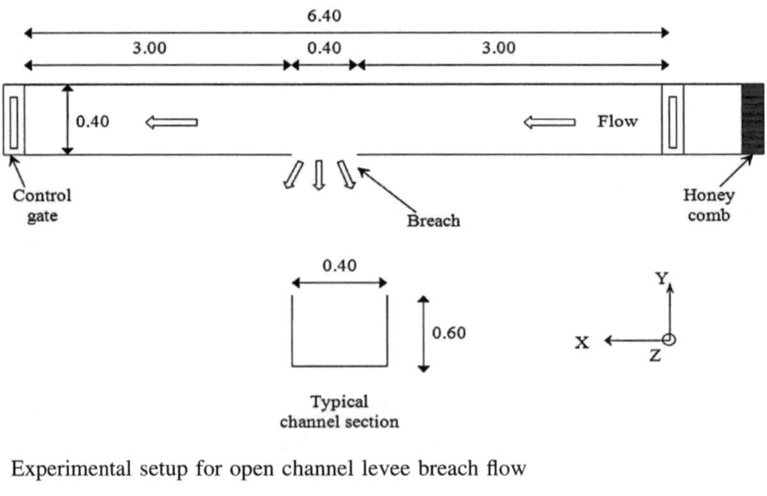

Fig. 2 Experimental setup for open channel levee breach flow

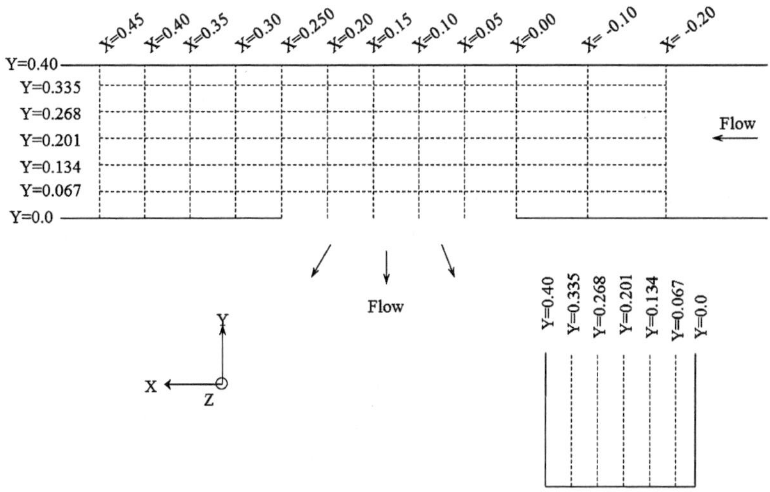

Fig. 3 Sections for measurements of flow velocity and turbulence

side containing the breach and the positive z-axis is upward in vertical direction. The origin of this coordinate system is taken to be the bed of the flume at the upstream point of the breach opening, as shown in Fig. 3. Distances are non-dimensionalized using the channel width of 0.40 m. A steady flow rate is maintained in all experiments and water level at the downstream is kept constant using a control gate. Measurements are taken after steady conditions are attained in the channel and Reynolds number is ensured to be high for all tests.

Table 1 shows the flow conditions for the set of experiments conducted, where Q_b is the breach discharge, Q_u is the main channel discharge upstream the breach, Q_r is the flow ratio as defined before, and L_b is the breach width. Two breach widths, L_b are considered and for each, the downstream channel gate is altered to change flow through the breach and thus breach flow ratio, Q_r is changed from 1 to 0.80 as shown in Table 1. In all experiments, water exiting the breach is drained quickly in order to prevent any accumulation and interference with flow through breach.

Table 1 Breach flow experimental conditions

$L_b = 0.25$ m			$L_b = 0.40$ m		
Q_u (m^3/s)	Q_b (m^3/s)	Q_r	Q_u (m^3/s)	Q_b (m^3/s)	Q_r
0.16	0.16	1	0.16	0.16	1
0.16	0.1282	0.80	0.16	0.1282	0.80

3 Breach Hydraulics

Figure 4 shows the water surface mapping around the breach location for $Q_r = 1$, where $Q_u = Q_b = 0.16$ m^3/sec. Water depth in the main channel upstream the breach is 0.38 m, it increases longitudinally across the main channel to reach 0.42 m downstream the breach and decreases rapidly in transverse direction at the breach location to reach 0.26 m at breach, which is the lowest flow depth inside the main channel. Due to the free-overfall conditions at the breach section, variations in water depth are more pronounced than in case of open channel dividing flow. Moreover, the free-overfall conditions lead to a rapid decrease in water depth across the channel at the breach section with velocities being the highest at breach section. As observed from Fig. 4, the water depth contours are parallel to the breach line due to the curvature of the incoming streamlines to exit from the breach. Flow depth at the stagnation zone is the highest along the breach section. Due to the flow circulation and significant decrease in velocities at the separation zone, the water depth in this zone is higher than that in main channel and equal to that in the stagnation zone downstream the breach. At $Q_r = 1$, the recirculation effects and the separation zone are the highest compared to other zone of $Q_r < 1$. Unlike open channel dividing flow, the breach flow separation is observed to occur at the near side of the breach in the main channel causing flow contraction, which can be clearly observed from the velocity vectors across the breach.

To obtain a clear picture on the distribution of x and y-velocity components around the breach section, experimental data are collected at a finer mesh at Z = 0.20 m from flume bottom, and results are plotted in Fig. 5. It is observed that the X-velocity decreases at breach mid-section and reaches zero by the end of the breach due to the complete curvature of streamlines exiting the breach in the y-direction. Experimental results showed that at the tip of the stagnation zone—at breach downstream end—streamlines tend to take a steep exit path in negative x-direction. On the other hand, the velocity in y-direction clearly shows the distribution of the exit velocities perpendicular to breach section for $Q_r = 1$, Fig. 5. These velocities give a better view for the separation and contraction zones, than the water depth mapping. In the separation zone (see Figs. 5, 6) that starts to form

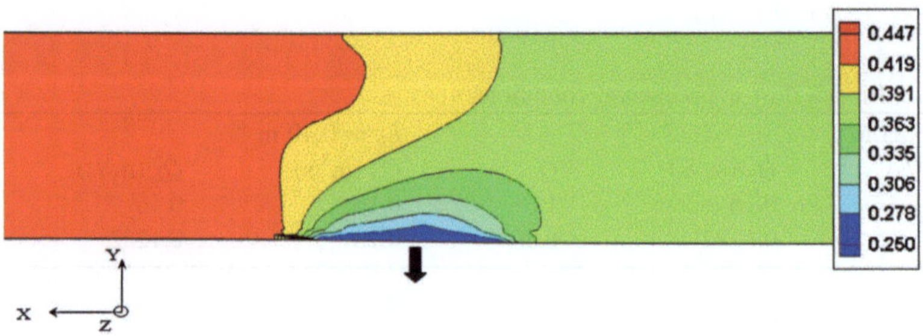

Fig. 4 Water depth around the breach section, $Q_u = 0.16$ m^3/s, $Q_r = 1$, $L_b = 0.40$ m

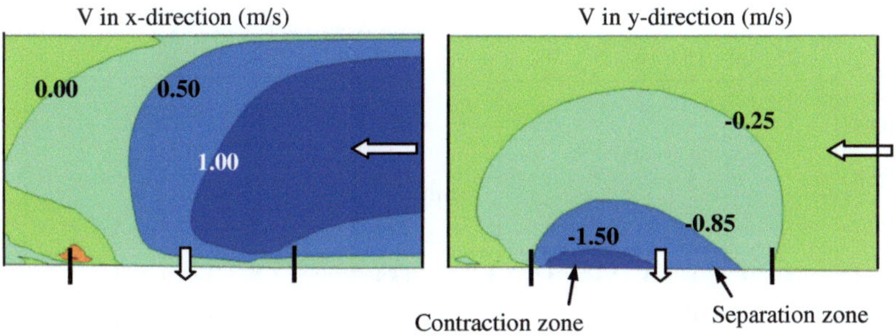

Fig. 5 Velocity x- and y-components around the breach section, $Q_u = 0.16\ m^3/s$, $Q_r = 1$, $L_b = 0.40$ m, Z = 0.20 m (grid refined to 2.5 cm mesh)

in the main channel before flow exit through the breach, the exit velocities in the y-direction are the lowest along the breach section. The highest values of velocity in y-direction indicate the zone where flow contraction (see Figs. 5, 6) occurs and were observed to occur in the far section of the breach after the separation zone. In general, the magnitude of both velocity components along the breach section is almost the opposite, where maximum values of one velocity component corresponded to minimum values of the other component, which is caused by the curvature in the streamlines to exit the breach section.

Figure 8 shows the x-velocity vectors around the breach section in the main channel at Z = 0.04 m and Z = 0.11 m for $Q_r = 1$. In Fig. 8, the increase in value of x-velocity is clearly observed with the increase in water depth. Except at the breach section, the x-velocity vectors in main channel around the breach show similar patterns a both Z = 0.04 m and Z = 0.11 m. Higher values are found to

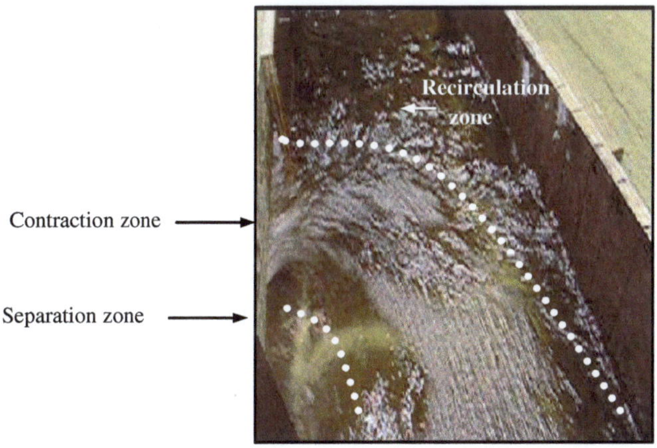

Fig. 6 Streamlines at the surface at the breach section, $Q_u = 0.16\ m^3/s$, $Q_r = 1$, $L_b = 0.40$ m, $Z = Z_{surface}$

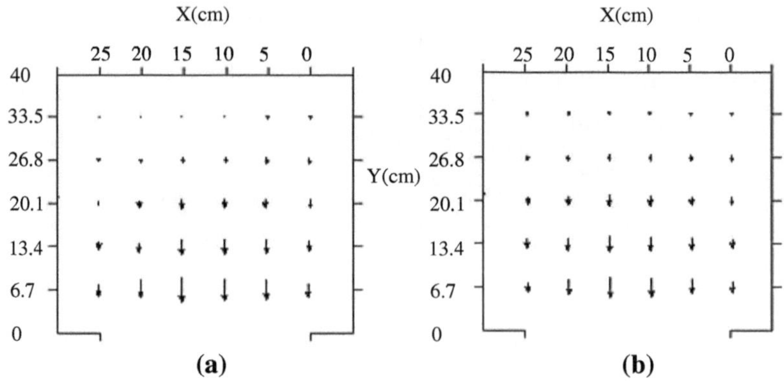

Fig. 7 Y-velocity vectors around the breach, $Q_u = 0.16$ m^3/s, $Q_r = 1$, $L_b = 0.25$ m, (**a**) Z = 0.04 m and (**b**) Z = 0.11 m

occur near the breach section and tend to decrease across the channel. Similar behavior is observed for the y-velocity vectors, as shown in Fig. 7. However, magnitude of velocity tends to decrease as we move away from flume bottom.

Figures 7 and 8 show clearly the impact of the breach on flow in man channel (for $Q_r = 1$), where streamlines are forced to take a curved path to exit breach section, starting a small distance approaching the breach causing a change in the distribution of velocities along the channel width. The magnitude of x- and y-velocity components shows a drastic decrease in value in main channel at breach location with highest values at breach section. The breach forced the streamlines in the main channel to take a curved path to exit at its location, where curvature is observed to increase at lower sections near the flume bottom as previously shown in Fig. 7.

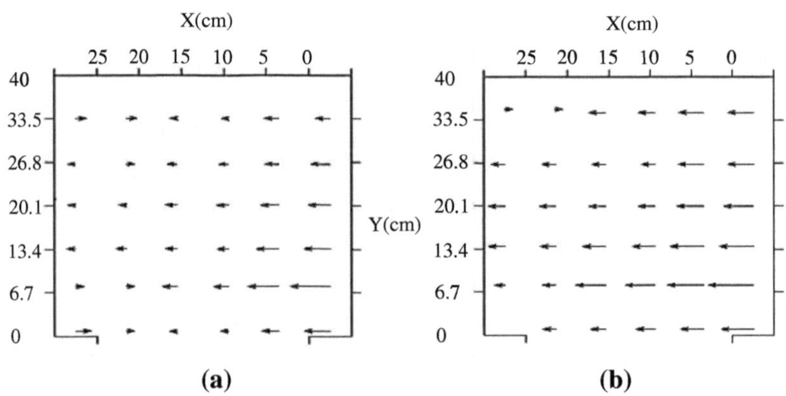

Fig. 8 X-velocity vectors around the breach, $Q_u = 0.16$ m^3/s, $Q_r = 1$, $L_b = 0.25$ m, (**a**) Z = 0.04 m and (**b**) Z = 0.11 m

Fig. 9 Streamlines direction and rough relative velocity at exit section at flume side breach, $Q_u = 0.16$ m^3/s, $Q_r = 0.8$, $Z = Z_{surface}$

Velocity magnitude and direction of streamlines exiting the breach section are shown in Fig. 9 at the surface for $Q_r = 0.80$. It clearly shows the exit angles of flow from the breach and the increase in velocity in the contraction zone and the reverse flow at the far point of the breach section. Such characteristics are confirmed by experimental results to have occurred along various sections in vertical direction with variable magnitude. Moreover, main characteristics of breach flow, such as contraction zone, separation zone, curvature of streamlines, and significant change in velocity across breach, are also observed for various breach widths (see Fig. 10, which shows magnitude of velocity vectors for $L_b = 0.25$ m and $L_b = 0.40$). For the same Q_r, the higher the breach opening, the larger the angle of entry of flow, and the bigger the extent of the flow separation that occurs in the main channel just before flow exit from breach and the smaller the contraction zone, which is a similar behavior to dividing open channel flow as observed by Ramamurth et al. (2007).

For $Q_r = 1$ and Lb = 0.25 m, Fig. 11 shows Y- and Z-velocity components at two locations, X = 0.10 m and X = 0.20. Both sections are after the region of the flow separation at the breach and they show clearly the impact of the vertical velocity component (Z-velocity) on the flow pattern near the breach section. This secondary downward current is very strong and obvious at section X = 0.20, near the end of the breach, while it is still present in other sections in the contraction zone but with less intensity. This secondary current caused the streamline to fall rapidly causing the rapid decrease in water depth at the vicinity of the breach. The Z-velocity component is observed to decrease with decrease in vertical distance at the contraction zone, at X = 0.10 m and 0.20 m.

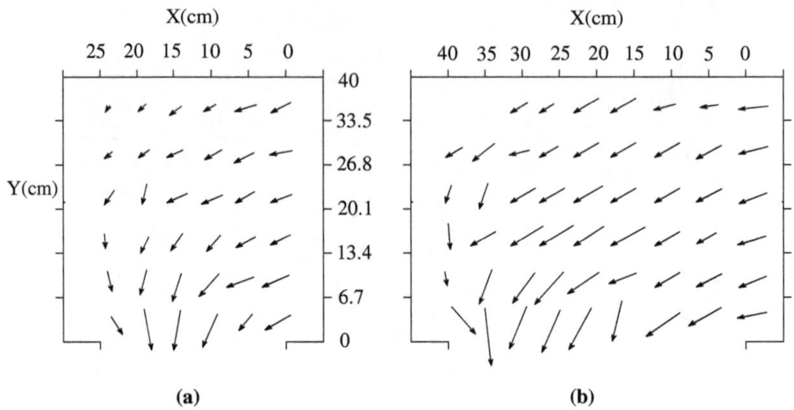

Fig. 10 X-Y vector velocity field around the breach, $Q_u = 0.16$ m^3/s, $Q_r = 1$, Z = 0.15 m for (a) $L_b = 0.25$ m, and (b) $L_b = 0.40$ m

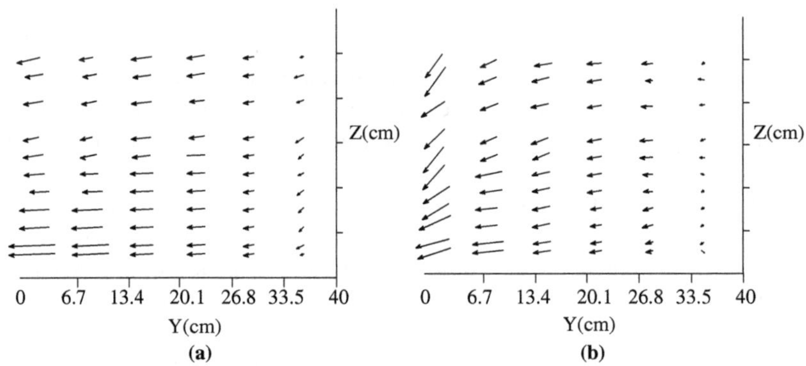

Fig. 11 Y-Z vector velocity profile along the breach, $Q_u = 0.16$ m^3/s, $Q_r = 1$, for $L_b = 0.25$ m, at **a** X = 0.1 m, and **b** X = 0.2 m

4 Incipient Motion of Sandbags

One of the most efficient and economic ways to close open channel breaches and quickly stop inundation of surrounding areas is to use sand bags. However, the size of the dumped sand bag has to be chosen such that the breach flow velocity will not sweep it away, which was the main problem that caused the inundation of the neighborhood at the vicinity of the 17th Street Canal in New Orleans (see Sattar et al. 2008). The U.S.Army Corps failed to choose the correct sand bag size and closure attempts failed. The size of the dumped sandbags can be predetermined by relating the critical flow conditions that will cause the onset of bag instability to bag size. Therefore, the sandbags could be sized so that they settle on reaching the channel bed and thus breach closure can be progressed successfully. These critical conditions are known in literature to be well described by either critical bed shear stress or critical flow velocity, where the latter approach is used in this section of the paper for the onset of stability of sandbags and relation to breach hydraulics.

For describing the shape of large solitary particles, Gogus and Defne (2005) suggested an imaginary circumscribing rectangular prism with dimensions, a = width of frontal area of the particle perpendicular to the flow direction; b = particle height; and c = length of the particle parallel to the flow direction. The dimensions of this prism can be used to describe the shape of a particle in terms of a dimensionless Shape Factor,

$$SF = ((a + b)/2c)(\forall/abc)^{\frac{1}{3}} \tag{1}$$

where $\forall$ is the particle volume. Vanoni (1975) suggested that the Shape Factor might take the form $SF = b/\sqrt{ac}$. However, according to Dey (2004), these shape factors encounter a major drawback as they do not account for the distribution of the surface area and the volume of the particles. Therefore, two particles can have the same shape factor and totally different hydrodynamic properties, e.g. a sphere and a cube. Therefore, the shape factor proposed by Alger and Simons (1968) can be used as a better indication for the particle hydrodynamics, which can be written as;

$$\beta = SF\frac{(S/\pi)^{\frac{1}{2}}}{(6\forall/\pi)^{\frac{1}{3}}} \tag{2}$$

where β is the modified shape factor; SF is the Shape Factor and we will use that proposed by Gogus and Defne (2005); and S is the sandbag total surface area. It is to be noted that now, for a sphere, $\beta = SF$.

The sandbags were made of sewed cloth containing sand with an average specific weight of 1.85 g/cm^3. Sandbags are grouped in four parts according to their dimension and orientation to flow-cubes, rectangular and irregular prisms, as shown in Fig. 12. Rectangular prisms are studied into two groups according to their orientation to flow direction.

Dimensions of sandbags studied are presented in Table 2, together with their corresponding shape factor (Gogus and Defne 2005) and the modified shape factor (Alger and Simons 1968).

In each experiment, a fixed discharge was maintained until uniform flow conditions are developed in main channel upstream the breach section, then a

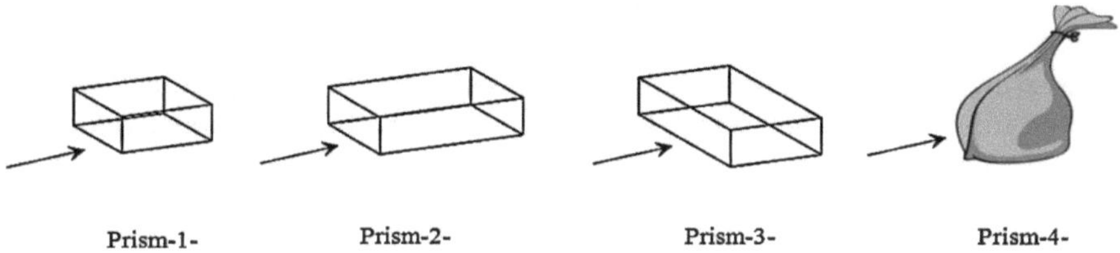

Fig. 12 Sandbags shapes used in experiments

Table 2 Shape factor of sandbags used in experiments

Shape of sandbag	*a (cm)*	*b (cm)*	*c (cm)*	*SF*	*β*
Prism-1-cube	3	3	3	1	1.11
	5	5	5	1	1.11
	7	7	7	1	1.11
	9	9	9	1	1.11
	12	12	12	1	1.11
Prism-2-rectangle	6	1.5	8	0.47	0.60
	3.5	1	15.5	0.15	0.21
	9	2	10	0.55	0.71
	3	1.2	9.5	0.22	0.29
	5	1.2	6.5	0.33	0.44
	3.5	1.75	14.5	0.18	0.24
	6	3	9	0.50	0.59
	4	2	15.5	0.19	0.25
	5	2.75	17	0.22	0.28
	7.5	3.5	11	0.50	0.59
	6	3.75	18	0.27	0.33
	16	5.5	17	0.63	0.75
Prism-3-rectangle	10	2	9	0.67	0.86
	6.5	1.2	5	0.77	0.99
	17	5.5	16	0.70	0.84
Prism-4-sphere	4	4	4	0.80	0.80
	5	5	5	0.80	0.80
	6	6	6	0.80	0.80
	8	8	8	0.80	0.80
	12	12	12	0.80	0.80

sandbag with selected dimension is placed on the bottom of a flume at breach section. Flow is altered in small steps and this procedure is repeated until motion of sandbag was observed. Afterwards, the flow value is decreased for three steps and the whole test is repeated to make sure that it reflected real critical conditions for motion of sandbag. Satisfying the criteria for sandbag motion, a complete water surface is mapped in flume and around the breach in addition to acquiring 3D velocity profiles along selected locations shown in Fig. 3 in addition to velocity profiles along the breach section. All previous studies on incipient motion of sandbags or large particles used the average channel section velocity at the initiation of the motion to determine the stability condition. However, the average velocity might not be a very good indication for flow critical conditions through a breach due to the velocity distribution along the breach section, which tends to increase along the breach to reach its maximum in the end of the contraction zone. As seen from Fig. 13, the average breach velocity is half the maximum velocity at the end of the contraction zone. Therefore, the depth averaged velocity at the location of the sand bag shall be taken as the critical velocity for sand bag incipient motion.

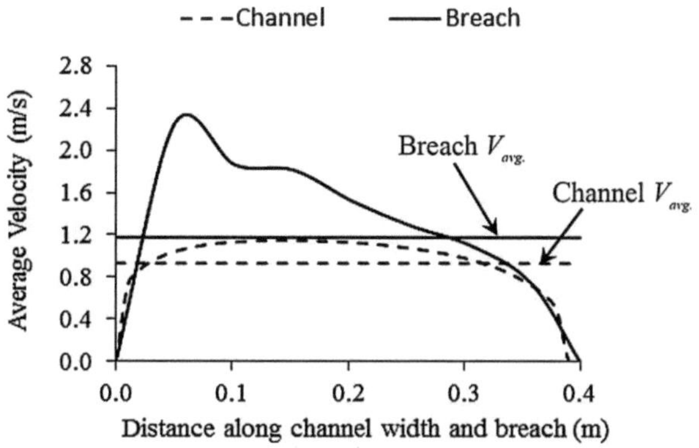

Fig. 13 Velocity profile across flume and along breach section, $Q_u = 0.16\ \mathrm{m^3/s}$, $Q_r = 0.8$, for $L_b = 0.40$ m

Assuming that vertical velocity distribution along each breach section roughly follows the sixth power law, the following can be used to relate critical velocity causing the instability of sandbags dumped in running flow, as (Izbash 1936) for spherical sand bags

$$V_{cr} = C_I\sqrt{2gD(\rho_s - \rho_w/\rho_w)} \tag{3}$$

and as (Zhu et al. 2004) for prism shaped sand bags

$$V_{cr} = C_{II}(H/b)^{\frac{1}{6}}\sqrt{gb(\rho_s - \rho_w/\rho_w)} \tag{4}$$

where V_{cr} is the critical velocity; D is the sandbag diameter; b is the height of the sandbag as defined before; ρ_s is the sand density; ρ_w is the water density; H is the water depth above the sandbag; and C_I, C_{II} are constants that are assumed in the current study to be related to sandbag shape factor β, and orientation in flow.

Equations 2 and 3 are plotted in the Fig. 14 using the experimental results for breach flow for $Q_r = 1$, and the constants are obtained for spherical sandbags and for prism sand bags. For spherical sandbags, results showed that $C_I = 0.4876 = 1.64\ \beta$ for all orientations of the spherical sandbags. On the other hand, results for prism sandbags showed that C_{II} can be roughly approximated by $1.222(H/b)^{1/6}$, with a relatively larger error than for spherical bags.

5 Conclusions

The open channel side or levee breach is very similar to open channel dividing flow in some aspects, and to dam-break flow in other aspects. In the current study, the hydraulic characteristics of an open channel side breach are investigated experimentally, while considering the problem as 3D. This would contribute to understanding levee breach hydraulics and thus the downstream stability in case of erodible bed downstream the breach and would help finding efficient ways to close

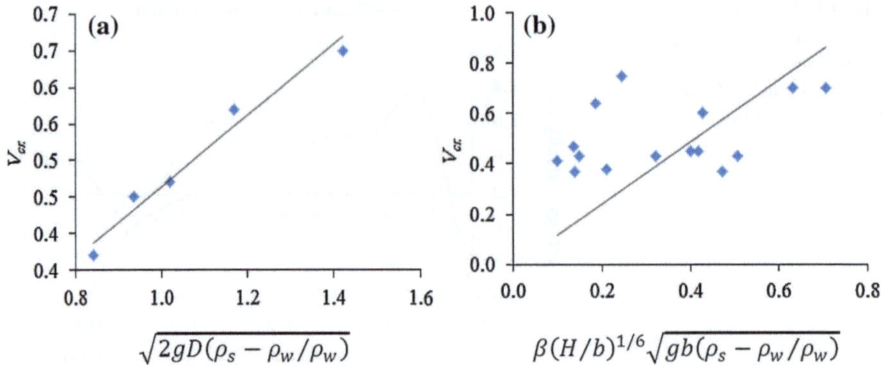

Fig. 14 Critical velocity for incipient sandbag motion for (**a**) spherical sandbags $\sqrt{2gD(\rho_s - \rho_w/\rho_w)}$, and (**b**) prism sandbags $\beta(H/b)^{1/6}\sqrt{gb(\rho_s - \rho_w/\rho_w)}$

the breach and prevent extensive inundation for adjacent areas. In this context, this work tested various sizes and shapes of sand bags and related them to critical velocity through the breach. While good results were obtained for spherical sand bags (which are the ones implemented practically), rough results were obtained for prism shaped sandbags despite using a shape factor in previously developed equations. However, the results can still be used as a guidance for choosing the suitable sandbag shape and size to help closing the breach.

Acknowledgments This work has been carried out under the financial support of Egyptian Science and Technology Development Fund (STDF), Egyptian State Ministry for Scientific Research, Project ID39.

References

Alger GR, Simons DB (1968) Fall velocity of irregular shaped particles. J Hydraul Div, ASCE 94(3):721–737

Apel H, Merz B, Thieken A (2008) Influence of dike breaches on flood frequency estimation. Comput Geosci 35(5):907–923

Dey S (2004) Incipient motion of bivalve shells on sand beds under flowing water. J Eng Mech, ASCE 129(2):232–240

Gogus M, Defne Z (2005) Effect of shape on incipient motion of large solitary particles. J Hydraul Eng, ASCE 131(1):38–45

Izbash SV (1936) Construction of dams by depositing rock in running water, vol 5. 2nd Congress on Large Dams, Washington

Jaffe D, Sanders B (2001) Engineered levee breaches for flood mitigation. J Hydraul Eng, ASCE, 127(6)

Jia Y, Zhu T, Rihai C, Zhang Y (2010) Numerical modeling of flow through a breached levee and during levee closure. In: Proceeding of world environmental and water resources congress 2010, Challenges of Change, ASCE

Ramamurth AS, Qu J, Vo D (2007) Numerical and experimental study of dividing open-channel flow. J Hydraul Eng, ASCE 133(10):1135–1144
Sattar AM, Kassem A, Chaudhry MH (2008) Case Study: 17th street canal breach closure procedures. J Hydraul Eng, ASCE 134 (11)
Vanoni VA (1975) Sedimentation engineering. ASCE Manual No. 54, ASCE, New York
Yen C (1995) Hydraulics and effectiveness of levees for flood control. In: Proceedings US Italy research workshop on the hydrometeorology, impacts and management of extreme floods, Perugia, Italy
Yu M, Deng Y, Qin L, Wang D, Chen Y (2009) Numerical simulation of levee breach flows under complex boundary conditions. J Hydrodyn, Ser N\B, 21(5):633–639
Zhu L, Wang J, Cheng N, Ying Q, Zhang D (2004) Settling distance and incipient motion of sandbags in open channel flows. J Waterw Port Coast Ocean Eng, ASCE 130(2):98–103
Zolghad M, Hashemi M, Hosseiipour E (2010) Modeling of flood wave propagation through levee breach using MIKE21, A case study in Helleh River, Iran. In: Proceeding of water environment and water resources congress 2010, Challenges of Change, ASCE pp 2683–2693

The Influence of Large Roughness Elements on Natural Morphological Changes in a Mountain River Bed

Mateusz Strutyński, Andrzej Strużyński and Krzysztof Kulesza

Abstract The presence of large roughness elements in a river bed influences the existing fluvial processes. Large roughness elements change the flow regime, continuity of river and morphological conditions, such as bed material size, water depth, and bed slope. Large roughness elements (for example boulders) are also often used to improve fish habitat and to change the hydraulic conditions of water flow (velocity magnitude, water flow direction). As long as the riverbed stays under their influence, one can observe that on the border between liquid and solid phases there is an extremely high turbulence. This causes the high erosion potentials in the zone of large roughness area. Determination of influence of large roughness elements on flow conditions is of crucial importance with regard to a proper engineering approach when designing flow channels. In the chapter, the results of investigations carried out in Carpathian mountain rivers: Raba, Porębianka and Czarny Dunajec are presented. Performed measurements included cross-section and longitudinal leveling, bed load sampling and hydrometric measurements. Differentiation of flow velocity in the region of large roughness element and its influence of the bed load size in the stream channel was shown. The aim of the chapter was to find the changes of fluvial processes due to the presence of large

M. Strutyński (✉) · A. Strużyński
Department of Water Engineering and Geotechnics, University of Agriculture in Krakow, al. Mickiewicza 24/28, 30-059 Kraków, Poland
e-mail: mstrutynski@ar.krakow.pl

K. Kulesza
Institute of Meteorology and Water Management—National Research Institute, Board of Water-Management Systems, ul. Piotra Borowego 14 30-215 Kraków, Poland
e-mail: k.kulesza@poczta.onet.pl

P. Rowiński (ed.), *Experimental and Computational Solutions of Hydraulic Problems*, GeoPlanet: Earth and Planetary Sciences, DOI: 10.1007/978-3-642-30209-1_16,

roughness elements and to connect it with the fish habitat conditions for different species. This improves the ecological state of water flow which is the requirement of the Water Framework Directive of the European Union.

Keywords Large roughness elements · River morphology · River bed stability

1 Introduction

Current water policy of the European Union posed new requirements for tasks realized in the sphere of engineering and water management. At present one of the crucial objectives of water management, pursuant to the EU-Water Framework Directive (2000/60/EC), is counteracting worsening of the state of some waters and achieving good condition of all water bodies by the end of 2015. For surface waters, the "good condition" is determined by "good ecological state" which depends on conducted hydromorphological assessment of a watercourse reach. Each European Union member state developed its own methodology of hydromorphological assessment. However, a vast majority of methods used for the hydromorphological assessment takes into consideration the size and shape of material forming the riverbed (Adynkiewicz and Piragas 2006). Diversification of riverbed material is not only favourable for aquatic organisms, but it also determines conditions of hydrodynamic balance of a river part.

Investigations conducted by Starkel (1991) demonstrate that the erosion processes are most intensified in southern Poland. Małopolska region is one of the most diversified regions in Poland concerning its natural environment. Numerous strongly changed water segments occur in this area. On the basis of analyses conducted in the area of Board of Water Management in Krakow operations distinguished were 321 strongly changed uniform surface water bodies and 21 artificial surface water bodies. At present, endeavours are made to minimize human pressure on the natural environment through the so-called "close to nature" river training. It is a response to progressing degrading of the aquatic environment and gradual worsening of flowing water quality. It is a compromise between traditional training (e.g., gradual consolidation) and activities aimed at creating conditions for restoration of watercourse ecological potential. Large Roughness Elements are natural components of the bed material. They may originate from riverbed or lateral erosion of the watercourse channel but may be also transported by the watercourse from upper parts where the bed slopes and energy of the river bed is higher. The presence of Large Roughness Elements in a river bed positively affects a change of fluvial conditions in it and the hydromorphological conditions in the channel bed. A zone of erosion (where processes of riverbed material washing out take place) and a turbulence (with greater turbulence caused by water flowing round large roughness element), may be distinguished in the zone of their influence. The size of these zones and range of their influence depend on the degree of the grain protruding above the riverbed covering and on the flow.

2 Objective and Methods

Large Roughness Elements laying in a riverbed create new hydrodynamic conditions. They diversify the flow regime (its size, dynamics and max. discharge), river continuity (transport and sedimentation of riverbed material) and morphological parameters (mean depth, mean velocity, slope, aerial view, cross-sections).

The presence of Large Roughness Elements in a riverbed diversifies the size of sediments forming the riverbed covering. Whirlpools and turbulence caused by the presence of Large Roughness Element leads to deposition in its shadow of small sediment which is not washed out. The impact of Large Roughness Elements on river bed is shown in Fig. 1.

Yagger et al. (2007) stated that Large Roughness Elements present in a river bed reduce the stresses which may cause small sediment transport intensity. Turbulence behind Large Roughness Element causes that fine material sedimenting directly behind a Large Roughness Element is raised only locally but is not transported down the river. Velocity pulsation may cause a temporary detaching of bed material grains from the riverbed (Byczkowski 1966). For individual bed material grains, the size and weight of a single particle are crucially important because of the bed material movement and its transport in the bed channel (Raudkivi 1998). Bed material movements cause morphological changes of the bed channel. Therefore, they are important for an assessment of the watercourse ecological state, according to the requirements of the EU-Water Framework Directive. Horizontal arrangement of riverbeds depends on many factors. It is shaped mainly by the size of material forming the bed covering and the adjoining area, the size and amount of bed material supplied from the upper parts of the catchment, longitudinal slope and flow regime.

Diversification of morphological conditions of riverbed is particularly important for aquatic organisms. Individual fish species living in Polish rivers require diverse material forming bed cover. Some of them, like, e.g., salmon, whose restitution has been attempted lately in Poland, requires forming a mound of fine grained material with diameter from 0.02 m. Trout, very popular in Polish rivers requires on its spawning site materials from 0.008 to 0.03 m in diameter. Large

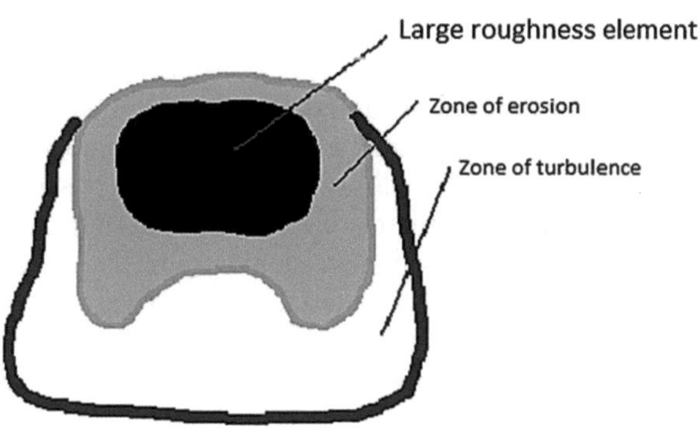

Fig. 1 Impact of large roughness element on a riverbed

Roughness Elements as natural components of bed material diversify the size of bed covering; therefore, they may contribute to an improvement of spawning conditions in rivers.

The aim of the chapter is to determine the influence of Large Roughness Elements on conditions shaping the morphology of a riverbed and their influence on the deposition of riverbed material. The chapter presents initial results of research conducted on the Czarny Dunajec River segment in Chochołów village and Raba River segment in Zarabie village.

Field measurements comprised granulometric analysis of riverbed material in the zone of Large Roughness Elements influence. Granulometric analyses were conducted using a traditional method (sieving). Geodesical measurements of the cross sections and longitudinal profiles were carried out by means of GPS-RTK device. Hydrometric measurements were made using magnetohydrodynamic current meter (Nautilus).

The size and dynamics of morphological processes depend on the flow discharge and dominant discharge during flood. The conditions of hydrodynamic balance are mostly affected by the quantity and dynamics of bed load transport. If the shear stresses on the river riverbed are lower than critical stresses for a given fraction, then the river material remains stable.

Bed shear stresses may be computed according to the following formula:

$$\tau_0 = \gamma YI[N \cdot m^{-2}] \quad (1)$$

where γ is the density of water ($N \cdot m^{-3}$), Y is the water depth (m), and I is the slope (–).

In order to determine the conditions of riverbed material forming the riverbed covering, boundary stresses for individual fractions were computed from the following formula:

$$\tau_{cr} = f_m d_m \Delta\gamma_s [N \cdot m^{-2}] \quad (2)$$

where f_m is the dimensionless stresses for mean fraction (–), d_m is the mean fraction (m), and $\Delta\gamma_s$ is the difference between bed load and water density ($N \cdot m^{-3}$)

Values of critical stresses for individual fractions d_i depend mainly on the values of dimensionless stresses f_i:

$$\tau_{cr\,i} = f_i \Delta\gamma_s d_i [N \cdot m^{-2}] \quad (3)$$

where f_i is the dimensionless stresses for fraction d_i (–), d_i is the fraction (m), and $\Delta\gamma_s$ is the difference between bed load and water density ($N \cdot m^{-3}$)

Results of field measurements made it possible to determine the boundary conditions for movement of riverbed material, dynamics of erosion processes in the zone of large roughness element and the effect of large roughness elements shape on the volume of material forming riverbed cover. On the basis of conducted measurements, also changeability of riverbed morphology was assessed within their range regarding ichtiofauna requirements for living and spawning purposes.

3 Field Investigations

3.1 The Czarny Dunajec River

The investigated reach of Czarny Dunajec River is characterized by a high flow dynamics. Frequent floods which happen in the watercourse catchment cause a considerable intensification of erosion processes. Because of big channel incision, lateral erosion processes are dominant along the analysed reach. Figure 2 shows a cross-section of the Dunajec River on the researched segment. It is characterised by a considerable slope of the bank, which is caused by washing out of the riverbed material from the right side of the bed.

Table 1 shows characteristic flows of the Czarny Dunajec River with given corresponding bankfull discharge and bed shear stresses for the river reach covered by the measurements. Roughness coefficient 0.03 and hydraulic gradient 0.006 were assumed for the computations (on the basis of field measurements).

There are two large roughness elements on the studied part of the Czarny Dunajec River changing flow distribution differ with size and shape. In Table 2 the dimensions (a, d and c) and shape factors of oversize grains are collected. The shape factors are calculated by the use of the following formula:

$$SF = \frac{c}{\sqrt{ad}} \tag{4}$$

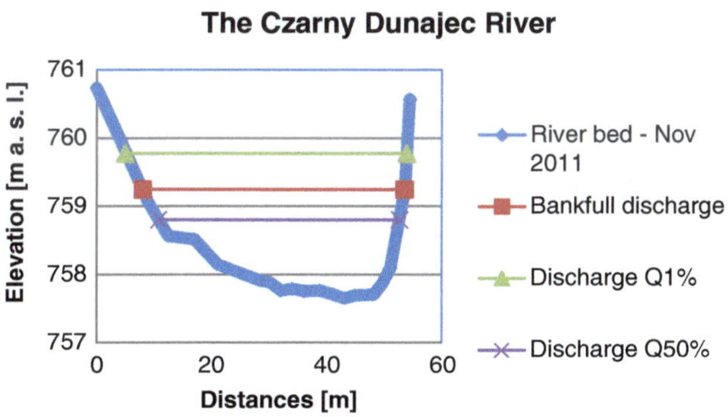

Fig. 2 Cross section of the Czarny Dunajec River, November 2011

Table 1 Hydraulic parameters on the Czarny Dunajec River

P (%)	Flow Q ($m^3 \cdot s^{-1}$)	Bankfull discharge (m)	Shear stresses ($N \cdot m^{-2}$)
1	252	2.0	117.72
5	162	1.80	105.95
10	129	1.70	100.06
25	78	1.33	78.28
50	49	0.99	58.27

Table 2 Compilation of sizes of large roughness elements on the Czarny Dunajec River

Number of LRE	Size			Shape coefficient SF
	a (m)	d (m)	c (m)	
1	0.65	0.35	0.21	0.75
2	0.41	0.36	0.22	0.58

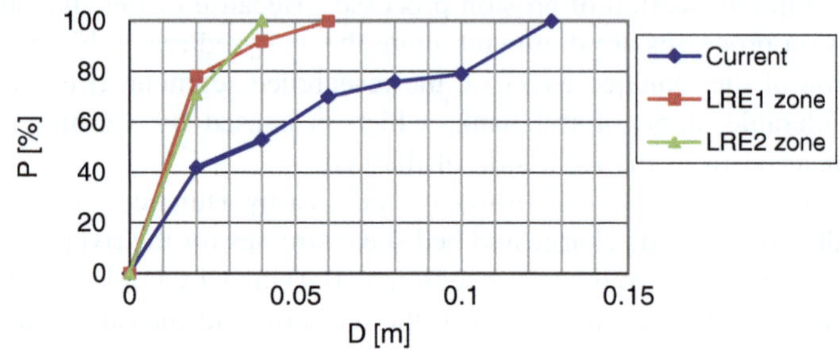

Fig. 3 The Czarny Dunajec River, grain size distribution curves

According to the classification given by Ksiazek (2000) after Novak (1988), the Large Roughness Element number 1 belongs to a group of spherical, while the Element number 2 to ellipsoidal shape. Both boulders protrude 0, 18 m above bed level.

Sieve curves of bed material sampled in the oversize grains interacting area are shifted up indicating higher amount of fine fractions in comparison to the free flow bed sections (Fig. 3).

3.2 December 2011

On the basis of conducted measurements it may be stated that Large Roughness Elements diversify the size of material forming the riverbed cover. Small material washed out during floods, sediments in the zone behind Large Roughness Element. Diameter d_{50} is about 0.016 m while in the current it is twice larger. The biggest size of grains from the current reaches almost 0.13 m and behind the Large Roughness Element number 1 is 0.06 m. Standard deviation of grain size distribution curve in the current is 2.94 in comparison to the value of 1.3, which characterises the bed armour. It points high mobility of fresh bed material. Diameter d_{50} of a sample collected from the current corresponds to d_{90} for the riverbed material behind the Large Roughness Elements.

The presence of Large Roughness Elements diversifies also hydrodynamic conditions which depend on hydraulic gradient. A single Large Roughness Element which protrudes above the riverbed covering causes water damming and a change of its gradient. In result of water streams flowing around a Large

Table 3 The Czarny Dunajec River, hydraulic parameters measured behind the large roughness elements

	LRE 1		LRE 2
Hydraulic gradient (–)		0.006	
Riverbed slope downstream LRE (–)	0.049		0.012
Hydraulic gradient behind LRE (–)	0.055		0.055
Area of fine sediment deposition (m^2)	0.267		0.338

Roughness Element, the point velocity and intensity of turbulence increase. Hydrometric measurements in the Czarny Dunajec River indicated that the increase of velocity and turbulence is observed especially on the both sides of Large Roughness Elements. Results of hydraulic gradient measurements for individual Large Roughness Elements were compiled in Table 3.

On the basis of conducted measurements it may be stated that the presence of Large Roughness Elements causes a local change of hydraulic gradient brought about by water damming by the protruding Large Roughness Element. Average hydraulic gradient in the current on the analyzed water reach is 0.006, while in the large roughness element region it increases about 9 times and is 0.055. This situation leads to creating new hydrodynamic conditions in their influence zone and diversifies the conditions of riverbed material transport.

3.3 The Raba River

Field measurements carried out on the Raba River include granulometric analysis of bed load. Due to the mountainous character of the Raba River and large average slope of the valley, flooding processes of the first terrace appears here every 2 year. This events cause a big change in river bed morphology and mixing of bed material. During field measurements performed between 2008 and 2011 also the large floods have passed. The biggest one occurred in May 17, 2010 with the discharge of 758 $m^3 \cdot s^{-1}$ and 4.58 water depth. After the flood in 2010 the d_{50} diameter of the bedload has increased, as shown in Fig. 4.

Measurements performed before and after the event indicate that small fractions of the river bed were washed out and then the armoring processes started again on the test section of the Raba River.

There are three Large Roughness Elements on the test section of Raba River. While one of them is located naturally, the two others have been artificially introduced by the Fishing Association for morphological differentiation of the river bed. Granulometric measurements carried out bellow the first Large Roughness Element with a diameter of 2.52 m indicate that size of hidden material was stable (Fig. 5).

Performed measurements indicate that bed material which is deposited below the large roughness element becomes stable and its diameter does not change even

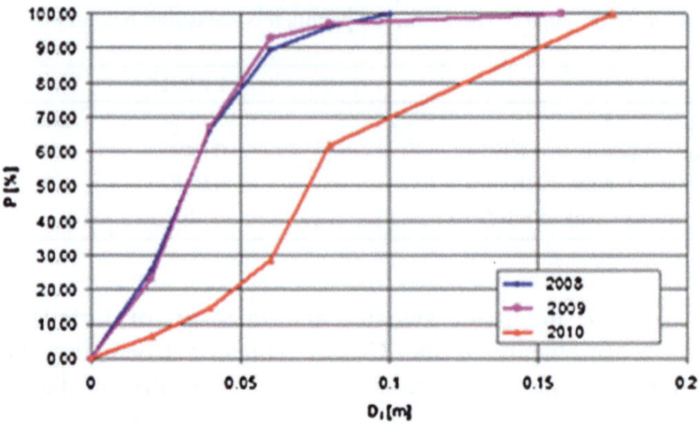

Fig. 4 Grain size distribution curves, the Raba River

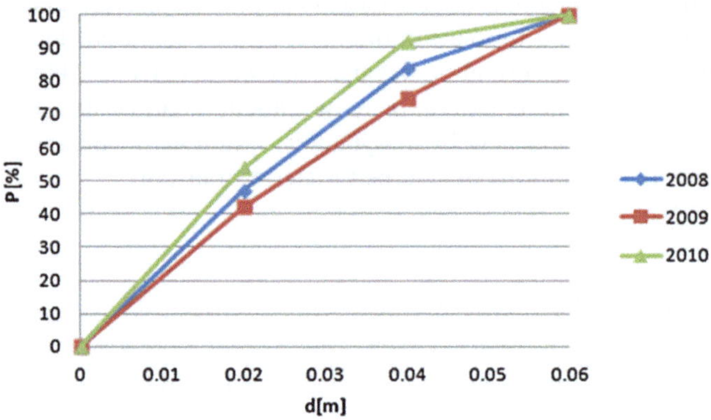

Fig. 5 Grain size distribution curves in the shallow zone of large roughness element, Raba River

after passing flood discharges. The shallow water area was approximately 18.56 m^2 and the shadow length from the boulder was 5.87 m. This caused bigger grains to be flushed out from the shallow zone. Only small particles (up to 0.06 m) could be fortified in this zone.

4 Analysis of Results

Shields (1936) conducted investigations aimed at determining dimensionless stress values. On the basis of conducted laboratory tests he found that the values of dimensionless stresses for uniform grain were 0.06. Research carried out by Michalik (1990) to assess the intensity of bed load transport in the Carpathian tributaries of the Vistula River revealed the value of dimensionless stresses 0.033 (tracer tests performed by Michalik in Raba River with using isotopes) and this

Table 4 Bed critical stresses for individual riverbed material fractions, Czarny Dunajec River

Diameter (m)	Critical stresses ($N \cdot m^{-2}$)	Critical depth (m)
0.02	10.89	0.19
0.04	21.78	0.37
0.06	32.67	0.56
LRE 1	190.60	3.24
LRE 2	196.02	3.33

value was adopted for further computations. Table 4 shows values of critical stresses for individual fractions of the riverbed material (Czarny Dunajec River).

Large roughness elements are regarded as stable when the bedshear stresses during bankfull flow are lower than boundary stresses. Along the analysed segment of the Czarny Dunajec River bankfull discharge for bankfull flow was estimated on the level of 2.47 m (Strużyński et al. 2012), which corresponds to bed shear stresses for bankfull flow 145.38 $N \cdot m^{-2}$. Therefore, it may be stated that the large roughness elements situated on the investigated Czarny Dunajec River reach are stable (Table 4).

To determine the hydrodynamic conditions on the test section of the Raba River the armoring prognosis was performed. It was made using the ARMOUR, PC program developed in the Department of Water Engineering and Geotechnics, University of Agriculture in Cracow. The armoring prognosis for the river bed on the test section of Raba River is shown in Fig. 6.

On the basis of the performed analysis it can be concluded that in the investigated part of the Raba River the armoring process exist up to the water depth of 3.4 m (Strutyński 2011). Then the bed cover starts to be transported. When the water depth exceeds 3.4 m, mass transportation of all fractions in river bed starts. After the flood in 2010 the armor layer was removed completely. Despite this the

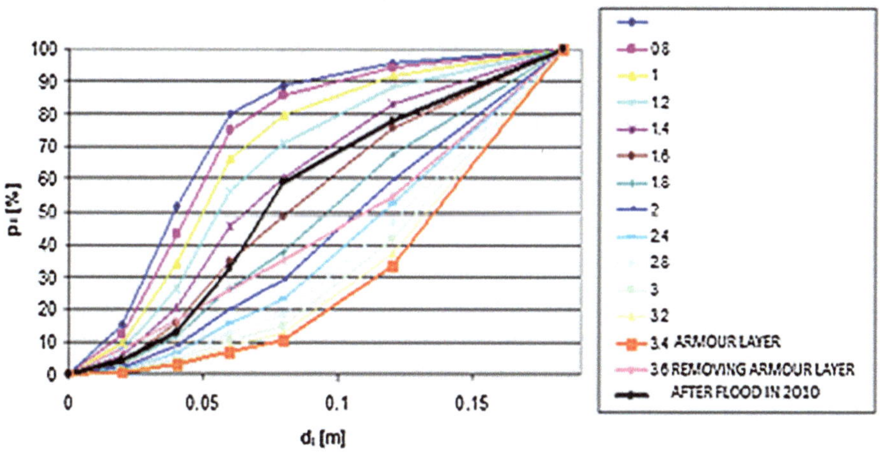

Fig. 6 Bed stability prognosis—The Raba River

mean diameter of river bed in the zone of large roughness element has not been changed. This happens due to the presence of vortex formed by the flow around the Large Roughness Elements. No Large Roughness Elements were shifted back or lowered into the bottom after the 2010 event.

5 Conclusions

On the basis of conducted investigations it may be stated that despite the exceeded boundary stresses values for fine material sedimenting in the shalow of Large Roughness Elements, they are stable. Such a situation has been caused by clogging of the cross-section by the Large Roughness Elements and reduction of stresses in their region. Despite a sudden change of the gradient (increase in bed shear stresses) fine riverbed material is not transported down the river. It should be also mentioned that in case of fine material behind the Large Roughness Elements no clogging of small grains occurs or their later entering the transport because of their constant dimensions.

Large Roughness Elements cause formation of local sandbanks whose diameter depends on the Large Roughness Element diameter and its protruding above the riverbed covering. Presence of the Large Roughness Elements leads to formation of local bed overdeepenings in the zone of their influence. Fine grained material with diameters up to 6 cm which becomes deposited immediately behind the Large Roughness Elements is not washed out during floods.

The presence of Large Roughness Elements on the analysed Czarny Dunajec River reach considerably diversifies the size of riverbed material. It is particularly important for creating spawning sites for trout which requires bed material diameters between 0.008 and 0.03 m. Such diversified morphological conditions lead to improvement of the watercourse ecological state, which is required by the implemented EU-Water Directive.

The investigations were financed by the National Science Centre research grant No. 1865/B/P01/2009/37: Assessment of the sustainable state of mountains rivers and streams on the basis of natural morphological conditions

References

Adynkiewicz-Piragas M (2006) Hydromorfologiczna ocena cieków wodnych w krajach Unii Europejskiej jako element wspierający ocenę ekologicznego stanu rzek zgodnie z wymogami Ramowej Dyrektywy Wodnej (Hydromorphological assesment of rivers in European Union countries as an element supporting ecological assessment of river conditions in accordance with water framework directive). Infrastruktura i Ekologia Terenów Wiejskich 4/3:7–15 (in Polish)

Byczkowski A (1966) Hydrologia tom I (Hydrology vol I), Wydawnictwo Szkoły Głównej Gospodarstwa Wiejskiego, Warszawa (in Polish)

Directive of the European Parliament and the Council 2000/60/EC (2000) Establishing a framework for community action in the field of water policy

Książek L (2000) Proces deformacji dna koryta potoku górskiego i jego związek z siłą wleczenia pojedynczych ziaren rumowiska. (Deformation process of bed in mountains stream and its connection with drag force of bed particles). PhD thesis Wydział Inżynierii Środowiska. Akademia Rolnicza w Krakowie (in Polish)

Michalik A (1990) Badania intensywności transportu rumowiska wleczonego w rzekach karpackich (Investigations of bed load intensity in Carpathian Mountain rivers). Zesz. Nauk. AR w Krakowie, Habilitation Dissertation 138 (in Polish)

Novak L (1988) Stabilita dna bystrinnych toku (Bed stability of mountain rivers). In: Vysoka Skola Zamedelska, Praha (in Czech)

Shields A (1936) (in German). Anwendung der Aehnlichkeitsmechanik under Turbulenzforschung auf die Geschiebebewegung [Application of similarity mechanics and turbulence research on shear flow]. Mitteilungen der Preußischen Versuchsanstalt für Wasserbau. 26. Berlin: Preußische Versuchsanstalt für Wasserbau

Starkel M (1991) Geografia Polski (Geography of Poland). Wydawnictwo Naukowe PWN, Warszawa (in Polish)

Strutyński M (2011) Ocena warunków hydrodynamicznych cieku w korycie z ziarnami ponadwymiarowymi (Estimation of hydrodynamic conditions in a river bed with large roughness elements) PhD thesis. Manuscript, Uniwersytet Rolniczy w Krakowie (in Polish)

Strużyński A, Strutyński M, Kulesza K (2012) Bed stability as parameter describing the hydromorphological balance of mountain river. Geoplanet (in print)

Raudkivi AJ (1998) Loose boundary hydraulics. Balkema Publishers, Rotterdam

Yagger EM, Kirchner JW, Dietrich WE (2007) Calculating bed load transport in steep boulder bed channels. Water Resour Res 43, W07418. doi:10.1029/2006WR005432

Bed Stability as a Parameter Describing the Hydromorphological Balance of a Mountain River

Andrzej Strużyński, Krzysztof Kulesza and Mateusz Strutyński

Abstract The hydromorphological parameters are used as abiotic descriptors for the classification system introduced by the Water Frame Directive. When studying the eco-literature, authors point out that the introduction of vulnerable species into restored river reach is often not effective as long as the ecosystem has been adapted to the changed river parameters. In fact, as being well-explored, hydromorphological parameters can be successfully used as reference factors of rivers. While bedload granulometry changes reflect the variation of many fluvial processes, bed stability is interpreted here as the key parameter. This first attempt of the evaluation is performed for three selected parts of the Carpathian rivers. The bed stability prognosis for different discharges is done by Armour PC program on the basis of the Gessler method which assumes the probabilistic nature of the bedload movement. The threshold of the bed transportation is given by the critical Shields parameter. The discharges causing bed movement are compared to bankfull and dominant ones as well as to $Q_{10\,\%}$, $Q_{25\,\%}$, $Q_{50\,\%}$ and $Q_{75\,\%}$.

Keywords Bankfull · Dominant discharge · Bed stability · Hydromorphological balance · Mountain river

---

A. Strużyński (✉) · M. Strutyński
Department of Hydraulic Engineering and Geotechnics, University of Agriculture in Kraków, Mickiewicza Ave. 24/28, 30-059 Kraków, Poland
e-mail: rmstruzy@cyf-kr.edu.pl

K. Kulesza
Institute of Meteorology and Water Management, National Research Institute, Board of Water-Management Systems, Piotra Borowego 14, 30-215 Kraków, Poland
e-mail: krzysztof.kulesza@imgw.pl

P. Rowiński (ed.), *Experimental and Computational Solutions of Hydraulic Problems*, GeoPlanet: Earth and Planetary Sciences, DOI: 10.1007/978-3-642-30209-1_17,

1 Introduction

1.1 Description of the Problem

The main division of hydromorphological parameters of river reaches appears due to the development of separate runs (upper, middle and lower) in which the erosion of banks and beds or material deposition exist. An accurate hydromorphological classification can be found in literature connecting fluvial processes to the actual profiling and trace development (Radecki-Pawlik 2006; Church 1992; Brice 1975; Thorne et al. 1997; Rosgen 1994; Church and Jones 1992). While flowing over miscellaneous substratum, the mountain river channel and corridor vary in width, depth and slope (Mueller and Pitlick 2005). This is caused by high energies of water dissipated in riverbed during peak flows. After stormy rainfalls, these rivers tend to develop their bed. These processes, however, do not appear regularly in natural mountain rivers as long as their channels are highly variable at time and space (Bartnik and Strużyński 1997; Strużyński et al. 2011). As long as the mountain river dynamics is high during flood events, they move their currents, change their beds, move their bars or transform terraces even after river training (Florek et al. 2007; Strużyński 2006). After a flood, the mountain river continuum can be disturbed not only by anthropopressure but also by natural factors (Grzybkowska 2007; Lane 1955; Vannote et al. 1980). In addition, the cultivation of the valleys changes continuously (Bartnik et al. 2009). The forested area, the use of agriculture or the expansion of urbanized areas with whole infrastructure influence these changes. Long term observations of the river systems often tend to identify the channel incision and transformation of riverbeds from alluvial to rocky systems (Korpak et al. 2008; Wyżga and Radecki-Pawlik 2011). For this reason, subsequent reaches can be hydromorphologically mixed. This could make it difficult to transform reference parameters of rivers to the other riches upstream or downstream. The problem of restoration of river systems becomes even bigger when ecological continuity would be taken into consideration when all pressures tends to develop the unique river ecosystem (Poole 2002). What also appears problematic is that it is still not obvious that after the restoration of physical, hydromorphological and biological conditions of the regulated river has been accomplished, the resulting ecosystem will go back to its proper and more valuable conditions (Allan 2004; Lake et al. 2007). These problems, however, cannot be solved without proper engineering methodology. The first step of every investigation is to describe representative hydromorphological parameters, quantify them and draw the scheme and borders of possible fluvial processes which can run in the system (Czoch and Kulesza 2006).

The goal of the study is to describe current hydromorphological dynamics of the measured river reaches and evaluate the conditions which would be considered as morphologically and ecologically well balanced river flowing through the rural areas.

1.2 The Propose of the Description of Hydromorphological Reference Parameters on the Basis of Bed Stability Prognosis

Parametric analysis of bed stability bases on the in situ measurements (geodetic, bed granulometry). The longitudinal profiling as well as cross-section data of the river channel and the terrace are used for flow capacity description. This data is used as an input to bed stability/armoring prognosis. The analysis presented in the chapter is done with the use of Armour program (Bartnik and Strużyński 1999). Calculations are done on the basis of the comparison of the actual bed shear to bedload critical stresses. As the bed material is multifractional and rough (diameter > 5 cm), the Wang formula is used (Wang 1977):

$$f_1 = \frac{f_{m1}}{1.786\left(\frac{d_i}{d_m}\right)^{0.946}} \quad \text{for } d_i/d_m = < 0.6, \tag{1}$$

$$f_2 = \frac{f_{m2}}{\left(\frac{d_i}{d_m}\right)^{0.314}} \quad \text{for } d_i/d_m > 0.6, \tag{2}$$

where f_1, f_2 are critical Shields stresses, d_i is the diameter of the fraction i, and d_m is the mean diameter.

Critical Shields stresses divide the bed material into bigger and smaller fractions. Basing on the direct measurements performed in mountain rivers, Bartnik proposed an additional separation of f_m into f_{m1} and f_{m2} parameters regarding the change of sieve curve distribution while during the transportation of bed material, smaller fractions hide in the holes created by the grains which stabilize the bed (Bartnik et al. 1992):

$$f_{m1} = 0.039\delta^{0.26} \quad \text{and} \quad f_{m2} = 0.028\delta^{0.26} \tag{3}$$

where $\delta = (d_{84.13}/d_{15.87})^{0.5}$ is the standard deviation of grain size distribution (Bartnik 1992).

The bedload transportation does not sustain with the same intensity during a flood. Before armour coat removal, the amount of transported material differs dramatically when compared to the massive transportation appearing with multifractional bed (Bartnik and Strużyński 2002). When the process of creating or removing armour coat appears, the grain size distribution changes. As long as there are no separate conditions of grain movement, the prognosis done by Armour program uses stochastic analysis of this phenomenon (Gessler 1970). Bed roughness depends on the characteristic diameters (i.e., d_{50}, d_{90}) (Bray 1982). Bartnik (1992) presents also the proportional dependency of Shields stresses on the shape factor of the grains:

$$- d_i < 0.04 \text{ m}, \quad f_m = 0.0058e^{2.37\,SF} \tag{4}$$

$$- d_i > 0.04 \text{ m}, \quad f_m = 0.0127e^{1.67\,SF} \tag{5}$$

where the shape factor SF = $c/(a\,b)^{0.5}$, and a, b and c are grain dimensions.

The program calculates the critical Shields stresses basing on the characteristic diameters and standard deviation of grain size distribution for input discharges (Bartnik and Strużyński 1997).

1.3 The Hydrological Foundation for Morphological Processes

Different bed transformation appears during various flows. This was graphically presented by Wolman and Miller (1960) as a relationship between transport ratio, applied stress and the frequency of stress application. Bed shear stresses developed by flowing water are hydraulic radius and slope dependent. For this reason, it is hard to connect bed transformation processes to the specific discharge. The same changes of bed material granulometry appearing for different characteristic discharges would indicate the difference of the hydromorphological processes found in the chosen river reach. This may probably be the cause of the existence of similar fluvial processes during bankfull discharge, dominant discharge or various characteristic discharges ($Q_{10\,\%}$, $Q_{25\,\%}$, $Q_{50\,\%}$, and $Q_{75\,\%}$) (Table 1).

In the chapter, bed stability prognosis was done for bankfull discharges estimated with the help of Riley formula (Radecki-Pawlik 2011):

$$BI = \frac{W_i - W_{I+1}}{D_i - D_{i+1}} \tag{6}$$

where BI is the Riley index, W and D are the channel width and depth on the chosen ("i") levels.

The analysis of the changes of BI index (Riley 1972) as well as hydraulic radius (R_h) were performed for the indication of the trends of the cross-sectional shape changes. The R_h was chosen instead of Wolman's (1955) method as it appears to be more sensitive than the channel width/depth ratio analysis. Although the BI and R_h analyses were intended for the search of elevation of the bankfull discharge, it can be also very helpful for the search of the terraces as well as low flow banks. The analysis of both parameters allows also for searching other alluvial forms, i.e., higher elevated branches or banks. This methodology allows to characterize hydromorphological state and current tendency of the fluvial processes. The bankfull discharge was first searched within the values from $Q_{10\,\%}$ to $Q_{25\,\%}$, as proposed by Pickup and Warner (Radecki-Pawlik 2011) but if not found the near range values are considered. The other parameter that has been checked was the dominant discharge which was calculated after Lambor (1971) as the maximal value of

Table 1 Hydrological data

Hydrological data	$Q_{p\,\%}$	1	2.5	5	10	25	50	75
Biały Dunajec	($m^3\,s^{-1}$)	499	357	284	210	121	74	47
Czarny Dunajec	($m^3\,s^{-1}$)	239	192	162	129	78	49	23
Skawa	($m^3\,s^{-1}$)	284	227	191	148	85	48	24

product of probable flow multiplied by the frequency of its appearance. Pickup and Warner (Copeland et al. 2000) propose to calculate the dominant discharge as $Q_{75\ \%}$. Another flow chosen was $Q_{50\ \%}$ which is often taken into consideration and described by Bray (Radecki-Pawlik et al. 2005) as a dominant discharge. Last but not least, the analysis was performed of the longitudinal profile of the river as it influences the hydromorphological dynamics with the slope and channel capacity as the main parameters (Thorne et al. 1997; Church and Jones 1992).

2 Study Site

The researched objects are localized in Southern Poland in the highest, Carpathian region. The river localizations were chosen in two steps. The first was to search for a natural run from orthophotomaps and then the field studies were performed. The rivers Biały Dunajec and Czarny Dunajec are ones which make the Dunajec River in borders of town Nowy Targ. Both of the tributaries start in the Tatra Mountains. The third river, Skawa, flows in the Beskidy Zachodnie Mountains. The investigated 500 m long reaches are identified in Fig. 1. The closing cross-sections for the investigated rivers are located in: southern suburbia of town Nowy Targ (Biały Dunajec) and also in the villages: Chochołów (Czarny Dunajec) and Bystra (Skawa).

The area of Biały Dunajec catchment is 224 km^2 with the river length of 31 km. The research area is located between 3 + 300 and 3 + 870 km. In the researched part Biały Dunajec stays a braided river. The length of Czarny Dunajec is 48 km and the catchment area equals 456 km^2. In the researched part of Czarny Dunajec it is a braided or curvy river. As far as both rivers are concerned, the Biały and Czarny Dunajec beds are incised. The length of the Skawa River is 96.4 km and the catchment area is 1,200 km^2. In its upper run, Skawa is a mountain river and below the

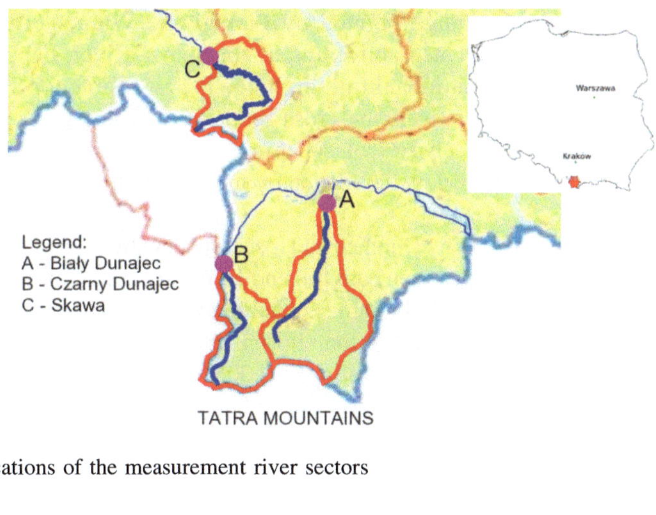

Fig. 1 Locations of the measurement river sectors

Table 2 The land use of sub-basins closed by investigated reaches

River	Urbanized areas (%)	Agricultural areas (%)	Forests (%)
Biały Dunajec	13	44	43
Czarny Dunajec	10	28	62
Skawa	2	50	48

Osielec gorge it becomes a highland river. The research area is located between 61 + 500 and 61 + 700 km, where Skawa stays a meandering river. The land use of sub-catchments close to the chosen locations varies as shown in Table 2.

The rivers flow across mountain areas which are mostly subject to agricultural or forested use, whereas urbanized areas are in the minority. The sub-basin of Czarny Dunajec is mostly forested and less under cultivation. There is a similarity of afforestation and agricultural use between the basins of rivers Biały Dunajec and Skawa with the difference in urbanized areas. The beds and banks of all the three rivers are not fortified. Upstream the measurement reach of Biały Dunajec, the hydropower plants are build. As they cross the river with dams, they block the movement of bed material downstream causing lowering of the bed elevation. The cause of the bed cutting down of Czarny Dunajec is different. The hypothetical cause is the land use where most forests are located (Krzemień 1976). In both Dunajec rivers there exist an exploitation of bed material from the channel. Many publications indicate that this procedure speeds up incising of rivers.

3 Presentation of Measurement Results

Basing on the Chezy formula, the discharge curves and other hydraulic parameters were automatically calculated by the author's computer program for the whole cross-section data (Fig. 2a, b and c). Moreover, the levels of chosen characteristic discharges are shown on the graphs. Only in the case of Czarny Dunajec, bankfull discharge is placed in the range pointed by Pickup and Warner (1976). As for Biały Dunajec River, it is located above and as for Skawa River, below the range of 4–10 years flow. The level of bankfull discharge is best visible for Skawa cross-sections. For Biały Dunajec River, the cross-sections evidence gives much more noticeable level for the discharge of 41 m^3s^{-1} and for Czarny Dunajec the 350 m^3s^{-1} flow. The bankfull level evidence by the use of chosen methods did not give reliable results.

In the measured reach, Czarny Dunajec had lost contact with its terrace even during 100-year flood. The distance between bankfull and 1st terrace elevation equals 3.5, 0.8, and 0.3 m in Czarny Dunajec, Biały Dunajec and Skawa rivers, respectively.

The slopes of the bed measured in the chosen cross-sections of Biały Dunajec, Czarny Dunajec and Skawa were as follows: 0.0057, 0.006 and 0.006 (–). While there are many similarities in Biały and Czarny Dunajec, Skawa varies in the plan view, bed cross-sections, arrangement of riffle-pool sequences and bed granulometry (Table 3).

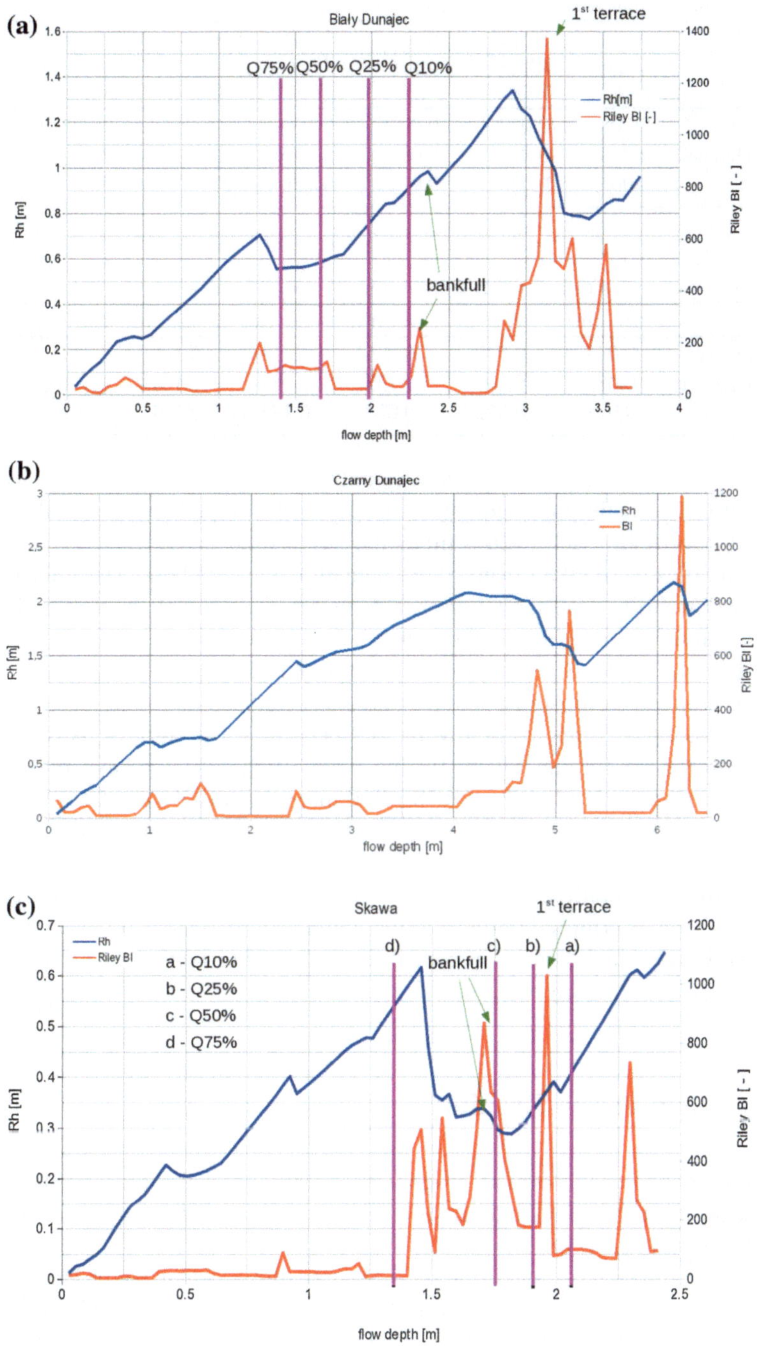

Fig. 2 Hydraulic characterization of the chosen cross-sections. **a** Biały Dunajec, **b** Czarny Dunajec, and **c** Skawa

Table 3 Hydraulic parametrization of investigated reaches in longitudinal

River	Average slope (−)	Slope riffles (−)	Slope pools (−)	I_{riffle}/I_{pool} (−)	Length riffles (m)	Length pools (m)	L_{riffle}/L_{pool} (−)
Biały Dunajec	0.008	0.012	0.004	3.32	60	99	0.61
Czarny Dunajec	0.008	0.015	0.004	3.75	192	276	0.70
Skawa	0.005	0.025	0.002	12.50	25	166	0.15

Table 4 Characteristics of bed material collected for bed stability analysis

River	d_{16} (m)	d_{50} (m)	d_m (m)	d_{84} (m)	d_{90} (m)	δ (−)	SF (−)
Biały Dunajec	0.045	0.090	0.095	0.151	0.162	1.83	0.51
Czarny Dunajec	0.081	0.101	0.135	0.181	0.187	1.49	0.51
Skawa	0.021	0.072	0.077	0.133	0.137	2.55	0.38

The differences are well indicated by both presented riffle-pool ratios of slope and length. The riffle-pool sequences are most developed in Skawa River. The length of pools can reach even 8 times the riffles one.

Bed samples were taken from river channel, banks and bars. The interpretation of the bed stability was then performed on the basis of channel samples. The measured and calculated parameters are shown in Table 4.

The Dunajec rivers bed granulometry shape (Novák 1988) is mostly ellipsoidal; however, elongated ellipsoids, spheres and disks can also be found there (SF in the range from 0.2 to 0.9, the average value of 0.5). The substratum of Skawa channel consists of the Carpathian Flysh. In this river, grains usually have the shape of elongated boards and rods (SF = 0.38). Bed material in Biały and Czarny Dunajec Rivers is mostly granite and quartzite. The finest grains are found in Skawa River (d_{90} = 0.14 m) and the biggest in Czarny Dunajec (d_{90} = 0.19 m). In Biały Dunajec, the bed stabilizing fraction (d_{90}) dimension is 0.16 m. The bed cover of Czarny Dunajec is armored (δ close to the value of 1.3) as stays opposed to the bed of Skawa River which can be indicated as mixed. The riverbed material in Biały Dunajec is also relatively well-sorted.

4 Discussion of the Results

The scenario of the performed prognosis contributes to the development of the armored bed. After the bed material is armored, the higher discharges for removing cover are calculated (Fig. 3).

Standard deviation of sieve curve δ changes the most in Biały Dunajec. The difference between δ value characterizing bed under $Q_{1\,\%}$ flow conditions and the maximum armored state equals 1.72, 1.35 and 0.43 for Skawa, Biały Dunajec and Czarny Dunajec, respectively. This reflects the range of sieve curve changes during the chosen characteristic flows.

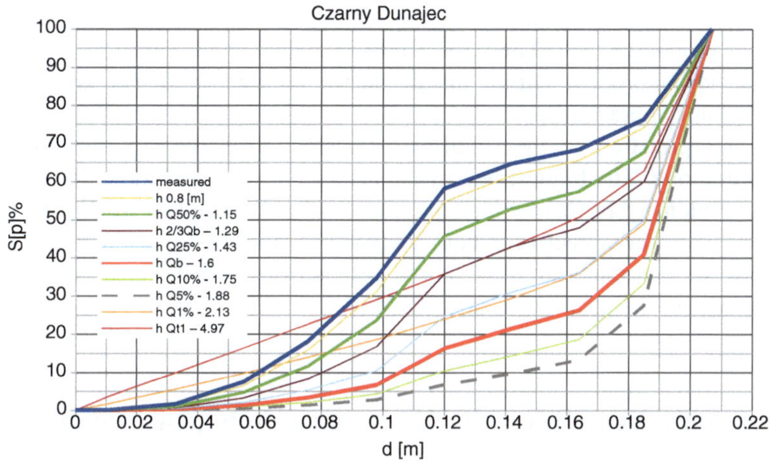

Fig. 3 An exemplary bed stability prognosis presented for the Czarny Dunajec river

The morphological parameters of chosen river reaches differ seriously which is indicated by all the characteristic discharges, i.e., by bankfull discharges which were placed in the range of different probable discharges (Fig. 4). ARMOUR simulation showed a high stability of Czarny Dunajec and Biały Dunajec bed material. Armoring processes last until $Q_{5\ \%}$ for Czarny Dunajec River and until $Q_{25\ \%}$ for Biały Dunajec. In the Skawa River, the bed stability exists up to bankfull discharge which is slightly lower than $Q_{50\ \%}$. The Skawa River is one of the three

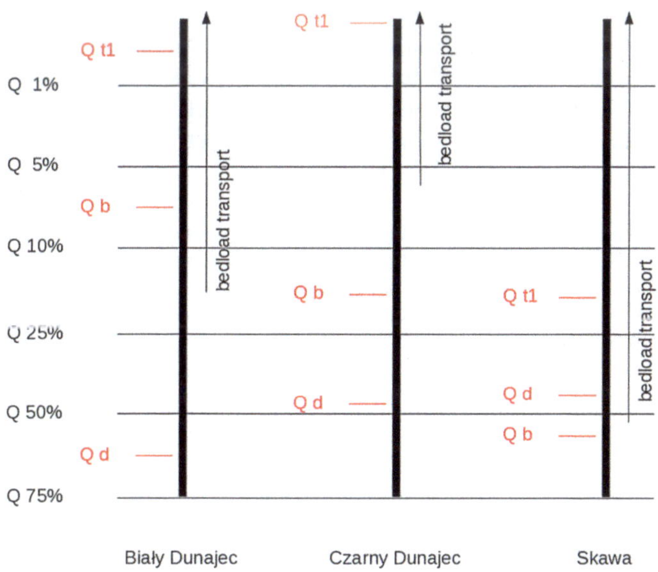

Fig. 4 Track record of the chosen discharges versus the bed transportation process

chosen in which the dominant discharge is bigger than the bankfull discharge. The dominant discharge calculated for Biały Dunajec and Czarny Dunajec rivers causes armoring process. During this discharge in the Skawa River, all fractions of the bed material are transported.

The interpretation of the smallest calculated discharge confirms the results presented above. The highest bed dynamics appears for the Skawa River. Fluvial processes intensity in Czarny Dunajec can be characterized as weak. The erosion processes were very effective in both Biały and Czarny Dunajec due to the reduced supply of the bed material. This caused incision of the bed which is found especially strong in the Czarny Dunajec River.

5 Summary Describing the Hydromorphological Balance of Chosen River Reaches

Biały Dunajec

Bed is stable up to $Q_{25\ \%}$. Bed erosion processes observed in the field. Incision processes exist. Connection with the terrace is lost. Riffle-pool sequences poorly developed. Hydromorphological balance possible after the restoration of delivery of bed material.

Czarny Dunajec

Bed is stable up to $Q_{5\ \%}$. Possible bank erosion processes. Connection with the terrace is lost. Incision processes poor. Riffle-pool sequences weak. River developed new hydromorphological balance in this reach.

Skawa

Bed is not stable but bedload transportation processes do not flush material from the reach. Bed erosion and accumulation processes exist. Connection with the terrace good. No incision processes found in this reach. Riffle-pool sequences well developed. This river is hydromorphologically attractive as an ecosystem. Active transportation hydromorphological balance exists.

Acknowledgments Financial support for this study was provided by Polish National Science Center. Project No N N305 186537: The description of balanced mountain rivers and streams state on the basis on natural morphological parameters. Authors would also send special thanks to Piotr Madej, Paweł Muciek, Miłosz Nitka and Grzegorz Zawadzki who were indispensable in the performed field works.

References

Allan D (2004) Influence of land use and landscape setting on the ecological status of rivers. Limnetica 23(3–4):187–198

Bartnik W (1992) Hydraulika potoków i rzek górskich z dnem ruchomym—początek ruchu rumowiska wleczonego (Fluvial hydraulics of mountain river with mobile bed—beginning of bed load motion). Zeszyty Naukowe Akademi Rolnicza w Krakowie 171:122 (in Polish with English summary)

Bartnik W, Strużyński A (1997) Deformacja koryta cieku po przejściu fali powodziowej na przykładzie potoku Tenczyńskiego (The example of riverbed deformation after the pass of the flood wave on basis of the Tenczyński stream), Konferencja Naukowa "Powódź w dorzeczu górnej Wisły w lipcu 1997 roku" PAN 1998, pp 155–167 (in Polish)

Bartnik W, Strużyński A (1999) Determining hydrodynamic balance in Mountain stream floods. In: 3rd International symposium on ecohydraulics, IAHR and Utah State University, on CD_ROM, Salt Lake City

Bartnik W, Strużyński A (2002) Estimation of hydraulic parameters of armored layer forming in mountain rivers and streams, Advances in hydro-science and engineering. ICHE and Warsaw University of Technology, published on CD-ROM

Bartnik W, Madeyski M, Michalik A (1992) Suspended load and bed load transport in mountain streams determied by different methods, in Erosion and Sediment transport monitoring programs in river basins. In: Proceedings of the Oslo symposium, Aug 1992 IAHS Publ. No 210, pp 3–9

Bartnik W, Bonnenberg J, Florek J (2009) Wpływ utraty naturalnej retencji zlewni na charakterystykę morfologiczną zlewni i cieku (The influence of the loos of natural water storage capacity of a river basin on the morphological characteristics of the river and its basin). Infrastruktura i Ekologia Obszarów wiejskich, Nr 2, PAN, Kraków, pp 69 (in Polish with English summary)

Bray DI (1982) Flow resistance in gravel-bed rivers. In: Hey RD, Bathurst JC, Thorne CR (eds) Gravel-bed rivers: fluvial processes, engineering and management. Wiley, New York, pp 109–137

Brice JC (1975) Air photo interpretation of the form and behaviour of alluvial rivers. Final report to the US Army Research Office

Church M (1992) Channel morphology and typology. In: Carlow P, Petts GE (ed) The rivers handbook, Blackwell Science, Malden

Church M, Jones D (1992) Channel bars in gravel-bed rivers. In: Hey RD, Bathurst JC, Thorne CR (eds) Gravel-bed rivers. Wiley, Chichester, pp 291–338

Copeland RR, Biedenharn DS, Fischenich JC (2000) Channel-forming discharge, US ARMY Corps of Engineers, ERDC/CHL CHETN-VIII-5, pp 1–10

Czoch K, Kulesza K (2006) Warunki referencyjne specyficzne dla typów cieków w Polsce jako podstawa do prac nad oceną ekologicznego stanu wód płynących (Specific reference conditions for surfaces water body (rivers) types in Poland as a base of works on assessment of ecological status of rivers), Infrastruktura i Ekologia Obszarów wiejskich, Nr 4/3/2006, PAN, Kraków, pp 25–36 (in Polish with English summary)

Florek J, Strużyński, A Mucha J (2007) Hydrodynamic effects of flood wave travel along Targaniczanka stream. Acta Scientiarum Polonorum, Formatio circumiectus, 6(4):39–50

Gessler J (1970) Self stabilizing tendency of alluvial channels. J Waterw Harbours Coast Eng Div ASCE 96(2):239–249

Grzybkowska M (2007) Zależności troficzne w wodach słodkich (Trophic relationships in freshwater ecosystems). In: Gwoździński K (ed) Bory Tucholskie i inne obszary leśne Ochrona, monitoring, edukacja (Tucholskie Forest and other forest areas. Protection, monitoring, education). Wydawnictwo Uniwersytetu Łódzkiego, Łódź, pp 213–231

Korpak J, Krzemień K, Radecki-Pawlik A (2008) Wpływ czynników antropogenicznych na zmiany koryt cieków karpackich (Influence of anthropogenic factors on changes of Carpathian stream channels) Infrastruktura i Ekologia Obszarów wiejskich, Nr 4, PAN, Kraków, pp 88 (in Polish with English summary)

Krzemień K (1976) Współczesna dynamika koryta potoku Konina w Gorcach (Current dynamics of the Konina Stream riverbed). Folia Geographica, Series Geographica-Physica 7:87–122 (in Polish with English summary)

Lake PS, Bond N, Reich P (2007) Linking ecological theory with stream restoration. Freshw Biol 52:597–615

Lambor J (1971) Hydrologia inżynierska (Hydrology for engineers). Arkady, Warszawa (in Polish)
Lane EW (1955) The importance of fluvial morphology in hydraulic engineering, In: Proceedings American society of civil engineering, 81, paper 745, pp 1–17
Mueller ER, Pitlick J (2005) Morphologically based model of bed load transport capacity in a headwater stream. J Geophys Res 110:F02016. doi:10.1029/2003JF000117
Novák L (1988) Stabilita dna bystřinných toků (Bed stability of the steep streams). Vysoká Škola Zemědělska, Praha, pp 174 (in Czech)
Pickup G, Warner RF (1976) Effects of hydrologic regime on magnitude and frequency of dominant discharge. J Hydrol 29:51–75
Poole A (2002) Fluvial landscape ecology: addressing uniqueness within the river discontinuum. Freshw Biol 47:641–660
Radecki-Pawlik A (2006) Podstawy hydrogeomorfologii cieków górskich (Basics mountain rivers of hydrogeomorphology). ISBN 83-89968-66-5, Bel studio, Kraków, pp 125 (in Polish)
Radecki-Pawlik A (2011) Hydromorfologia rzek i potoków górskich (Hydro-morphology of the mountain rivers and streams). Wydawnictwo Uniwersytetu Rolniczego w Krakowie, Kraków, pp 288 (in Polish)
Radecki-Pawlik A, Tyrka M, Radecki-Pawlik B, Zasępa P (2005) Wyznaczenie wielkości przepływu brzegowego w dwóch zróżnicowanych zlewniach górskich (Bankfull discharge determination in the two diverse mountainouss catchments) Infrastruktura i Ekologia Terenów Wiejskich, Komisja Technicznej Infrastruktury Wsi PAN/o Kraków, Nr 1, ISSN 1732-5587, pp 69–82 (in Polish with English summary)
Riley SJ (1972) A comparison of morphometric measures of bankfull. J Hydrology 17(1–2): 23–31
Rosgen DL (1994) A classification of natural rivers. Catena 22:169–199
Strużyński A (2006) Skutki powodzi roztopowej w roku 2006 w uregulowanym odcinku delty śródlądowej rzeki Nidy (Consequences of the spring flood in year 2006 on the regulated part of middle delta in the Nida River), Infrastruktura i Ekologia Terenów Wiejskich, Komisja Technicznej Infrastruktury Wsi PAN/o Kraków, Nr 4/1-monografie, ISSN 1732-5587, pp 235–247 (in Polish with English summary)
Strużyński A, Wyrębek M, Strutyński M, Kulesza K (2011) Cross-section changes in the lower part of a mountain river after the flood in spring 2010, as presented by means of CCHE2D program, Experimental methods in hydraulic research, geoplanet: earth and planetary sciences, vol 1, pp 287–297. doi: 10.1007/978-3-642-17475-9_21
Thorne CR, Hey RD, Newson MD (1997) Applied fluvial geomorphology for river engineering and management. Wiley, Chichester
Vannote RL, Minshall GW, Cummins KW, Sedell JR, Cushing CE (1980) The river continuum concept. Can J Fish Aquat Sci 37:130–137
Wang FY (1977) Bed load transport in open channels. In: Proceedings of IAHR, Baden–Baden, vol A9, pp 63–69
Wolman M (1955) The natural channel of Brandywine River, Pensylvania. US Geol Survey 271:56
Wolman MG, Miller JP (1960) Magnitude and frequency of forces in geomorphic processes. J Geol 68(1):54–74
Wyżga B, Radecki-Pawlik A (2011) Jak zmniejszyć zagrożenie i ryzyko powodziowe w dorzeczu górnej Wisły? (How to decrease flood risk in Upper Vistula basin?), Gospodarka Wodna, 10/2011, pp 414–421

Part II
Computational Solutions of Hydraulic Problems

An Analysis of the Impact of Most Important Processes on the Concentration Reduction Values of Chosen Contaminants in a Groundwater (Based on Computational Solutions)

Andrzej Aniszewski

Abstract This chapter presents a general overview of 2D mathematical models for inorganic and organic contaminants moving in an aquifer, taking into consideration the most important processes that occur in the ground. These processes affect, to a different extent, the concentration reduction values for the contaminants moving in a groundwater. In this analysis, the following processes have been taken into consideration: reversible physical non-linear adsorption, chemical and biological reactions (as biodegradation/biological denitrification) and radioactive decay (for moving radionuclides). Based on these 2D contaminant transport models it has been possible to calculate numerically the dimensionless concentration values with and without all the chosen processes in relation to both the natural sites (piezometers) and the chosen contaminants.

1 Introduction

A general description of all the most important mechanisms (processes) affecting the movement of contaminants in a groundwater along with mathematical equations in relation to these processes are presented. It is to be noted that all interactive mechanisms (processes) overlapping in the ground strongly depend on climatic and ground conditions. This is connected, among other things, with negative or positive charging of the solid phase of natural soils (as an "anionic" or "cationic" exchange) (Zheng and Wang 1999; Chiang 2005; Weiss and Cozzarelli

A. Aniszewski (✉)
West Pomeranian University of Technology,
Piastów Street 50a, 70-310 Szczecin, Poland
e-mail: andrzej.aniszewski@zut.edu.pl

P. Rowiński (ed.), *Experimental and Computational Solutions of Hydraulic Problems*, GeoPlanet: Earth and Planetary Sciences, DOI: 10.1007/978-3-642-30209-1_18,

2008; Kraft et al. 2008; Aniszewski 2009, 2011; Taniguchi and Holman 2010). The climatic and ground conditions selected for the presented transport models are described in greater detail by Zheng and Wang (1999), Chiang (2005), Aniszewski (2009, 2011), Taniguchi and Holman (2010).

It should be noted that in various climatic and ground conditions, physico-chemical properties of moving groundwater differ considerably one from another, impacting on all the major processes (parameters) used in the transport models presented here.

The main topic of this chapter is a comparison between the concentration values that have been calculated numerically according to the equations presented below with and without all the chosen processes acting in the groundwater (as Δ values in Table 1 later in the text). The differences Δ allow also to estimate the dimensionless values of concentration reductions in relation to the numerically calculated concentration values without all the chosen processes, treated in the further analysis as the standard errors $(\Delta/C^*_{\max c})$ given in Table 1 (where $C^*_{\max c} = C_{\max c}/C_o$.

The reduced 2D contaminant transport equations (models) are presented in the further part of this paper in relation to both the Polish temperate (cold) climate and the chosen natural sandy aquifer with fine sand filling.

Based on the calculated standard errors given in Table 1 one can assess the importance (hierarchy) of the particular processes affecting the concentration reduction during contaminant movement in ground media in relation to the chosen climatic and ground conditions.

2 General Mathematical Description of Contaminant Transport in a Groundwater without and with the Chosen Processes and Contaminants

The well-known most simplified 2D advection–dispersion equation without adsorption process is as follows (Zheng and Wang 1999; Chiang 2005):

$$\frac{\partial C}{\partial t} + u_x \frac{\partial C}{\partial x} = D_x \frac{\partial^2 C}{\partial x^2} + D_y \frac{\partial^2 C}{\partial y^2} \tag{1}$$

where C is the solute concentration in flowing groundwater in aqueous phase (in the local equilibrium conditions); u_x is the component of the so-called pore groundwater velocity in pore space along the x axis; D_x, D_y are the components of the longitudinal and transverse dispersion coefficients along the x and y axes that depend on the longitudinal and transverse dispersivities (α_L, α_T); t is the co-ordinate of time; (x, y) are the Cartesian co-ordinates of the adopted reference system.

In Eq. (1) the adsorption process was neglected theoretically in the first stage of numerical calculations in relation to analyzed chloride and sulfate indicators. So,

Table 1 The maximum dimensionless values of the chosen concentrations ($C^*_{\max c}$) calculated according to respective equations along with the standard error values ($\Delta/C^*_{\max c}$)

Chosen piezometers				III (8)	IX (4)	VII (6)	X (5)
Dimensionless and dimensional distances from leakage source in lagoon 4 [total distance L to piezometer X(5)]				0.38(x ≈ 40.0 m)	0.66(x ≈ 70.0 m)	0.86(x ≈ 90.0 m)	1.00(x(L) ≈ 105.0 m)
Initial dimensionless concentrations in lagoon 4 (Aniszewski 2009, 2011)		Chlorides (NaCl) [adsorption process]		1.00	1.00	1.00	1.00
		Sulfates (Na_2SO_4) [adsorption process]		1.00	1.00	1.00	1.00
		Nitrates (NO_3^-) [biodegradation process]		1.00	1.00	1.00	1.00
		BOD indicator [biodegradation process]		1.00	1.00	1.00	1.00
		Chosen radionuclide [radioactive decay]		1.00	1.00	1.00	1.00
Dimensionless values of contaminant concentrations $C^*_{\max c}$		Chlorides	Eq. (2)	0.4394[1]	0.2536[1]	0.1837[1]	0.0939[1]
			Eq. (5)	0.4035[2]	0.2383[2]	0.1758[2]	0.0910[2]
		Sulfates	Eq. (2)	0.3846[1]	0.1795[1]	0.1129[1]	0.0743[1]
			Eq. (5)	0.3276[2]	0.1584[2]	0.0978[2]	0.0634[2]
		Nitrates	Eq. (5)	0.4983[2]	0.2171[2]	0.1040[2]	0.0335[2]
			Eq. (9)	0.4979[3]	0.2145[3]	0.1031[3]	0.0333[3]
		BOD indicator	Eq. (5)	0.6524[2]	0.5252[2]	0.3534[2]	0.2126[2]
			Eq. (9)	0.6504[3]	0.5242[3]	0.3524[3]	0.2116[3]
		Chosen radionuclide	Eq. (5)	0.4035[2]	0.2383[2]	0.1758[2]	0.0910[2]
			Eq. (12)	0.3879[4]	0.2308[4]	0.1708[4]	0.0895[4]
Values of standard errors $\Delta/C^*_{\max c}$	Chlorides	$\Delta/C^*_{\max c}$ Eq. (5) × 100 % $\Delta = \lvert C^*_{\max c}$ [Eqs. (2)–(5)]$\rvert$		6.5[5]	5.3[5]	3.9[5]	2.7[5]
	Sulfates	$\Delta/C^*_{\max c}$ Eq. (5) × 100 % $\Delta = \lvert C^*_{\max c}$ [Eqs. (2)–(5)]$\rvert$		14.4[5]	10.7[5]	7.8[5]	6.0[5]
	Nitrates	$\Delta/C^*_{\max c}$ Eq. (9) × 1000 ‰ $\Delta = \lvert C_{\max c}$ [Eqs. (5)–(9)]$\rvert$		9.8[6]	8.7[6]	6.2[6]	0.8[6]
	BOD Indicator	$\Delta/C^*_{\max c}$ Eq. (9) × 1000 ‰ $\Delta = \lvert C^*_{\max c}$ [Eqs. (5)–(9)]$\rvert$		4.7[6]	3.3[6]	2.5[6]	2.0[6]
	Chosen radionuclide	$\Delta/C^*_{\max c}$ Eq. (12) × 100 % $\Delta = \lvert C^*_{\max c}$ [Eqs. (5)–(12)]$\rvert$		4.0[7]	3.2[7]	2.9[7]	1.8[7]

[1] MDC (maximum dimensionless concentrations) acc. to Eq. (2) without adsorption ($R = 1.00$ for chlorides and sulfates). [2] MDC acc. to Eq. (5) with adsorption ($R_a \approx 1.06$ for chlorides, $R_a \approx 2.18$ for sulfates, $R_a \approx 1.82$ both for nitrates and for BOD—same as for nitrates and $R_a \approx 1.06$ for chosen radionuclide—same as for chlorides). [3] MDC acc. to Eq. (9) both with adsorption ($R_a \approx 1.82$ both for nitrates and for BOD indicator—same as for nitrates) and biodegradation ($k_1^* \approx 0.17$ for nitrates and $k_1^* \approx 0.62$ for BOD). [4] MDC acc. to Eq. (12) with adsorption ($R_a \approx 1.06$ for chosen radionuclide—same as for chlorides) and with radioactive decay ($\lambda^* \approx \lambda \cdot t$). [5] SE (standard errors) between MDC acc. to Eqs. (2) and (5) with ($R_a \approx 1.06$) and without ($R = 1.00$) adsorption process for chlorides and with ($R_a \approx 2.18$) and without ($R = 1.00$) adsorption process for sulfates [%]. [6] SE between MDC acc. to Eqs. (5) and (9) with ($k_1^* \approx 0.17$ for nitrates and $k_1^* \approx 0.62$ for BOD) and without ($k_1^* = 0$) biodegradation (with adsorption $R_a \approx 1.82$ both for nitrates and BOD—same as for nitrates) [‰]. [7] SE between MDC acc. to Eqs. (5) and (12) with ($\lambda^* \approx \lambda \cdot t$) and without ($\lambda = 0$) radioactive decay (with adsorption $R_a \approx 1.06$ for chosen radionuclide—same as for chlorides) [%]

for such an assumption, the well-known retardation factor resulting from the adsorption process was neglected in Eq. (1), as

$$R = 1 + \frac{\rho}{m} \cdot \frac{\partial S}{\partial C} = 1.00$$

The dimensionless form of Eq. (1) is

$$\frac{\partial C^*}{\partial \tau} + \frac{\partial C^*}{\partial \xi} = D_x^* \frac{\partial^2 C^*}{\partial \xi^2} + D_y^* \frac{\partial^2 C^*}{\partial \eta^2} \tag{2}$$

in which the following auxiliary dimensionless parameters were taken into account:

$$\frac{x}{L} = \xi; \quad \frac{y}{L} = \eta; \quad \frac{C}{C_o} = C^*; \quad \frac{u_x\, t}{L} = \tau\,; \quad \frac{D_x}{u_x L} = D_x^*; \quad \frac{D_y}{u_x\, L} = D_y^* \tag{3}$$

where L is the measured distance from the source of the contaminant outflow (injection) to the last groundcross-section (piezometer) in the chosen ground, treated as the known length of the solution area in the numerical solution (calculations) of the transport Eqs. (2, 5, 8 and 12) later in the text ($L \cong 105.0$ m); C_o in Eq. (3) is the initial measured contaminant concentration in the source of the outflow (injection) into the chosen natural aquifer, along the vertical slot with assumed infinitesimal width, treated as the known initial concentrations in the numerical calculations of the transport Eqs. (2, 5, 8, and 12).

The detailed description of the field research in natural site with the measurements of both C_o and maximum dimensional concentrations $C_{\max c}$ in piezometers for the chosen contaminants is given in Aniszewski (2009) along with the illustrative map of this natural site (piezometers), given in Aniszewski (2011).

The initial and boundary conditions used in numerical calculations, for all the indicators chosen in this chapter, are given as Eqs. (16) and (17) in the further part of the chapter. These conditions were also related to the considered semi-confined plane of groundwater. Taking into account the symmetry, the area of flow was considered for $0 \langle\ \xi \leq \infty$ and $0 \langle\ \eta \leq \infty$. The numerical calculations for all the indicators we chose were realized based on modified computational program PCCS-2. This program allows obtaining the dimensionless values of the contaminant concentrations (without and with all the processes chosen) in the range <0, 1>.

The well-known 2D advection–dispersion equation with the adsorption process is the following (Zheng and Wang 1999; Chiang 2005):

$$\frac{\partial C}{\partial t}\left(1 + \frac{\rho}{m}\frac{\partial S}{\partial C}\right) + u_x \frac{\partial C}{\partial x} = D_x \frac{\partial^2 C}{\partial x^2} + D_y \frac{\partial^2 C}{\partial y^2} \tag{4}$$

where S is the mass of the solute species adsorbed on the grounds per unit bulk dry mass of the porous medium (in the local equilibrium conditions); ρ is the bulk density of the porous medium; m is the effective porosity of the porous medium;

$[1 + (\rho/m) \cdot (\partial S/\partial C)]$ is the constant in time retardation factor ($R \neq 1.00$) resulting from sorption process (Chiang 2005).

The dimensionless form of Eq. (4) is:

$$\frac{\partial C^*}{\partial \tau} + A_1^* \frac{\partial C^*}{\partial \xi} = A_1^* \cdot D_x^* \frac{\partial^2 C^*}{\partial \xi^2} + A_1^* \cdot D_y^* \frac{\partial^2 C^*}{\partial \eta^2} \tag{5}$$

in which the following auxiliary dimensionless parameters were taken into account:

$$\begin{gathered} \frac{x}{L} = \xi; \quad \frac{y}{L} = \eta; \quad \frac{C}{C_o} = C^*; \quad \frac{u_x\, t}{L} = \tau; \quad \frac{D_x}{u_x L} = D_x^* \\ \frac{D_y}{u_x\, L} = D_y^*; \quad \frac{m}{m + \rho\, N\, K\, C_o^{(N-1)}\; C^{*(N-1)}} = A_1^* \end{gathered} \tag{6}$$

where K, N are the parameters of the Freundlich non-linear isotherm accepted for mathematical description of adsorption process for all the indicators chosen in this chapter (K is the Freundlich constant and N is the Freundlich exponent, that depend on the solute species, the nature of the porous medium and some other conditions of the system).

So, the general equation describing retardation factor for all the indicators adopted here takes the form:

$$R = 1 + \frac{\rho}{m} \cdot \frac{\partial S}{\partial C} = 1 + \frac{\rho}{m} \cdot N \cdot K \cdot C^{(N-1)} \tag{7}$$

Numerical values of the calculated retardation factors based on Eq. (7) (treated as average values R_a) for all the chosen indicators are given in the explanation under Table 1 (in footnote 2).

The well-known 2D advection–dispersion equation with biodegradation/biological denitrification and adsorption processes is the following (Zheng and Wang 1999; Chiang 2005):

$$\frac{\partial C}{\partial t}\left(1 + \frac{\rho}{m}\frac{\partial S}{\partial C}\right) + u_x \frac{\partial C}{\partial x} = D_x \frac{\partial^2 C}{\partial x^2} + D_y \frac{\partial^2 C}{\partial y^2} - k_1\left(C - \frac{\rho\, S}{m}\right) \tag{8}$$

where k_1 is the first-order reaction rate for the kinetically-controlled biodegradation process (biological denitrification).

However, in this analysis, for biodegradation process, the same parameter, k_1, was assumed for both the dissolved (aqueous) and the sorbed (solid) phases (with a possibility of certain differences between the values of this parameter in relation to the two above-mentioned phases) (Zheng and Wang 1999; Chiang 2005; Weiss and Cozzarelli 2008; Kraft et al. 2008; Joekar-Niasar et al. 2008; Aniszewski 2011).

The nitrate ions (NO_3^-) and BOD indicator (biochemical oxygen demand, which can also be used for determining indirectly the concentration of substances susceptible to biodegradation process) were chosen in the present analysis (Chiang 2005; Weiss and Cozzarelli 2008; Kraft et al. 2008; Joekar-Niasar et al. 2008).

Basing on the analyzed literature, in the case of nitrate biodegradation, one can rather speak of biological transformation (reaction), treated in a saturated zone mainly as biological denitrification. More detailed explanations concerning the complex biodegradation process in a natural groundwater is given by (Zheng and Wang 1999; Chiang 2005; Weiss and Cozzarelli 2008; Kraft et al. 2008; Joekar-Niasar et al. 2008; Aniszewski 2011).

The dimensionless form of Eq. (8) is:

$$\frac{\partial C^*}{\partial \tau} + A_1^* \frac{\partial C^*}{\partial \xi} = A_1^* \, D_x^* \frac{\partial^2 C^*}{\partial \xi^2} + A_1^* \, D_y^* \frac{\partial^2 C^*}{\partial \eta^2} - k_1^* \, C^* (A_1^* + A_2^*) \tag{9}$$

in which the following auxiliary dimensionless parameters were taken into account:

$$\begin{gathered}
\frac{x}{L} = \xi; \ \frac{y}{L} = \eta; \ \frac{C}{C_o} = C^*; \ \frac{u_x \, t}{L} = \tau; \ \frac{D_x}{u_x L} = D_x^*; \ \frac{D_y}{u_x \, L} = D_y^* \\
k_1^* = k_1 \, \frac{L}{u_x} = k_1 \, t; \quad \frac{m}{m + \rho \, N \, K \, C_o^{(N-1)} \, C^{*(N-1)}} = A_1^* \\
\frac{\rho \, K \, C_o^{(N-1)} \, C^{*(N-1)}}{m + \rho \, N \, K \, C_o^{(N-1)} \, C^{*(N-1)}} = A_2^*
\end{gathered} \tag{10}$$

The well-known 2D advection–dispersion equation with radioactive decay and adsorption processes (Zheng and Wang 1999; Chiang 2005) is the following:

$$\frac{\partial C}{\partial t} + u_x \frac{\partial C}{\partial x} = D_x \frac{\partial^2 C}{\partial x^2} + D_y \frac{\partial^2 C}{\partial y^2} - \frac{\rho}{m} \frac{\partial S}{\partial t} - \lambda \left(C - \frac{\rho \, S}{m}\right) \tag{11}$$

where λ is the first-order decay rate usually expressed as a half-life ($t_{1/2}$) (assuming that, generally, if the reaction is radioactive decay, the first-order decay constant, λ, for the dissolved phase should be equal to such a constant for the sorbed solid phase.

The numerical calculations were carried out for the theoretical radionuclide taken from the literature (following Aniszewski 2009), with the first-order decay rate constant $\lambda \cong 0.01 \ s^{-1}$ (for the relatively short half-life time $t_{1/2} \cong 70 \ s$).

However, based on the analyzed literature, we also see that the numerical value of the first-order decay rate constant λ (expressed as a half-life $t_{1/2}$) can be in a huge range for various kinds of radionuclides moving in a groundwater stream (Zheng and Wang 1999; Chiang 2005; Aniszewski 2009).

The dimensionless form of Eq. (11) is:

$$\frac{\partial C^*}{\partial \tau} + A_1^* \frac{\partial C^*}{\partial \xi} = A_1^* \cdot D_x^* \frac{\partial^2 C^*}{\partial \xi^2} + A_1^* \cdot D_y^* \frac{\partial^2 C^*}{\partial \eta^2} - \lambda^* \cdot C^* (A_1^* + A_2^*) \tag{12}$$

in which the following auxiliary dimensionless parameters were taken into account:

$$\frac{x}{L} = \xi;\ \frac{y}{L} = \eta;\ \frac{C}{C_o} = C^*;\ \frac{u_x\,t}{L} = \tau;\ \frac{D_x}{u_x L} = D_x^*;\ \frac{D_y}{u_x\,L} = D_y^*$$
$$k_1^* = k_1\,\frac{L}{u_x} = k_1\,t;\quad \frac{m}{m + \rho\,N\,K\,C_o^{(N-1)}\,C^{*(N-1)}} = A_1^* \tag{13}$$
$$\frac{\rho\,K\,C_o^{(N-1)}\,C^{*(N-1)}}{m + \rho\,N\,K\,C_o^{(N-1)}\,C^{*(N-1)}} = A_2^*$$

The numerical values of the calculated dimensionless and maximum ($\eta = 0$) concentrations ($C^*_{\max c} = C_{\max c}/C_o$) based on Eqs. (2), (5), (9) and (12) are given in Table 1 and referred to the footnotes: (1) (without adsorption), (2) (with adsorption), (3) (with biodegradation and adsorption) and (4) (with radioactive decay and adsorption) in the previously chosen piezometers, as described by Aniszewski (2009, 2011).

3 Final Results of Numerical Calculations

The numerical calculations of the dimensionless maximum concentrations ($C^*_{\max c}$) for all the indicators chosen in this chapter were carried out based on the well-known "upwind" scheme that is described, in detail, e.g., in Szymkiewicz (2010). In the adopted numerical solution of Eqs. (2, 5, 9 and 12) the values of dimensionless steps of the difference scheme grid ($h^* = \Delta x/L \approx 0.01$, $k^* = \Delta y/L \approx 0.01$) were determined in relation to the real range scale ($L \cong 105.0$ m), under the assumption that the numerical value of the Peclet number Pe should not be larger than two ($Pe \leq 2$):

$$Pe = \frac{u_x \cdot \Delta x\ (\Delta y)}{D_x\ (D_y)} = \frac{u_x \cdot \Delta x\ (\Delta y)}{\alpha_L \cdot u_x\ (\alpha_T \cdot u_x)} = \frac{\Delta x\ (\Delta y)}{\alpha_L\ (\alpha_T)} \leq 2 \tag{14}$$

For such an assumption, the difference scheme grid should also be no larger than twice the dispersivities (α_L, α_T) as the characteristic length of heterogeneities of the ground systems. In practice, the Peclet number constraint is often relaxed outside the area of interest, where a lower predictive accuracy is acceptable (Szymkiewicz 2010). However, the dimensionless time step ($w^* = u_x \Delta t/L \approx 0.001$) was adopted to minimize also the so-called "numerical dispersion" and "artificial oscillations" (under the assumption that the numerical value of the Courant number C_a should not be larger than one):

$$C_a = u_x \cdot \Delta t/\Delta x \leq 1 \tag{15}$$

In these numerical calculations (for the adopted dimensionless steps of the difference scheme grids) the important consistency, stability and convergence conditions were also preserved to minimize the “numerical dispersion” and “artificial oscillations” in relation to all the above-mentioned equations describing particular processes (Szymkiewicz 2010). More details concerning the adopted “upwind” scheme is given also in Aniszewski (2009, 2011). The initial and boundary dimensionless conditions for numerical solution of Eqs. (2), (5), (8) and (12) were adopted in the following form:

- initial condition:

$$\begin{aligned} C^*(0 \langle \xi \le \infty,\ 0 \langle \eta \le \infty,\ \tau = 0) = 0 \\ C^*(\xi = 0,\ \eta = 0,\ \tau = 0) = 1 \end{aligned} \tag{16}$$

- boundary conditions:

$$\begin{aligned} C^*(\xi = 0,\ 0 \langle \eta \le \infty,\ \tau \rangle\ 0) = 0 \\ C^*(\xi = 0,\ \eta = 0,\ \tau \rangle\ 0) = 1 \\ C^*(\xi \to \infty,\ 0 \langle \eta \le \infty,\ \tau \rangle\ 0) = 0 \end{aligned} \tag{17}$$

Recapitulating all the results (in % and ‰) of the standard error calculations resulting from the above-mentioned computational (numerical) solutions as $(\Delta / C^*_{\max c})$ are the following (see Table 1):

- The standard errors between the maximum dimensionless concentrations based on Eqs. (2) and (5) (without and with adsorption processes) are: from 2.7 to 6.5 for chlorides and from 6.0 to 14.4 for sulfates (as the maximum errors) (values referred to footnote 5, in %);
- The standard errors between the maximum dimensionless concentrations based on Eqs. (5) and (9) (without and with biodegradation process and with adsorption in both cases) are: from 0.8 to 9.8 for nitrates and from 2.0 to 4.7 for the BOD indicator (as the minimum errors) (values referred to footnote 6 in ‰);
- The standard errors between the maximum dimensionless concentrations based on Eqs. (5) and (12) (without and with radioactive decay and with adsorption in both cases) are: from 1.8 to 4.0 for the chosen radionuclide (values referred to footnote 7, in %).

The numerical calculations and standard errors presented in Table 1 of this paper confirm the importance (hierarchy) of the chosen processes occurring in the considered groundwater in relation to concentration reduction values (from sulfates causing the biggest reduction to nitrates causing the lowest one).

4 General Conclusions

Basing on all the numerical calculations presented here and the earlier author’s research, the final conclusions can be summarized as follows:

- Basing on the numerical calculations, one can conclude that the non-linear adsorption process is of very great importance in terms of the concentration reduction, especially for the sulfates and to a lesser extent for the chlorides moving in a natural aquifer (see standard errors referred to footnote 5 in Table 1) (Aniszewski 2009, 2011).
- Basing on the numerical calculations, one can say that the biodegradation of the chosen compounds (nitrates and the BOD indicator) moving in a natural groundwater is of a very small importance in terms of the concentration reduction (see standard errors referred to footnote 6 in Table 1) (Aniszewski 2011).
- Basing on the numerical calculations, one can say that the radioactive decay is of a small importance in terms of the concentration reductions (see standard errors referred to footnote 7 in Table 1) (Aniszewski 2009).
- Practical engineering applications of the presented equations should also be underlined in relation to various kinds of assessments of environmental impacts and their usefulness as a relatively simple and cheap mathematical tool related mainly to definition of groundwater quality treated as a fundamental source of drinking water reservoirs.

It should also be noted that it is very hard to define concentration values based on the numerous more-or-less similar literature models (equations) and then to compare them with the concentration values obtained with the author's models presented here. This is connected with different climatic and ground conditions and some other important factors, namely:

- Different scales (lengths) of contaminant propagation plumes used each time, with the necessity of determining geometric scale-dependent similarity for the contaminant plumes (containing various micro- or macroscopic ground heterogeneities).
- Different selection or calculation of all the required parameters occurring in the equations presented here and literature ones, based on laboratory of field measurements (as various numerical values ranging over large intervals, also as concerns different climatic and ground conditions.

References

Aniszewski A (2009) Mathematical modeling and practical verification of groundwater and contaminant transport in a chosen natural aquifer. Acta Geophys 57:435–453

Aniszewski A (2011) Particular application of a mathematical transport model incorporating subsurface reactive pollutants. Acta Geophys 59:110–123

Chiang WH (2005) 3D-groundwater modeling with PMWIN. A simulation system for modeling groundwater flow and pollution. Springer, Heidelberg

Joekar-Niasar V, Ataie-Ashanti B, Sattari Z (2008) Large scale modeling of nitrogen transformation in the unsaturated zone—a case study of Tehran City, Iran. Taylor & Francis Group, London, Chap. 17, doi: 10.1201/9780203894569. ch17, 169–179

Kraft GJ, Browne BA, De Vita WM, Mechenich DJ (2008) Agricultural pollutant penetration and steady state in thick aquifers. Ground Water 46:41–50

Szymkiewicz R (2010) Numerical modelling in open channel hydraulics. Water science technology library. Springer, Dordrecht

Taniguchi M, Holman IM (2010) Groundwater response to changing climate. CRC Press, Boca Raton. doi:10.1201/b10530-18

Weis JV, Cozzarelli IM (2008) Biodegradation in contaminated aquifers: incorporating microbial/molecular methods. Ground Water 46:305–322

Zheng CH, Wang PP (1999) MT3DMS: a modular transport three-dimensional multispecies transport model for simulation of advection, dispersion, and chemical reactions of contaminants in groundwater systems; documentation and user's guide. Department of Geological Sciences, University of Alabama, Tuscaloosa, AL 35487, Contract Reports SERDP-99

Some Properties of Lagrangian Modeling of Saltating Grains Over Movable Bed

Włodzimierz Czernuszenko and Robert J. Bialik

Abstract The two-dimensional Lagrangian model of saltating grains is developed to simulate the particles' movement over movable bed and for provision of some properties of this kind of modeling. Stochastic method of the particle's collision with the channel bed is applied to identify and quantify the effects of the so-called cohesion coefficient on the particles behavior. The relationships between this coefficient and the coefficients of friction, restitution and an angle of collision are also presented. It is shown that the cohesion coefficient may be used to determinate the amount of momentum transferred by moving particles to the bottom and vice versa and controlling of its value may allow for the calculation of the velocity of moving bed.

1 Introduction

The sediment transport is usually divided into three parts (Graf and Altinakar 1998): wash load, suspended load, and bed load. In the bed load layer, particles may slide, roll or move in the form of shorter or longer jumps and the latter form of bed load is usually called saltation. On the other hand, Gao (2008) claimed that particles in the bed load are mostly transported into two regimes: saltation or sheet- flow. In the saltation regime, as previously defined, the particles roll, slide or

W. Czernuszenko · R. J. Bialik (✉)
Institute of Geophysics, Polish Academy of Sciences, Ks. Janusza 64, 01–452 Warsaw, Poland
e-mail: rbialik@igf.edu.pl

W. Czernuszenko
e-mail: wczer@igf.edu.pl

P. Rowiński (ed.), *Experimental and Computational Solutions of Hydraulic Problems*, GeoPlanet: Earth and Planetary Sciences, DOI: 10.1007/978-3-642-30209-1_19,

jump on the bed. On the other hand, in the sheet flow regime the particles are transported due to the collisions with other grains and in this form the bed is completely mobile. Modelling of sheet flow and saltation has attracted the attention of a number of researchers since the 1980s: (1) regarding the sheet flow (i.e., Sayed and Savage 1983, Hanes and Bowen 1985, Drake and Calantoni 2001, Gao 2008 or Yeganeh-Bakhtiary et al. 2009); and (2) regarding the saltation (i.e., Wiberg and Smith 1985, Lee et al. 1994, Nino and Garcia 1994, Lukerchenko et al. 2006 or Bialik et al. 2012). However, in each of these works the saltation is considered as a form of the bed load transport as defined by Graf and Altinakar (1998) rather than the bed load regime and, moreover, the movement of particles is mostly modeled over a non-moving bed. Therefore, the main objective of this work is to show some properties of modeling of saltating particle motion over a movable bed. The academic problem of the influence of the cohesion coefficient on saltating grains is deeply investigated. Moreover, this chapter provides some possible solution and the connection between modeling of saltation and sheet flow.

2 A Theoretical Model for Saltating Grain in Water

The motion of a saltating spherical particle in a fluid may be described by Newton's equation and presented in the one of the forms proposed, for example, by Wiberg and Smith (1985), Nino and Garcia (1994), Lee and Hsu (1994) or Bialik (2011a). Generally, this expression is as follows (Hinze 1975):

$$m_s \frac{\mathrm{d}\boldsymbol{v}_s}{\mathrm{d}t} = \boldsymbol{F}_D + \boldsymbol{F}_L + \boldsymbol{F}_M + \boldsymbol{F}_v + \boldsymbol{F}_B + \boldsymbol{F}_g + \boldsymbol{F}_{pc} \tag{1}$$

where m_s stands for the mass of solid particle (sediment) in water flow, $\boldsymbol{v}_s$ denotes the velocity of solitary particle (grain) along the bed in water stream. $\boldsymbol{F}_D$ is the drag force that acts on the particle in a uniform pressure field when there is no acceleration of the relative velocity between the particle and the conveying fluid. $\boldsymbol{F}_L$ describes the lift force, i.e., the Saffman lift force that is due to the pressure distribution developed on a particle due to rotation induced by velocity gradient. The higher velocity on the top of the particle gives rise to a low pressure, and the high pressure on the low velocity side gives rise to a lift force. On the other hand, $\boldsymbol{F}_M$ denotes the Magnus force, that is, the lift developed due to rotation of the particle. The lift is caused by a pressure difference between the two sides of the particle resulting from the velocity differential due to rotation that may be caused by a source other that the velocity gradient. Next, two unsteady forces are included in the model. $\boldsymbol{F}_v$ is the force representing virtual or apparent mass effect. This force relates to the force required to accelerate the surrounding fluid. The other one of unsteady forces is the Basset force $\boldsymbol{F}_B$. This force accounts for the viscous effects due to acceleration. The value of the Basset force depends on the acceleration history up to the present time. $\boldsymbol{F}_g$ denotes the force due to the gravity acceleration.

The last force on the right hand side of Eq. (1) that should be taken into the model account is the force $\boldsymbol{F}_{pc}$ exerted on the saltating particle from another particle during the collision between them (for details about the particle–particle collision force, see Czernuszenko 2009).

Equation (1) has to be supplemented with (a) the trajectory equation; (b) description of the flow velocity field that in the considered situation fulfils the logarithmic law; and (c) the procedure for the calculation of the post-collision particle's velocity with the channel bottom (Fig. 1) and then is easily numerically solved using the fourth-order Runge-Kutta method. The particles' post-collision velocities in the case of sliding of particles during the collision with the channel bed are easy to calculate with the use of impulse equation and thus, the post-collisional velocities of the flying and laying particles may be obtained from the following expressions (Czernuszenko 2009):

$$\boldsymbol{v}_1 = \boldsymbol{v}_1^0 - (\boldsymbol{n} - f\boldsymbol{t})(\boldsymbol{n} \cdot \boldsymbol{V}^0)\frac{(1+e)m_2}{m_1 + m_2} \tag{2}$$

$$\boldsymbol{v}_2 = \boldsymbol{v}_2^0 - (\boldsymbol{n} - f\boldsymbol{t})(\boldsymbol{n} \cdot \boldsymbol{V}^0)\frac{(1+e)m_1}{m_1 + m_2} \tag{3}$$

where $\boldsymbol{v}_1$ and $\boldsymbol{v}_2$ are the post-collision velocities vectors of particles 1 and 2, respectively, and the m_1 and m_2 stand for the masses of these particles, $\boldsymbol{V}^0$ is the relative particles' velocity vector and superscript '0' stands for the particles' properties before the collisions. Moreover, coefficient f describes friction and coefficient e denotes restitution. The post-collisional velocities in the case of particle-wall interaction are easy to get from Eqs. (2) to (3) by assuming that the

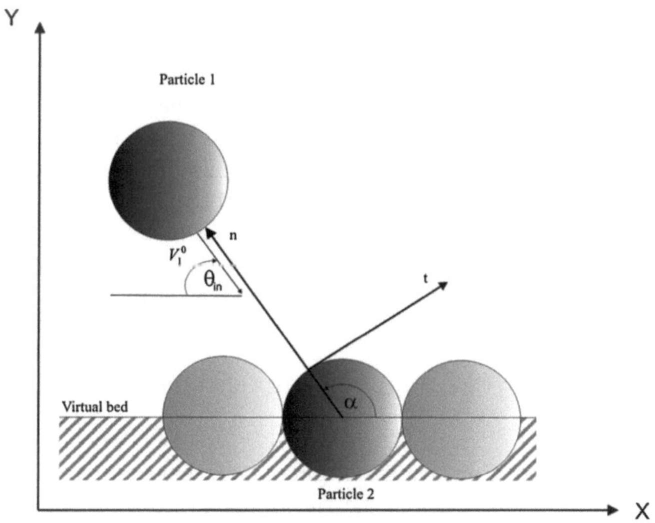

Fig. 1 Scheme of particle collision with the channel bed

bed is formed by uniformly packed spheres with the same size as the transported particles. Moreover, it is assumed that particles' velocity mostly depend on the ratio m_2/m_1 and the investigation of this ratio and its influence on the particles movement close to the channel bed is the main goal of the presented paper.

3 Interpretation of the Ratio m_2/m_1

In general, the objective of all researchers involved in the analysis of the bed load transport is to describe the field of the average sediment concentration that moves with a random velocity. The field of concentration may be identified with the probability distribution of the sediment particle position that in the initial moment of time t_0 was located in the point x_0. The problem of determining the average concentration field is reduced to seeking the probability density function of location of sediment particles $X(x, t)$, where x is Lagrangian coordinate and X is Euler coordinate. Finding the analytical form of this function is very difficult, if at all possible, because of the difficulty in determining the particle velocity $\boldsymbol{V}(\boldsymbol{x}, t)$, which depends on many random collisions with other grains. Let us determine the location of the sediment particles, i.e., define a function $\boldsymbol{X}(\boldsymbol{x}, t)$ for one particular case when the first particle is moving with a known velocity $\boldsymbol{v}_1^0$ and then hits another particle that lays on the bed, i.e., the velocity of the second particle is $v_2^0 = (0{,}0)$. Then, the particle movement depends on the angle of collision; after the collision the particle is reflected at further or closer distance from the bed. The permissible range of collision angles depends on the angle of impact and the geometry of the particles; for example, for the uniformly packed, spherical particles this range is (0°, 120°). When the angle of impact is in the range (0°, 30°), then one can show that the minimum collision angles vary linearly with the angle of impact and the maximum angle of collision can be computed from the formula proposed by Rowiński and Czernuszenko (1999). For the impact angles higher than 30°, the collision angle is chosen randomly with the uniform probability from the range (60°, 120°). Recently, for the particles with non-uniform diameters analogous expressions for the maximum and minimum values of the angle of collision surface have been derived by Bialik (2011b).

Let us now consider all possible trajectories of our particle in a space-time $(\boldsymbol{r}, t)$ by which it can move. This will be the set of all possible functions $r(\boldsymbol{x}(t), y(t))$ that after Rowiński et al. (2005) may have the following form:

$$r(\mathbf{x}(t), y(t)) = \int_0^t v(\mathbf{x}(\tau), y(\tau))d\tau \tag{4}$$

such that their initial position is at the origin of coordinate system, i.e., at the point (0,0). To remind, Eq. (4) is called the functional expression as the independent variable is a function describing all the trajectories in which the particle can move.

We are interested in the distribution of values of $y(t)$. This distribution describes the probability that the particle will never travel at a greater distance (height) from the channel bottom than the size of a:

$$P(y(t) < a) = \int_0^t v(x(\tau), y(\tau))\, d\tau < a \tag{5}$$

Let us now consider a special case of particle motion, i.e., as before, the particle starts to move from the point $x = 0$, but further, if the particle in time t is in point $\boldsymbol{x}$, it is likely that the particle stops. In theory, this is called colloquially 'accidental motion with holes'. It is possible that at the point $\boldsymbol{x}$ there is a hole in which the particle falls into and will never move again. This is the case quite often found in the studies of saltating particles. In the probability distributions of particle trajectories, which does not move to a greater distance than the arbitrarily chosen number a, it is assumed that the particle selected for the simulation does not fall into the hole. This can be done properly by selecting the range of collision angles as a function of incidence angle and depending on the coefficient of cohesion (m_2/m_1). To remind, this coefficient is defined as the ratio of the mass of the particle (2) resting on the bottom to the mass of flowing particle (1), i.e., the ratio m_2/m_1. For very large values of this ratio, the particles that lay on the channel bottom are somewhat more compact, still and focused. On the other hand, for small values of this ratio, these particles are less dense and more mobile. This ratio behaves similarly to the restitution coefficient e, which is defined as the ratio of the particles' velocity after the collision to the particles velocity before the collision. Figure 2 schematically shows the movement of saltating particles over the

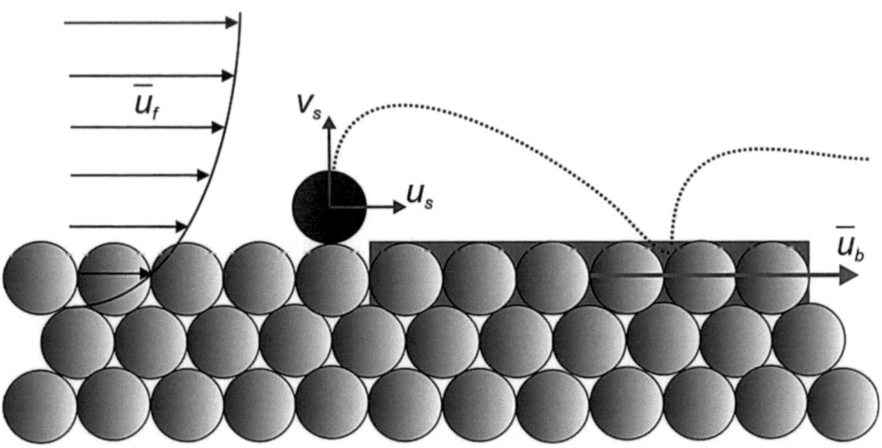

Fig. 2 Scheme of the particles movement over movable bed, v_s and u_s are the longitudinal and vertical velocities of particles, respectively, u_f is the flow velocity, and u_b is the mean velocity of the channel bed

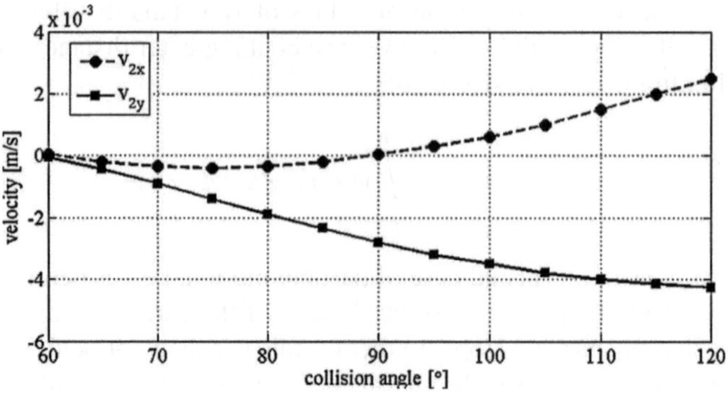

Fig. 3 Particle velocity of the surface layer vs. collision angle, for the following data: $V_1 = (0.074, -0.043\ \text{m/s})$, $\theta_{in} = 30°$, $m_2/m_1 = 60$

movable bed. Due to the compact arrangement of the particles it is assumed that in the layer of the moving bed the velocity of all particles is the same.

Let us examine the role of the ratio m_2/m_1 (cohesion coefficient) on the concentration of the bottom particles, for the case when the flying particle strikes the particles lying on the bottom at an angle of 30° with the velocity (0.074, −0.043 m/s) and for different possible angles of collision from the range (60°, 120°). Firstly, let us assume that the ratio $m_2/m_1 = 60$. Figure 3 shows that, depending on the angle of collision, the particle that lies on the channel bottom receives more or less momentum, and so for the impact angles from the range (60°, 90°) the momentum causes the particle horizontal velocity to be approximately zero. However, for the angle of 120° the momentum is the greatest, causing the particle horizontal velocity to be equal to 2.4 mm/s. Moreover, averaging horizontal velocity within all possible angles of collision, we get $v_{2x} = 0.49$ mm/s. While averaging the vertical velocity, we obtain $v_{2y} = -2.54$ mm/s.

The velocity of bottom particles also depends on the angle of impact of collisions. Figure 4 shows the relation between the longitudinal particle velocity and the angle of collision for incidence angle equal to 30° and the ratio m_2/m_1 from the range of 5 to 40. One can see that the greater the value of the angle of collision, the greater the average longitudinal particle velocity.

Figure 5 shows the dynamics of the changes of the particle velocity components for larger values of m_2/m_1, up to 1,000, and for the impact angle equal to 30°. For m_2/m_1 of 1,000, the values of velocity are approximately equal to 0. This result is interpreted in this way that the particles are arranged on the bottom so concisely that their movement is almost impossible, i.e., their concentration is maximum and for the uniform spherical particles is equal to $\pi/6 = 0.5236$. It is generally assumed that the maximum possible concentration is 0.6 or even 0.65 (see Bagnold 1973). Determination of the minimum particles velocity can be linked to the determination of the maximum concentration, and so, for $v_1^0 = (0.074, 0.043\ \text{m/s})$ that it can be assumed on the basis of Fig. 5 is equal to (0.00006, 0.00031 m/s) for

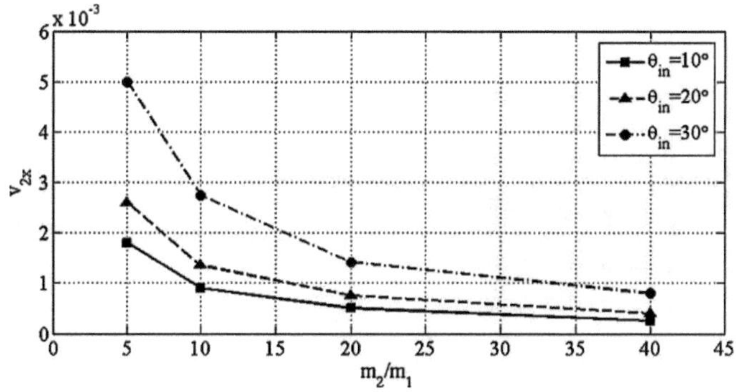

Fig. 4 Dependence of the average longitudinal particle velocity on the ratio m_2/m_1 and on the angle of incidence θ_{in}

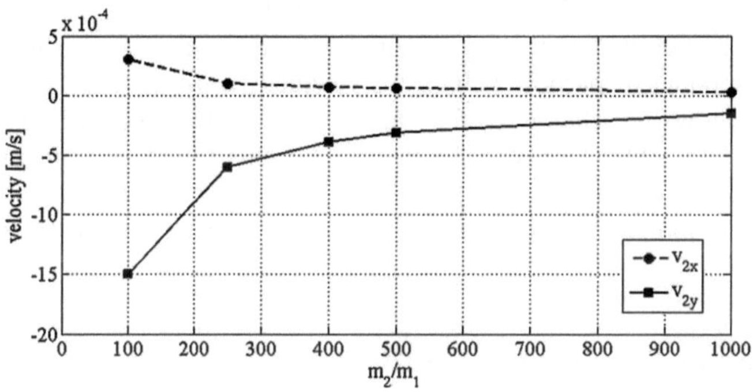

Fig. 5 Estimation of the minimum velocity for the particles lying on the bed for different values of m_2/m_1 and $\theta_{in} = 30°$ with the assumption that $v_1^0 = (0.074, -0.043 \text{ m/s})$

$m_2/m_1 = 500$, as 0.075 % of the initial value, i.e., the maximum concentration corresponds to the cohesion coefficient equal to 500.

We will now consider the relation between the coefficient of restitution and the ratio m_2/m_1. In analyzing of the bed load sediment transport process, these coefficients are dependent on each other. It is sufficient to operate a single factor in the analysis of sediment movement in the longitudinal direction, such as adopting a coefficient $e = 0.01$ and a ratio $m_2/m_1 = 100$. The simulation assuming these values of the coefficients does not cause any changes in the value of the longitudinal velocity component and only affects the transverse component.

To show the relation described above, we will rewrite Eqs. (2) and (3) in the following forms:

$$v_2 = v_2^0 - (\boldsymbol{n} - f\boldsymbol{t})(n \cdot V^0)[\mathrm{w}_2] \tag{6}$$

$$v_1 = v_1^0 - (\boldsymbol{n} - f\boldsymbol{t})(\boldsymbol{n} \cdot V^0)[\mathrm{w}_2] \tag{7}$$

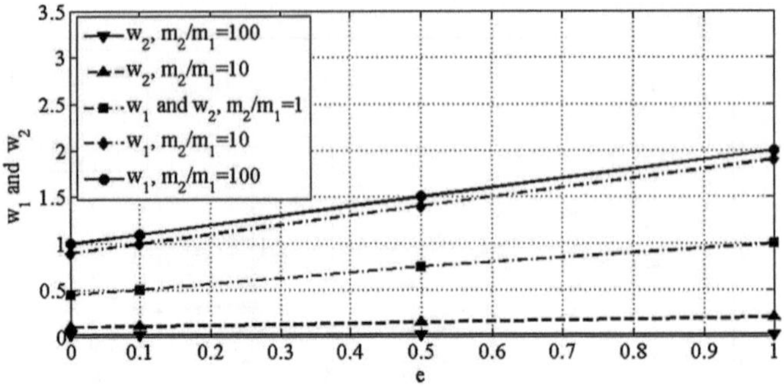

Fig. 6 Values of the factors $[w_1]$ and $[w_2]$ as a function of friction and restitution coefficients for the three selected ratios $m_2/m_1 = 1$, 10 and 100

where $[w_1] = (1+e)\ m_2/(m_1 + m_2)$, $[w_2] = (1+e)\ m_1/(m_1 + m_2)$

Any combination of selected coefficients e and m_2/m_1 does not affect the value of $[w_1]$, i.e., do not change the longitudinal velocity of flowing particle $\boldsymbol{v}_1$ and the vertical component of particle velocity is reduced $[w_2]$-fold.

Figure 6 shows the change of the values of w_1 and w_2 for e of 0.01–1 and for m_2/m_1 from 1 to 100. Moreover, for $m_2/m_1 = 1$ this figure presents symmetry in the values of the coefficients $[w_1]$ and $[w_2]$, while for larger values of the ratio m_2/m_1, two components of the particle's velocity are constrained to zero.

4 Preliminary Numerical Results

Let us recall in a few sentences how the simulation of a single trajectory is carried out: firstly the flying particle comes with a known velocity and hits the second particle that lies on the bed. Shortly before the collision, the angle of incidence is calculated. Next, the angle of collision is generated with uniform probability of its value in the range of allowable angles, as described before. The permissible range of collision angles for spherical particles depends only on the angle of impact

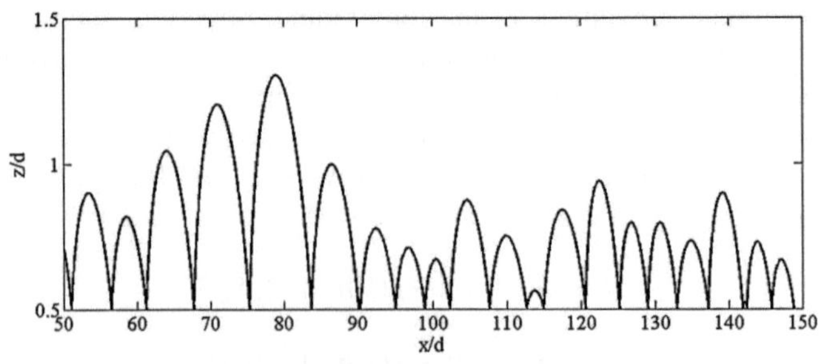

Fig. 7 Particle trajectory for $d = 2.47$ mm, $u_* = 0.075$ m/s, $e = 1, f = 0$, $m_2/m_1 = 100$

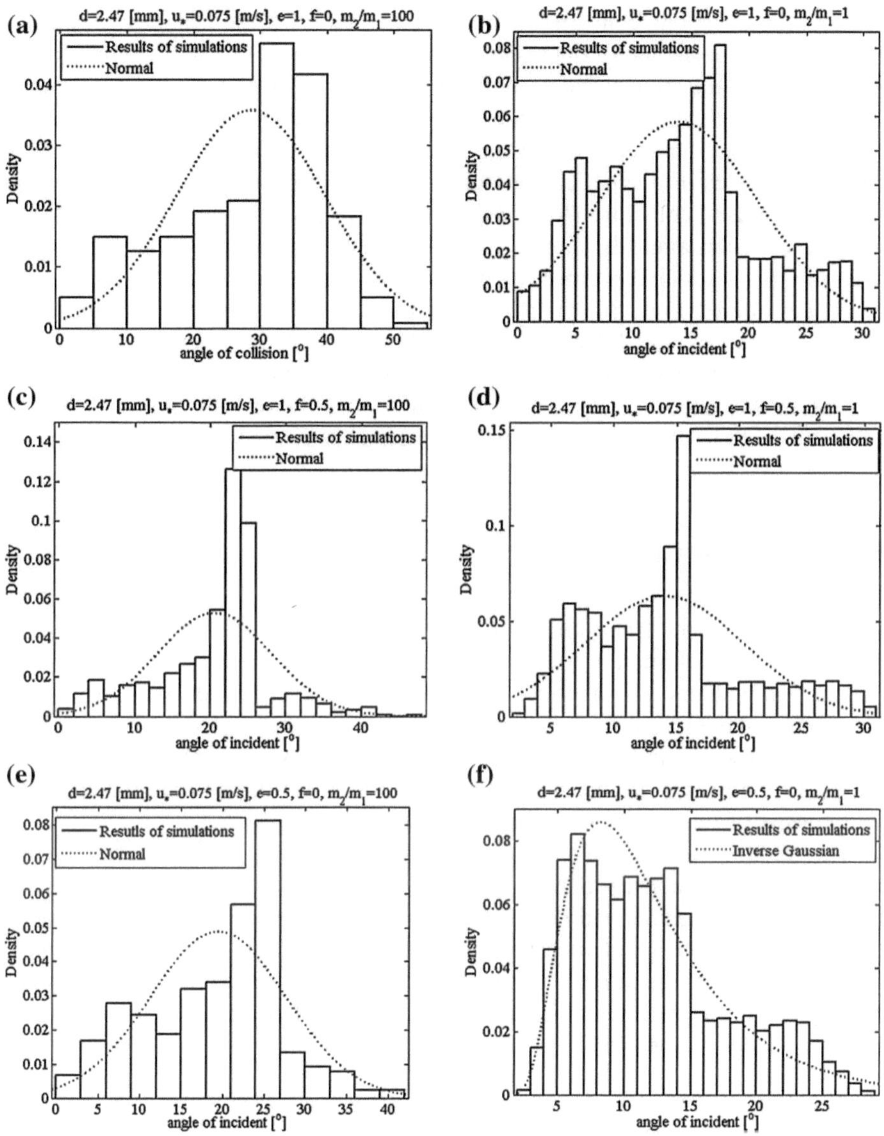

Fig. 8 Probability distributions of angles of incidence for particles of diameter $d = 2.47$ mm, $u_{*} = 0.075$ m/s for different ratios m_2/m_1

(incidence). Given the impact angle and the angle of collision we implement the following collision resulting from the simulation. Figure 7 presents one possible trajectory that was obtained based on the presented above model of saltating particle.

Now the questions concerned with the simulations of the saltating particles is whether the impact angle depends on the ratio m_2/m_1. We will search for the answer to this question through multiple simulations of particles' movement for different coefficients of restitution e, friction f and the ratio m_2/m_1.

Figures 8(a–f) present the results of numerical simulations of the impact (incident) angles of a spherical saltating particle. Comparing the results of simulations for elastic collisions (Figs. 8a, b) it is easy to see that for $m_2/m_1 = 100$ the average collision angle is approximately 35° and is significantly larger than for $m_2/m_1 = 1$, for which it is equal to 16°. Besides, it is important to notice that peaks in all considered Figs. 8(a–f) in relation to the normal distribution are shifted towards larger values. All presented distributions have a larger left asymmetry, with greater left asymmetry of the distribution for the $m_2/m_1 = 1$. Moreover, Fig. 8f shows that for $m_2/m_1 = 1$ the results of numerical simulation fulfill the Inverse Gaussian distribution rather than Normal one.

5 Concluding Remarks

The analysis of the cohesion coefficient m_2/m_1 and its relationships with the coefficients of restitution e, friction f and an angle of incident θ_{in} was made based on the simulations of spherical saltating particles obtained with use of the Lagrangian model. The main conclusion is that the ratio m_2/m_1 may be successfully used to set the links between the saltating grains and the moving bed and that the presented results are the first step to build a model of bed load transport that will consist of two regimes: saltation and sheet flow.

The authors would like to stress that they are aware that the accurate link between the cohesion coefficient m_2/m_1 and (1) momentum transfer to the bed from the moving grains and (2) the particles concentration in the surface layer of the bed requires the actual measurement data. However, these data have been still unavailable and the next step forward for all researchers is to provide the enough number of the experimental data that may be used for the verification of all existing models of bed load.

Acknowledgments The research was supported by The National Sciences Centre Grant No. N N306 658140. Professor W. Czernuszenko is also indebted to many colleagues with whom he has been fortunate to work and discuss all research topics for the last 50 years.

References

Bagnold RA (1973) The nature of saltation and of 'bed-load' transport in water. Proc R Soc Lond A 332:473–504

Bialik RJ (2011a) Particle–particle collision in lagrangian modeling of saltating grains. J Hydraul Res 49(1):23–31. doi: 10.1080/00221686.2010.543778

Bialik RJ (2011b) Numerical study of saltation of non-uniform grains. J Hydraul Res 49(5):697–701. doi: 10.1080/00221686.2011.598025

Bialik RJ, Nikora VI, Rowiński PM (2012) 3D lagrangian modelling of saltating particles diffusion in turbulent water flow. Acta Geophys. 60(6):1639–1660 doi: 10.2478/s11600-012-0003-2

Czernuszenko W (2009) Model of particle-particle interaction for saltating grains in water. Arch Hydro-Eng Environ Mech 56(3–4):101–120

Drake TG, Calantoni J (2001) Discrete particle model for sheet flow sediment transport in the nearshore. J Geophys Res 106(C9):19,859–19,868. doi: 10.1029/2000JC000611

Graf WH, Altinakar MS (1998) Fluvial hydraulics: flow and transport processes in channels of simple geometry. Wiley, Chichester

Gao P (2008) Transition between two bed-load transport regimes: saltation and sheet flow. J Hydraul Eng 134(3):340–349. doi: 10.1061/(ASCE)0733-9429(2008)134:3(340)

Hanes DM, Bowen AJ (1985) A granular-fluid model for steady intense bed-load transport. J Geophys Res 90(C5):9149–9158. doi: 10.1029/JC090iC05p09149

Hinze JO (1975) Turbulence. McGraw-Hill, New-York

Lee HY, Hsu IS (1994) Investigation of saltating particle motion. J Hydraul Eng 120(7):831–845. doi: 10.1061/(ASCE)0733-9429(1994)120:7(831)

Lukerchenko N, Chara Z, Vlasak P (2006) 2D numerical model of particle-bed collision in fluid-particle flow over bed. J Hydraul Res 44(1):70–78. doi: 10.1080/00221686.2006.9521662

Nino Y, Garcia M (1994) Gravel saltation: 2 modeling. Water Resour Res 30(6):1915–1924. doi: 10.1029/94WR00534

Rowiński PM, Czernuszenko W (1999) Modelling of sand grains paths in a turbulent open channel flow. In: Proceeding of the 28th IAHR Congress, Graz, Austria (CD), 1999

Rowiński PM, Czernuszenko W, Krukowski M (2005) Migration of floating particles in a compound channel, Water Quality Hazards and Dispersion of Pollutants: 121-141

Sayed M, Savage SB (1983) Rapid gravity flow of cohesionless granular materials down inclined chutes. J Appl Math Phys (ZAMP) 34:84–100. doi:0044-2275/83/001084-17

Wiberg PL, Smith JD (1985) A theoretical model for saltating grains in water. J Geophys Res 90(C4):7341–7354

Yeganeh-Bakhtiary A, Shabani B, Gotoh H, Wang SSY (2009) A three-dimensional distinct element model for bed-load transport. J Hydraul Res 47(2):203–212. doi: 10.3826/jhr.2009.3168

Numerical Verification of Log-Law in Flows with Pressure Gradient

Włodzimierz Czernuszenko and Alexey Rylov

Abstract The chapter deals with 3D rough turbulent flows with pressure gradient in a straight open channel with regular bed roughness. The bed of the channel is characterized by roughness elements which are supposed to be uniform in size and form a regular surface of the bed. Turbulence structure above viscous sub-layer for such a roughness type is assumed to be homogenous in horizontal plane. Due to numerous experiments, this type of flow with zero pressure gradient is known to match the logarithmic law (log-law). So a question arises if it can be extended to flows with non-zero pressure gradient, and second, what are parameters of the log-law. An open channel turbulent flow is described by the Reynolds equations with simple turbulent model, which has eddy viscosities described by an enhanced mixing length hypothesis. The Reynolds equations with the continuity equation for steady, parabolic 3D turbulent flow in an open channel are solved for accelerating and decelerating flows. For these types of flow the additive parameter B of the log-law is calculated and results are discussed.

1 Introduction

The mean velocity vertical profile in an open channel flow is of great interest to civil engineers, particularly for both uniform and non-uniform flows over gravel-bed. The channel boundaries are usually rough and the roughness elements disturb

W. Czernuszenko
Institute of Geophysics Polish Academy of Sciences, 01-452 Warsaw, Ks. Janusza 64, Warsaw, Poland
e-mail: wczer@igf.edu.pl

A. Rylov (✉)
Trianon Co, Stantsionnaya 30a, Novosibirsk 630032, Russia
e-mail: raa05@yandex.ru

P. Rowiński (ed.), *Experimental and Computational Solutions of Hydraulic Problems*, GeoPlanet: Earth and Planetary Sciences, DOI: 10.1007/978-3-642-30209-1_20,

the flow. The disturbances produce a complicated flow pattern even in the region away from the boundary especially in the case of non-uniform flows. This effect is most intensive when size of roughness elements is relatively large, for example, in the case of gravel or pebble-bed. Nikuradse's concept of logarithmic velocity profile with the absolute size of roughness elements k and the constant B, can be used only for sand roughness (Yalin 1977). When the relative spacing, size and shape of the roughness elements are particularly important, Nikuradse's concept becomes inaccurate. Nevertheless, Nikuradse's sand grain roughness has become a standard method of describing the roughness of boundary in both pipes and open channel flows.

Two types of large relative roughness are distinguished. First, there are large-scale roughness elements such as boulders spread on the river bed which might protrude to the water surface. Second, the bed is composed of coarse but somewhat uniform materials. The second case is similar to sand roughness, however the relative roughness here is much larger and the whole flow may feel directly the influence of the bed materials (Perry et al. 1969). Present investigation aims at the second case, i.e., the riverbed is covered by coarse but somewhat uniform materials.

Kamphuis, based on the laboratory measurements of velocity over several roughness heights (diameter of river gravel and pebbles) from $D = 0.5$ till 40 mm, concluded that the diameter was a poor roughness criterion (Kamphuis 1974). The logarithmic velocity distribution is valid with the average size of roughness elements ranged from $k = 1.5\,D$ till $2.5\,D$, and the log-law constant is equal to 8.5 [this constant is denoted by B, see Eq. (9)]. The actual point where the velocity is zero (effective velocity origin) has been found 0.3 D below of the crest of roughness elements.

Determination of the theoretical (virtual) bottom location where $y = 0$ becomes important when the roughness size is large in ratio to flow depth. Bayazit in his experiments with hemispheres of diameter $D = 23$ mm, found that the theoretical bed would lie at the distance 0.35 D below the tops of the hemispheres in order for velocity distribution to agree with logarithmic law (Bayazit 1976). He showed that the usual log-law remains valid for velocity distribution as long as the flow depth is larger than the roughness height. Equivalent sand roughness in this case is equal to 2.5 D and the constant $B = 8.5$.

Kirkgoz deals with the velocity distribution in rough boundary flows with different sizes of roughness elements, namely, $k = 1$, 4, 8 and 12 mm (Kirkgoz 1989). The experimental values of position of the reference level where the mean velocity is zero lie between 0.25 and 0.75 mm below the tops of roughness elements. Finally, he concluded that the size of roughness elements is reflected in the values of the friction velocity and the value of U_* increases with increasing roughness size.

It is common to distinguish two regions for the velocity profile: the inner or the wall region, and the outer or the core region. Velocity distribution in the inner region follows the log-law whereas in the wake region there are some deviations from it. To reduce these deviations some authors suggest using an additive constant increasing with Reynolds number in this region (Guo et al. 2005). However, in practice the log-law is frequently used from the bottom to the surface, i.e., to the whole outer region.

The main aims of the work are:

1. To develop the basic equations for the vertical velocity profiles in non- uniform flows over rough bed, i.e., the log-law for the flow with non-zero pressure gradient;
2. to show that B-constant is subject to changes and depends on the pressure gradient in non-uniform flows.

2 Fundamental Physics and Associated Equations

Three-dimensional, steady turbulent flow in an open channel is governed by the Reynolds-averaged Navier-Stokes equations. The continuity and momentum equations for incompressible turbulent flows may be written in the Cartesian tensor notation in the forms:

continuity equation

$$\frac{\partial U_i}{\partial x_i} = 0, \tag{1}$$

momentum equations

$$U_j \frac{\partial U_j}{\partial x_i} + \frac{1}{\rho} \frac{\partial P}{\partial x_i} = F_i - \frac{\partial}{\partial x_j} \left(\overline{u_i u_j} - \nu \frac{\partial U_i}{\partial x_j} \right), \tag{2}$$

where U_i and u_i are the i-th components of the time average velocity and turbulent velocity, respectively, so that full velocity $\tilde{U}_i(x, y, z, t) = U_i(x, y, z) + u_i(x, y, z, t)$, P is the time average pressure, ρ is the density, ν is the kinematic viscosity, g is the acceleration of gravity and $\mathbf{F} = (g \sin \alpha, g \cos \alpha, 0)$. Also notation x for longitudinal, y for vertical and z for lateral coordinates as well as U, V, W or u, v, w for the corresponding velocity components will be used in the chapter.

Wide open channel flow conditions, i.e., flows with large aspect ratio, allow us to assume that all terms in the lateral, z-momentum equation are zero and for uniform flow the convective inertia term is negligibly small compared to others. Also, y-momentum equation obviously gives vertical pressure distribution very close to hydrostatic one. At last, x-momentum equation is simplified to

$$\frac{\partial}{\partial y} \left(-\overline{u'v'} + \nu \frac{\partial U}{\partial y} \right) = g S_0, \tag{3}$$

where $S_0 = \tan \theta$ is the bed slope.

Integration of Eq. (3) across the boundary layer (above roughness tops) gives the formula for distribution of the total shear stresses

$$\frac{\tau_{total}(y)}{\rho} = -\overline{u'v'} + \nu\frac{\partial U}{\partial y} = ghS_0\left(1 - \frac{y}{h}\right), \tag{4}$$

Hence, the total shear stresses vary linearly with y, they become equal to zero at free surface ($y = h$) and assume to reach their maximum value at the roughness height but not at the virtual bed ($y = 0$), because Eq. (4) is not valid at $y = 0$.

The total shear stresses can be expressed as a sum of two components, namely, the mean Reynolds τ_t stresses, and the mean viscous τ_v stresses. Their bed values correspond to a vertical level somewhere below the roughness height. Starting at this level, the viscous shear stresses become negligibly small compared to the turbulent stresses. These stresses are essential only in viscous sub-layer, where the stresses of other type are negligibly small compared with the viscous ones. The thickness of viscous sub-layer is hard to estimate but approximately it is less than 15% of k_s for the rough sand bed (Yalin 1977). Quantitatively, the viscous shear stress due to the molecular viscosity μ in viscous sub-layer is given by

$$\tau_l(y) = \mu\frac{\partial U}{\partial y}, \tag{5}$$

Outside viscous sub-layer in fully turbulent flow the viscous stress in Eq. (4) is negligibly small compared to the other one, thus the total shear stress (in wall region, $y < 0.2h$) can be modeled by Prandtl mixing length hypotheses (MLH) as follows

$$\tau_{\mathrm{total}}(y) = \rho\ell^2\left(\frac{\partial U}{\partial y}\right)^2, \tag{6}$$

where ℓ is Prandtl's mixing length.

Following Prandtl's assumption of constant total stress the average velocity profile can be obtained from Eq. (6). This equation with assumption $\ell = \kappa y$, where κ is the Karman constant related to the mixing length ℓ, can easily be integrated over the turbulent core outside the viscous sub-layer, i.e., in the interval $[y_k, y]$

$$\frac{U(y)}{U_*} = \frac{1}{\kappa}\ln\frac{y}{y_k} + \frac{U_k}{U_*}, \tag{7}$$

where U_k is the value of U at the top of averaged roughness elements, i.e., at $y_k = k_{ef}$, where k_{ef} is the height of averaged roughness which interferes the flow. Shear velocity U_* is the most fundamental scale and there are several methods to estimate this scale. For the zero pressure-gradient case we usually use the formula $U_* = \sqrt{ghS_0}$. The Clauser's method is also applicable (Clauser 1954). It relies on the validity of the logarithmic law for the inner (wall) region of the boundary layer holding $y/h < 0.2$.

The above description is adequate only for small distances from the wall, i.e., in the inner layer where the mixing length rises linearly with the distance to the wall, and the shear stress is constant. The second term on the right-hand side of Eq. (7) is called a roughness function.

In case of hydraulically rough flow and sufficiently large Reynolds number, the velocity distribution does not depend on viscosity, and expression U_k/U_* is constant. Therefore, Eq. (7) becomes

$$\frac{U(y)}{U_*} = \frac{1}{\kappa}\ln\frac{y}{k_{ef}} + B, \qquad \left(B = \frac{U_k}{U_*}\right) \tag{8}$$

where y is the depth above the channel bottom. It is easy to see that the effective velocity origin is located over the theoretical bed at $y_0 = k_{ef}e^{-\kappa B}$. For sand roughness $B = 8.5$ at $Re = k_{ef}U_*/\nu > 70$ and $y_0 = k/30$.

Previous conclusion requires some comments. Very precise measurements of Nezu and Rodi showed that our data for uniform flow, though formally valid only in the inner region, can be applied throughout the channel depth, with two constants: κ (Karman's constant) and B (Nezu and Rodi 1986). Close examination of velocity measurements shows a slight deviation from the logarithmic law in the region close to the water surface ($0.7 < y/h < 1$). To adapt Eq. (8) to the available measurement data, Coles' law was usually adopted (for details see Cordoso 1989; Nezu and Rodi 1986).

Equation (8) is widely used for zero pressure gradient (ZPG) rough turbulent flows, when the relative roughness is small—$k_{ef}/h < 0.05$ (Nikora et al. 2004). In case of a non-uniform flow over rough bed, when the relative roughness is large, coefficient B is no longer constant. Generally, the B-value depends on the size, shape and distribution of the roughness elements. It is smaller than 8.5 in rough flows with small relative roughness, e.g., Bayazit found that B should be smaller than 8.5 when $h/k_{ef} < 3$ (Tu 1991). B decreases with increase of k_{ef}/h till 0.25. Then its value remains roughly about 5.0 in experiments with bed materials $D_s = 23.5$ mm.

A safe way to determine the equivalent sand roughness for the bed grain diameter is still a problem. Various reseachers proposed different values. Nikuradse in his tests with uniform sands came to $k_{ef} = D_s$; Einstein and El-Sami took

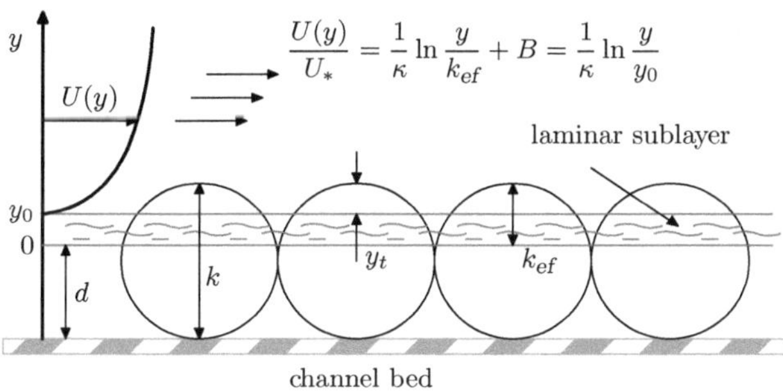

Fig. 1 Basic definitions of log-law for flows over spherical roughness elements with diameter k: d is the zero-plane displacement, k_{ef} is the height of roughness interfered with the flow

$k_{ef} = D_{65}$ for non-uniform sands, Mayer-Peter chose $k_{ef} = D_{90}$, Lane and Carlson—$k_{ef} = D_{75}$ [for details, see Tu (1991)]. Location of reference-level, zero-bed or theoretical bed, defined by $y = 0$, as it is called by various researches (McLean et al. 2002), also varies among a series of works. For uniform roughness elements (hemispheres) Einstein and El-Sami found $y_0 = 0.20\,D$; also for hemispheres Bayazit got ($D = 23$ mm) $y_0 = 0.35\,D$, Grass et al. for rounded pebbles—$D = 9$ mm, $y_0 = 0.18\,D$ (Grass et al. 1993) (see Fig. 1).

3 Flows in Gravel-Bed Channels with Pressure-Gradient

For environmental engineers, the knowledge of the vertical distribution of mean velocity in open-channel flows is significantly important, especially for non-uniform flows, which are more natural in rivers or open channel flows. However, efforts for assessing this velocity distribution have been mainly devoted to uniform flow conditions. Thus, the non-uniform flow is of interest here, and first, the basic equation of two-dimensional vertical velocity profile should be formulated for flows with pressure gradient.

Non-uniform flows are produced by changes in the channel geometry, while transiting from one uniform-flow state to another, and in accelerating and decelerating flows. Depending on a sign of the pressure gradient we can divide velocity profiles into groups

$\partial P = \partial x < 0$	accelerating flow,
$\partial P = \partial x = 0$	zero-pressure gradient flow, this case has been already discussed in Sect. 2,
$\partial P = \partial x > 0$	decelerating flow.

Here P stands for free stream pressure. Some authors have studied the influence of $\partial P/\partial x$ on velocity distributions. Most of them dealt with the case of decelerating flows. The formulation of the velocity distribution in equilibrium turbulent boundary layers is subjected to decelerating flows, how it was reviewed by Yaglom in Yaglom (1979). He showed that, in the overlapping region the velocity distribution follows the logarithmic law, it also holds for accelerating flows as long as they are in local equilibrium. However, logarithmic law can be expected not to hold up for arbitrarily high values of the pressure gradient. In case of accelerating flows, Cordoso gave a criterion for break-down of the universal law-of-the wall [see (Cordoso 1989)]

$$\frac{\nu}{\rho U_*^3}\frac{\partial P}{\partial x} < -0.02. \tag{9}$$

Based on experimental works, impact of spacial acceleration on deformation of the mean velocity field can be described as (Cordoso 1989; Tu 1991)

- Velocity distributions could no longer be represented entirely by the universal log-law. The deviations from log-law are to be attributed mainly to acceleration. The existence of the inner layer being in equilibrium and following the law-of-wall and of an outer layer deviating from that law can be postulated. The inner log-law seems to extend up to about 0.05 of the flow depth;
- Experiments show that in accelerating and decelerating flows all mean velocity profiles follow a logarithmic law in the inner layer. However, differences in the profile form are observed, i.e., the accelerating flow profile has bigger shear velocity than the decelerating one;
- For the equal water depths the point velocity in the rising branch (for accelerating flow) is generally larger than the one in the falling branch (for decelerating flow);
- Vertical shear-stress distributions for the rising branch are usually concave, implying accelerating flow ($\partial P/\partial x < 0$); those for falling branch are convex, implying decelerating flow ($\partial P/\partial x > 0$). If the longitudinal pressure gradient is zero, the constant shear stress layer is observed.

3.1 Basic Turbulence Relations

Non-uniform flows are produced by changes in the channel geometry. If the flow depth varies in flow direction, the flow is called gradually varied. Equations describing the gradually varied flow, in accelerating and decelerating variants, are derived under the following simplifying assumptions :

1. the slope of the channel bottom is small,
2. pressure distribution at a channel cross-section is hydrostatic,
3. head losses in gradually varied flow may be determined by using equation for head losses in uniform flows, i.e., $U_*^2 = ghS_f$.

For wide open channel flows we can quite naturally assume all terms in the lateral z-momentum equation to be zero. Then, y-momentum equation from Eq. (2) for steady flows takes the form

$$U\frac{\partial V}{\partial x} + V\frac{\partial V}{\partial y} + \frac{1}{\rho}\frac{\partial P}{\partial y} = g_y - \frac{\partial \overline{uv}}{\partial x} - \frac{\partial \overline{v^2}}{\partial y} + \frac{\partial}{\partial x}\left(\nu\frac{\partial V}{\partial x}\right) + \frac{\partial}{\partial y}\left(\nu\frac{\partial V}{\partial y}\right), \tag{10}$$

where $g_y = g\cos\theta$. Further, we will need the following additional assumptions:

4. flow is unidirectional with the velocity vector equal to $(U + u, v)$,
5. x-derivatives of turbulent stresses can be neglected.

Under above assumptions the previous equation is simplified to

$$\frac{1}{\rho}\frac{\partial P}{\partial y} = g_y - \frac{\partial \overline{v^2}}{\partial y}. \tag{11}$$

After integration of Eq. (11) over interval $[y, h]$ we get

$$P(y) = \rho g h \cos\theta\left(1 - \frac{y}{h}\right) + \left(\overline{v^2}(h) - \overline{v^2}(y)\right), \tag{12}$$

Equation (12) allows us to conclude that only for small bed slope and vertically homogeneous turbulence the pressure distribution is hydrostatic.

The x-momentum equation from Eq. (2) under similar assumptions takes the form

$$U\frac{\partial U}{\partial x} + \frac{1}{\rho}\frac{\partial P}{\partial x} = g_x - \frac{\partial \overline{uv}}{\partial y} - \frac{\partial \overline{u^2}}{\partial x} + \nu\frac{\partial^2 U}{\partial y^2}, \tag{13}$$

where $g_x = g\sin\theta = gS_0$. Assume components of turbulent velocity to vary negligibly in flow direction, integrate Eq. (13) over flow vertical interval $[\varepsilon, h]$ and denote $\bar{U} = 1/h\int_\varepsilon^h U\,dy$, so that $U(y) = \bar{U} + \delta_u(y)$. Assumptions $\delta_u(y) \ll U$, $\partial\delta_u/\partial x \ll U$ lead to

$$\bar{U}\frac{\partial \bar{U}}{\partial x} + \frac{g}{2}\frac{\partial h}{\partial x} = g(S_0 - S_f), \tag{14}$$

where symbol S_f is defined by $gS_f = \tau_b/\rho h$.

In order to deduce a formula for friction velocity U_*, defined by $U_*^2 = \tau_b/\rho = ghS_f$, for flows with non-zero pressure gradient, substitute velocity derivative from continuity equation $h\bar{U}_x + \bar{U}h_x = 0$ into Eq. (14) and make use of relation $P_x = \rho g h_x/2$ for depth-averaged pressure P in hydrostatic case; then

$$U_*^2 = ghS_0 - \frac{\partial P}{\partial x}\frac{h}{\rho}\left(1 - 2\,\mathrm{Fr}^2\right), \qquad \mathrm{Fr} = \bar{U}/\sqrt{gh}, \tag{15}$$

The above equation is applicable for steady non-uniform flows. However, it is also valid for unsteady flows in a case of kinematic wave. But it cannot be extended to unsteady flows with dynamic wave. Examination of Eq. (15) shows that it is unconditionally valid only if gravitational force prevails over streamwise pressure, i.e., in a case of acceleration flow, where $\partial P/\partial x < 0$. Decelerating flows imply a possibility for us to fail finding U_* from Eq. (15). This occurs when the pressure force raises to some threshold, and starts to dominate over gravitation. Below we are about to apply the MLH for describing the velocity profile. We can certainly assume the total shear stress to vary linearly

$$-\overline{uv}(y) = U_{*p}^2\left(1 - \frac{y}{h}\right), \tag{16}$$

Considering Eq. (16) for $y \ll h$ and assuming validity of Prandtl's MLH with $\ell = \kappa y$ gives the log-law in a form

$$\frac{U(y)}{U_{*p}} = \frac{1}{\kappa}\ln\frac{y}{k_{ef}} + B, \qquad B \equiv \frac{U(k_{ef})}{U_{*p}}, \tag{17}$$

where U_{*p} is defined by Eq. (15) and k_{ef} is the effective roughness height.

Further we are going to verify previous results by a numerical approach, i.e., to solve the hydrodynamic equations with turbulence model for accelerating and decelerating flows and try to compare computed vertical velocity distribution against that one from Eq. (17).

3.2 Turbulence Model

The applied turbulence model is a simplified version of models presented by the authors before (Czernuszenko and Rylov 2000, 2003). To solve motion Eq. (2) the total turbulent stress tensor must be known, i.e., normal and shear turbulence stresses at any point of the cross-section should be prescribed. The normal turbulent stresses and the shear turbulent stresses are defined according to Czernuszenko and Rylov (2003). The model describes turbulence structure in the case of anisotropic normal turbulent stresses and isotropic shear stresses in the form

$$-\rho\overline{u_i u_j} = \begin{cases} -\rho\overline{u_i^2} & \text{for } i = j \\ \rho\ell^2 D_{ij} S & \text{else,} \end{cases} \tag{18}$$

where D_{ij} is the deformation rate tensor, and S is defined by Czernuszenko and Rylov (2000)

$$S = \sum_{i,k} \left| \frac{\partial U_i}{\partial x_k} \right|. \tag{19}$$

For open channel flows the right-hand sum in Eq. (19) may be reduced to two terms containing derivatives of streamwise velocity in lateral and vertical directions, respectively. The other terms in the sum above containing the lateral and vertical components of the mean velocity vector do not exceed 1–2 %

3.3 Hydrodynamic Model

As pointed out in Sect. 2 three-dimensional steady turbulent flow in an open channel is governed by the Reynolds-averaged Navier–Stokes Eqs. (1, 2). To solve these equations we need to take into account the turbulence model from Sect. 2 and give the initial and boundary conditions as well. We are about to write these equations for the tilted open channel with bed slope defined by angle θ. Turbulent transport of momentum in x-direction is assumed to be negligible. Therefore, the terms involving second derivatives with respect to x may be omitted. This kind of flow is usually called parabolic in longitudinal direction. Substitution of Eq. (18) into (1, 2) produces

$$\frac{\partial U}{\partial x} + \frac{\partial V}{\partial y} + \frac{\partial W}{\partial z} = 0, \tag{20}$$

$$\frac{\partial U^2}{\partial x} + \frac{\partial VU}{\partial y} + \frac{\partial WU}{\partial z} + \frac{1}{\rho}\frac{\partial P}{\partial x} = g \sin\theta + \frac{\partial}{\partial y}\left(l^2 S \frac{\partial U}{\partial y}\right) + \frac{\partial}{\partial z}\left(l^2 S \frac{\partial U}{\partial z}\right), \tag{21}$$

$$\begin{aligned} \frac{\partial UV}{\partial x} + \frac{\partial V^2}{\partial y} + \frac{\partial WV}{\partial z} + \frac{1}{\rho}\frac{\partial P}{\partial y} &= g \cos\theta + \frac{\partial}{\partial y}\left(-\overline{v^2}\right) \\ &+ \frac{\partial}{\partial y}\left(2Sl_y^2 \frac{\partial V}{\partial y} + Sl_{yz}^2\left(\frac{\partial W}{\partial y} + \frac{\partial V}{\partial z}\right)\right) \\ &+ \frac{\partial}{\partial z}\left(\frac{1}{2}S\left(l_y^2 + l_z^2\right)\left(\frac{\partial W}{\partial y} + \frac{\partial V}{\partial z}\right)\right), \end{aligned} \tag{22}$$

$$\begin{aligned} \frac{\partial UW}{\partial x} + \frac{\partial VW}{\partial y} + \frac{\partial W^2}{\partial z} + \frac{1}{\rho}\frac{\partial P}{\partial z} = \frac{\partial}{\partial z}\left(-\overline{w^2}\right) + \frac{\partial}{\partial y}\left(Sl^2\left(\frac{\partial W}{\partial y} + \frac{\partial V}{\partial z}\right)\right) + \\ + \frac{\partial}{\partial z}\left(2Sl_z^2 \frac{\partial W}{\partial z}\right), \end{aligned} \tag{23}$$

Details on the enhanced mixing length model can be found in Czernuszenko and Rylov (2003).

Boundary conditions that accompany Eqs. (20–23) need to be specified along solid boundaries, water surface and up-stream cross-section bounding the calculation domain. Since parabolic flows are considered, boundary conditions do not need to be given at the downstream end of the calculation domain. It is assumed that the flow is uniform with constant width and depth [rigid lid approximation, (Rastogi and Rodi 1978)]. The conditions at the solid boundaries were specified using wall functions technique proposed by Launder and Spalding (1974), i.e., the conditions are specified at a point near a wall which lies outside the laminar sublayer and satisfy the logarithmic portion of the universal law of the wall. To do so we use Eq. (8).

Normal velocity components at the solid boundaries and free surface are set to zero. The free surface boundary conditions were specified following the approach of Rastogi and Rodi, which considers free surface acting as a plane of symmetry (Rastogi and Rodi 1978). Therefore, the gradients of U and V in y-direction are zeros. The condition at the initial cross section $x = 0$ for longitudinal velocity U was taken along with logarithmic distribution. Components V and W were set equal to zero. The essential problem with the boundary conditions is in specifying the logarithmic velocity profile, which in the case of rough irregular bed surface is not clear.

3.4 Numerical Simulations

The model described above was applied to simulate the 3D mean stream velocity distribution of turbulent flow in a rectangular channel. The reference laboratory data are chosen from measuments by Mansour-Tehrani Mansour-Tehrani (1992). To solve the above set of equations the numerical parabolic procedure known as Patankar-Spalding algorithm is used (Patankar and Spalding 1972). It solves the set of above equations for three components of velocity U, V, W and pressure P, at each forward step in longitudinal direction. At each such a step cross-section pressure field is tuned to force velocity field to satisfy continuity equation. The scheme has HYBRID pattern approximation of convective terms (Leschziner 1980). However, due to insignificant secondary velocities it has actually the second order of accuracy. The program Canal of version 4.6.2, which allows simulating 3D turbulent parabolic flows along with the specified approach, was used in all computations.

A series of numerical simulations has been performed in order to verify the possibility of reproducing some of measured flow parameters. Also, we would like to find out if the model with rigid-lid apporoximation is applicable to handle free surface open channel flows. To achieve the last purpose, computational runs for all the laboratory cases with 3 bed sphere diameters have been done (see Table 1).

Each run was targeted to deliver $\partial P/\partial x = 0$ state. The coefficient B in the run was treated as the relevant computed one (B_1 in Table 1, and B_0 corresponds to the parameter value computed from measured data). There are some discrepancies between B_0 and B_1 columns. First, they can exist due to difficulties in measuring subtle quantities like slope, friction velocity or the height of averaged roughness k_{ef}. Second, experimental work was carried out at small enough Reynolds numbers (case K1Q3 gives value of about 5,000), and turbulence-generating sublayer near bed roughness gives even lesser values. So flows tend to become partly laminar.

The procedure of seeking B appears to be essentially iterative. It consists of a series of steps. Each step starts at given B and produces value of $\partial P/\partial x$ after

Table 1 Channel data in experiments by Mansour-Tehrani (1992)

Case	Roughness height, mm	Depth, mm	Mean velocity, mm/s	Shear velocity, mm/s	Karman	Effect. rough. height, mm	B_0	B_1
	k	h	U_m	U_*	κ	k_{ef}	B	B
K1Q1	1.15	49.8	227.8	12.48	0.392	0.37	7.28	9.45
K1Q2	1.15	49.7	164.8	9.18	0.398	0.47	7.81	9.91
K1Q3	1.15	49.9	106.0	6.46	0.387	0.48	6.79	8.13
K6Q1	6	49.0	219.4	15.91	0.37	1.47	5.43	6.97
K6Q2	6	49.1	154.6	13.15	0.395	1.475	5.34	5.36
K6Q3	6	48.9	111.0	8.46	0.384	1.37	5.23	6.33
K12Q2	12	46.9	155.9	14.82	0.42	2.72	5.35	5.7
K12Q3	12	49.7	106.0	9.21	0.382	2.73	5.30	6.11

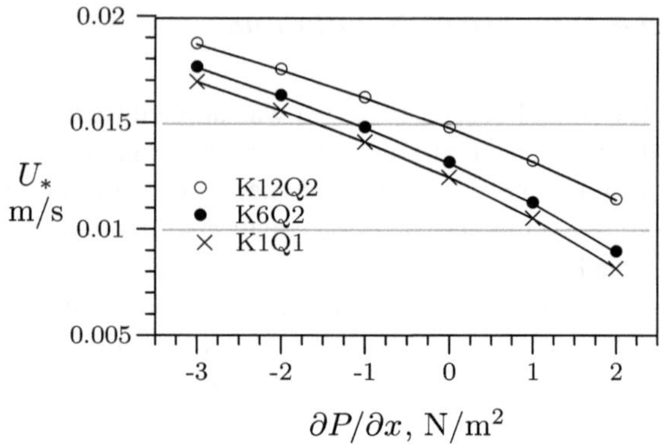

Fig. 2 U_* values at various $\partial P/\partial x$

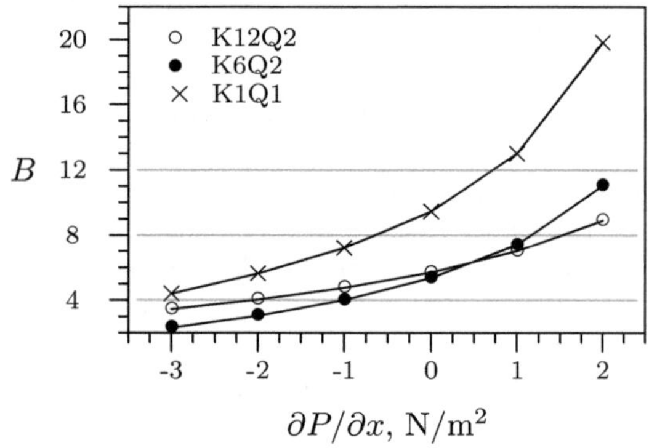

Fig. 3 B values at various $\partial P/\partial x$

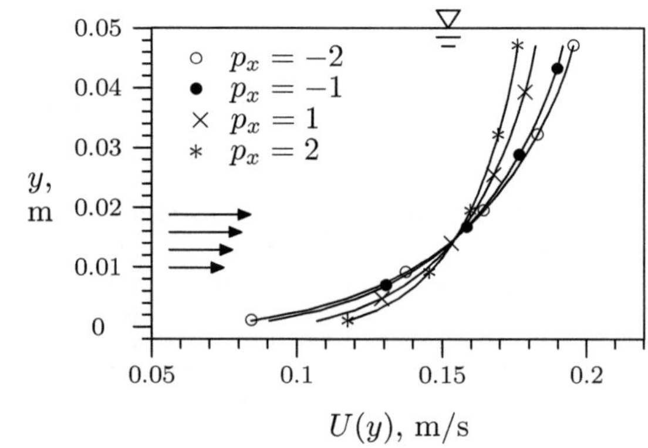

Fig. 4 U profiles at channel centerline for various $\partial P/\partial x$

computing. Achieved pairs $(B, \partial P/\partial x)$ give a chance to make a prediction what the next value of B should be in order to get to $\partial P/\partial x = a$ state. This procedure converges for reasonable a values (not too large ones).

The B values in zero pressure grandient flow cases are summarized in Table 1. In order to demonstrate flow evolution for non-zero pressure grandients, three cases—one at each sphere diameter—have been chosen. The above-mentioned iterative procedure was applied to build a series of $(\partial P/\partial x, U_*, B)$ triples. Figures 2 and 3 show friction velocity and parameter B values, respectively.

Formula (15) poses no constraints on negative values of $\partial P/\partial x$. On the other hand, U_* solution to (15) within domain $\partial P/\partial x > 0$ exists only for a bounded range of $\partial P/\partial x$. When $\partial P/\partial x$ tends to its upper bound, Eq. (15) loses validity since its inference is based on small enough pressure grandient values. Moreover, under such conditions (hard decelerating flows) we cannot assume that rigid-lid approximation is still valid. Therefore, positive part of $\partial P/\partial x$ should expose less stability in behaviour compared to negative one. Figure 3 demonstrates exactly such effect. Already $\partial P/\partial x = 2$ invokes suspiciously steep rise of B for K1Q1 case. Perhaps, the case of the least diameter spheres is a frontier zone to laminar flow regimes since K6Q2 case shows more moderate rise of B, and the largest diameter spheres provide the most gradual B increase.

Also, for acceleration flows it is worthy to note that abnormal evolution of flow parameters starts an order of magnitude below the edge established by relation (9), which was elaborated for this type of flows.

Illustration of changes in vertical profiles of streamwise velocity is placed in Fig. 4. The move of U profile after changes in $\partial P/\partial x$ should be reckoned quite expectable. Really, due to the preserved total discharge decelerating pressure gradient ($\partial P/\partial x > 0$) retards the flow and enforces U front to flatten. When this gradient alters the sign, accelerating force appears. It acts in reverse direction and the profile goes forth in its upper part.

3.5 Conclusions

- The numerical model with rigid-lid approximation of free surface is applicable for modelling open channel flows with non-zero pressure gradients.
- The logarithmic pattern of vertical profile for streamwise velocity component, which is approved to be valid in stationary unidirectional flows, can also be applied to accelerating as well as decelerating open channel flows. However, friction velocity should be corrected along with (15).
- Velocity distributions could no longer be represented entirely by the universal log-law. Existence of two flow strata—inner layer in equilibrum at the wall, which follows the law-of-the-wall, and outer layer deviating from that law should be postulated. The inner layer seems to extend up to about 0.25 of the flow depth; in other words, it is five times larger than Cordoso mentioned Cordoso (1989).
- Tuning parameters proved that computational model reproduces experimental data adequately, namely, the state of $\partial P/\partial x = 0$ was gained at quite close values

of k_{ef} and B for cases K6 and K12. The case K1Q1 seems to be less representative since lower sizes of grains lead to lower accuracy in measuments of grain-dependent variables, and second, at such a small Reynolds number the departure from the law-of-the-wall is associated with the tendency for the turbulent boundary layer to relaminarize (*Relaminarization* means reversed transition from turbulent back to laminar flow).

- The additive constant B in log-law formula for accelerating and decelerating flows depends on the streamwise pressure gradient. When $\partial P/\partial x$ rises, also B rises. The dependence is close to linear at small enough pressure gradient values.

Acknowledgments This work was supported by grant No. N306 658140 from The National Science Centre Grant, Poland. Authors are grateful to Dr P. Rowinski for reviewing an early draft of the results and for his helpful criticism.

References

Bayazit M (1976) Free surface flow in a channel of large relative roughness. J Hydr Res 14(2):115–126

Cordoso A (1989) Spacially accelerated flow in an smooth open channel. These No.813, Ecolé Polytechnique Federale de Lausanne.

Clauser FH (1954) Turbulent boundary layers in adverse pressure gradients. J Aero Sci 21:91–108

Czernuszenko W, Rylov AA (2000) A generalisation of Prandtl's model for 3D open channel flows. J Hydraul Res 38(2):133–139

Czernuszenko W, Rylov AA (2003) A Modeling of shear and normal turbulent stresses. In: Open channel flows. XXX Congress IAHR Thessaloniki, Greece. Proceedings Theme C, vol 1:135–142

Gimenez-Curto LA, Corniero Lera MAC (1996) Oscillating turbulent flow over rough surfaces. J Geophys Res 101(C9):20, 745–20, 758.

Grass AJ, Stuart RJ, Mansour-Tehrani M (1993) Common vertical structure of trbulent flows over smooth and rough boundaries. AIAA J 31(5):837–847

Guo J, Julien PY, Meroney RN (2005) Modified log-wake law for zero-pressure-gradient turbulent boundary layers 43(4):421–430

Kamphuis JW (1974) Determination of sand roughness for fixed beds. J Hydr Res 12(2):193–203

Kirkgoz SM (1989) Turbulent velocity profiles for smooth and rough open channel flow. J Hydr Eng ASCE 15(11):1543–1561

Launder BE, Spalding DB (1974) The numerical computation of turbulent flows. Comput Meth Appl Mech Eng 3:269

Leschziner MA (1980) Practical evaluation of three finite difference schemes for the computation of steady-state recirculation flows. Comput Meth Appl Mech Eng 23:293–312

Mansour-Tehrani M (1992) Spatial distribution and scaling of bursting events in boundary layer turbulence over smooth and rough surfaces. University of London, Ph.D. Dissertation.

McLean S, Dittrich A, Aberle J (2002) Zero-plane displacement for rough-bed open-channel flows. In: Proceedings of the international conference on fluvial hydraulics river flow 2002, Louvain-la-Neuve, Belgium, pp 83–92.

Nezu I, Rodi W (1986) Open-channel flow measurements with a laser doppler anemometer. J Hydr Eng 112(5):335–355

Nikora VI, Koll K, McLean S, McEwan I, Dittrich A (2004) Velocity distribution in the roughness layer of rough-bed flows. J Hydr Eng 130(10):1036–1042

Patankar SV, Spalding DB (1972) A calculation procedure for heat, mass and momentum transfer in three dimensional parabolic flows. J Heat Mass Trans 15:1787–1806

Perry AE, Schofield WH, Joubert PN (1969) Rough Wall Turbulent Boundary Layers. J Fluid Mech 37:383–413

Rastogi AK, Rodi W (1978) Predictions of heat and mass transfer in open channels. J Hydr Div 104(HY3):397–420.

Raupach MR, Antonia RA, Rajagopalan S (1991) Rough-wall turbulent boundary layers. Appl Mech Rev 44(1):1–25

Tu H (1991) Velocity distribution in unsteady flow over gravel beds. These No. 911, Ecolé Polytechnique Federale de Lausanne, Lausanne EPFL.

Yaglom AM (1979) Similarity laws for constant-pressure and pressure-gradient turbulent wall flows. Ann Rev Fluid Mech 11:505–540

Yalin MS (1977) Mechanics of sediment transport. Pergamon Press, Oxford

Yang S-Q (2009) Veloicty distribution and wake-law in gradually decelerating flows. J Hydr Res 47(2):177–184

Application of the RMA2/RMA4 Models to Simulate Pollution Transport in a Retention Reservoir

Monika Gałek and Paweł S. Hachaj

Abstract In this chapter the authors present an application of the computational two-dimensional depth-averaged models RMA2 and RMA4 to simulate water flow and pollution propagation in the Dobczyce retention reservoir. The bases of the models, the methodology used and the obtained results are all presented along with the justification of the proposed solution. Several simulated events are shown as examples. Possible future application of the presented research to solve cases of economical and administrative interest is outlined.

1 Introduction and Study Area

The main aim of this chapter is to present the RMA2/RMA4 models as a tool supporting analysis of propagation of mineral and organic pollution in a water reservoir. This analysis contains contamination events that come as both constant and temporary leakage to the Dobczyce lake. The discussed reservoir is situated in southern Poland in Małopolskie Voivodeship. The lake was created by building a dam on the sixtieth kilometer of the Raba River, which is one of Vistula river tributaries. The knowledge about pollution propagation in this reservoir is as significant as a function of the Dobczyce Lake itself: the lake is the primary water source for the nearby city of Cracow.

M. Gałek (✉) · P. S. Hachaj
Institute of Water Engineering and Water Management, Cracow University of Technology, Warszawska 24, 30-962 Kraków, Poland
e-mail: monika.galek@iigw.pl

P. S. Hachaj
e-mail: pawel.hachaj@iigw.pl

P. Rowiński (ed.), *Experimental and Computational Solutions of Hydraulic Problems*, GeoPlanet: Earth and Planetary Sciences, DOI: 10.1007/978-3-642-30209-1_21,

The RMA models (the name of that model family comes from the Resource Management Associates, where it was developed) have been created to simulate water flow in rivers and estuaries, however in this chapter they are used to calculate this phenomenon in a mountain reservoir. The RMA2 model has already been successfully applied to a retention reservoir (Zając 2012).

2 Model Description

The former chapter by the same authors (Gałek and Hachaj 2008) addressed a similar issue but for the lake working under catastrophic flood conditions. In that case, dispersive movement could be neglected as much smaller than advective one and the FESWMS (Finite Element Surface Water Modeling System) model has been used. In this chapter the lake working under average conditions is considered (12 m^3/s total discharge) and both transport processes are taken into account. The RMA2/RMA4 models suit this purpose better.

RMA2 (Donnel et al 2009) is a two-dimensional depth-averaged hydrodynamic model, provided by the AQUAVEO company (AQUAVEO). To obtain the results of steady-state and dynamic simulations, RMA2 uses: the Reynolds form of Navier–Stokes equations for turbulent flows, the Manning or Chezy equations to calculate the bed and banks friction, and eddy viscosity coefficients to define characteristics of turbulence. The forms of fluid mass and momentum conservation equations are:

$$h\frac{\partial v_x}{\partial t} + hv_x\frac{\partial v_x}{\partial x} + hv_y\frac{\partial v_x}{\partial y} - \frac{h}{\rho}\left[E_{xx}\frac{\partial^2 v_x}{\partial x^2} + E_{xy}\frac{\partial^2 v_x}{\partial y^2}\right] + gh\left[\frac{\partial a}{\partial x} + \frac{\partial h}{\partial x}\right] + \frac{gv_x n^2}{\left(1.486h^{\frac{1}{6}}\right)^2}\left(v_x^2 + v_y^2\right)^{\frac{1}{2}} - \zeta V_a^2 \cos\psi - 2hv_y\omega\sin\Phi = 0 \tag{1}$$

with an analogous formula for the y direction, and

$$\frac{\partial h}{\partial t} + h\left(\frac{\partial v_x}{\partial x} + \frac{\partial v_y}{\partial y}\right) + v_x\frac{\partial h}{\partial x} + v_y\frac{\partial h}{\partial y} = 0 \tag{2}$$

where: h is the water depth; ρ the fluid density; t time; g the acceleration due to gravity; x,y the Cartesian coordinates; v_x, v_y the velocities in Cartesian directions; E_{xx}, E_{xy} the turbulent (eddy) diffusivity parameter described below; Va the wind speed; a the bottom elevation; n the Manning roughness coefficient; ζ the empirical wind shear coefficient; ψ the wind direction; ω the rate of the Earth angular rotation; and Φ the local latitude.

Subsequent terms represent: depth-averaged velocity change in the x direction during the time; velocities change in the x direction along the x coordinate; velocities change in the x direction along the y coordinate; hydrodynamic loss due to eddy viscosity; impact on acceleration due to gravity; influence of the bottom roughness; effect of wind speed and direction; impact of the Coriolis force.

In the simulations used in this chapter, both the wind (analysis containing wind impact would be too extensive for this short chapter) and the Coriolis force impacts (because of small size of the reservoir) are neglected.

One of the important parameters for RMA2 is the bottom roughness. Changes of the bed friction may result in different effects of flow direction as well as velocity magnitude. The bottom shear stress is represented by the following equation:

$$\tau = \rho g R S \tag{3}$$

where: R is the hydraulic mean radius; S the bed slope.

The Manning velocity formula for uniform flow is:

$$\tau_x = \rho g \left(\frac{n}{1.49}\right)^2 \frac{v_x \sqrt{v_x^2 + v_y^2}}{h^{1/3}} \tag{4}$$

where τ_x is the bottom shear stress component in x direction; the formula for the y direction is analogous.

The turbulent exchange is the momentum transfer of fluid particles due to their chaotic motion. The turbulence effect occurs in time scales shorter than the iteration time step. The value of turbulent diffusivity parameter used in this model to overcome this problem may exceed the physical value of this parameter for water. The exact value of this coefficient could require empirical verification.

$$E_{xx} \frac{\partial^2 v_x}{\partial x^2} = \mu \frac{\partial^2 v_x}{\partial x^2} + \frac{\partial}{\partial x} \frac{\delta \overline{v_x'^2}}{\partial x} \tag{5}$$

$$E_{xy} \frac{\partial^2 v_x}{\partial y^2} = \mu \frac{\partial^2 v_x}{\partial y^2} + \frac{\partial}{\partial x} \frac{\delta \overline{v_x' v_y'}}{\partial y} \tag{6}$$

(analogous formulas for E_{yx} and E_{yy}), where: E_{yx}, E_{yy} are the eddy viscosity tensor components; μ the molecular viscosity; v_x' the turbulent fluctuations velocity in the x direction; and v_y' the turbulent velocity fluctuations in the y direction.

The final solution may be obtained in two types of simulations: *Steady-state simulation*—boundary conditions are constant between the time steps, and therefore the solution is independent of time. The solution is given as one result only. *Dynamic simulation*—the results depend on time. The boundary conditions also have to be dependent on time. The choice of the simulation type depends on the desired application.

The solution is obtained using a finite element method thanks to the Galerkin algorithm of weighted residuals. The equations are being solved on a mesh containing one- or two-dimensional elements.

RMA4 (Letter et al. 2003) is a model related to RMA2. It is designed to simulate the depth-averaged advection–diffusion processes. This water quality model can be used for estimation of concentration of any constituent dissolved in an aquatic environment. The RMA4 application can be used to identify potential critical areas for pollutants spread, e.g., oil spills. It may also be used to determine the horizontal salinity distribution, or other water quality criteria.

The form of the depth-averaged transport equation used by RMA4 is as follows:

$$h\left(\frac{\partial c}{\partial t} + v_x\frac{\partial c}{\partial x} + v_y\frac{\partial c}{\partial y} - \frac{\partial}{\partial x}D_x\frac{\partial c}{\partial x} - \frac{\partial}{\partial y}D_y\frac{\partial c}{\partial y} - \sigma + kc + \frac{R(c)}{h}\right) = 0 \quad (7)$$

where: c is the concentration of a given constituent; D_x, D_y the turbulent mixing coefficient (non-diagonal diffusion terms are assumed to be zero); k the first order decay coefficient of a given pollutant; σ the source/sink of constituent; and $R(c)$ the rainfall/evaporation rate.

Subsequent terms represent: local storage; advection term (x); advection term (y); dispersion (x); dispersion (y); local sources; exponential decay; rainfall and/or evaporation effects.

3 Methodology

The models calculate on two-dimensional meshes built of triangular and quadrilateral elements. The method of creating meshes is discussed in (Hachaj 2007) and the mesh built previously for (Gałek and Hachaj 2008) is also used in this chapter after slight modifications due to specific RMA2 requirements. The boundary conditions are located on the tributaries (as the water inflow discharges) and on the outflow (as the water surface level).

The simulation methodology is based on injecting virtual pollutants into the chosen parts of the lake. To present the possibility of calculating different scenarios, the authors considered three different cases, as shown in Fig. 1:

- Persistent inflow of organic pollution from Raba (1) and Wolnica (2);
- 12 h organic leakage near tributaries (3), (4) and (5);
- 3-hours leakage of industrial waste from haven (6) and power station (7);

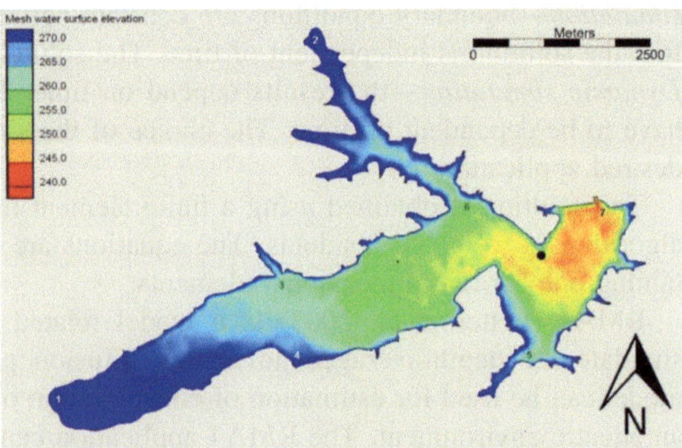

Fig. 1 The bathymetry of the Dobczyce Lake for 272.6 m a.s.l. with positions of contaminants injection points (1–7) and the water supply inlet position (•)

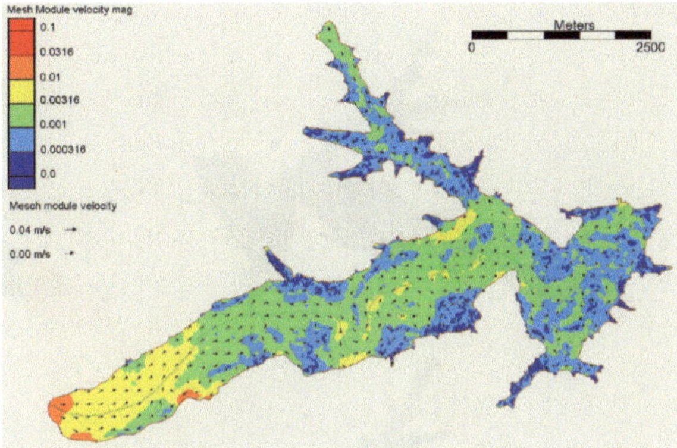

Fig. 2 The planar velocity field obtained from the RMA2 simulation for water surface level of 272.6 m a.s.l. and average flow in the lake. The *grey line* denotes the position of the main current

The danger exerted on the water supply inlet by all these events has been considered.

All the simulations were conducted for static velocity field shown in Fig. 2 obtained for windless conditions and average inflow discharges. The calculations may as well be performed for any reasonable wind, even one changing in time.

The crucial point of the pollution transport simulation setup was choosing the diffusivity parameter. In principle it may vary for different types of flow by the order of magnitude and more (Roberts and Webster 2002). Following the guidelines contained in (Roberts and Webster 2002) backed by earlier results (Fisher et al. 1979; Holley and Jirka 1979) the value of 0.1 m^2/s has been used for "organic" type pollutants and 0.05 m^2/s for "industrial" type ones. However reasonable they are, it should be stressed now that those values require experimental verification before using them for practical application purposes. Fortunately, comparing simulations performed for various values of that parameter indicates that the model is not extensively susceptible to its changes.

4 The Results

The result of the analysis is a simulation file presenting the process of contamination expansion in the lake. It can be presented graphically as a set of contaminant concentration maps scaled in parts per million (or, roughly, g/m^3). In the cases presented below, the processes have been observed for two weeks each, and all the maps are drawn using a logarithmic scale along the concentration axis.

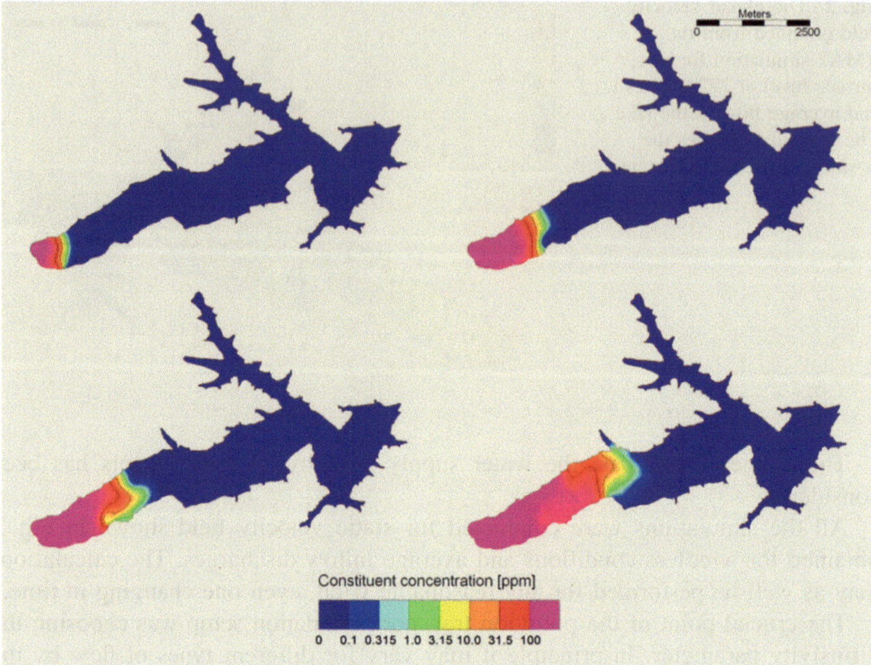

Fig. 3 Process of transporting the simulated contamination from the Raba river after 14 h (*upper left*), 2 d (*upper right*), 4 d (*lower left*), and 10 d (*lower right*) after the *zero* moment

4.1 Persistent Organic Contaminants Inflow

For the first pair of simulations, the following assumption has been chosen: Raba (the main inflow) and Wolnica (the most important side tributary) have been polluted by a constituent of organic origins. Starting from the *zero* moment the rivers inject 100 ppm polluted water into the lake.

The results for Raba are shown in Fig. 3. The contaminant flows along the main current and spreads around. The area of exponential concentration growth (from 0 to 100 ppm) becomes wider and wider as the time passes and the polluted area takes the full width of the lake. It should be pointed out that the outmost contour is drawn for 0.1 ppm, i.e., for 10^{-3} of the original concentration. It can be seen that after 10 days the contaminant does not pass even a half of the way to the water supply inlet. This gives the water station staff and appropriate protective services more than enough time to react.

For the Wolnica tributary a similar situation is observed (Fig. 4). Due to the fact that the channel is narrower and shallow, but the current is slower (see Fig. 2 again), diffusive effects take the upper hand over advection: the contour lines are slightly further from each other than they were for Raba-originating pollutants.

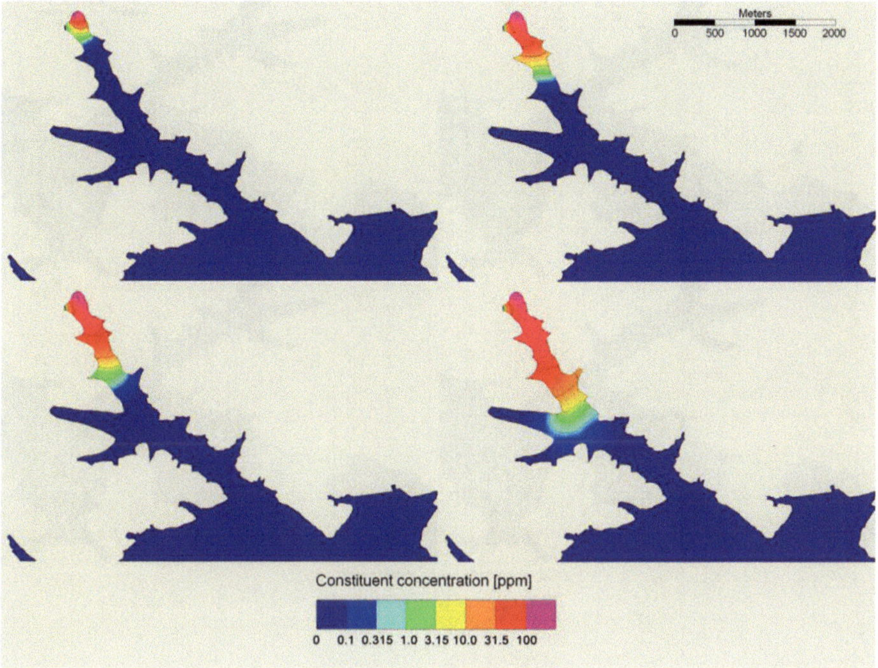

Fig. 4 Process of transporting the simulated contamination from Wolnica tributary after 14 h (*upper left*), 2 d (*upper right*), 4 d (*lower left*), and 10 d (*lower right*) after the *zero* moment

Both results suit appropriate theoretical anticipations and both of them indicate that there is more than two weeks to react if a real pollution event like that happens. This contrasts the results of (Gałek and Hachaj 2008), where for a high flow four days were enough for Wolnica-originating virtual pollutants to reach the water supply inlet even without considering any diffusive movements.

4.2 Twelve-Hour Leakages of Organic Substances

The next set of three simulations carry the assumption that 10^4 kg of an organic pollutant entered the lake near the outflow of one of selected tributaries. Such a leakage lasted for 12 h (i.e., its intensity was 231 g/s). The *zero* moment for each of these simulations is set to be the end of this leakage time. Subsequent Figs. (5, 6 and 7) show the simulated contaminant behavior—again through a period of 10 days.

For the leakage starting in point **3** the contaminant disperses and start to flow along with the main current. It does not threat the water supply inlet during the first

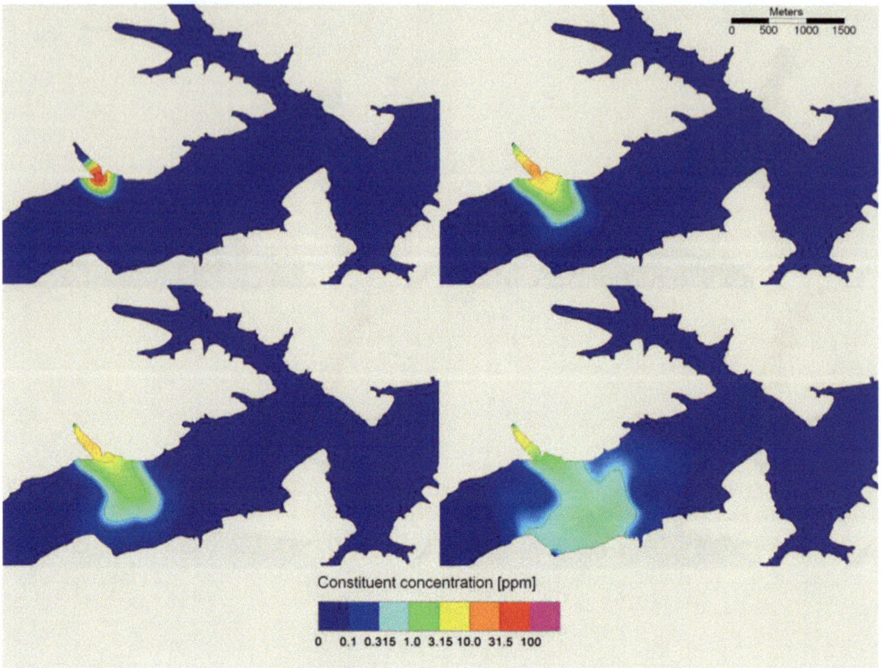

Fig. 5 Process of transporting simulated 12 h leakage from position **3** shown in Fig. 1 after 14 h (*upper left*), 2 d (*upper right*), 4 d (*lower left*), and 10 d (*lower right*) after the *zero* moment

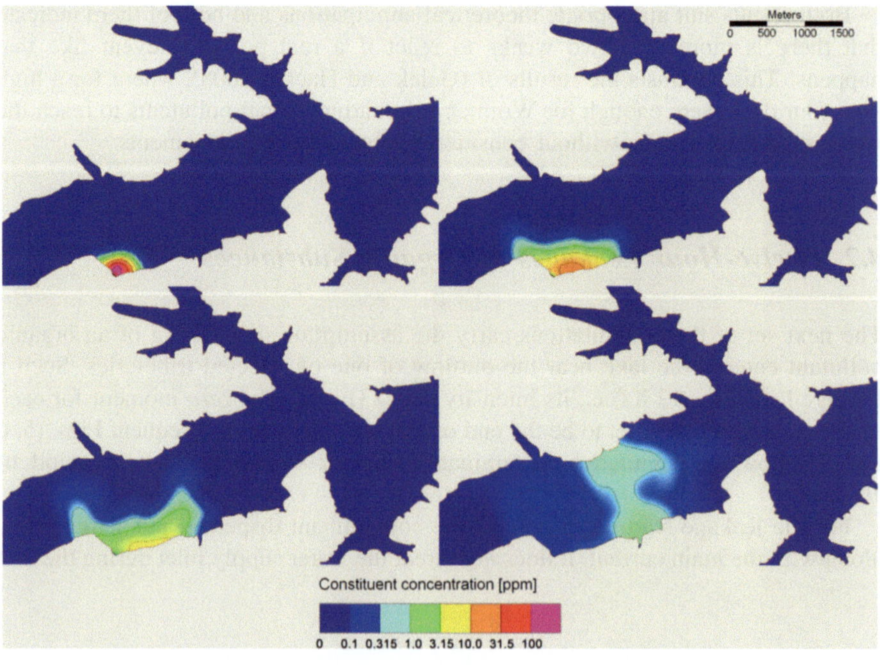

Fig. 6 Process of transporting simulated 12 h leakage from position **4** shown in Fig. 1 after 14 h (*upper left*), 2 d (*upper right*), 4 d (*lower left*), and 10 d (*lower right*) after the *zero* moment

two weeks after the event. Its concentration drops by three to four orders of magnitude as it diffuses along the passing time of 10 days (see Fig. 5).

Constituent **4** drop zone (Fig. 6) has been located opposite to the one of constituent **3**. Firstly, the substance dissipates along the right hand bank (along with the current) as well as northwards. Next, a fraction of the contaminant flows downstream still dissipating while the rest stays in stagnant areas slowly leaving them through turbulent diffusion processes. No threat for the water supply inlet exists during the whole simulation time.

Simulated contamination starting from the zone **5**, in the shallow southern bay, stays out of the reach of the main lake current for almost two days. During that time it spreads in a radial way driven mainly by the turbulent diffusion. It is also pushed northwards by two minor tributaries that enter the bay but it is a weak effect. After two days the simulated constituent is taken by the current towards the outflows of the lake, i.e., the main outflow located at the dam and the water supply inlet. Although most of it should reach the dam outflow along with the current, some fraction of considerable concentration may reach the water supply inlet in a time of one to two weeks. It still leaves some time to react after the leakage event is detected (see Fig. 7).

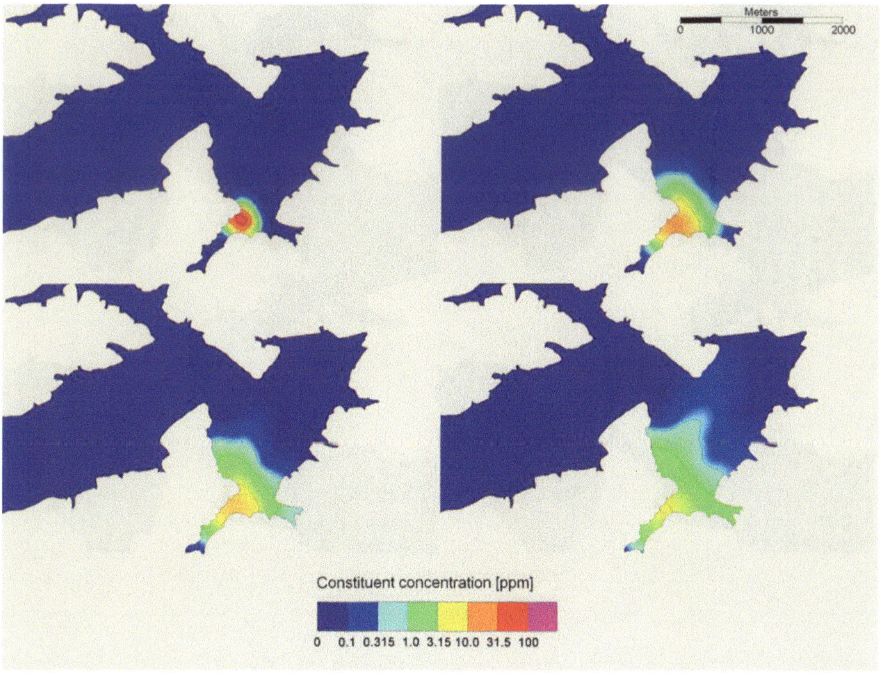

Fig. 7 Process of transporting simulated 12 h leakage from position **5** shown in Fig. 1 after 14 h (*upper left*), 2 d (*upper right*), 4 d (*lower left*), and 10 d (*lower right*) after the *zero* moment

4.3 Three-Hour Leakage of Industrial Pollution

The last two simulations presented in this chapter were performed under an assumption of rapid leakage of constituents that have lower eddy diffusivity parameter than the ones used before (0.05 m^2/s instead of 0.1 m^2/s). For short they are called *industrial* constituents and they origin from the zones where industrial-originated contamination may eventually appear: a haven located near the end of the Wolnica bay (**6** in Fig. 1), and the power plant inlet located eastwards from the dam (**7** in Fig. 1).

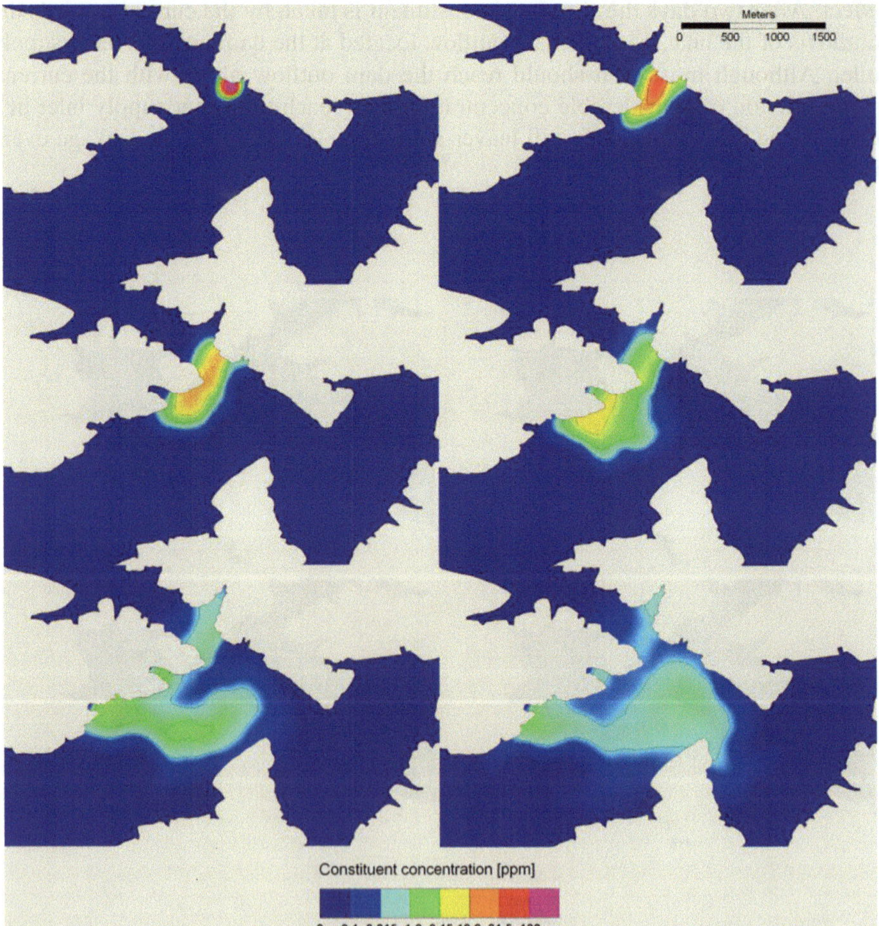

Fig. 8 Process of transporting simulated 3 h leakage from position **6** shown in Fig. 1 after 5 h (*upper left*), 1 d (*upper right*), 2 d (*middle left*), 4 d (*middle right*), 8 d (*lower left*) and 12 d (*lower right*) after the *zero* moment

As before, the total mass of the leaked pollutant is assumed to be ten tons, but this time it enters the lake during only three-hour event. It results in 926 g/s mass load. The end of the leakage marks the *zero* moment for subsequent Figs. 8 and 9.

The RMA2 hydrodynamic model predicts that a weak backward current should exist for the considered state of the lake near the mouth of the Wolnica bay. Such a setup causes the simulated constituent to flow west first, then to turn south and east. Effectively, the contaminated area splits firstly into two and then even into three regions. This makes the way of pollutant particles to the water supply inlet longer. However it should be beard in mind that both the model predictions (taken either from FESWMS or RMA2 models) as well as observation performed in the

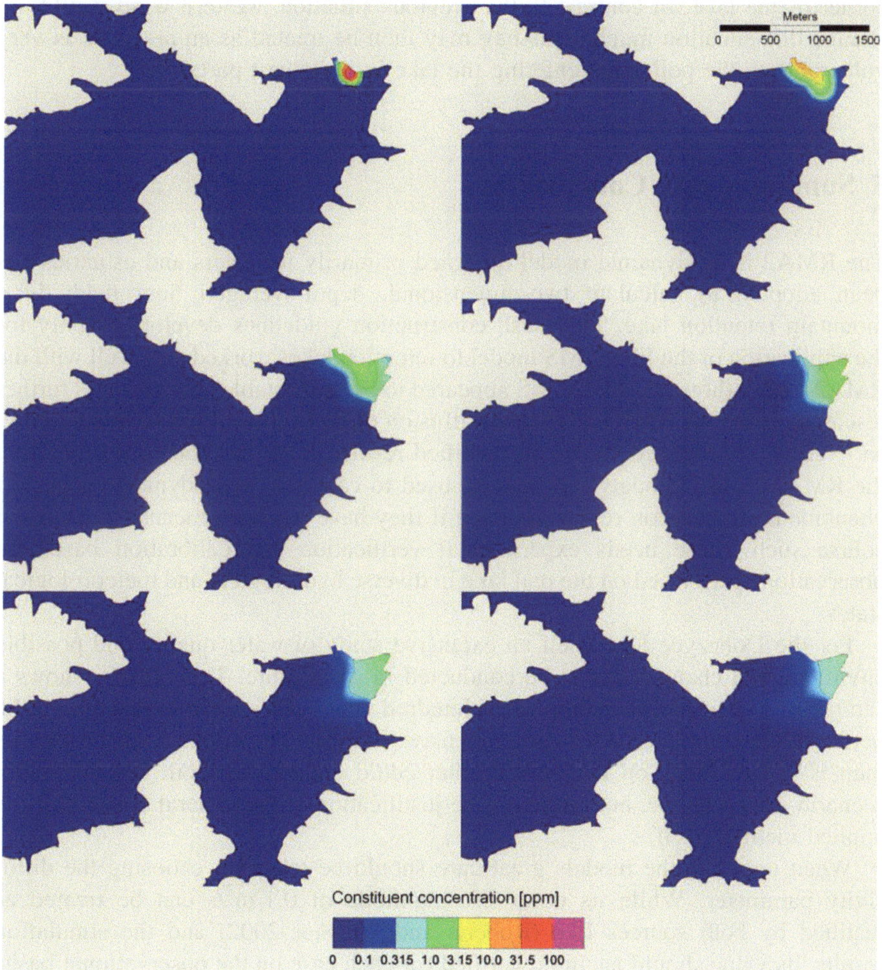

Fig. 9 Process of transporting simulated 3 h leakage from position **7** shown in Fig. 1 after 5 h (*upper left*), 1 d (*upper right*), 2 d (*middle left*), 4 d (*middle right*), 8 d (*lower left*) and 12 d (*lower right*) after the *zero* moment

lake exhibit strong dependence of the flow field in that region on the wind speed and direction, which will be addressed in subsequent publications. In general, when pushed by a western wind (which is most common in southern Poland) the constituent would flow along the northeastern bank and enter the proximity of the water supply inlet much quicker than it is shown in Fig. 8.

The situation shown in Fig. 9, where the pollutant starts from the zone of the power plant inlet, is quite different. It gives two clues encouraging a further analysis. Firstly, most of the contaminant is quickly flushed out of the lake through the main outflow. It does not threaten the water supply inlet at all but it has a big impact on the Raba water quality below the reservoir. Secondly, one can also observe accumulation of the pollutant in a dead zone located in the easternmost corner of the lake. In contrast to the previous situation, western wind would not change this situation much. This bay may then be treated as an area that is very vulnerable to the pollutants entering the lake in its lowest part.

5 Summary and Conclusions

The RMA2 hydrodynamic model designed primarily for rivers and estuaries has been adopted to calculate two-dimensional, depth-averaged flow fields in a mountain retention lake. The mesh construction guidelines developed firstly for the application of the FESWMS model to retention lakes worked also well with the RMA2 model (however FESWMS appeared to be more stable; this requires further research). Then the RMA4 advection/diffusion transport model has been applied as an overlay. It produced physically justified results. It can be concluded then that the RMA2/RMA4 models can both be used to calculate hydrodynamic transport phenomena in retention reservoirs even if they have not been meant for them. Of course such a use needs experimental verification and calibration based on observations performed on the real lake in diverse hydrological and meteorological states.

For the Dobczyce lake itself an extensive study of water quality and possible environmental changes has to be conducted in near future. This chapter shows a sample of five out of probably a few hundred simulated scenarios that should be performed to complete such a comprehensive analysis. The intention of the authors then is not to focus on conclusions that could be derived from any particular scenario shown above, but rather on the justification of the general concept of the applied method itself.

When applying the models great care should be taken on choosing the diffusivity parameter. While its order of magnitude of 0.1 m^2/s can be treated as justified by both sources like (Roberts and Webster 2002) and the simulation results, its value should again be verified for each case on the observational basis. Fortunately the model is not very susceptible to the changes of this parameter so there is no need to find its exact value very precisely (which may be virtually impossible).

Finally: The dynamics of the simulated density fields is difficult to present in a chapter. The frames have been chosen to be as representative as possible, but they are still 4 to 6 out of over 300 ones made for each of the scenarios. Interested people are welcome to contact the authors of this chapter to obtain animation files containing all the frames in their proper sequence.

References

AQUAVEO (online) Water Modeling Solutions, http://www.aquaveo.com/

Donnel BP et al (2009) Users Guide to RMA2 WES Version 4.5

Fisher HB et al (1979) Mixing in inland and coastal waters. Academic, New York

Gałek M, Hachaj PS (2008) Tracking of modelled pollution originating from tributaries in the Dobczyce lake working under flood conditions. Publs Inst Geophys Pol Acad Sc E-10(406):39

Hachaj PS (2007) Modelling of a two-dimensional velocity field for the water flow in the lake of Dobczyce. In: Transport phenomena in hydraulics, p 87

Holley ER, Jirka GH (1979) Mixing in Rivers, Tech Report No. E-86-11, U.S. Army Corps of Engineers, Vicksburg, Miss

Letter JV et al (2003) Users Guide to RMA4 Version 4.5

Roberts PJW, Webster DR (2002) Turbulent diffusion. In: Shen HH (ed) Environmental fluid mechanics: theories and applications. American Society of Civil Engineers, New York

Zając K (2012) Symulacje przepływu wody w zbiorniku retencyjnym Tresna za pomocą modelu numerycznego RMA2 (Simulating water flow in the Tresna retention reservoir using the RMA2 model), Engineer degree thesis, Politechnika Krakowska (in Polish)

The "River Memory" Effect: An Attempt to Understand and Model it

Paweł S. Hachaj

Abstract Observations made by people working with retention reservoirs indicate that for standard, most common hydro-meteorological conditions, the majority of water in those lakes remains stagnant while almost all the discharge is carried by a relatively narrow current. This phenomenon is commonly called "the memory of the river" as that current appears often in locations close to the original bed of the river as it had been before the lake dam has been built. Such a behavior seems to appear not only when there are remnants of the old bed in the lake bathymetry, but also when there is no trace of the old bed. To the author's knowledge, no theoretical study of this phenomenon exists so far. Hardly any articles about it can be found. The common intuition is that the formula for turbulent viscosity should be changed from Newtonian, linear one to something more sophisticated. In this article, the use of the Stribeck viscosity formula is proposed to explain water behavior for very small flow velocities ($v \leq 10^{-4}$ m/s) that normally occur in retention lakes. Physical justification, computational formulas changes proposals, and simplified simulation results are all presented.

1 River Memory: Known but Overlooked

Observations made by people working with retention reservoirs indicate that for the most common hydro-meteorological conditions, when the total discharge is relatively low and the wind is moderate, the majority of water in those lakes

P. S. Hachaj (✉)
Institute of Water Engineering and Water Management,
Cracow University of Technology, Warszawska 24, 31-155 Kraków, Poland
e-mail: pawel.hachaj@iigw.pl

P. Rowiński (ed.), *Experimental and Computational Solutions of Hydraulic Problems*, GeoPlanet: Earth and Planetary Sciences, DOI: 10.1007/978-3-642-30209-1_22,

remains stagnant, while almost all the discharge is carried by a relatively narrow current. This phenomenon has been observed frequently for many lakes by both applied scientists who were to conduct their measurements there, as well as the members of staffs of hydrotechnical facilities connected with those lakes. There is no standard name for that effect so far, but it is most commonly called "the River Memory". This is because the current appears often in locations close to the original bed of the river as it had flown in times before the dam has been built. Such a behavior tends to appear not only when there are remnants of the old bed in the lake bathymetry, like it is for, e.g., Goczałkowice or Porąbka reservoirs, but also when there is no trace of the old bed, like for, e.g., Dobczyce or Tresna lakes.

Although widely known, this effect has been mostly overlooked. According to the author's knowledge, no theoretical study of this phenomenon exists so far. Judging from conversations the author has had with a number of fluid mechanics as well as with hydrology specialists, that effect, even if noticed, has been treated as a curiosity, a local oddity, an interesting but meaningless observation or just something that exist and we should live with it, taking the assumption of its existence. For example, an analysis of the Pakość lake behavior by Nachlik et al. (1980) proved to be futile without that empirical assumption.

In times when most numerical simulations of hydrodynamics in retention lakes were in fact one-dimensional, recognizing and proper understanding the River Memory effect was not very important. But nowadays scientists are equipped with two-dimensional, depth-averaged two-dimensional and even three-dimensional models that can be applied to simulate flow fields in retention reservoirs: ELCOM (Hodges and Dallimore 2007), FESWMS (Froelich 2003), RMA2 (Donnel et al. 2009), TUFLOW (Syme et al. 2010)—to name just a few. When using such models it is important to re-create real behavior of water in retention lakes with as many distinguishable features as possible. Yet current models fail to show the River Memory effect unless they are manually forced to do so—as it will be shown in Sect. 4.

A politician could say that if the facts do not suit the theory—the worse for the facts. On a contrary, a scientist should feel an urge to alter the theory. The new theory should agree with the old one for most cases but the ones containing the new observations. In the considered instance, the laws of free-surface flow need to be altered for very low velocities (order of magnitude of a millimeter per second and less) that appear in the retention lakes in a way that do not harm them for higher velocities. Natural candidates to modify are components responsible for motion resisting forces in the governing equations. *In extenso*: the viscosity terms.

2 On Various Models of Friction

Viscosity is the internal friction in liquids—so says the definition of viscosity. In this section, various approaches to friction will be briefly discussed. This should help us to understand better what the desired form of viscosity terms is. Starting from very simple approaches, the sequence of models will end at the one proposed to solve the secret of the River Memory.

2.1 Amontes-Coulomb Model; Dynamic and Static Friction

The simplified Amontes-Coulomb formula is the most basic approach to friction. It assumes that when two surfaces slide one on another the friction force does not depend on velocity:

$$F_f = \lambda N, \tag{1}$$

where F_f is the friction force; λ is the friction coefficient; and N is the normal force between both surfaces.

However, it was always known that formula (1) is not compete, as common observations show that it is easier to keep a body moving than to let it start to move. Thus, the friction equation needed to be supplemented by a static variant that adds some sort of threshold velocity dependence to this model:

$$F_f = \begin{cases} F_0, & \textit{if } v = 0 \textit{ and } F_0 < T \\ \lambda N + 0v, & \textit{if } v \neq 0 \end{cases}, \tag{2}$$

where F_0 is the external tangential force; v is the velocity; and T is the threshold tangential force also known as traction (Bayer 2004).

There are two issues to be recorded, both important for the main goal of this chapter.

First: Traction exceeds sliding friction: $T > \lambda N$. So for any value of F_0 such that $T > F_0 > \lambda N$, two possible states exist: in one of them the bodies do not slide one on another, in another one they do—with acceleration:

$$a = \frac{F_0 - \lambda N}{m}, \tag{3}$$

where m is the mass of the moving body.

Second: If the slide velocity is very small, some effects of static friction appear even if in motion (Rice and Ruina 1983; Åström and Canudas-de-Wit 2008), which can be written in a simplified form as:

$$F_f = \lambda N + (T - \lambda N)e^{v/v_0} \tag{4}$$

where v_0 is a parameter.

These phenomena were investigated at the beginning of 20th century by Richard Stribeck, a key person for Sect. 2.4 of this chapter (Stribeck 1902; Jacobsen 2003).

2.2 Newtonian Model of Viscosity and its Ramifications

When it comes to calculating internal friction in liquids, i.e., viscosity, there are no longer two fixed surfaces that slide one on another. Instead, a continuum of layers is considered. Thus, forces and velocities are replaced by their first order spatial derivatives: shear stresses and shear rates, respectively. In the original Newton model of viscosity it is assumed that the shear stress grows linearly with the shear rate. In general, the equation of Newtonian and quasi-Newtonian viscosity can be expressed as an increasing function with no residuum, in a simplified form:

$$S = \mu\left(\frac{\partial v^{\alpha}}{\partial h}\right), \tag{5}$$

where S is the shear stress; μ is the viscosity coefficient; h is the direction normal to v; and α is a parameter.

The value of parameter α determines the model type inside the family:

- $\alpha = 0$—superliquidity;
- $0 < \alpha < 1$—shear thinning, also called pseudo-plastic (Dunn 2005) or fluidizing (Locat and Demers 1988) models;
- $\alpha = 1$—original Newton formula;
- $\alpha > 1$—shear thickening, also called as dilatant (Dunn 2005) models.

The Newtonian family of viscosity models applies in principle to the laminar flow regime only. Yet event for turbulent flows a variant of the Newton linear formula is commonly used in practical applications (e.g., Donnel et al. 2009; Hodges and Dallimore 2007 and others), but with scalar values of shear stress and viscosity coefficient replaced with appropriate tensors. In principle, for turbulent flows the eddy viscosity coefficient takes the form of a tensor of the third rank:

$$\mu = \begin{pmatrix} E_{xx} & E_{xy} & E_{xz} \\ E_{yx} & E_{yy} & E_{yz} \\ E_{zx} & E_{zy} & E_{zz} \end{pmatrix}, \tag{6}$$

where all the E_{ab} values are scalars.

In principle, such a tensor can have nine independent terms: three diagonal ones and six mixed ones. In practical applications, some simplifications are often used, e.g., two-dimensional models discard both the third row and column; some models use a symmetrical tensor, other assume all the mixed terms to be equal or even zero; some other assume that all the components are equal. Certain models require higher values of those parameters than physically justified ones in order to maintain numerical stability. But these simplifications are not responsible for the fact that those models do not predict the River Memory phenomenon: there is just no place for such a nonlinear behavior in residue-less Newtonian calculations.

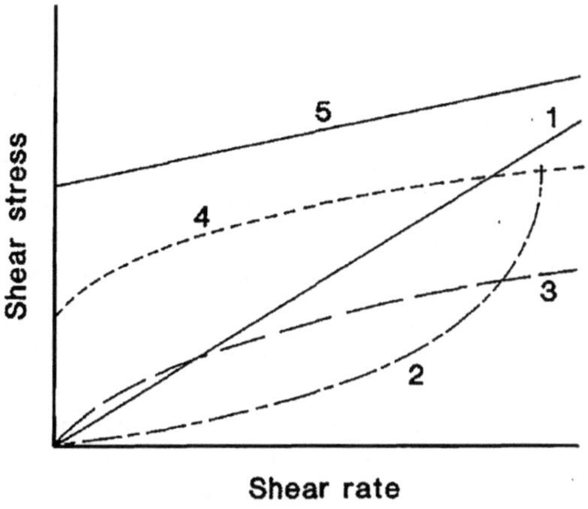

Fig. 1 Major types of fluids distinguished by their viscosity (Locat and Demers 1988): *1* Newtonian, *2* Thickening, *3* Thinning, *4* Casson, *5* Bingham

2.3 Bingham and Casson Models

The simplest introduction of a static term to the concept of viscosity has been proposed by Eugene Bingham (Bingham 1916); see also line 5 in Fig. 1. This approach assumes that a small constant term should be added to the standard Newton viscosity formula:

$$S = Y + \mu \frac{\partial v}{\partial h}, \tag{7}$$

where Y is the yield stress, the minimal shear stress needed to start the flow.

Later experiments indicated that the transition between no-flow and flow regimes is smooth for many liquids; this leads to the Casson formula (Casson 1959) depicted as line 4 in Fig. 1

$$S = \left(Y + \sqrt{\mu \frac{\partial v}{\partial h}} \right)^2. \tag{8}$$

The value of the yield stress for most "standard" liquids is small enough to be neglected, leading back to the Newton formula. The Bingham and Casson models are commonly used to describe fluids of high viscosity like pitch or liquid chocolate, but also for blood (Kim 2002) and water-clay mixtures (Komatina and Jovanović 1997; Locat and Demers 1988). Moreover, it turned out that pure water in capillaries or porous media can be treated as a Bingam plastic (Kovacs 1981). It may be tempting then to give the Bingam viscosity formula a chance to explain the River Memory phenomenon. Unfortunately it is not enough.

The first objection one could raise is whether it was justified to use any static terms to describe water in large scales. When considering a retention reservoir one can assume a size like: kilometers of length, hundreds of meters of width and at least several meters of depth. For the vast majority of water particles in there, there is no solid object nearby, unlike in capillaries or porous media. The particles are in continuous, chaotic, thermal movement. Moreover, a lake is by no means an isolated system: various factors, starting from wind blows, through boats, ending at fish and water birds, disturb the water inside the lake continuously. The sum of those disturbances is usually greater than any collective current that may appear due to fixed inflow and outflow of a few cubic meters per second. Local velocities caused by those random effects can also be much higher than the collective flow velocity which is around a millimeter per second or less.

Is there any place for a static picture then? According to common understanding of lake hydraulics there is not. But the observations of flow regimes in retention reservoirs including most recent observations on the Goczałkowice lake being performed within the ZiZOZap project indicate that local instabilities are damped, stagnant areas are much larger than any computational model predicts, and a relatively narrow current appears inside the lake. (*ZiZOZap*: *The Integrated System of Aiding the Management and Protection of a Dam Lake*; *an ongoing project*, http://www.zizozap.pl/. *Papers and reports to be published.*) It does not seem possible to explain these phenomena without taking into account static viscosity effects.

While the first objection can be overcome by admitting that the observations contradict common assumptions, the second objection is more substantial. Let us assume that for very slow velocities water really behaves as Bingham or Casson equations predict. If so, one should expect that the velocity field follows the scheme observed for Bingham plastic: a relatively wide core moving with constant speed surrounded by border zones where the velocity changes gradually from zero near the banks up to the velocity of the core, see e.g., (Estellé and Lanos 2006) or (Komatina and Jovanović 1997). This is not our case! (Fig. 2).

Summing up, the two objections may be written using words that the author has heard a few times: *We are not sure if we are allowed to use the Bingham formula for lakes. And even if we do that, it will not work there.*

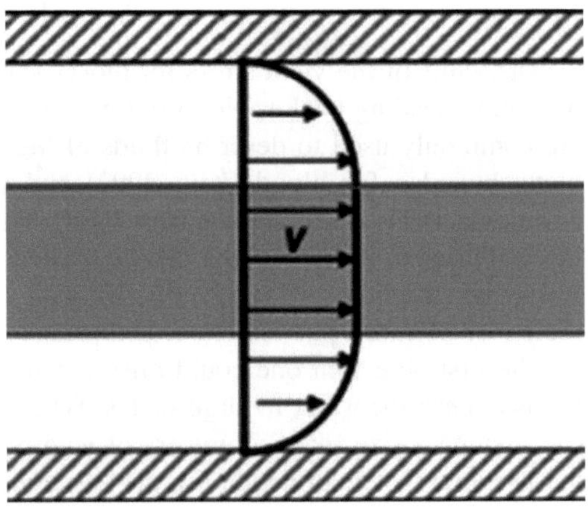

Fig. 2 Planar velocity field of a Bingham liquid. The central core is surrounded by border zones. They together fill the whole channel width. Based on (Estellé and Lanos 2006)

2.4 LuGre (Stribeck) Model: A Possible Solution

The LuGre model name originates from the names of two scientific centers: Lund and Grenoble. It is also called the Stribeck model (Åström and Canudas-de-Wit 2008; Andersson et al. 2007), as it derives from original works of R. Stribeck. It has been developed as a tool to describe friction of two lubricated surfaces sliding one on another and it is widely known and commonly used in tribology. As such, it contains:

- a static friction term that drops rapidly but not instantly with velocity (like the original Stribeck formula);
- a term proportional to shear stress (like the Newton equation);
- a gradual transition between static and kinetic regimes (like the Casson model).

The friction equation has the following form:

$$F_f = F_C + \left(Y_f - F_C\right) e^{-|v/v_0|^\alpha} + \mu_f v, \tag{9}$$

where F_C is the Coulomb term responsible for the velocity-independent friction; Y_f is the Yield friction responsible for the strength of static effects; v_0 is a parameter responsible for the length of the static exponential tail; α is a parameter responsible for the slope of the static tail; and μ_f is an analogue of the Newton viscosity coefficient.

Formula (9), presented graphically in Fig. 3, should be adjusted before applying it to slow flows of water. First, as there are no fixed surfaces in that case, the Coulomb component may be discarded. Second, following Tustin (1947) and, to an extent, Bo and Pawlescu (1982), it can be assumed that $\alpha = 1$. Then v and v_0 can be replaced by appropriate shear rates $\partial v/\partial h$ and R_0. So in terms of viscosity the equation is:

$$S = Y e^{-\left|\frac{1}{R_0}\frac{\partial v}{\partial h}\right|} + \mu \frac{\partial v}{\partial h}. \tag{10}$$

It is noteworthy that for higher velocities Eq. (10) is closer to the Newton formula than (7) and (8) are, as the static term does not remain constant in the LuGre formula but it quickly drops to negligible values.

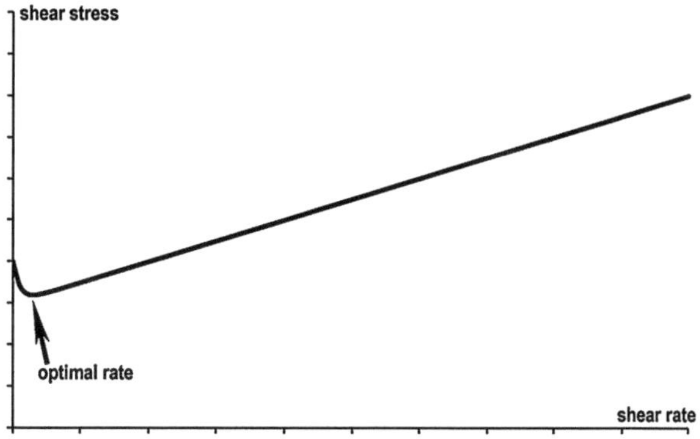

Fig. 3 LuGre friction graph. There exists the optimal velocity (*shear rate*) for which the friction (*shear stress*) is minimal

The adjusted LuGre viscosity model has a potential to explain the River Memory effect. If the viscosity really increases as the shear rate is close to zero, then for small discharges that flow in beds of large cross-sections, making all the water flow with very low shear rates, requires more energy than carrying the total flow by a current of the optimal shear rate and minimal shear stress. The rest of the channel shall then remain stagnant. This model has also another feature, shared in fact by Bingham and Casson ones that is absent in the Newtonian approach: local disturbances in the discharge filed, as they disperse and their velocity decreases, are damped by the static friction component. This prevents occasional wind blows, boats, fish and water birds mentioned in Sect. 2.3 disturbing the global picture. On the other hand, constant wind, even if small, may alter the velocity field much. These findings seem to suit the facts known from observations.

3 LuGre Viscosity Terms in Computer Modeling: A Proposal

In depth-averaged two-dimensional hydraulic models the turbulent viscosity effects are usually contained in the momentum conservation equations as appropriate terms (Donnel et al. 2009; Froelich 2003; Syme et al. 2010). For example, for the x direction momentum transfers those terms have forms like:

$$\cdots - \frac{h}{\rho}\left(E_{xx}\frac{\partial^2 v_x}{\partial x^2} + E_{xy}\frac{\partial^2 v_x}{\partial x\,\partial y}\right)\cdots, \tag{11}$$

where h is the water depth; ρ is the water density; and E_{ab} are appropriate components of the eddy viscosity tensor.

For a 3-D ELCOM model (Hodges and Dallimore 2007) the off-diagonal eddy viscosity terms are set to zero, and the overall momentum transport equation has the following viscosity-related term:

$$\cdots - \frac{\partial\left(E_{xx}\frac{\partial v_x}{\partial x}\right)}{\partial x} - \frac{\partial\left(E_{yy}\frac{\partial v_y}{\partial y}\right)}{\partial y} - \frac{\partial\left(E_{zz}\frac{\partial v_z}{\partial z}\right)}{\partial z}\cdots. \tag{12}$$

Exact handling of E_{ab} values depends on the actual model, but in principle those components have the following generalized form:

$$E_{ab} = \tau(1 + m_{ab}), \tag{13}$$

where τ is a scalar parameter interpreted as the viscosity coefficient, and m_{ab} is the dimensionless parameter responsible for sub-time-scale velocity fluctuations (mixing) between directions a and b or fluctuations *in* the given direction if $a \equiv b$.

Users of these models have always given an access to the τ parameter or its equivalent. They may also have access to the m_{ab} parameters. Yet no altering of any of them can lead those models to automatically apply the LuGre viscosity approach instead of the Newtonian one. It is possible, however, to proceed iteratively, increasing τ manually after each run of the program in the regions where shear rates are small. Taking into account that such a procedure requires selective modifying boundary conditions on the computational mesh it is easy to see that such a process is time consuming, not very effective, and that is very vulnerable to human mistakes. A sample solution obtained using such an iterative method is shown in the next section.

On the other hand, it is possible to modify the models themselves—their creators have the access to the source codes, so the Newton formula could be replaced by the LuGre one. As the static viscosity component depends on total water velocity, the appropriate nine terms of three-dimensional momentum transfer equations should have ELCOM-like forms:

$$\ldots - \frac{\partial\left(Ye^{-\left|\frac{R}{R_0}\right|} + \tau(1 + m_{ab})\frac{\partial v_b}{\partial a}\right)}{\partial b} \ldots, \tag{14}$$

where:

$$R = \sqrt{\left(\frac{\partial v}{\partial x}\right)^2 + \left(\frac{\partial v}{\partial y}\right)^2 + \left(\frac{\partial v}{\partial z}\right)^2}, \tag{15}$$

and where two new parameters appear:

Y strength of the static viscosity; its unit should be the unit of τ divided by the unit of time, setting Y to zero disables the static component totally;

R_0 characteristic shear rate responsible for positioning of the optimal rate point.

It may also be necessary to modify bed and banks friction terms accordingly, in order to avoid artificial strengthening the flow near the banks; but that requires further research that is far beyond the scope of this chapter.

For two-dimensional depth-averaged models there are only four eddy viscosity terms. In order to take the third dimension into account the equation of shear rate, Eq. (15) should be rewritten as:

$$R = \sqrt{\left(\frac{\partial v}{\partial x}\right)^2 + \left(\frac{\partial v}{\partial y}\right)^2 + \left(\varepsilon\frac{v}{h}\right)^2}, \tag{16}$$

where h is the local depth; *and* ε is a parameter of the order of magnitude of 1.

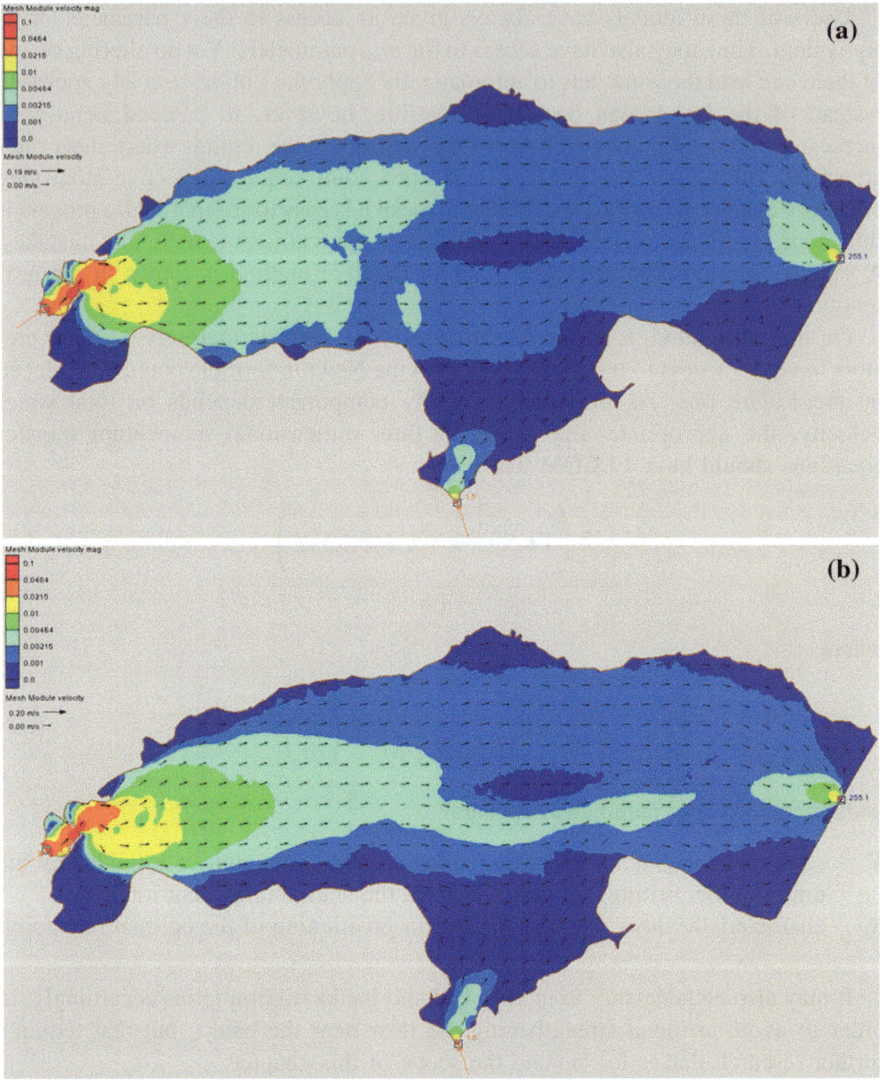

Fig. 4 Planar velocity maps as obtained from the FESWMS model. The scale along the v axis is logarithmic, three contours per order of magnitude. **a** unaltered model outcome, **b** outcome after six repetitions of the procedure described

In some cases it may also be assumed that the horizontal scale of the phenomena that appear in a lake exceeds much their vertical scale, thus R is in fact determined by the vertical component. So it can be written that:

$$\frac{R}{R_0} = \frac{v}{\mathrm{P}_0 h}, \tag{17}$$

where the parameter P_0 is equal to R_0 or slightly greater.

4 Forcing a Program to Simulate the River Memory Effect

In order to show an example of emulating a model containing the River Memory effect, the FESWMS program provided by the Aquaveo company has been used. Regarding the usage of this program for retention lakes, see Hachaj (2007), Bronczakowska (2012), Trzewik and Hachaj (2012) and references therein. The simulations were conducted for the Goczałkowice lake with the water surface level of 255.1 m a.s.l., the main inflow (Vistula) of 20 m^3/s and side inflow (Bajerka) of 1.5 m^3/s. The algorithm worked as follows:

1. Following a standard procedure and keeping the turbulence coefficient parameter at its default value, listed in the software user interface as “100”, the first water velocity map has been obtained, as shown in Fig. 4a.
2. A threshold form of eddy viscosity has been introduced: for areas with relatively higher velocities and higher shear rates, the internal turbulent viscosity parameter has been assumed to be “80”, while for stagnant areas it has been assumed to be “240”.
3. Starting from the upper part of the lake, the potential stagnant areas have been identified on the velocity map. The rest of the lake has been marked as non-stagnant. The coefficients have been set as described above and the model has been re-run, giving the second velocity map.
4. It turned out that some areas previously stagnant changed to non-stagnant and vice versa. The shapes of high and low viscosity areas have then been adjusted. Then, following the current down the lake, new potential stagnant areas have been marked.
5. Steps 3 and 4 have been repeated. After four repetitions stagnant areas have been marked all the way down to the dam outflow. After another two repetitions no significant changes on the velocity map occurred. That map has then been declared as the final one (Fig. 4b).

While very rough, arbitrary, and inaccurate, the procedure was able to generate a picture containing a pattern resembling the River Memory phenomenon.

5 Summing-Up Remark

This chapter is just a first step towards solving the River Memory phenomenon. As very little nature-taken quantitative data exists so far for this effect, it is much too early to get to any definite conclusions. Yet the author strongly suspects (in hope that these suppositions are justified by the reasoning presented above) that our understanding of water viscosity for large scales and very low shear rates should be redeveloped. In fact, it is another opportunity for water to prove that it is the most marvelous substance in our Universe.

Acknowledgments The author would like to express his immense gratitude to Prof. dr hab. eng. Elżbieta Nachlik and dr hab. eng. Krzysztof Wojciech Książyński for long and inspiring discussions he have had with them. The author also wishes to thank all the Friends and Colleagues, who provided him with a lot of valuable information and were kind enough to listen to his hypotheses as they evolved.

References

Andersson S et al (2007) Friction models for sliding dry, boundary and mixed lubricated contacts. Tribol Int 40:580

Åström KJ, Canudas-de-Wit C (2008) Revisiting the LuGre model. IEEE Control Syst Mag 28(6):101

Bayer RG (2004) Mechanical wear fundamentals and testing. CRC Press, Boca Raton

Bingham EC (1916) An investigation of the laws of plastic flow. US Bur Stan Bull 13:309

Bo LC, Pawlescu D (1982) The friction–speed relation and its influence on the critical velocity of the stick-slip motion. Wear 82(3):277

Bronczakowska B (2012) Symulacje przepływu wody w zbiorniku retencyjnym Tresna za pomoca modelu FESWMS (*Simulating water flow in the Tresna retention reservoir using the FESWMS model*), IIGW Politechnika Krakowska (in Polish)

Casson N (1959) Rheology of disperse system. Pergamon Press, Oxford

Donnel BP et al (2009) Users Guide to RMA2 WES Version 4.5. Engineering research and development center, US Army

Dunn DJ (2005) The viscous nature of fluids, in fluid mechanics. Free Study

Estellé P, Lanos C (2006) Squeeze flow of Bingham fluids under slip with friction boundary condition. Rheol Acta 46(3):397

Froelich DC (2003) Two-dimensional depth-averaged flow and sediment transport model. US Department of Transportation

Hachaj PS (2007) Modelling of a two-dimensional velocity field for the water flow in the lake of Dobczyce. In: Transport Phenomena in Hydraulics, p 87

Hodges B, Dallimore C (2007) Estuary, Lake and Coastal Ocean Model ELCOM. Centre for Water Research, University of Western Australia

Jacobsen B (2003) The Stribeck memorial lecture. Tribo Int 36(11):781

Kim (2002) A study of non-newtonian viscosity and yield stress of blood in a scanning capillary-tube rheometer, Drexel University

Komatina D, Jovanović M (1997) Experimental study of steady and unsteady free surface flows with water-clay mixtures. J Hydr Res 35(5):579

Kovacs G (1981) Seepage hydraulics, developments in water science. Elsevier, Amsterdam

Locat J, Demers D (1988) Viscosity, yield stress, remolded strength, and liquidity index relationships for sensitive clays. Can Geotech J 25:799

Nachlik E et al (1980) Implementation study of the Pakość Lake. IIGW Politechnika Krakowska (in Polish)

Rice JR, Ruina AL (1983) Stability of steady frictional slipping. J Appl Mech Trans ASME 50(2):343

Stribeck R (1902) Die wesentlichen Eigenschaften der Gleit- und Rollenlager (*The key qualities of sliding and roller bearings*) Zeitschrift des Vereines Seutscher Ingenieure 46(38, 39) pp 1342 and 1432 (in German)

Syme WJ et al (2010) TUFLOW user manual GIS Based 1D/2D Hydrodynamic modeling. BMT WBM

Trzewik M, Hachaj PS (2012) Flow simulations in the Porąbka lake with the FESWMS model. *In this volume*

Tustin A (1947) The effects of backlash and of speed-dependent friction on the stability of closed-cycle control systems. J IEEE Pt 1(94):143

Application of Lattice Boltzmann Method for Generation of Flow Velocity Field Over River Bed-Forms

Mikołaj Karpiński, Robert J. Bialik and Paweł M. Rowiński

Abstract In this chapter a Lattice Boltzmann Method (LBM) was presented as a relatively simple tool for generation of flow velocity field over river sand waves. The special attention was paid to the way of discretization of the presented equations. The choice of this method was conditioned by its advantages for modeling of complex geometries. Moreover, the computations with use of this method are easy to parallelize in contrast to the widely used models. The numerical results were compared with the available laboratory data and a reasonable agreement was found among the mean fluid velocity over the stoss, crest and trough of the considered fixed, triangular dune.

1 Introduction

Bed-forms are crucial features influencing the flow resistance; they also steer the flow into certain patterns and frequently produce flow structures. To make the picture more complex they also respond to the flow characteristics by changing their shape and the speed of their movement. In this study a relatively simple case is considered; the bed-forms are assumed to be immobile and only one-way coupling towards understanding of their influence on the bulk flow velocities will be studied herein.

M. Karpiński (✉) · R. J. Bialik · P. M. Rowiński
Institute of Geophysics, Polish Academy of Sciences, Ks. Janusza 64
01-452 Warsaw, Poland
e-mail: mkarpin@igf.edu.pl

R. J. Bialik
e-mail: rbialik@igf.edu.pl

P. Rowiński (ed.), *Experimental and Computational Solutions of Hydraulic Problems*, GeoPlanet: Earth and Planetary Sciences, DOI: 10.1007/978-3-642-30209-1_23,

There is a number of experimental studies, including laboratory and field research on river bed-forms, its statistical description and motion (Aberle et al. 2010; Coleman and Melville 1994, 1996; Coleman et al. 2011; Hino 1968; Julien and Klaassen 1995; Nikora et al. 1997 or Singh et al. 2011) or structure of flow over ripples and dunes (Bennett and Best 1995; Best and Kostaschuk 2002; Fernandez et al. 2006; Nelson and Smith 1989; Nelson et al. 1993; Parsons et al. 2005 or Sukhodolov et al. 2006). On the other hand, to date, several numerical models for investigation of flow velocity field over bed-forms have been developed. Recently, most of them have been based on the Large Edgy Simulation (LES) (i.e. Grigoriadis et al. 2010, 2012; Stoesser et al. 2008; Yue et al. 2005, 2006) due to the fact that in comparison to Reynolds-averaged Navier–Stokes equation (RANS) this kind of modeling is useful to investigate the instantaneous flow and coherent structures.

The purpose of this paper is to develop and apply the Lattice Boltzmann Method (LBM) for generation of flow velocity field over sand waves. It is important to note here that in contrast to the widely used Large Eddy Simulation (LES) or Reynolds-averaged Navier–Stokes (RANS) models, the Lattice Boltzmann Method is less popular in geophysical and hydraulics communities and thus, the second goal of this work is to show the advantages of this kind of modeling. This aim will be fulfilled through the comparison of results of flow velocity field over bed-forms obtained from the proposed model with the available data from the measurements of flow over artificial bed-forms of Lyn (1993).

2 Description of the Model

The fundamental principles on which the LBM is based and which must be met are: (1) to construct simplified molecular dynamics that incorporates the essential characteristics of physical microscopic processes; and (2) to satisfy the desired macroscopic equations by the macroscopic averaged properties. In this section, the LBM is briefly introduced, with particular attention paid to the way of discretization of the presented equations. Moreover, the boundary conditions used in the numerical simulations are described in the end of this section.

The starting point for derivation of LBM is the Boltzmann equation that in the general form may be written as:

$$\frac{\partial f}{\partial t} + \mathbf{v} \cdot \nabla f = C(f), \tag{1}$$

where $f = f(\mathbf{x}, \mathbf{v}, t)$ is the distribution function of water particles with the velocity $\mathbf{v}$ at the time t around the position $\mathbf{x}$ and $C(f)$ is the collision operator describing the effect of collisions between particles.

Next, the set of possible velocity vectors $\{\mathbf{c}_i\}_{i=0\ldots b}$ has to be restricted to the finite number of values $b+1$ where the index i specifies the allowable velocity

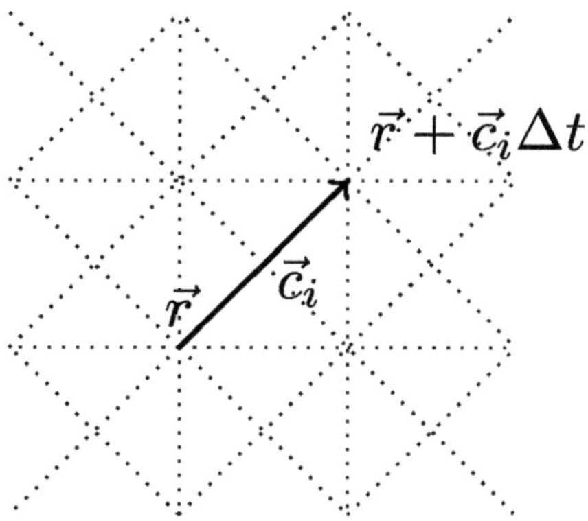

Fig. 1 Rectangular lattice used in the LBM for generation of flow velocity with the basic parameters of Eq. (2)

directions. Thus, the Lattice Boltzmann Equation in the discrete form reads as follows (Chen and Doolen 1998):

$$f_i(\mathbf{r} + \mathbf{c}_i\Delta t, t + \Delta t) = f_i(\mathbf{r}, t) + \Omega_i\big(\mathbf{f}(\mathbf{r}, t)\big), \tag{2}$$

where $f_i(\mathbf{r}, t)$ represents the probability of finding the particle at the position $\mathbf{x} = \mathbf{r}$ at the time t with the velocity of particle $\mathbf{v} = \mathbf{c}_i$ and Ω_i denotes here the function known as the collision operator.

Figure 1 schematically shows the particle movement from the position $\mathbf{r}$ to another position $\mathbf{r} + \mathbf{c}_i\Delta t$, which takes place on a rectangular lattice used in solving the LBM. It is important to note here that in the LBM the lattice may be selected in the various ways, depending on the complexity of the problem being solved. In the literature, i.e., He and Luo (1997) or Qian et al. (1992), they are usually classified using the following term $D\alpha Q\beta$, where α stands for the dimensionality of the space and β denotes the number of discrete velocities. The most popular lattices, which are usually used are: $D2Q9$ or $D3Q19$. Due to the fact that in our work we consider the 2-D case, we choose the $D2Q9$ lattice (as is presented in Figs. 1, 2).

Moreover, to use the LBM, the main fluid quantities (density and velocity) are obtained by simple summation upon the discrete velocities and then are defined in the following forms

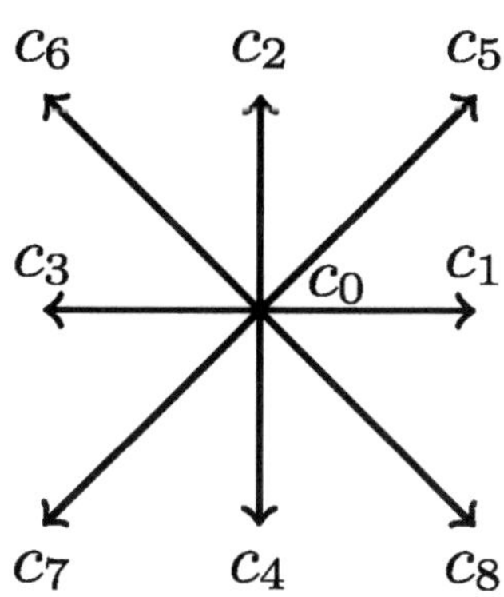

Fig. 2 Scheme of velocity vector directions for $D2Q9$ lattice

- macroscopic fluid density $\rho(\mathbf{r}, t)$:

$$\rho(\mathbf{r}, t) = \sum_i f_i(\mathbf{r}, t), \tag{3}$$

- macroscopic fluid velocity $\mathbf{u}(\mathbf{r}, t)$:

$$\mathbf{u}(\mathbf{r}, t) = \frac{1}{\rho(\mathbf{r}, t)} \sum_i f_i(\mathbf{r}, t)\mathbf{c}_i. \tag{4}$$

As has been already mentioned in Eq. (2), Ω_i denotes the function which is known as the collision operator. In practical applications this term is usually approximated using the Bhatnagar-Gross-Krook method (see Bhatnagar et al. 1954) and then Eq. (2) may be rewritten in the following form:

$$f_i(\mathbf{r} + \mathbf{c}_i \Delta t, t + \Delta t) = f_i(\mathbf{r}, t) + \underbrace{\frac{1}{\tau}\left(f_i^{eq}(\mathbf{r}, t) - f_i(\mathbf{r}, t)\right)}_{\Omega_i\left(\mathbf{f}(\mathbf{r}, t)\right)} \tag{5}$$

where τ is the relaxation time which is a function of kinematic viscosity, and f_i^{eq} is the local equilibrium distribution function, which can be obtained from (He and Luo 1997)

$$f_i^{eq} = w_i \rho \left(1 + \frac{u_a c_{ia}}{\mathbf{c}_s^2} + \frac{u_a u_b Q_{iab}}{2\mathbf{c}_s^4}\right) \tag{6}$$

where $\mathbf{c}_s$ is the speed of sound $c_s = \frac{1}{\alpha}\sum_i w_i c_i^2$, $Q_{iab} = c_{ia} c_{ib} c_s^2 - \delta_{ab}$ is the parameter, $\mathbf{c}_i = (c_{ix}, c_{iy})$ is the discrete velocity vector and $\mathbf{u}_i = (u_{ix}, u_{iy})$ is the macroscopic velocity vector. Moreover, in Eq. (6) Einstein's summation convention is used and $\{w_i\}_i$ denotes the set of weights normalized to unity that for the lattice $D2Q9$ are:

$$w_i = \left\{ \begin{array}{lll} 4/9, & for & i = 0 \\ 1/9, & for & i = 1, 2, 3, 4 \\ 1/36, & for & i = 5, 6, 7, 8 \end{array} \right\} \tag{7}$$

and, in addition, the $D2Q9$ discrete velocity vectors have the following values (see Fig. 2)

$$\left\{ \begin{array}{lll} c_0 = (0, 0), & c_3 = (-1, 0), & c_6 = (-1, 1) \\ c_1 = (1, 0), & c_4 = (0, -1), & c_7 = (-1, -1) \\ c_2 = (0, 1), & c_5 = (1, 1), & c_8 = (1, -1) \end{array} \right\} \tag{8}$$

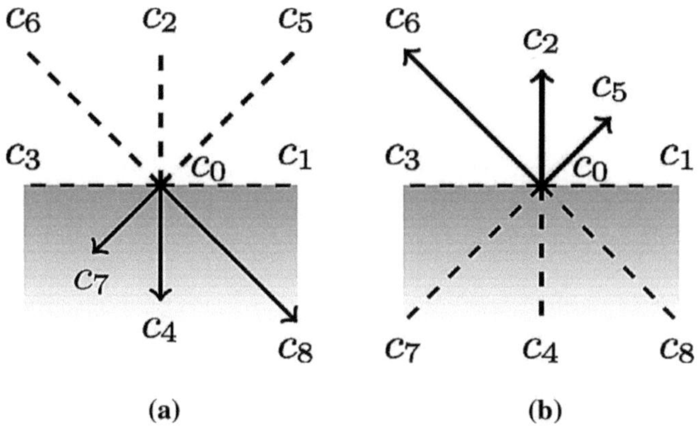

Fig. 3 No-slip boundary condition (bounce-back boundary condition) for the collision with the channel bottom. **a** Before collision. **b** After collision

2.1 Boundary Conditions

To apply the Lattice Boltzmann Method (LBM) for generation of the fluid velocity field over the bed-forms, the boundary conditions have to be determined. Typically in such a case the following conditions are required: (1) on the channel bottom; (2) on the free surface; (3) at the inflow to the channel; and (4) at the outflow from the channel.

No-slip boundary conditions are used in the model (see Fig. 3). The basic rule is that the particle is reflected in the lattice node with the same velocity, but in the opposite direction. This condition is regarded as the greatest advantage of the Lattice Boltzmann Method due to the simplicity of its application to complex geometries.

Figure 4 shows the boundary condition for the collision with the free surface that is called *Free-slip boundary condition*. It is important to note here that inclusion of the free surface is usually omitted in the LBM and the free surface is treated as the channel bed. The basic rule of *Free-slip boundary condition* is that the particle is reflected in the lattice node in such a way that all the normal to the free surface velocities should be zero and tangential components have the same value as before the collision.

The inflow and the outflow can be described by the periodic boundary conditions. Boundary conditions of this kind will be applied in the comparison between the numerical results and the experimental data.

2.2 Effect of Gravity

Let us consider a non-trivial bed slope. The acceleration of flow due to gravity may be taken into account by means of the procedure described in Guo et al. (2002). That procedure modified Eq. (4) to the following form:

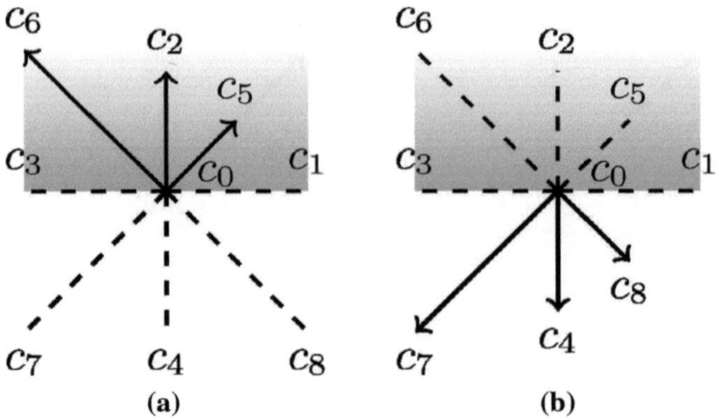

Fig. 4 Free-slip boundary condition for the collision with the free surface. **a** Before collision. **b** After collision

$$\mathbf{u}(\mathbf{r},t) = \frac{1}{\rho(\mathbf{r},t)} \sum_i f_i(\mathbf{r},t)\mathbf{c}_i + \frac{\Delta t}{2\rho(\mathbf{r},t)} \mathbf{F}. \tag{9}$$

and Eq. (5) reads:

$$f_i(\mathbf{r} + \mathbf{c}_i \Delta t, t + \Delta t) = f_i(\mathbf{r},t) + \frac{1}{\tau}\left(f_i^{eq}(\mathbf{r},t) - f_i(\mathbf{r},t)\right) + \Delta t F_i \tag{10}$$

where

$$F_i = \left(1 - \frac{1}{2\tau}\right) w_i \left[\frac{\mathbf{c_i} - \mathbf{u}}{c_s^2} + \frac{(\mathbf{c_i} \cdot \mathbf{u})}{c_s^4}\mathbf{c_i}\right] \cdot \mathbf{F}. \tag{11}$$

3 Numerical Results

In this section, the results of numerical simulations are presented, showing a reasonable agreement among the mean fluid velocity over the stoss, crest and trough of the dune and the experimental data. Moreover, the obtained results confirm the theoretical considerations that anticipate the reverse flow in the lee part of the dune.

The mean flow velocity distributions at three specific points of dune, i.e., stoss, crest and trough obtained from the above LBM were compared with the experimental data of Lyn (see Lyn 1993), conducted in an adjustable-slope recirculating flume, which was approximately 13 m long and 26.67 cm wide. The channel consisted of *periodic array of* 45° *triangular elements of amplitude* $h = 1.2$ cm *and wavelength* $L = 15$ cm (see Figs. 5, 6). The conditions in the channel were as

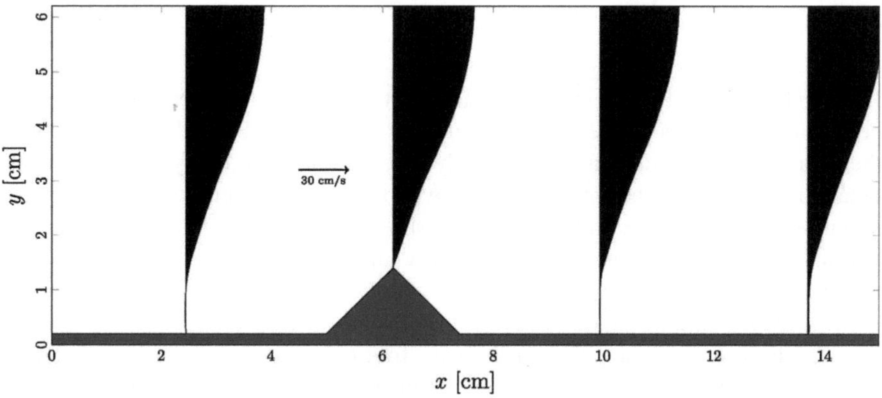

Fig. 5 Distributions of the mean stream-wise velocity in the stoss, crest and trough of the fixed triangular dune

follows: flow bed depth $H = 6.1$ cm, energy slope $S = 0.00525$, mean velocity $U = 25.4$ cm/s, Reynolds number $Re = UH/\nu = 15.5 \times 10^3$. In addition, the bed-form along its length and width was two-dimensional so the comparisons with the above model results are well justified.

Figure 5 shows the distributions of the mean flow velocity along four verticals ($x = 0$, $x = L/4$, $x = L/2$ and $x = 3L/4$), as was done by Lyn (1993). In the crest region, $x = 0$, the mean velocity fulfills the logarithmic distributions, whereas in the trough of the dune, $x = L/4$, and in the reattachment region, $x = L/2$, it may be noticed that the mean velocities indicate the existence of the reverse flow, especially, close to the bottom. However, showing the vector velocity field above this part of the considered artificial dune more accurately and clearly (as it is presented in Fig. 5), one can see that the recirculating flow region also occurs in the fourth case, $x = 3L/4$ (Fig. 6).

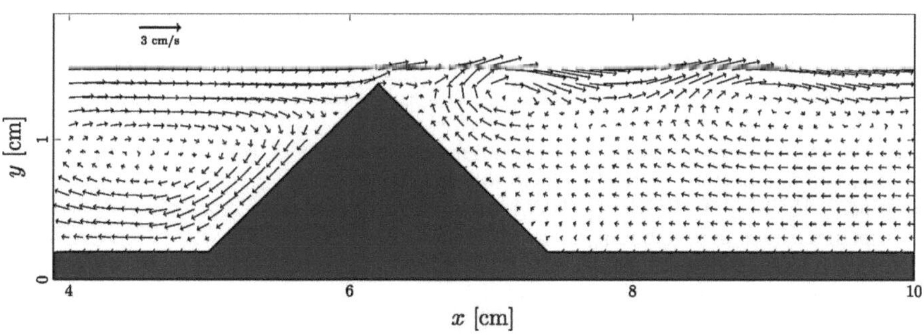

Fig. 6 Velocity vector field in the reattachment region

4 Conclusions

In this chapter we briefly introduced the Lattice Boltzmann Method for the generation of flow velocity field over fixed river bed-forms. Both the comparison of the results of numerical simulations with experimental data of Lyn (1993) and the detailed analysis of the velocity field over the artificial dune, with the special attention paid to the lee part of this dune show that the presented method may be alternative to other, widely used to study the velocity field: LES and RANS models. The main advantages of this method are its ease of numerical implementation as well as the ease of application to complex geometries, like river beds. We also believe that the LBM in the near future can be successfully applied not only to description of the velocity field over the fixed dunes, but will be used for analysis of the dynamics and evolution of river bed-forms as well. This study is of preliminary character and we do realize that future in-depth research is needed to justify the usefulness of the method for modeling of flow in complex river geometries.

Acknowledgments This work was supported by the Institute of Geophysics of Polish Academy of Sciences through the project for Young Scientists No. 16/IGF PAN/ 2011/Mł. The authors are also grateful to V. Nikora, A. Rajwa, and A. Sukhodolov for stimulating discussions on this topic and constructive comments on the manuscript.

References

Aberle J, Nikora V, Henning M, Ettmer B, Hentschel B (2010) Statistical characterization of bed roughness due to bed forms: a field study in the Elbe River at Aken. Germany. Water Resour Res 46:W03521

Bennett SJ, Best JL (1995) Mean flow and turbulence structure over fixed, 2-dimensional dunes - implications for sediment transport and bedform stability. Sedimentology 42(3):491–513

Best J, Kostaschuk R (2002) An experimental study of turbulent flow over a low-angle dune. J Geophys Res 107(C9):3135

Bhatnagar PL, Gross EP, Krook M (1954) A model for collision processes in gases I: small amplitude processes in charged and neutral one-component systems. Phys Rev 94(3):511–525

Chen S, Doolen GD (1998) Lattice Boltzmann method for fluids flows. Annu Rev Fluid Mech 30:329–364

Coleman SE, Melville BW (1994) Bed-form development. J Hydraul Eng ASCE 120(4):544–560

Coleman SE, Melville BW (1996) Initiation of bed forms on a flat sand bed. J Hydraul Eng ASCE 122(6):301–310

Coleman SE, Nikora VI, Aberle J (2011) Interpretation of alluvial beds through bed-elevation distribution moments. Water Resour Res 47:W11505

Fernandez R, Best J, Lopez F (2006) Mean flow, turbulence structure, and bed form superimposition across the ripple-dune transition. Water Resour Res 42(5):W05406

Grigoriadis DGE, Balaras E, Dimas AA (2010) Large-eddy simulations of unidirectional water flow over dunes. J Geophys Res 114:F02022

Grigoriadis DGE, Dimas AA, Balaras E (2012) Large-eddy simulations of wave turbulent boundary layer over rippled bed. Coast Eng 60:174–189

Guo Z, Zheng C, Shi B (2002) Discrete lattice effects on the forcing term in the lattice Boltzman method. Phys Rev E 65:046308

He X, Luo LS (1997) Theory of the lattice Boltzmann method: from the Boltzmann equation to the lattice Boltzmann equation. Phys Rev E 56:6811–6817

Hino M (1968) Equilibrium-range spectra of sand waves formed by flowing water. J Fluid Mech 34(3):565–573

Julien PY, Klaassen GJ (1995) Sand-dune geometry of large rivers during floods. J Hydraul Eng ASCE 121(9):657–663

Lyn DA (1993) Turbulence measurements in open-channel flows over artificial bed forms. J Hydraul Eng ASCE 119(3):306–326

Nelson JM, Smith JD (1989) Mechanics of flow over ripples and dunes. J Geophys Res 94(C6):8146–8162

Nelson JM, McLean SR, Wolfe SR (1993) Mean flow and turbulence fields over 2-dimensional bed forms. Water Resour Res 29(12):3935–3953

Nikora VI, Sukhodolov AN, Rowinski PM (1997) Statistical sand wave dynamics in one-directional water flows. J Fluid Mech 351:17–39

Parsons DR, Best JL, Orfeo O, Hardy RJ, Kostaschuk R, Lane SN (2005) Morphology and flow fields of three-dimensional dunes, Rio Parana, Argentina: results from simultaneous multibeam echo sounding and acoustic Doppler current profiling. J Geophys Res 110(F4): F04S03

Qian YH, D'Humieres D, Lallemand P (1992) Lattice BGK models for Navier-Stokes equation. Europhys Lett 17(479)

Singh A, Lanzoni S, Wilcock PR, Foufoula-Georgiou E (2011) Multiscale statistical characterization of migrating bed forms in gravel and sand bed rivers. Water Resour Res 47:W12526

Sukhodolov AN, Fedele JJ, Rhoads BL (2006) Structure of flow over alluvial bedforms: an experiment on linking field and laboratory methods. Earth Surf Process Landforms 31:1292–1310

Stoesser T, Braun C, Garcia-Villalba M, Rodi W (2008) Turbulence structures in flow over two-dimensional dunes. J Hydraul Eng ASCE 134(1):42–55

Yue WS, Lin CL, Patel VC (2005) Large eddy simulation of turbulent open-channel flow with free surface simulated by level set method. Phys Fluids 17(2):025108

Yue WS, Lin CL, Patel VC (2006) Large-eddy simulation of turbulent flow over a fixed two-dimensional dune. J Hydraul Eng ASCE 132(7):643–651

Modelling the Flow Conditions Over a Side Channel Spillway of the Pilchowice Storage Reservoir on the Bóbr River

Jerzy Machajski and Dorota Olearczyk

Abstract In the chapter, the results of numerical modelling of flow conditions by flume of side-channel spillway of Pilchowice storage reservoir on the Bóbr River are presented, for which some reconstruction of outlet installations are necessary. Modelling was preceded by an analysis of functioning principles of devices of this type, given in currently available literature. Modelling was also preceded by analytical calculations, determining a capacity ability of side-channel spillway as a function of valid damming levels in reservoir and capacity ability of a flume which takes over the waters overflowing spillway and conveys them to multistage cascade, connecting the spillway with a downstream river section. Analytical analysis of functioning principles of such devices allowed us to derive a differential equation describing the course of water line along a flume of side-channel spillway. Numerical calculations were carried out using the authors' computer program. Calculations were made for actual state of object and for several proposed solutions of its reconstruction, assuming that due to hard workability of rock foundation there was a necessity of executing the building works in essentially minimum range. The results of calculations confirm the design principles of devices of this type, underlined in the literature, extorting their working in every operating conditions as not submerged. They confirm also an evident necessity of flume reconstruction, that ensures a safety of Pilchowice reservoir in terms of computational discharges, significantly exceeding a nominal capacity of outlet installation in the current state of its technical design. In the numerical analysis,

J. Machajski (✉)
Institute of Geotechnics and Hydrotechnics, Wrocław University of Technology, Wyspiańskiego 27, ul. Wybrzeże 50-370 Wrocław, Poland
e-mail: Jerzy.Machajski@pwr.wroc.pl

D. Olearczyk
Institute of Environmental Engineering, Wrocław University of Environmental and Life Sciences, Plac Grunwaldzki 24, 50-363 Wrocław, Poland
e-mail: dorota.olearczyk@up.wroc.pl

P. Rowiński (ed.), *Experimental and Computational Solutions of Hydraulic Problems*, GeoPlanet: Earth and Planetary Sciences, DOI: 10.1007/978-3-642-30209-1_24,

the attention is paid to possibility and outright necessity of leading a flume with a longitudinal bottom slope greater than zero, showing a profit from such a solution.

Keywords Storage reservoir · Side-channel spillway · Flume · Numerical modelling

1 Introduction

In accordance with the Polish Building Law, technical conditions should be determined for every structure, in respect, among other things, to their location. For hydro-engineering structures, first conditions were determined in technical regulations from 1967, changed in 1973 into the technical conditions. Successive conditions appeared in 1996 and were valid till 2007, when the currently valid regulations came out (Directive 2007). In regulations of 1996, significant changes have been done related to the requirements of objects of a given class of importance and a technical class of some of such objects has been changed. As a consequence, among other things, computational discharges have to be changed too.

According to the regulations of 1996, the Pilchowice storage reservoir class of importance has been changed from II to I, and, as a result, the design and control discharges have increased. In such a situation, the question appeared whether under new exploitation conditions a safe passage of increased discharges through the existing reservoir outlet devices will be possible. The authors of this chapter undertook a trial to solve that problem.

2 Object Characteristics

The main tasks of the Pilchowice storage reservoir, located in km 196 + 700 of the Bóbr River course, is water storing for energetic purposes and flood protection of the downstream Bóbr River Valley. The dam of 62 m height, built in 1904–1912, makes a storage reservoir with the maximum capacity of 55.12 mln m^3. Its arch body was made of concrete mixed with broken stones. During the period of most frequent occurrence of flood freshet wave, July–October, a flood control storage equal to 26.0 mln m^3 is kept in the reservoir (Machajski and Olearczyk 2006; Machajski and Rędowicz 2008).

3 Outlet Devices

Surplus of water is carried away through outlet devices, including a diverse channel, two bottom outlets and side channel spillway. The side channel spillway,

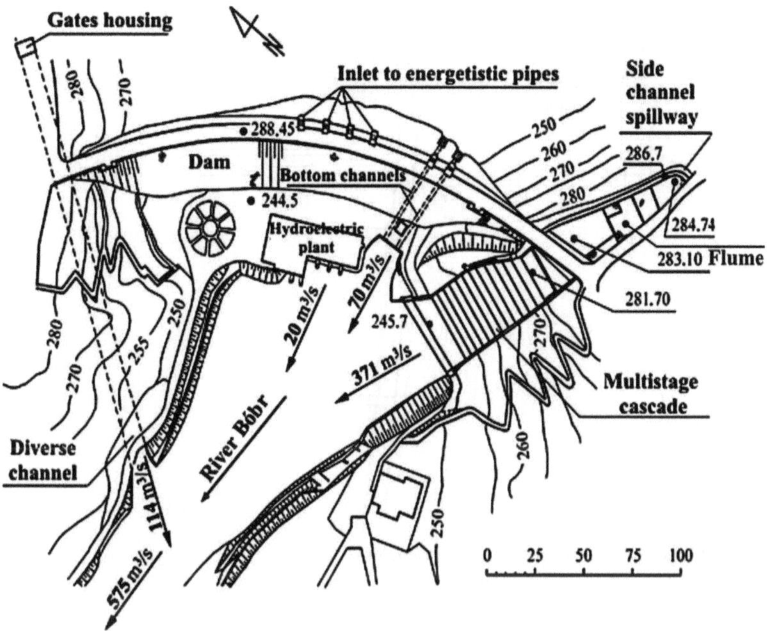

Fig. 1 Functional plan of the Pilchowice dam

of 84 m crest length, starts its operation when water level in the reservoir reaches its crest equal to 286.70 m a.s.l., which corresponds to reservoir capacity of 50.50 mln m^3. Water from the spillway by flume, flows downstream through multistage cascade, founded on the rock massif, which closes the Bóbr River Valley from its left bank. Diverse channel of 387 m length was hewn in the rock mass as a tunnel with oval cross-section of 6.40 m height and 7.80 m width. Diverse channel at half of its whole length is divided by vertical shaft, in the lowest part of which, three conduits, each 1,500 mm in diameter, are built in, equipped with valves.

Bottom outlets consist of two conduits, of 1,500 mm diameter each, equipped with vertical lift gates from headwater and valves from tailwater. At the beginning of reservoir exploitation, in the water power station four turbine sets were installed, which were supplied by independent water intakes located in dam body, closed from the headwater with vertical lift gates. The total capacity of all mounted turbines is about 37 $m^3\ s^{-1}$ (Machajski and Rędowicz 2008) (Fig. 1).

4 Capacity Ability of Outlet Devices

The capacity ability of outlet devices is given for chosen several reservoir water levels, up to dam crest equal to 288.75 m a.s.l. (Fig. 2). The capacity ability presented in Table 1 is given for individual devices, and as the total sum in reference to reservoir exploitation conditions (Machajski and Rędowicz 2008).

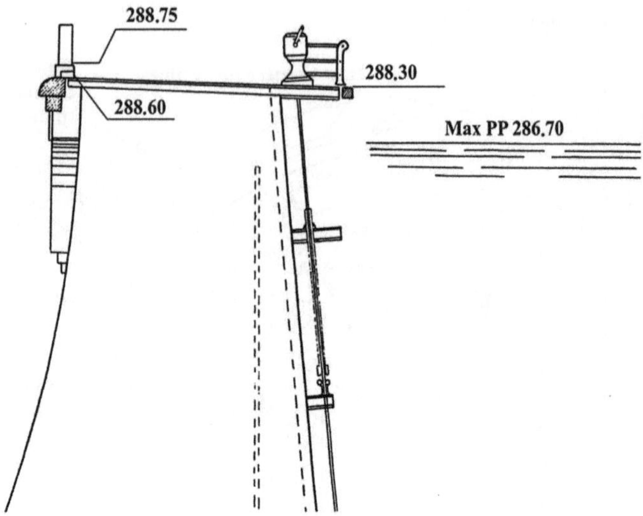

Fig. 2 The Pilchowice dam crest height position

Table 1 Capacity ability of storage reservoir Pilchowice outlet devices

Water level	Elevation m a.s.l.	Diverse channel $m^3\ s^{-1}$	Bottom outlet $m^3\ s^{-1}$	Spillway $m^3\ s^{-1}$	Sum $m^3\ s^{-1}$
Min. PP	258.00	66.7	42.3	–	109
NPP (summer)	272.40	91.6	58.4	–	150
NPP (winter)	278.18	100	64	–	164
Max. PP	286.70	111	71	–	182
Dam crest from upstream	288.30	113 Σ 185	72	292	477
Dam crest from downstream	288.75	113.5 Σ 186	72.5	372	558

5 Analytical Calculations

Side channel spillway fulfils two parallel functions, i.e., that of overflow part and that of flume, which is conducted parallel to the overflow edge. The flume takes over the water overflowing spillway and conveys it to devices that connect side channel spillway with downstream river section. The analysis of water motion conditions within the side channel spillway has changed gradually. At present, the analysis is based on the law of linear momentum conservation (quantity of motion) assuming that the only force ensuring flow of water in the flume results from the water level slope in the direction of water routing. Next, energy of the water overflowing the spillway is dissipated due to its mixing with water in the flume without influence on the flow conditions in the flume. Thus, the purpose of hydraulic computation of side channel spillway, assuming the reservoir exploitation conditions, is determination of a necessary length of the spillway edge, and determination of the free water surface in the flume at the assumed parameters of its cross-section, constant bottom slope and uniform increase of discharge along flume length (Khatsuria 2005; Novak et al. 2007; Şentürk 1994; Tančev 2005).

The spillway discharge is determined assuming that it works in the whole range of expected discharges as not submerged. Calculations were carried out using the formula for free spillway with straight insert on its crest (Rogala et al. 1991; Şentürk 1994; Tančev 2005):

$$Q = \varepsilon\, 2/3\, \mu\, B\, \sqrt{2\, g}\, H_o^{3/2} \tag{1}$$

where ε is the coefficient of weir side contraction, $\varepsilon = 1.0$; μ is the discharge coefficient, $\mu = 0.654$; B is the spillway crest length, $B = 84.0$ m; H_o is the energy height calculated in relation to spillway crest.

For the calculations, three characteristic reservoir water levels were taken, regarding the weir crest (286.70 m a.s.l.), dam crest (Fig. 2), the determined water layer thickness, and the resultant energy of the stream. The results of calculations are as follow:

- 288.75 (dam crest from downstream)—286.70 = 2.05 m → Q = 476.15 $m^3\ s^{-1}$
- 288.60 (dam crest from downstream)—286.70 = 1.90 m → Q = 424.86 $m^3\ s^{-1}$
- 288.30 (dam crest from upstream)—286.70 = 1.60 m → Q = 328.32 $m^3\ s^{-1}$.

In further analysis for flume parameters determination, the discharge equal to 424.86 $m^3\ s^{-1}$ was assumed. In professional literature, according to classification of water motion in open channels (Jain 2001; Rogala et al. 1991), water motion in the side channel spillway flume is defined as spatially variable steady flow. Hence, in the calculations for determination of the flume parameters, an impact of external forces, in compliance with Newton's second dynamic principle, should be considered in two cross-sections, 1–1 and 2–2, at distance dx, including gravity force, material roughness of the flume cross-section and resultant force of hydrostatic pressures in two cross-sections mentioned above (Fig. 3).

The gravity force is represented by horizontal component of water weight W between two cross-sections, 1–1 and 2–2, given in the form (Ghosh 2006; Khatsuria 2005; Novak et al. 2007; Şentürk 1994):

$$W\ \sin\theta = \gamma\, S_o\, A\, dx \tag{2}$$

where S_o is the longitudinal slope of the flume bottom, A is the wetted area between cross-sections 1–1 and 2–2, γ is the weight density of water and θ is the inclination angle of the flume bottom.

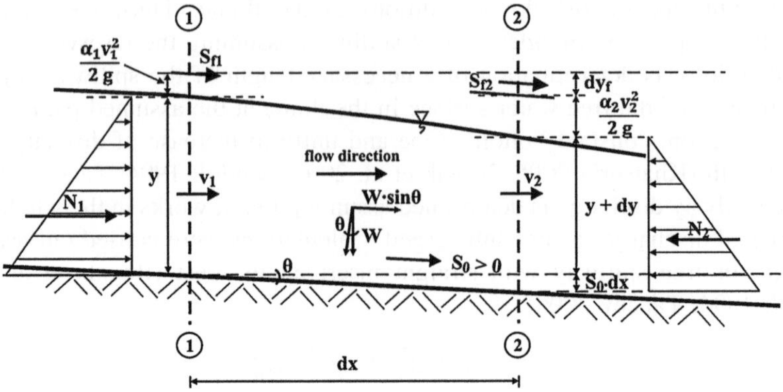

Fig. 3 Calculating scheme for the side channel spillway flume dimensioning

The force resulting from an impact of roughness of the flume material can be expressed as (Ghosh 2006; Khatsuria 2005; Novak et al. 2007; Şentürk 1994):

$$F_s = \gamma A S_f dx \tag{3}$$

where S_f is the energy line slope.

The force resulting from differences of hydrostatic pressures in cross-sections 1–1 and 2–2 can be written as (Ghosh 2006; Khatsuria 2005, Novak et al. 2007; Şentürk 1994):

$$N_1 - N_2 = -\gamma A dy \tag{4}$$

where N_1 is the force of hydrostatic pressure in cross-section 1–1, N_2 is the force of hydrostatic pressure in cross-section 2–2, and dy is the depth difference in cross-sections 1–1 and 2–2.

The resultant change of momentum in time unit between cross-sections 1–1 and 2–2 is equal to the sum of external forces and is described by the equation (Ghosh 2006; Khatsuria 2005; Novak et al. 2007):

$$\frac{dM}{dt} = N_1 - N_2 + W \sin\theta - F_s \tag{5}$$

After substitution of $A = Q/v$ and $q = dQ/dx$ and transformations, the following equation is derived (Ghosh 2006; Khatsuria 2005; Novak et al. 2007):

$$\frac{dy}{dx} = \frac{S_o - S_f - 2 \alpha Q q / g A^2}{1 - \alpha Q^2 / g A^2 D} \tag{6}$$

where α is the kinetic energy correction coefficient (Saint–Venant), $\alpha = 1.10$; q is the rise of flow per length of flume dQ/dx; D is the hydraulic depth, $D = A/B$ and B is the width of flume.

The above differential equation can be solved in two ways: either by applying numerical methods available for equations of this type or by applying simplified methods, which in general consist in introduction of finite differences in place of differentials. The equation determining a change of water levels Δy per length Δx is (Ghosh 2006; Khatsuria 2005, Novak et al. 2007, Şentürk 1994):

$$\Delta y = -\frac{\alpha\, Q_1\,(v_1 + v_2)}{g\,(Q_1 + Q_2)}\left[(v_2 - v_1) + \frac{v_2\,(Q_2 - Q_1)}{Q_1}\right] + S_o\Delta x - S_f\,\Delta x \quad (7)$$

Applying the above formula, the flume parameters were determined to ensure a possibility of whole discharge passage in amount of 424.86 $m^3\ s^{-1}$ with simultaneous guarantee that spillway works as not-submerged.

The analytical calculations reveal that a free flow of computational discharge through the existing side channel spillway is not possible due to too shallow and too narrow flume in its upper part. Therefore, the operating conditions of the spillway are changing quickly into conditions of submerged weir. For that reason, the flume should be deepened with simultaneous set of its width at 22.50 m. Only under such conditions the assumed discharge of spillway can be obtained for reservoir water level elevation equal to 288.60 m a.s.l. These requirements put under a big question the possibility of the flume rebuilding, mainly in terms of economic calculations. The reconstruction would force the necessity of constructional changes of the weir—new static conditions of its work and also a necessity of constructional changes of outer wall on the opposite side. In this situation, it is assumed that by numerical analysis a minimum of works necessary to improve spillway functioning, i.e., increasing its capacity ability, will be determined. Verified by numerical analysis, side channel spillway characteristic for its existing state is shown in Fig. 4.

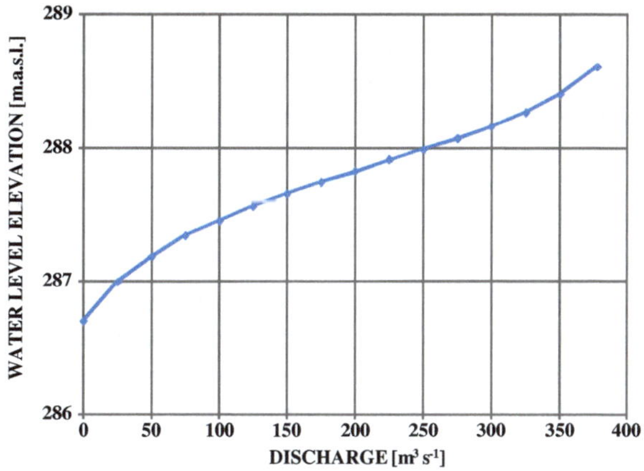

Fig. 4 Side channel spillway characteristic for the existing state (analytical)

6 The Bóbr River

The Bóbr River is the left-sided tributary of the Odra River. The catchment area to the Pilchowice dam cross-section is 1209.0 km^2. The catchment average slope is about 5.5 %, whereas longitudinal slope of the river to the Pilchowice dam is 11.94 ‰. The river length on that section equals to 79.7 km. A hydrographical network in the catchment is well developed and a drainage density ranges from 0.87 km^{-2} in the upper part of catchment to 0.77 km^{-2} in the Pilchowice cross-section. The mountainous nature of tributaries of the Bóbr River significantly influences a velocity of water mass motion, concentration time of freshet wave and favours the freshets and floods occurrence. Rivers and watercourses in the upper Bóbr River catchment are supplied with water mainly from rain in summer season and from snow in winter season. The nature of water supply causes two maximum discharges from the catchment, summer and spring. Summer freshets are the biggest in size and magnitude, whereas spring freshets are suitably lower. Significant for freshet waves formation on the Bóbr River are its left-sided tributaries in Karkonosze Mountains: the Kamienna and Łomnica Rivers.

On the Bóbr River, upstream the Pilchowice reservoir, there are five gauging stations. Hydrological data were determined at the Jelenia Góra gauging station, key for the reservoir, and were transferred into reservoir cross-section according to the catchment area growth. Characteristic discharges in Pilchowice dam cross-section are as follows (Polish abbreviations): minimum NNQ = 0.500 $m^3\ s^{-1}$, average low flow SNQ = 2.32 $m^3\ s^{-1}$, mean SSQ = 14.8 $m^3\ s^{-1}$, average maximum flow SWQ = 109 $m^3\ s^{-1}$, maximum WWQ = 494 $m^3\ s^{-1}$ (1997). Maximum discharges with a given probability of exceedance are: $Q_{50\ \%} = 145\ m^3\ s^{-1}$, $Q_{10\ \%} = 311\ m^3\ s^{-1}$, $Q_{5\ \%} = 380\ m^3\ s^{-1}$, $Q_{1\ \%} = 538\ m^3\ s^{-1}$, $Q_{0.5\ \%} = 605\ m^3\ s^{-1}$, $Q^{\alpha}_{0.1\%} = 877 m^3 s^{-1}$ (Machajski and Rędowicz 2008).

According to the regulations (Directive 2007) valid in Poland, Pilchowice dam is classified into hydro-engineering structures of the first class of importance. For structures of this class, founded on a rock, not subjected to destruction due to their overflow, a probability of exceedance for computational discharges should be set at 0.5 and 0.1 %, but for the higher discharges—with the upper extension at $t_\alpha = 1$ and confidence level 0.84. For Pilchowice dam cross-section, the discharges are, respectively:

- design $Q_m = Q_{0.5\%} = 605\, m^3 s^{-1}$,
- control $Q_k = Q^{\alpha}_{0.1\%} = 877\ m^3 s^{-1}$.

Computational discharges significantly exceed the reservoir outlet devices capacity ability, so numerical analyses were carried out for the choice of optimum solution for their possible reconstruction. First, a side channel spillway was analyzed in respect of the flume parameters (Machajski and Rędowicz 2008).

7 Outlet Devices Rebuilding Concept

In the concept of reconstruction of the Pilchowice reservoir's outlet devices, the reservoir safety was taken into account due to computational discharges and capacity ability of existing (corrected) outlet devices. First, conditions of computational discharges passage through the reservoir were determined, analyzing possibilities of their reduction to safe discharges both for downstream area and for reservoir. Analysis of reservoir flood routing was carried out using Puls method, often called inlet-storage-outlet method (Ghosh 2006). In this method, a flood wave transformation through a reservoir is described by continuity equation written in differential form (Jain 2001):

$$\frac{\partial Q}{\partial x} + \frac{\partial A}{\partial t} = 0 \tag{8}$$

where Q is the discharge [$m^3\ s^{-1}$], x is a co-ordinate consistent with flow direction [m], A is the flow cross-section area [m^2], t is time [s].

Equation (8), after integration from x_1 to x_2 and modification, takes the form in which an integral means a water capacity stored in reservoir or flowed out from reservoir:

$$Q(x_1) - Q(x_2) + \frac{d}{dt}\int_{x_1}^{x_2} A\ dx = 0 \tag{9}$$

and further:

$$Q(x_1) = Q(x_2) + \frac{\Delta V}{\Delta t} \tag{10}$$

where x_1 is the co-ordinate at inflow cross-section, x_2 is the co-ordinate at dam cross-section, $Q(x_1)$ is the reservoir inflow [$m^3\ s^{-1}$], $Q(x_2)$ is reservoir outflow [$m^3\ s^{-1}$], $\Delta V/\Delta t$ means changes of reservoir capacity in time.

Knowing a flood wave hydrograph, reservoir storage curve and characteristics of outlet devices, calculations of flood wave transformation through the reservoir were carried out, using Eq. (10), under the general assumptions:

- safe height differences between dam crest and water level elevations: 1.0 m and 0.10 m respectively for design and control discharges (Directive 2007),
- during control flood wave passage through the reservoir all outlet devices are opened except water power station turbine No V of 9.62 $m^3\ s^{-1}$,
- during design flood wave routing (Directive 2007), two bottom outlets of 1500 mm diameter and water power station turbine No V of 9.62 $m^3\ s^{-1}$, are excluded from exploitation,
- verified characteristic of side channel spillway elaborated on the basis of own analytical calculations was applied (Machajski and Rędowicz 2008),

Table 2 Results of hypothetical flood wave transformation through the Pilchowice reservoir

Type of wave $m^3 s^{-1}$	$Q_{red.}$ $m^3 s^{-1}$	Reduction percentage	Elevation of water level in reservoir a.s.l.		Time of overflowing h	Duration time of exceedance, h	
			Initial	Final		Permissible discharge	Allowable discharge
Design $Q_{0.5\%} = 605$	422	30	272.40	288.29	79	99	36
Control $Q_{0.1\%} = 877$	666	24	272.40	289.27	74	220	200

- maximum permitted water level elevation in the reservoir referring to the dam crest is 288.75 m a.s.l. (Fig. 2),
- calculations were carried out for initial water level in the reservoir equal to 272.40 m a.s.l. which corresponds to normal water level elevation NPP (Machajski and Rędowicz 2008).

In Table 2 the results of calculations are given in the form of reduced discharges, Q_{red}, during flood wave routing through reservoir, final water level elevations in the reservoir, time when spillway overflow started and duration time when permissible and allowable discharges were exceeded downstream the reservoir.

During design flood passage, there is no possibility of preserving a safe dam crest height above the maximum water level elevation in the reservoir, which reaches 288.29 m a.s.l., i.e., 0.46 m to the dam crest in an initial state of reservoir fulfillment of 272.40 m a.s.l.

During control flood passage, the water level elevation in the reservoir exceeded dam crest about tens of centimeters. According to calculations, water level in the reservoir reaches 289.27 m a.s.l., and exceedance is equal to 0.52 m above the dam crest in an initial state of reservoir fulfillment of 272.40 m a.s.l.

Results of water level elevation in reservoir for control discharge, given in Table 2, are somewhat higher because in calculations the amount of water overflowing dam crest is not considered. Results for design discharge for initial water level in the reservoir equal to 272.40 m a.s.l. are almost the same with those observed during flood routing in July 1997.

On the basis of numerical calculations, consisting in changing a depth and width of flume and in determining a free water level along flume length, a required range of works for spillway's flume rebuilding was determined. It is necessary to deepen the flume with simultaneous adjustment its bottom longitudinal slope with small widening its upper part. These requirements can be realized without significant interference in construction of the whole object; therefore, they can be accepted in economic respect. The proposed solutions of spillway rebuilding are shown in Fig. 5, whereas in Fig. 6 a free water surface and velocity along the flume is presented.

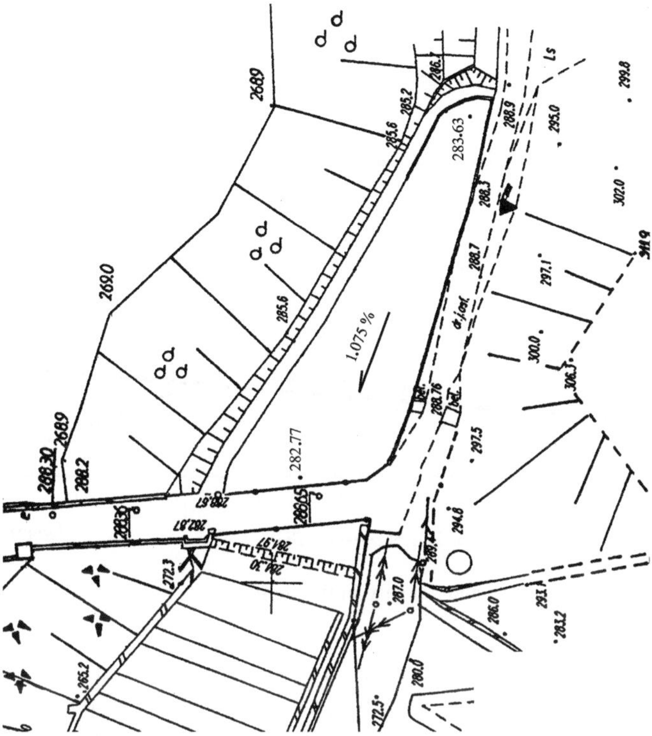

Fig. 5 Concept of reconstruction the side channel spillway

The spillway can be reconstructed without a necessity of reservoir emptying; also, its safety is ensured because flood waves can pass through the spillway without any negative consequences for the whole object. A significant drawback of spillway reconstruction is the lack of possibility of improving the operating conditions of freshet wave routing; in the existing state, the operation of outlet installation (diverse channel and bottom outlets) is carried out until a water level elevation in the reservoir will reach the dam crest, that is, the outflow will be somewhat over 200 $m^3 s^{-1}$.

Hydraulic characteristics of side channel spillway, corrected by analytical calculations, are determined for discharges in range from 0.0 to 425 $m^3 s^{-1}$ and in the range of water level elevations in reservoir from 286.70 m a.s.l. (spillway crest) to 288.75 m a.s.l. (dam crest from downstream—Fig. 2) and is shown in Fig. 7.

The rise of discharge curve, visible in Fig. 7, shows a moment of change in the character of spillway work from not-submerged to submerged. This is a situation which should never develop for this type of device—in every condition determined by weir parameters and water level elevations in the reservoir, this device should work as not-submerged.

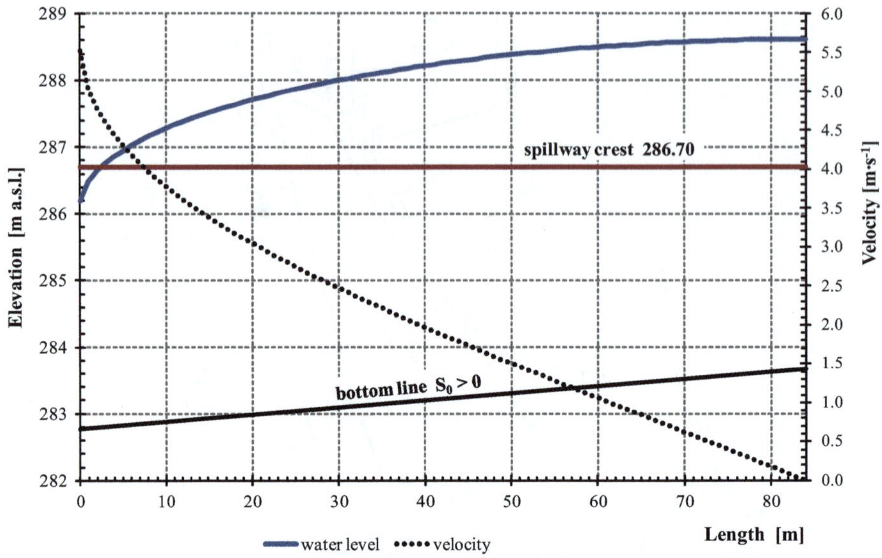

Fig. 6 The free water surface and velocity in the flume

Because investigations concerned an existing object, it was important to determine the conditions for the changes in character of the weir work and expected capacity ability. The spillway capacity ability obtained on the basis of numerical analysis is close to the required, that under given conditions can overflow by weir. This corresponds to discharge of about 425 $m^3\ s^{-1}$, which could be obtained for corrected spillway layout in plan, its crest length, weir shape and flume parameters.

Fig. 7 Hydraulic characteristics of side channel spillway for the proposed solution

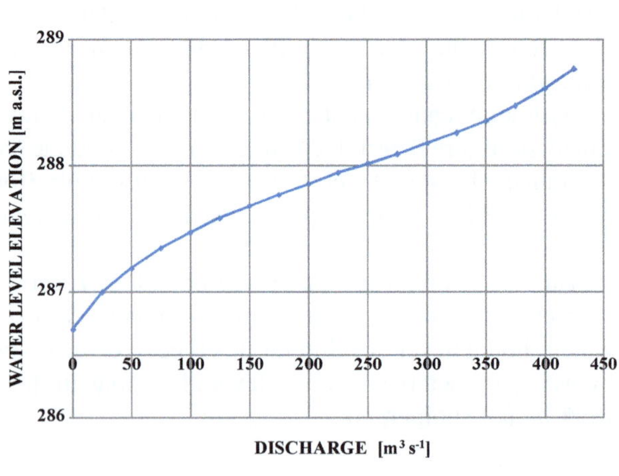

8 Recapitulation

Recapitulating the numerical analysis of side channel spillway of Pilchowice reservoir connected mainly with hydraulic characteristics of its basic elements, weir and flume, there is a possibility of improving the capacity ability of this device. The proposed solution that corrects the configuration of the flume and its width in the upper part are verified by numerical analysis. This confirmed a necessity of flume rebuilding, guarantying exploitation safety of reservoir mainly because of computational discharges, significantly exceeding the capacity ability of outlet devices in the existing state. In numerical analysis an attention was paid to possibility and outright necessity of drawing the flume with constant longitudinal slope bigger than zero. This allows drawing the following conclusions:

1. In order to improve the exploitation safety of the Pilchowice reservoir, it is necessary to introduce the proposed changes in the existing state of flume of the side channel spillway.
2. A change in side channel spillway solutions consists in the deepening of flume bottom with the maintenance of its longitudinal slope bigger than zero and the widening in its upper part.
3. The proposed changes of side channel spillway layout in plan, introduced in a minimal range, will increase its capacity ability by about 100 $m^3\ s^{-1}$.

References

Ghosh SN (2006) Flood control and drainage engineering. Taylor & Francis Group, London

Jain SC (2001) Open-channel flow. Wiley, New York

Khatsuria RM (2005) Hydraulics of spillways and energy dissipators. Marcel Dekker Publishers, New York

Machajski J, Olearczyk D (2006) Water management instruction on storage reservoir Pilchowice. Wroclaw University of Technology, Institute of Geotechnics and Hydrotechnics. Wrocław (in Polish)

Machajski J, Rędowicz W (2008) Analysis in technical and economical aspects of enlargement capacity ability of storage reservoir Pilchowice on the Bober river. Research report no 11/ 2008. Wrocław University of Technology, Institute of Geotechnics and Hydrotechnics. Wrocław (in Polish)

Novak P, Moffat AIB, Nalluri C, Narayanan R (2007) Hydraulic structures. Taylor & Francis Publishers, New York

Rogala R, Machajski J, Rędowicz W (1991) Applied hydraulics. Wroclaw University of Technology Publications, Wroclaw (in Polish)

Directive (2007) Directive of the Poland's minister of environment of April 20, 2007, on technological conditions of hydro-engineering structures. Dz.U. Nr 86/2007, poz. 579 (in Polish)

Şentürk F (1994) Hydraulics of dams and reservoirs. Water Resources Publications, Colorado

Tančev L (2005) Dams and appurtenant hydraulic structures. A.A. Balkema Publishers, London

Flood Quantile Estimates Related to Model and Optimization Criteria

Iwona Markiewicz, Witold G. Strupczewski and Krzysztof Kochanek

Abstract Flood frequency analysis (FFA) provides information about the probable size of flood flows. Empirical methods are more commonly employed in engineering design and planning, and among empirical methods the at-site frequency analysis is by far the most commonly used method. When applying the methods of at-site flood frequency analysis, it is clear that the role of hydrology seems minor at best and the role of statistics seems to be the lead one, whereas it should be the other way round. FFA entails the estimation of the upper quantiles of an assumed form of a probability density function of the annual or partial duration maximum flows, as the true function is not known. In the paper, the five two-parameter models and their three-parameter counterparts have been assumed successively for describing the annual peak flows for Nowy Targ gauging station on the Dunajec River. The 1 % quantile has been estimated by four optimization criteria. To find the best fitting model, three discrimination procedures have been applied. The best fitting model and, thus, hydrological design value depends on the optimization criterion and the procedure of discrimination. It is characteristic for hydrological size of samples. At the same time, the designers of the hydraulic structures want to have a unique value, not accepting the uncertainty. It seems essential that we should go back and start examining the way in which we have been doing the hydrological frequency analysis.

I. Markiewicz (✉) · W. G. Strupczewski · K. Kochanek
Institute of Geophysics Polish Academy of Sciences, Warsaw, Poland
e-mail: iwonamar@igf.edu.pl

W. G. Strupczewski
e-mail: wgs@igf.edu.pl

K. Kochanek
e-mail: kochanek@igf.edu.pl

P. Rowiński (ed.), *Experimental and Computational Solutions of Hydraulic Problems*, GeoPlanet: Earth and Planetary Sciences, DOI: 10.1007/978-3-642-30209-1_25,

1 Introduction

The assessment of the flood upper quantiles is the basic goal for the flood frequency analysis (FFA). The probable size of flood peaks obtained in this way has many practical applications. This information is required for dimensioning hydraulic structures such as culverts, dams, bridges, overflow channels, spillways, levees, and others. Moreover, it is essential in determining the limits of flood zones with varying degree of flood risk, estimating the risk of exploitation of floodplains as well as for the valuation of some insurance premiums. FFA plays an important role in water resources design and management, especially in reducing the flood risk.

Among the methods for the estimation of flood quantiles, empirical methods are more commonly employed in engineering design and planning. However, the physically oriented methods are in progress; there is the stochastic watershed modeling in conjunction with Monte Carlo simulation. The flood frequency analysis boils down to the estimation of the upper (flood) quantiles of an assumed form of a probability density function (PDF) of the annual or partial duration maximum flows, as the true function is not known. The at-site frequency analysis is the most commonly used approach. Hence, the estimation of flood quantiles refers to the choice of the probability model, i.e., the form of probability density function describing the annual peak flows for the investigated gauging station. Simultaneously, the optimization criterion is selected, i.e., the method of estimation of parameters, and thus quantiles of the assumed distribution. To find the best-fit model to empirical data, the chosen discrimination procedure is applied.

Many distributions have been suggested for fitting the flood extremes data, e.g., NERC (1975). According to the hydrological report of the World Meteorological Organization from 1989 (Cunnane 1989), the Gumbel and log-normal distributions were the most commonly used for the description of the annual maximum flows, while nowadays the heavy-tailed distributions are recommended for this issue, e.g., FEH (1999). There are many methods of parameter estimation (Singh and Strupczewski 2002), the most popular of which are the methods of moments, maximum likelihood, linear moments, probability weighted moments, least squares and entropy. Iacobellis et al. (2002) analysed a relation between the variation coefficient C_V of the series of annual maximum flows and the physically consistent quantities. They found that C_V is controlled mainly by the climate and by some water loss features. Then, the effects of climatic and geomorphologic parameters on the skewness coefficient C_S were investigated (Gioia et al. 2012). However, the scientists mainly have been putting effort into improving the techniques of the flood quantiles assessment by proposing the new types of PDFs and new or modified methods of estimation. The proliferation of the distributions and estimation methods employed is itself symptomatic of the weak theoretical bases of the model selection. At the same time, no hydrological theory has been formed yet for indication of the probability distribution which should be used under specific conditions and which estimation method should be employed at that time.

In hydrological practice, a number of problems with the assessment of flood quantiles arise. These are: unknown probability distribution function of annual peaks, short time series, error corrupted data, the simplifying assumptions of independent identically distributed (i.i.d.) data and, in particular, of the stationarity of relatively long series. All these may cause a high uncertainty of upper quantile estimate. The only effect of a sample size is widely documented for various distribution models and estimation methods (Landwehr et al. 1980; Strupczewski et al. 2005; Kochanek et al. 2005; Markiewicz et al. 2010). Inclusion of other sources of error into flood frequency analysis would result in increasing uncertainty of design flood estimate. This is a feature not appreciated by the hydrologists-designers as they want to have the only one value for designing flood related structures.

The aim of this study is to answer the question: where are we heading in flood quantiles estimation? Whether the theory meets the expectations of the practice? The paper is organized as follows. After providing some introduction to the topic, the models and the optimization criteria investigated in the paper are presented in Sects. 2 and 3. Next, the three discrimination procedures used to find the best fitting model are shown. The following section provides the issue of the estimation of flood quantiles for the example of Nowy Targ gauging station on the Dunajec River. The results are presented and discussed in two subsections. The paper is concluded in the final section.

2 Models

The choice of the most appropriate distribution to describe the maximum annual flows has been the object of wide interest for a long time (Jenkinson 1969; NERC 1975; Rao and Hamed 2000; Malamud and Turcotte 2006).The frequency distribution functions which are used in flood modeling should be without upper bound and with non-negative skewness (Kaczmarek 1977; Cunnane 1989; Rao and Hamed 2000). Since the hydrological samples of peak flows are usually of a relatively small size, both two- and three-parameter distributions are mainly used for FFA to get parameter estimates reliably and efficiently (Landwehr et al. 1980; Kuczera 1982; FEH 1999; Kochanek et al. 2005). Usually they are the scale and shape parameters with the lower bound parameter serving as the third one.

In this study, the probability density functions commonly used in FFA have been investigated. They are two-parameter gamma (Ga), Weibull (We), log-normal (LN), log-logistic (LL), log-Gumbel (LG), and they three-parameter counterparts, i.e., Pearson (Pe), Weibull (We), log-normal (LN), generalized log-logistic (GLL) and generalized extreme value (GEV). They are listed in Table 1.

The last two PDFs, i.e., LL, LG and their counterparts GLL, GEV are the heavy-tailed distributions. Models of this type are recommended by the researches of hydrological extreme events for modelling the annual maximum flows (FEH 1999; Katz et al. 2002; Griffis and Stedinger 2007; Adlouni et al. 2008).

Table 1 Distribution functions used in flood frequency analysis

Distribution	Probability density function (PDF)
Pearson (Pe) $\varepsilon = 0$: gamma (Ga)	$f(x) = \frac{\alpha^\lambda}{\Gamma\lambda}(x-\varepsilon)^{\lambda-1}\exp[-\alpha(x-\varepsilon)]$ $\alpha > 0$-scale, $\lambda > 0$-shape; $\varepsilon < x < \infty$
Weibull (We)	$f(x) = \frac{\kappa}{\alpha}\left(\frac{x-\varepsilon}{\alpha}\right)^{\kappa-1}\exp\left[-\left(\frac{x-\varepsilon}{\alpha}\right)^{\kappa}\right]$ $\alpha > 0$-scale, $\kappa > 0$-shape; $\varepsilon < x < \infty$
Log-normal (LN)	$f(x) = \frac{1}{(x-\varepsilon)b\sqrt{2\pi}}\exp\left[-\frac{(\ln(x-\varepsilon)-m)^2}{2b^2}\right]$ m-scale, $b > 0$-scale; $0 \le \varepsilon < x < \infty$
Generalized log-logistic (GLL) $\varepsilon = 0$: log-logistic (LL)	$f(x) = \frac{1}{\alpha}\left[-\frac{\kappa}{\alpha}(x-\varepsilon)\right]^{1/\kappa-1}\left[1+\left\{-\frac{\kappa}{\alpha}(x-\varepsilon)\right\}^{1/\kappa}\right]^{-2}$ $\alpha > 0$-scale, $\kappa < 0$-scale; $\varepsilon < x < \infty$
Generalized extreme values (GEV) $\varepsilon = 0$: log-Gumbel (LG)	$f(x) = \frac{1}{\alpha}\left[-\frac{\kappa}{\alpha}(x-\varepsilon)\right]^{1/\kappa-1}\exp\left\{-\left\{-\frac{\kappa}{\alpha}(x-\varepsilon)\right\}^{1/\kappa}\right\}$ $\alpha > 0$-scale, $\kappa < 0$-scale; $\varepsilon < x < \infty$

The heavy-tailed PDFs have conventional moments only in a certain range of the shape parameter values and the range decreases with growing moment order. Figure 1 illustrates the restrictions in existing conventional moments ratios for two-parameter distributions. However, the evidence of the heavy tail of hydrological variables is not sufficiently supported (Rowinski et al. 2002; Weglarczyk et al. 2002). Moreover, the Polish datasets of annual peak flows should be modeled using soft-tailed distributions, rather than heavy-tailed distributions (Strupczewski et al. 2011).

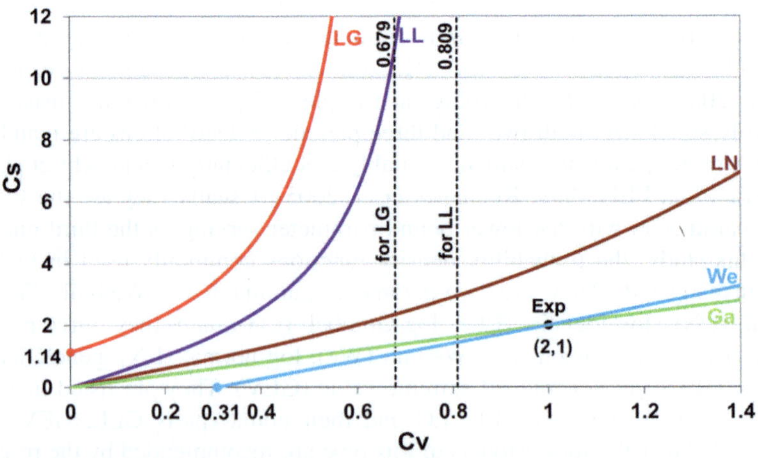

Fig. 1 Conventional skewness coefficient (Cs) versus conventional variation coefficient (Cv) for two-parameter distributions

3 Optimization Criteria

Several optimization criteria have been proposed for the assessment of the flood quantiles. The most popular are: the criterion of conventional moments (MO) (e.g., Kendal and Stuart 1969), the criterion of linear moments (LM) (e.g., Hosking and Wallis 1997) and the maximum likelihood criterion (ML) (e.g., Kendal and Stuart 1973). The first and second criteria are built on the summary statistics. Based on various principles, the summary statistics provide the measures of location, dispersion, skewness, kurtosis and so on. They are served for fitting the assumed PDF to the annual peak flow series. The definitions of dispersion and skewness measures for the MO and LM criteria are given in Table 2, while the location measure for them is expressed by the mean $(\mu \equiv \lambda_1)$. The linear moments have advantages in comparison with the classical moments, namely, their estimators are unbiased and the sampling L-moments ratios have very small biases for moderate and large samples (Hosking and Wallis 1997). In contrary to the MO and LM sample summary statistics, the ML criterion is distribution dependent and in fact it bases on the main probability mass of a sample. The ML estimators are asymptotically the most efficient but in the case of model misspecification, they can be extremely highly biased (Markiewicz et al. 2010).

The complement to the optimization criteria commonly used in FFA is the criterion built on the mean deviation (MD), proposed by Markiewicz et al. (2006), Markiewicz and Strupczewski (2009). The MD criterion applies the mean (μ) as the location measure, the mean deviation (δ_μ) as the dispersion measure and the skewness measure δ_S (see Table 2).

4 Discrimination Procedures

To find the best fitting model to empirical data from the set of competing models, some discrimination procedure is applied. It must define a test statistic as well as a decision rule indicating the action to be taken for the observed sample. One can also prioritize all competing models according to the values of the selection criterion. However, as shown in Strupczewski et al. (2004), the use of a discrimination procedure without the knowledge of its performance for the considered PDF may lead to erroneous conclusions. It seems reasonable to use and compare the results of several discrimination procedures.

4.1 Akaike Information Criterion

The Akaike information criterion with a correction for finite sample sizes bases on the quantity *AICc* defined by (Clifford and Chih-Ling 1989):

Table 2 Dispersion and skewness measures for MO, LM and MD optimization criteria

Optimization criterion	Dispersion measure		Skewness measure	
MO	Standard deviation	$\sigma = \left[\int_{-\infty}^{+\infty} (x-\mu)^2 dF(x)\right]^{1/2}$	Third moment	$\mu_3 = \int_{-\infty}^{+\infty} (x-\mu)^3 dF(x)$
LM	Second *L*-moment	$\lambda_2 = \int_{-\infty}^{+\infty} 2(x-\mu)F(x)dF(x)$	Third *L*-moment	$\lambda_3 = \mu + \int_{-\infty}^{+\infty} 6x\left[F^2(x) - F(x)\right]dF(x)$
MD	Mean deviation	$\delta_2 = \int_{-\infty}^{+\infty} \lvert x-\mu\rvert dF(x)$	Skewness measure	$\delta_S = \int_{-\infty}^{+\infty} (x-\mu)dF(x) - \int_{-\infty}^{\mu} (\mu-x)dF(x)$

$$AICc = -2\ln\left[L\left(f\left(x/\hat{\theta}\right)\right)\right] + 2K + \frac{2K(K+1)}{N-K-1} \tag{1}$$

where $L\left(f\left(x/\hat{\theta}\right)\right)$ is the likelihood function for the estimated model and K stands for the number of model parameters. The likelihood function is obtained by replacing model parameters with theirs estimates obtained by various optimization criteria (see Sect. 3). The best model among all competing models is this one with the lowest *AICc* value, which is provided in fact by the maximum likelihood (ML) criterion.

4.2 Daniels Characteristic

Next discrimination procedure bases on the Daniels characteristic D_{Max} (Kaczmarek 1977):

$$D_{\max} = \max_{i=1,\ldots,N}\left|\frac{F(x_{i:N}) - \hat{F}_{i:N}}{1 - F(x_{i:N})}\right| \tag{2}$$

where $F(x_{i:N})$ expresses the theoretical probability of the i-th element of the ordered random sample $x_{1:N} \geq \ldots \geq x_{N:N}$ and $\hat{F}_{i:N}$ is its empirical probability given by the Weibull formula:

$$\hat{F}_{i:N} = i/(N+1) \tag{3}$$

This discrimination procedure is particularly useful in flood frequency modeling, since it is especially sensitive within the area of probabilities of non-exceedance (cumulative probability) F approximating to one, i.e., in the upper quantiles area, which are of particular interest for the FFA.The model corresponding to the largest $D_{\max}$ value is selected as the best fitted to the data.

4.3 Difference Between the Quantile Estimates

Although the various differences between estimates of various methods have served to assess model fitting, the discrimination procedure has been explicitly proposed by Mitosek et al. (2006). Since the parametric methods of the PDF estimation are asymptotically unbiased, the estimate of any quantiles tends in infinity to its exact value for the sample withdrawn from the population of known distribution function. In Mitosek et al. (2006), the difference between 1 % quantile assessed by the criteria of moments and the maximum likelihood has been used as a measure of fitness of the model to the sample. The 1 % quantile, i.e., the quantile with the probability F equal to 0.99, is the most frequently used design value.

Here, three discrimination characteristics D_i, for $i = 1, 2, 3$, are proposed:

$$D_1 = \hat{x}^{ML}_{1\,\%} - \hat{x}^{MO}_{1\,\%} \quad (4)$$

$$D_2 = \hat{x}^{ML}_{1\,\%} - \hat{x}^{LM}_{1\,\%} \quad (5)$$

$$D_3 = \hat{x}^{ML}_{1\,\%} - \hat{x}^{MD}_{1\,\%} \quad (6)$$

where $\hat{x}^{MO}_{1\,\%}$, $\hat{x}^{LM}_{1\,\%}$, $\hat{x}^{MD}_{1\,\%}$, $\hat{x}^{ML}_{1\,\%}$ are the 1 % quantiles estimated by the criterion of moments, the criterion of linear moments, the criterion built on the mean deviation and the maximum likelihood criterion, respectively. For each discrimination characteristics D_i, the best model is the one with the lowest D_i value, for $i = 1, 2, 3$. Note that for the normal distribution the criteria of ML and MO are equivalent.

5 Case Study

For the estimation of flood quantile, the 90-year-long series of annual maximum flows for Nowy Targ gauging station on the Dunajec River has been selected as an example. The data and their characteristics are shown in Fig. 2.

The five two-parameter models, i.e., Ga, We, LN, LL and LG, and their three-parameter counterparts, i.e., Pe, We, LN, GLL and GEV, have been assumed as the hypothetical models, successively. The 1 % quantile has been estimated by the four optimization criteria, namely MO, LM, MD and ML. For each optimization criterion, three discrimination procedures have been applied to find the best fitting model among two-parameter models and then among three-parameter models. The discrimination procedures, i.e., *AICc*, $D_{\max}$ and D_i, for $i = 1, 2, 3$, are applied separately for two- and three-parameter distributions, since the results of comparison between different models with different numbers of parameters lose in clarity (Mitosek et al. 2006). For a set of competing models containing both two- and three-parameter PDFs, a discrimination procedure which takes into account the number of model parameters, should be used, such as, for example, the *AICc*, otherwise the three-parameter distribution would always be superior.

5.1 Flood Quantile Estimates for Two-Parameter Models

The values of 1 % quantile estimates are presented in Table 3. The assessed values differ considerably, from 552 to 929. The criterion of moments is the most stable in respect to various models and the maximum likelihood criterion is the most sensitive.

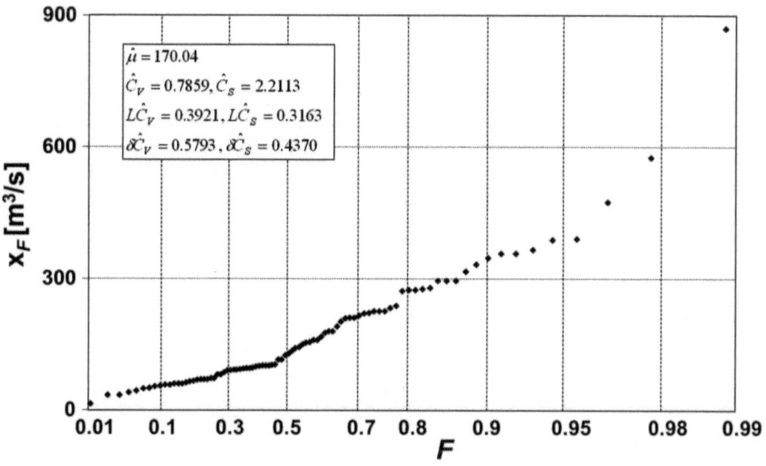

Fig. 2 Annual maximum flows for Nowy Targ on the Dunajec River from the period 1921–2010, with their characteristics

Table 3 The 1 % quantile estimates for two-parameter models

Optimization criterion	$\hat{x}_{1\,\%}$				
	Ga	We	LN	LL	LG
MO	620.64	604.11	671.10	638.76	630.68
LM	590.57	558.09	706.79	788.88	900.07
MD	603.07	570.88	726.41	827.07	914.32
ML	555.68	552.85	718.76	929.17	731.92

Which of the assessments of 1 % quantile is the closest to the true value? As one can see in Tables 4, 5 and 6, the discrimination procedures applied here don't give a clear answer. The italic font means that the optimization criterion is the best for particular model, while the bold font stands for the best model for particular optimization criterion. For example, following the discrimination procedure based on the Daniels characteristic (Table 5), the criterion of moments is the most appropriate for the assessment of the 1 % quantile, if the gamma distribution is assumed as the hypothetical model for describing the series of annual peak flows for Nowy Targ gauging station.

The *AICc* indicates the ML as the best estimation criterion for all considered models, which is obvious, since both the maximum likelihood optimization criterion and the Akaike discrimination procedure base on the maximized likelihood function for the model (Eq. 1). Following the *AICc*, the log-normal distribution fits to the data the most accurately (Table 4). The LN model is also chosen by discrimination characteristics D_1 and D_3, while D_2 points out the Weibull as the best one (Table 6). The Daniels characteristic gives different results. The LL distribution is indicated by the LM, MD and ML criteria of optimization as the best fit

Table 4 Akaike information criterion corrected for Nowy Targ record and two-parameter models

Optimization criterion	*AICc*				
	Ga	We	LN	LL	LG
MO	1088.7	1093.1	**1081.4**	1095.9	2157.2
LM	1086.7	1091.5	**1080.8**	1085.1	1209.4
MD	1087.4	1091.7	**1080.9**	1084.5	1198.0
ML	*1086.0*	*1091.0*	***1081.0***	*1084.0*	*1143.0*

Table 5 Daniels characteristic for two-parameter models

Optimization criterion	D_{max}				
	Ga	We	LN	LL	LG
MO	*8.8240*	*16.120*	2.1758	1.8015	**1.6900**
LM	15.210	55.578	1.4466	**0.5633**	0.7860
MD	12.022	37.690	*1.1560*	**0.5436**	*0.7820*
ML	31.978	69.362	1.2728	***0.3880***	1.9364

Table 6 Difference between the quantile estimates for two-parameter models

Characteristic	Ga	We	LN	LL	LG
D_1	64.960	51.265	**47.661**	290.40	101.24
D_2	34.886	**5.2418**	11.968	140.28	168.16
D_3	47.384	18.031	**7.6492**	102.10	182.40

model to the data and the LG distribution is chosen by the MO criterion (Table 5). The optimization criterion of conventional moments is superior for the Ga and We models, while the criterion based on the mean deviation is the best for the LN and LG. The ML criterion wins in the case of the LL model. In general, the results obtained from the discrimination procedures for two-parameter PDFs are inconclusive and it is difficult to conclude about the best model and the optimization criterion for the annual maximum flow series for the Nowy Targ gauging station.

5.2 Flood Quantile Estimates for Three-Parameter Models

The estimates of 1 % quantile for the data series for Nowy Targ, shown in Table 7, are scattered from 579 to 864. However, the scope of the assessed values is smaller than for two-parameter models. The results for the discrimination procedures investigated in the paper are presented in Tables 8, 9 and 10.

The *AICc* values for the Pearson and Weibull distributions are indeterminate by the MO, LM and MD optimization criteria (Table 8), since these three criteria yield the lower bound parameter ε (see Table 1) greater than the minimum

Table 7 The 1 % quantile estimates for three-parameter models

Optimization criterion	$\hat{x}_{1\%}$				
	Pe	We	LN	GLL	GEV
MO	665.95	665.02	649.79	619.46	641.51
LM	650.75	635.73	678.94	719.70	699.77
MD	663.57	667.89	737.73	830.26	801.02
ML	579.71	579.10	701.30	812.19	864.17

Table 8 Akaike information criterion corrected for Nowy Targ record and three-parameter models

Optimization criterion	*AICc*				
	Pe	We	LN	GLL	GEV
MO	*****	*****	**1086.8**	1103.0	1092.0
LM	*****	*****	**1083.6**	1090.9	1086.7
MD	*****	*****	1087.6	1087.0	**1084.5**
ML	1085.2	1087.7	***1082.7***	*1086.5*	*1084.4*

Table 9 Daniels characteristic for three-parameter models

Optimization criterion	D_{max}				
	Pe	We	LN	GLL	GEV
MO	*3.5700*	3.5053	**3.2613**	3.5584	3.3025
LM	4.6892	6.2390	2.1049	**0.9351**	1.3651
MD	3.8207	*3.4890*	*0.8730*	***0.5510***	*0.6590*
ML	17.717	25.306	1.5643	**0.6719**	0.6594

Table 10 Differences between the quantile estimates for three-parameter models

Characteristic	Pe	We	LN	GLL	GEV
D_1	86.245	85.918	**51.512**	192.72	222.66
D_2	71.046	56.635	**22.361**	92.486	164.40
D_3	83.864	88.790	36.427	**18.074**	63.150

observed value of annual maximum flows. The Akaike information criterion corrected for finite sample indicates the LN model as the best fitted to the data, if the MO, LM or ML optimization criterion is applied. In the case of using the MD criterion, the GEV distribution becomes preferable. The LN model is pointed out as the best one by two of the three discrimination characteristics D_i for $i = 1, 2$, while D_3 chooses the GLL distribution (Table 10). Following the Daniels characteristic, the GLL wins along with the LM, MD and ML criteria of optimization, while the LN model is superior for the criterion of moments, MO (Table 9).

We should note about the convective diffusion (CD) distribution, known also as the inverse Gaussian or Halphen type A. The CD represents flood frequency characteristics of Polish rivers quite well, in particular of lowland rivers (Strupczewski et al. 2011). The section in Nowy Targ represents the mountain river; nonetheless, according to the *AICc* discrimination procedure, the three-parameter CD model is the best compared to the five three-parameter models discussed in the following paper, using the optimization criteria of MO and ML. In the case of two-parameter models, the CD occupies the second position. The development of the LM and MD criteria for the convective diffusion distribution are in progress.

Similarly to the two-parameter models, the discrimination procedures do not overlap in the results for three-parameter PDFs. Hence, concluding about the superiority of one of the models and/or optimization procedures for the flood quantile estimation for selected gauging station is a difficult and risky task.

6 Conclusions

Flood frequency analysis has been used for design of hydrological structures for over the century. We have made great strides in hydrologic data collection and conceptualization of the processes surrounding floods. However, the main effort has been devoted to the improvement of the statistical techniques of flood frequency modeling. This resulted in a proliferation of models, optimization criteria and discrimination procedures. As shown in the paper, the choice of the best fitting model to data series and, thus, hydrological design value (i.e., 1 % quantile) depends on the optimization criterion and the procedure of discrimination. It is characteristic for hydrological size of samples. The heterogeneity of results causes an increase in our awareness of the uncertainty of flood quantile estimates instead of leading to clear solution. This stands in contrast with the expectation of engineers and hydrologists as they want to have a unique value, not accepting the uncertainty! So, answering the question where we are heading in flood quantiles estimation, it sees essential that we sholud go back and start examining the way in which we have been doing the hydrological frequency analysis.

Acknowledgments This work was financed by the Polish Ministry of Science and Higher Education under the Grant IP 2010 024570 titled "Analysis of the efficiency of estimation methods in flood frequency modeling".

References

Clifford MH, Chih-Ling T (1989) Regression and time series model selection in small samples. Biometrika 76(2):297–307.

Cunnane C (1989) Statistical distributions for flood frequency analysis. Operational hydrology Report No. 33, World Meteorological Organization, Geneva

FEH (1999) Flood estimation handbook. Institute of hydrology Wallingford, Oxfordshire OX10 8BB, U.K

Gioia A, Iacobellis V, Manfreda S, Fiorentino M (2012) Influence of Infiltration and soil storage capacity on the skewness of the annual maximum flood peaks in a theoretically derived distribution. Hydrol Earth Syst Sci 16:937–951. doi:10.5194/hess-16-937-2012

Griffis VW, Stedinger JR (2007) Evolution of flood frequency analysis with Bulletin 17. J Hydrol Eng ASCE 12(3):283–297

Hosking JRM, Wallis JR (1997) Regional frequency analysis. An approach based on L-moment. Cambridge University Press, Cambridge CH2 1RP

Iacobellis V, Claps P, Fiorentino M (2002) Climatic control on the variability of flood distribution. Hydrol Earth Syst Sci 6:229–238. doi:10.5194/hess-6-229-2002

Jenkinson AF (1969) Statistics of extremes. In: Estimation of maximum floods. WMO No 233, TP126, Tech. Note No. 98, 183-228, Secretariat of the World Meteorological Organization, Geneva

Kaczmarek, Z (1977) Statistical methods in hydrology and meteorology. Published for the Geological Survey, U.S. Department of the interior and the National Science Foundation, Washington, D.C., by the Foreign Scientific Publications Department of the National Centre for Scientific, Technical and Economic Information, Warsaw

Katz RW, Parlange MB, Naveau P (2002) Statistics of extremes in hydrology. Adv Water Resour 25:1287–1304

Kendall MG, Stuart A (1969) The advanced theory of statistics. Distribution theory, vol 1. Charles Griffin & Company Limited, London

Kendall MG, Stuart A (1973) The advanced theory of statistics. Inference and relationship, vol 2. Charles Griffin & Company Limited, London

Kochanek K, Strupczewski WG, Węglarczyk S, Singh VP (2005) Are parsimonious FF models more reliable than the true ones? II. Comparative assessment of the performance of simple models versus the parent distribution. Acta Geoph Pol 53(4):437–457

Kuczera G (1982) Robust flood frequency models. Water Resour Res 18(2):315–324

Landwehr JM, Matalas NC, Wallis JR (1980) Quantile estimation with more or less flood like distributions. Water Resour Res 16(3):547–555

Malamud BD, Turcotte DL (2006) The applicability of power-law frequency statistics to flood. J Hydrol 322:168–180

Markiewicz I, Strupczewski WG, Kochanek K, Singh VP (2006) Relations between three dispersion measures used in flood frequency analysis. Stoch Env Res Risk A 20:391–405. doi:10.1007/s00477-006-0033-x

Markiewicz I, Strupczewski WG (2009) Dispersion measures for flood frequency analysis. Phys Chem Earth 34:670–678. doi:10.1016/j.pce.2009.04.003

Markiewicz I, Strupczewski WG, Kochanek K (2010) On accuracy of upper quantiles estimation. Hydrol Earth Syst Sci 14:2167–2175. doi:10.5194/hess-14-2167-2010

Mitosek HT, Strupczewski WG, Singh VP (2006) Three procedures for selection of annual flood peak distribution. J Hydrol 323:57–73

NERC (1975) Flood studies report volume 1: hydrological studies. Natural Environment Research Council, London

Rao AR, Hamed KH (2000) Flood frequency analysis. CRC Press, Boca Raton

Rowinski PM, Strupczewski WG, Singh VP (2002) A note on the applicability of log-Gumbel and log-logistic probability distributions in hydrological analyses: I. Known pdf. Hydrolog Sci J 47(1):107–122

Singh VP, Strupczewski WG (2002) On the status of flood frequency analysis. Hydrol Process 16:3737–3740

Strupczewski WG, Mitosek HT, Kochanek K, Singh VP, Weglarczyk S (2004) Probability of correct selection from lognormal and convective diffusion models based on the likelihood ratio. Stoch Environ Res Risk Ass 18:1–11. doi:10.1007/s00477-004-0210-8

Strupczewski WG, Kochanek K, Singh VP, Weglarczyk S (2005) Are parsimonious FF models more reliable than the true ones? I. Accuracy of quantiles and moments estimation (AQME)—method of assessment. Acta Geoph Pol 53(4):419–436

Strupczewski WG, Kochanek K, Markiewicz I, Bogdanowicz E, Weglarczyk S, Singh VP (2011) On the tails of distributions of annual peak flow. Hydrol Res 42(2–3):171–192. doi:10.2166/nh.2011.062

Weglarczyk S, Strupczewski WG, Singh VP (2002) A note on the applicability of log-Gumbel and log-logistic probability distributions in hydrological analyses: II. Assumed pdf. Hydrol Sci J 47(1):123–137

Multi-Temporal Analysis for Mexico City Aquifer

Gerardo Ruiz and Rosio Ruiz

Abstract The Mexico City population growth demands an increase of water, the vital liquid that is obtained from the internal and external resources of exploitation. The dependency of the water on the groundwater extracted from wells makes it necessary to review the conditions of exploitation. These revisions are made from the depth of groundwater, providing measurement values that can be important for the exploitation of this source and its recovery. The main objective of the study reported in this chapter was to introduce a system that enables storage of information, analysis and visualization of measurements of static and dynamic levels and specific flow, as well as the use of Geographic Information System (GIS) to support a better understanding of the modeled aquifer and improve decision making in the operation of the network of wells in the Valley of Mexico. Multi-temporal analysis of the evolution of groundwater is presented up to 2009 of the aquifer of the Mexico City metropolitan area, for which the measurements of static and dynamic levels of 225 wells were made. Hydraulic balance of groundwater was performed to determine the degree of overexploitation, to check changes in the static levels according to the policy of operation of the entire system of wells of the aquifer.

G. Ruiz (✉)
Facultad de Ingeniería, Universidad Nacional Autónoma de México,
México 04510, Mexico
e-mail: gerardrui@hotmail.com

R. Ruiz
Facultad de Ingeniería, Universidad Nacional Autónoma de México,
México 04510, Mexico
e-mail: rosruur@yahoo.com.mx

R. Ruiz
Circuito Escolar s/n, División de Ingenierías Civil y Geomática,
Departamento de Hidráulica, Ciudad Universitaria, México 04510, Mexico

P. Rowiński (ed.), *Experimental and Computational Solutions of Hydraulic Problems*,
GeoPlanet: Earth and Planetary Sciences, DOI: 10.1007/978-3-642-30209-1_26,

Keywords Aquifer · Geographic Information System · Multi-temporal analysis · Drawdown

1 Introduction

Within the objectives of this chapter is to introduce a system that allows storing information, analysis and visualization of the measurements of static and dynamic levels of the outdoors measurements made during the year 2009, as well as the years following 1985. We use the Geographic Information System (GIS) as a support in the modeling of water-bearing layers and better decision making in the operation of the well network of the Valley of Mexico, based on the measurement data.

In order to carry out the multi-temporal study, the static and dynamic levels of 225 wells, located in the aquifer of Mexico City, were modeled. The hydraulic balance of the underground water to determine the overexploitation degree is presented.

Using the multi-temporal analysis through GIS, the effects of overexploitation on the Valley of Mexico were estimated; the areas and volumes in the periods 1985–2009 and 2005–2009 were quantified, trying to show the zones of drawdown or recharge. These results were processed through GIS with the intention of displaying two-dimensional graphs of the behavior of the aquifer in different stages.

1.1 Zone of Study and Hydrogeology

The study is made for a single system of the aquifer, namely the aquifer subsystem of Mexico City. Regionally, the subsoil of the Valley of Mexico is constituted by a sequence of alluviums and volcanic products, both permeable, which allows the infiltration and underground water circulation. The aquifer is confined by lacustrine clays in the east portion of the valley and in the western portion it works like a free aquifer (see Fig. 1).

The main recharge of the aquifer comes from the infiltration of water that precipitates on the mountains that surround it, especially in the south portion, due to the high permeability of rocks. The principal discharge of the aquifer is carried out by means of pumping of wells perforated in the valley, while the discharges to other basins do not exist, because the Valley of Mexico is an endorheic basin, without natural exits of superficial water and groundwater.

The hydrogeological behavior and hydraulic characteristics of rocks depend on their lithology, texture, structure and fracturing. Based on these parameters, the rocks of the area are classified in three hydrogeological units, taking into account their degree of permeability, compaction, granulometry and cementation. These are:

Permeable unit. The rocks grouped in this unit are volcanic products of the Quaternary, which are constituted by pyroclastic materials in all the variety of sizes, that consolidated slowly. There are basaltic rocks with high secondary

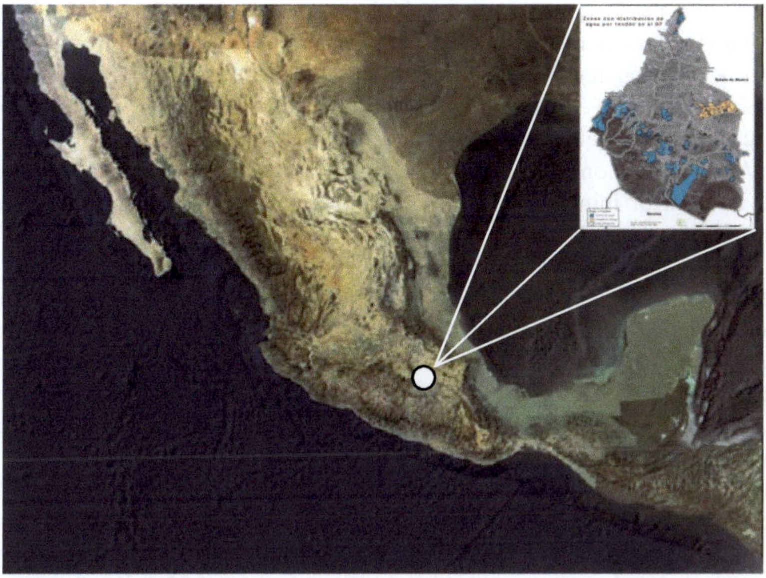

Fig. 1 Location of the zone of study, Mexico City, Mexico

permeability, that have intense fracturing and vesicular texture. They a form the main body of the mountain range of the Chichinautzin and Sta. Catarina, in the south and southwest portion of the area.

Semi-permeable unit. This unit includes rocks of predominantly andesitic composition, with secondary permeability due to fractures, that conform with the main bodies of the mountains of Cruces and Nevada, located to the west, respectively, and oriented in the river basin under study, as well as the andesitic rocks that constitute the mountain range of Guadalupe, located to the north. It also includes the extensive volcanic fans that form flanks of the mountain (Tarango formation), made up of pyroclastic flows of intermediate composition to acid layers of pumice, fluvial deposits and lahars.

Impermeable unit. Included in this unit are lacustrine materials that are located in the center of the river basin, constituted by predominantly argillaceous materials moderately consolidated, considered as an aquitard, because have considerable contribution to the aquifer volume in long periods of time.

1.2 Background and Distribution of the Water in Mexico City

The water for the zone of study is obtained from three main sources: aquifer mantles, Lerma-Cutzamala River and Magdalena River. Aquifers are the principal water supply source in the Metropolitan Zone of Mexico City (MZMC); the soil of this zone is volcanic and clay type. Rain plays an important role in the recharge of the aquifer mantles.

At present, the volume of water which we extract from the aquifer is greater than what is naturally recovered of the rain; the amount extracted from the subsoil is 45 m^3/s and that recovered 25 m^3/s. Consequently, the soil is compacted and causes subsidence of 0.10 m/year, although in certain places, like Xochimilco, Tláhuac, Ecatepec, Netzahualcóyotl and Chalco, the soil has been compacted up to 0.40 m/year in only one year; for that reason, the water that is extracted contains a greater amount of minerals, that makes it to be of lower quality. Statistical registries show annual subsidence ranging from 0.15 to 0.25 m/year in the region of the International Airport of Mexico City.

Mexico City is supplied by main principal and secondary networks. The main network of pipe, 690 km long, is formed by tubes of 0.5 and 1.73 m diameter. The secondary networks contains more than 10,000 km of pipes, with an interior diameter of 0.5 m, and have 243 tanks of storage with a capacity of 1,500,000.00 m^3, with 227 pumping station that increases the pressure on the network.

The water is led into Mexico City by means of 514 km of aqueducts and lines of conduction, which arrive at the takings of the users, by means of 910 km long primary network and 11,900 km long distribution networks. In this way, the Mexico City's inhabitants are provided with an average of 35 m^3/s of water. There also exist 27 water treatment plants. In the MZMC, there are basically three uses for water: 67 % are destined to the domestic sector, 17 % are used in the industry and 16 % are used in schools, hospitals and offices.

Water consumption minimum in Mexico City, as divided by social classes, is as follows: in some illegal establishments it is around 28 l/p/day (liters per person per day), the consumption estimation average in the sectors of medium income is between 275 and 410 l/p/day; and in the sectors of maximum income is between 800 and 1,000 l/p/day.

2 Measurement and GIS Analysis

The modeling data will provide a description of the system of the groundwater of the MZMC which are interconnected directly or with data modeled in the own interface of the Arc-GIS (Fürst and Nachtnebel 1993), providing a platform for the measured integration of the model and data. A model of the aquifer was made using GIS in two dimensions.

The models of two dimensions are generally applicable for greater-scale studies, whereas the specific studies of the site require a description of three dimensions of the inferior surface (Faunt 2009; Strassberg et al.2007).

The techniques of visualization available within ArcGIS help in the characterization and the visualization of the inferior surface and provide tools to define structures and location of different wells, as well as its characteristics (Maidment 2002).

The model allows us to make a better visualization of the zones in which, throughout the time, they have undergone drawdown of the aquifer and, in a few cases, a recharge, helping us to make better decisions in the future.

The present study was carried out in the period August–December 2009; it concerned the measurement and configuration of the static levels of 225 wells of the piezometric network of Mexico City, the position of the levels of the water with respect to the surface of the land.

The depths of the static levels vary fundamentally, between 15.15 and 176.15 m, controlled by the topographic configuration and the extraction of wells. The drawdown of greatest values is in the system in the west area of Mixcoac, San Ángel and Tacubaya. The depth diminishes in the northwest to the south, and the northeast to the east. The greatest depth, of 176.15 m, corresponds to the well denominated "Monte Sur N. 2", which is located in the town of Santa Cecilia located in the south system, the smallest depth, of 15.15 m, correspond to the well Per-9 "Periferico 9", that is located in the south system near Xochimilco.

In general, the wells perforated in the portion west of the delegations Gustavo A. Madero, Miguel Hidalgo, Alvaro Obregón and the south of Xochimilco, show static levels greater than 70.00 m, while in the rest of the basin, the levels vary between 30.00 and 70.00 m.

3 Results

The configuration of the elevation of the static level is such that the hydraulic load varies from 2,035 to 2,240 m above sea level. It is observed that the highest levels, amounting to 2,240 m a.s.l., are in the west zone and descend to the downtown of Mexico City. Also, in the south zone, in Santiago Tepalcatlapan, 2,225 m a.s.l. descend to the zone of Coyoacan, until reaching values of 2,175 m a.s.l. around Mount Estrella in Iztapalapa.

The groundwater flow settles down from the highest levels towards the lowest, that is, of hills of the west and south towards the center of the valley. In the north portion of the aquifer, Tlalnepantla, Azcapotzalco, Miguel Hidalgo and Cuauhtémoc, the groundwater flow is generated in the high parts, Mountains of Cruces, has a main direction from west to east and feeds the zone bordering the refinery of Azcapotzalco.

Watershed groundwater with hydraulic load of 2,180 m a.s.l. is located between the delegations Gustavo A. Madero and Venustiano Carranza, which could correspond to the limit between aquifers of Mexico City to the west and Texcoco to the east. From this zone, the contribution is drawdown to the west and to the east, in the lake of Texcoco.

In the average portion of the aquifer of the City, in the zone of Alvaro Obregón, Contreras, Benito Juárez and Coyoacán, the configuration of the elevation of the static level indicates that the groundwater flow is generated in the Mountains of Cruces, has a main direction from west to east and contributes to the low parts of

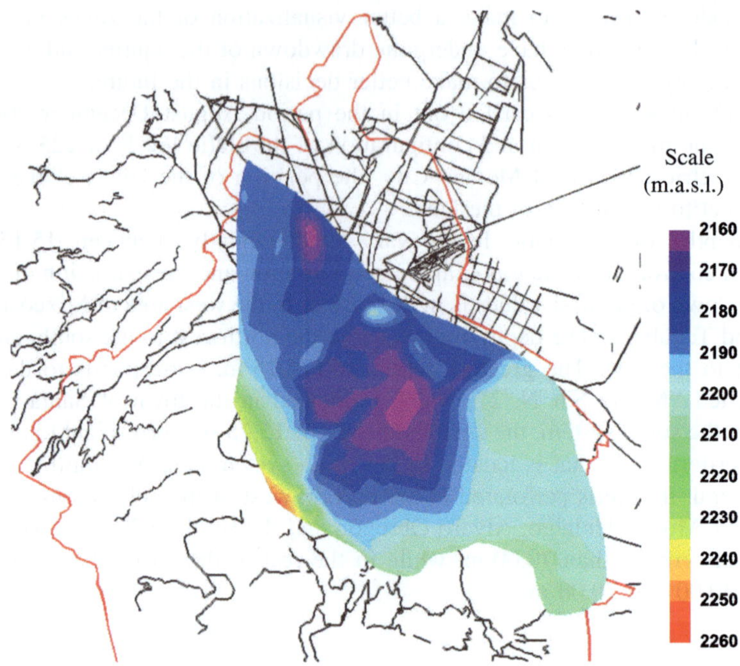

Fig. 2 Static level measured in 2009 in m a.s.l., aquifer of Mexico City, Mexico

the basin, where it delimits a cone of abasement that intercepts groundwater flow of the eastern part. In Fig. 2 we present the static level in 2009, in Fig. 3 the dynamic level measured in 2009 in m a.s.l.

The 1985–2009 evolution of static level, based on the piezometric data of the well network obtained in 1985, compared with the measured ones in the present year, is shown in Fig. 4, which contains the evolution experienced by water-bearing object of this study in period 1985–2009 (Pan et al. 1999).

In the analyzed period, the observed values of drawdown and recharge ranged from 36.93 to 33.87 m, respectively, both appearing in the south system, in the well 165 denominated "Nuevo Universidad" (greater drawdown) and Per-9 (greater recharge).

The principal recharge to the aquifer corresponds to the infiltration of rain that is generated by the topographic elevations, mainly in the south portion, due to the high permeability of rocks; which is next in importance is the Mountain of Cruces in the west, with an average permeability. The groundwater circulates to the mountains, to the center of the valley.

The total of values reported in 1985 (50 measurements) and compared to the values measured in 2009 (225 measurements) indicated that single three stations had recharged, which were: Santa Cruz Xochitepec, Periferico 9 and Periferico 19,

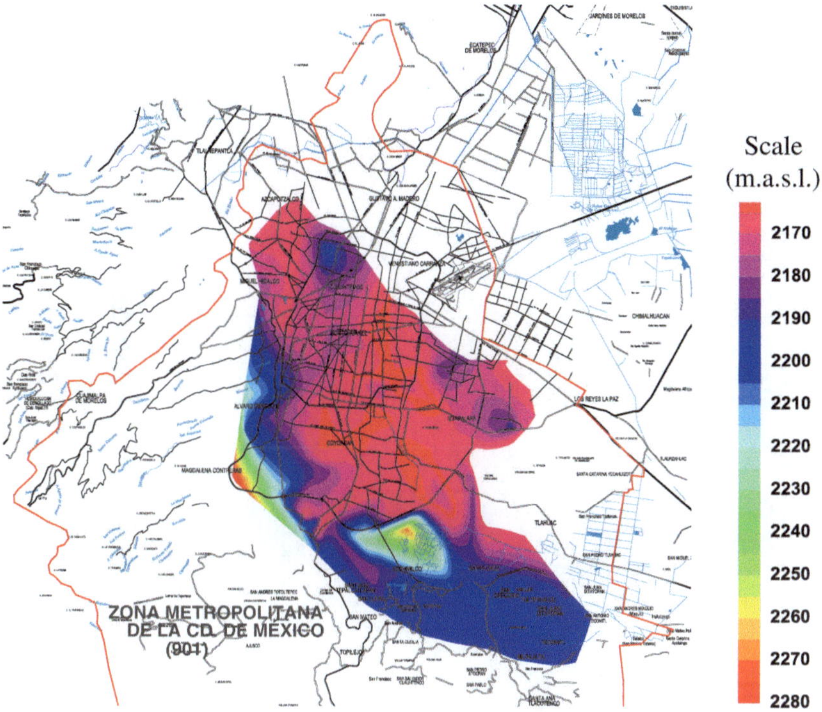

Fig. 3 Dynamic level measured in 2009 in m a.s.l., aquifer of Mexico City, Mexico

probably due to industries and growth of the population. In Fig. 4, configuration in GIS is shown.

Figure 5 shows the configuration of the evolution undergone by the static level in the period 2005–2009. In this period, there appeared a drawdown in the aquifer level of 44.10 m, located in the well Santa Catarina N. 12 of the east system and a maximum recharge in this period of 11.15 m in the well No. 69, Jardines del Pedregal, located in the center system.

Among all the systems that integrate the aquifer of Mexico City, the system to greatest recharge is the north, with recharges between 2 and 4 m. According to the information that appears in Fig. 4, it is observed that the recharge of the aquifer is in the southwestern zone and decreases until becoming a strong drawdown in the zones of Iztapalapa and Venustiano Carranza.

It is necessary to comment that a diminution of the discouragement is the drawdown in this period as compared to the period 1985–2005, and also a smaller recovery of the aquifer, although by a short period; if continues, the recovery of the aquifer of this form will have excellent results in a relatively short time.

Finally, although there were several wells that did not have measurements of dynamic depth, the results were obtained showing lines of equal depth, that gives a

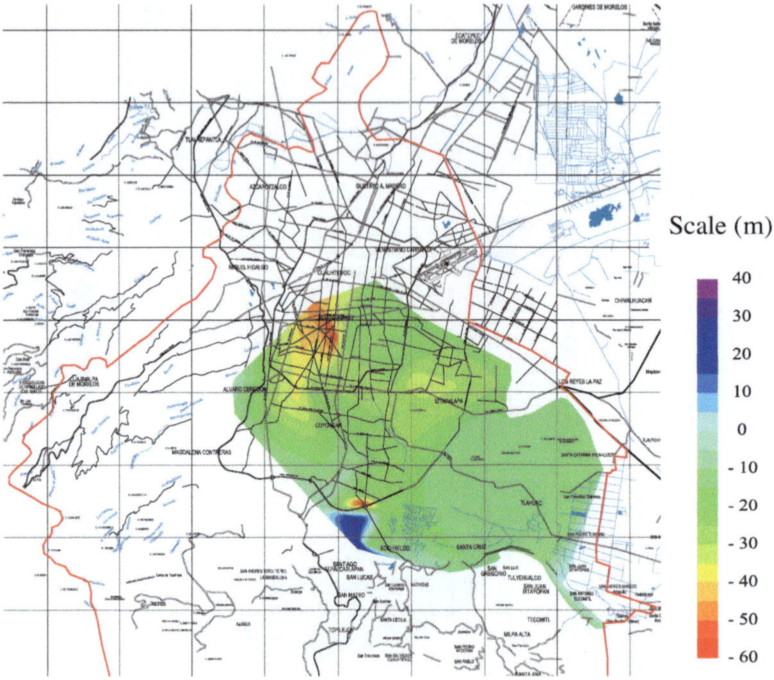

Fig. 4 Evolution of the static level 1985–2009, Mexico City, Mexico

clear idea of the behavior of the underground water in a dynamic phase. It is important to point out that the interpretation in the dynamic depth is based on the characteristics of its pumping equipment and its regime of operation, since they have a significant influence on the total volume of extraction.

The term of frequent use in wells is the specific discharge that expresses the amount of cubic meters per second that a well is able to provide by each meter of drawdown of its water level (drawdown is the same at the dynamic level except the static level). The specific discharge is expressed in $m^3/s/m$, the values oscillate between 0.01 and 0.06 $m^3/s/m$.

4 Conclusions

The depth of the static levels is controlled by the topographic configuration and the elevation of the land; in the aquifer of Mexico City it varies between 15.15 and 176.15 m. The measurement of the dynamic levels begins in this project; the depths of such a level vary between 15.74 and 181.70 m.

As concerns the average portion of the aquifer, the configuration of the elevation of the dynamic level indicates average elevations of the order 2,165 m a.s.l., fed mainly by the west zone and the south of Mexico City. In the analyzed period

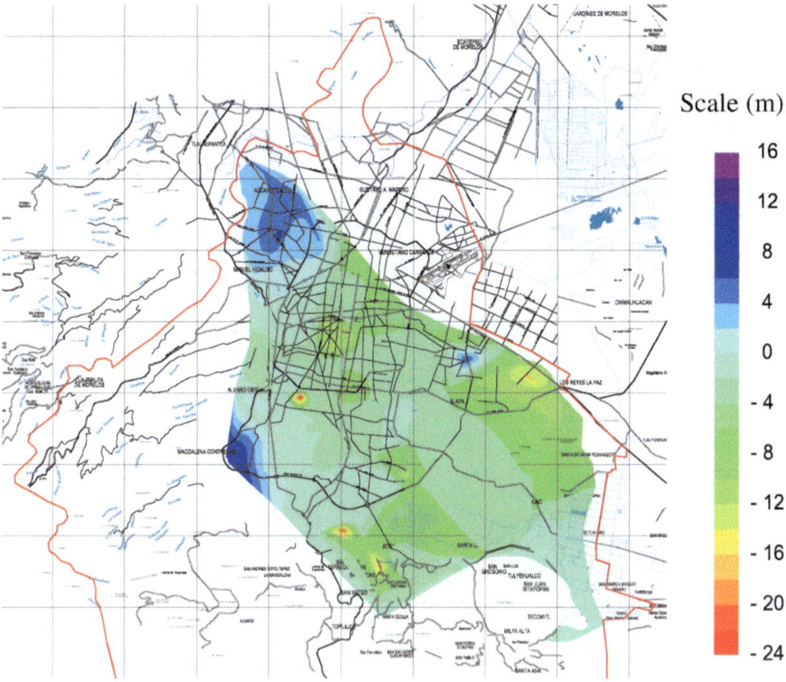

Fig. 5 Evolution of static level 2005–2009, Mexico City, Mexico

1985–2009, the observed values of drawdown and recharge are 36.93 and 33.87 m, respectively, both appearing in the south system.

It is necessary to mention that a diminution of the discouragement in the period 2005–2009 as compared to the period 1985–2005 is drawdown; also, there was a smaller recovery of the aquifer, although by a short period of time. If continued, the recovery of the aquifer will have excellent results in a relatively short time. The use of GIS is of the great help in this project.

Acknowledgments The authors express gratefulness to the Water System of Mexico City by the proportionate data and the sponsorship of the work. The study was made in the Faculty of Engineering of the National Autonomous University of Mexico.

References

Faunt CC (ed) (2009) Groundwater availability of the central valley aquifer. US Geol Surv Prof Calif 1766:225

Fürst GY, Nachtnebel HP (1993) Application of GIS in decision support systems for groundwater management. In: Proceedings of the HydroGIS **'93**, April 1993, Viena, Austria vol 211 pp 13–21

Maidment D (2002) Arc Hydro–GIS for water resources. ESRI Press, Redlands

Pan D, Domon G, De Bois S, Bouchard A (1999) Temporal (1958–1993) and spatial patterns of land use changes in Haut-Saint-Laurent (Quebec, Canada) and their relation to landscape physical attributes. Landsc Ecol 14:35–52

Strassberg G, Maidment D, Jones N (2007) A geographic data model for representing ground water systems. Ground Water J 4:515–518

Using Gene Expression Programming to Determine the Impact of Minerals on Erosion Resistance of Selected Cohesive Egyptian Soils

Ahmed M. A. Sattar

Abstract Cohesive sediment soils are encountered throughout Egypt at many locations, posing various physical and chemical characteristics in beds of lakes, estuaries and flash flood flows. The entire delta region is made up of clayey soil formed from various consecutive Nile floods before construction of the High Dam. Thus, it is very important to determine the erosional stability of such cohesive soils as a function of sediment chemical properties and mineral content. In the current research, 48 samples are collected from various locations throughout Egypt. All samples are subject to physical tests for grain size distribution, and X-ray diffraction analysis for mineral contents. Laboratory experiments are carried out on these samples for finding the difference in terms of erosion characteristics caused by different sediment composition among all samples. Assuming other properties of cohesive soils constant, the gene expression programming (GEP) algorithms are applied to relate the clay mineral content to experimental critical shear stress. Results show an excellent potentiality for the GEP for being applied on finding relations between complex parameters with nonlinear relationships with respect to soil erosion.

1 Introduction

One of the most important factors controlling the interaction of flow and the evolution/change of bed morphology as a result of such flow is the soil erodibility, i.e., the potential for soil surface to being transported by the moving water. This

A. M. A. Sattar (✉)
Department of Irrigation & Hydraulics, Faculty of Engineering,
Cairo University, Orman Street, Giza, Egypt
e-mail: ahmoudy77@yahoo.com

P. Rowiński (ed.), *Experimental and Computational Solutions of Hydraulic Problems*,
GeoPlanet: Earth and Planetary Sciences, DOI: 10.1007/978-3-642-30209-1_27,

erodibility is linked to the water velocity and, therefore to the shear stresses that erode soil. The soil erodibility is directly related to the soil grain size (especially for non-cohesive soil), i.e., larger soil particles need higher velocity for removal and transport. Non-cohesive soils represent sandy type soils, and Hydraulic Engineering Circular, HEC-18 (FHWA 2001) contains several established erodibility models for non-cohesive soils. Richardson and Davis (1995) presented various established scour predictors for sandy soils. However, sandy soils are only available on the shores of Egypt, while the entire delta region is made up of clayey soil formed from various consecutive Nile floods before construction of the High Dam (a pivot of Egypt industrialization, controlling flood waters and generating hydropower, www.en.wikipedia.org/wiki/Aswan_Dam) posing various physical and chemical characteristics in beds of lakes, estuaries and flash flood flow areas. The clayey soil found mostly in the delta region is considered to be from the cohesive type soil. Cohesive type soils follow an opposite regime with respect to erodibility than non-cohesive soils, where cohesive soils erosion resistance increase with the decrease in particle diameter and thus increase in plasticity. Many attempts have been made to correlate critical shear stress with various common parameters of cohesive soils (water content, unit weight, plasticity index, percent of silt and clay particles by weight passing sieve # 200, undrained shear strength) leading to the development of a database of 91 Erosion Function Apparatus (www.humboldtmfg.com/pdf2/hm4000ds.pdf) EFA tests, which was used to perform regression analyses and obtain correlation equations; however, all attempts failed to reach a model that describes the various contributing parameters with reasonable value for the coefficient of determination, *R-squared* (Cao et al. 2002). Other phenomena impact the erosion properties of cohesive soils, such as those developing when clay dries, and diagenetic bonds due to aging, such as those developing when clay turns into rock under pressure over geologic time. Due to the complexity of interaction between these parameters, it is very difficult to predict critical shear stress empirically on the basis of few properties. However, investigating the impact of such parameters individually on the erosion resistance of cohesive soils remained a topic that is frequently visited. Many researchers studied the impact of individual parameters on the erosion resistance of cohesive soils (mainly clay) and reported some empirical formulas, e.g., Dunn (1959); Enger et al. (1968), Hydrotechnical, Lyle and Smerdon (1965); Smerdon and Beasly (1959); Arulanandan (1975); Kelly and Gularte (1981) and Cao et al. (2002). Many of the related parameters were studied thoroughly except for the mineral content of clay, which did not receive such attention and thus in various reports the impact of many soil parameters on critical shear stress is documented, with no data on the impact of soil minerals as seen in Table 1 (Independent Levee Investigation Team 2006).

Recently, evolutionary algorithms have been used as a superior alternative for regression analysis and artificial neural networks, for finding relations between various parameters and producing a higher R-squared value and less mean error in prediction using a newly developed equation. Applications of evolutionary algorithms, especially gene expression programming (GEP) in water and environmental engineering, are not as numerous as the other soft computing tools of

Table 1 Impact of various soil parameters on critical shear stress of cohesive soils (clay)

When parameter increases	Erosion change
Unit weight	Decreases
Plasticity index	Decreases
Undrained shear strength	Increases
Void ratio	Increases
Swell	Increases
Percent passing sieve #200	Decreases
Clay minerals	——

artificial neural networks. They are restricted to finding functions in relatively fewer sub-areas including scour prediction downstream of hydraulic structures (Guven and Gunai 2008), stage discharge relationship prediction (Guven et al. 2009), predicting sediment transport in sewer pipe systems (Ghani and Azamathulla 2011), and prediction of bridge pier scour (Azmathulla et al. 2009).

This chapter aims at experimentally investigating the impact of various mineral constituents on the critical shear stress for soil erosion for selected Egyptian cohesive soils using experimental methods and novel evolutionary algorithms known as GEP, to relate various mineral constituents to the critical shear stress. First, an overview of the basic theory of gene expression programming is given, accompanied by implementation technique steps. Afterwards, a description for the experimental setup used to collect erosion data is given with the results. Key parameters considered are the mineral constituents of the soil samples with most of the other parameters either constant or have little variation.

2 Gene Expression Programming

Gene expression programming was invented by Ferreira in 1999, and is the natural development of genetic algorithms, GAs and genetic programming GP. GEP uses the same kind of diagram representation of GP, but the entities produced by GEP (expression trees) are the expression of a genome. Therefore, with GEP, the second evolutionary threshold—the Phenotype Threshold—was crossed, providing new and efficient solutions to evolutionary computation. So, the great insight of GEP consisted in the invention of chromosomes capable of representing any expression tree. GEP is a full-fledged genotype/phenotype systems, with the genotype completely separated from the phenotype, thus it surpasses the GP system by a factor of 100–60,000 (Ferreira 2001).

The fundamental difference between GP and GEP resides in the nature of the individuals forming the expression trees (Fig. 1). In GP, the individuals are nonlinear entities of different sizes and shapes called parse trees (or expression tree) that represent a program/function, while in GEP, the individuals are also nonlinear entities of different sizes and shapes, Expression Trees ET, but these complex entities are encoded as simple strings of fixed length (chromosomes).

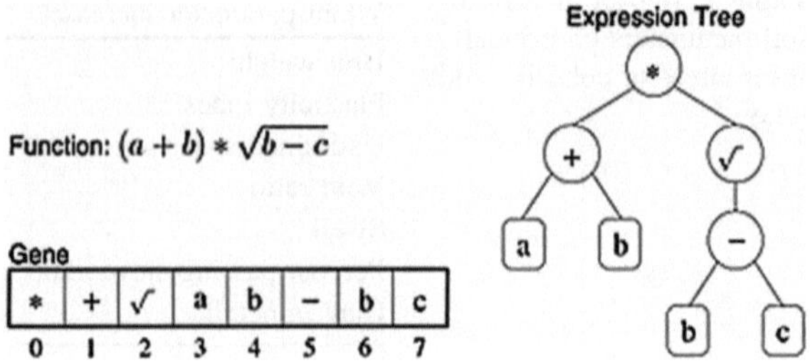

Fig. 1 Expression tree representation for a function

Initially, the chromosomes of each individual in the population are generated randomly. Furthermore, the structure of chromosomes was designed to allow the creation of multiple genes, each encoding a sub-expression tree. The genes are structurally organized in a head and a tail, and it is this structural and functional organization of GEP genes that always guarantees the production of valid programs, no matter how much or how profoundly the chromosomes are modified. The individuals in the new generations are subject to development processes such as expression of the genomes, confrontation of the selection environment, and reproduction with modifications. These processes are repeated for a predefined number of generations or until a solution is achieved as shown in the GEP flow chart in Fig. 2 (Ferreira 2001).

GEP evolves computer programs to solve problems by executing the following steps (Ferreira 2001);

*Step*1: One (or more) initial population of individuals is randomly generated with functions and terminals related to the problem domain.

*Step*2: The implementation of GEP iteratively performs the following steps until the termination criterion has been satisfied

(1) The fitness function for every individual is estimated to enable program to find solution by itself.
(2) Chromosomes are created by choosing a set of terminals and appropriate functions.
(3) The chromosome architecture is chosen including the length of the head and the number of genes.
(4) The linking function is set for Expression Trees, ETs.
(5) The next generation is produced using the genetic operations (reproduction, crossover and mutation).
(6) The termination criterion is checked. If it is not satisfied, the next iteration is performed; if satisfied, go to step 3.

*Step*3: The result may be a solution to the problem domain.

The GEP fitting for the experimental data is done using the commercial software GenXProTools, unconstrained and non-linear data mining software (www.gepsoft.com).

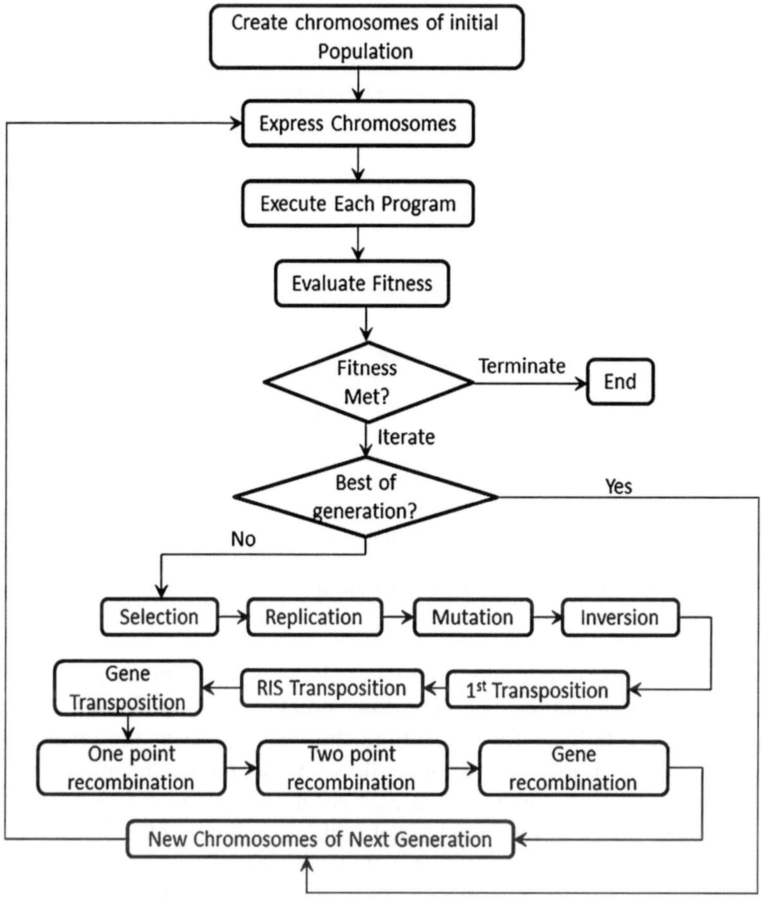

Fig. 2 Flow chart of gene expression programming (Ferreira 2001)

3 Experimental Setup

The experiments are performed in a straight flume (Fig. 3), which is located at the Irrigation and Hydraulics Department at Cairo University. The flume is composed of steel skeleton with steel sheeting for the bed and plexiglass on both sides; it is fed through a small centrifugal pump with a capacity of 150 L/s, through 8 inch UPVC pipes. Water is circulated through the flume and then to the below tanks beside the flume and mounted on the surface of the lab. The flume is 12 m long, 1.2 m deep and 0.30 m wide with adjustable bed slope that is set to 1/200 and an average water depth of 0.40 m. The average manning roughness coefficient of bottom and wall was calculated from velocity to water surface profile measurements and is found to be approximately 0.01. The discharge is measured by an orifice plate and a manometer installed on pipe on pump discharge side. Flow

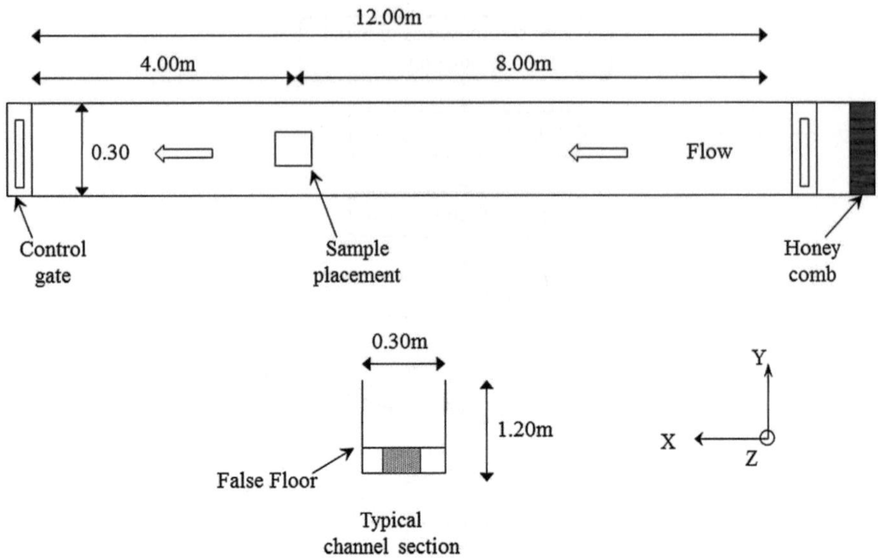

Fig. 3 Experimental setup for bed erosion experiments

velocity and depth are controlled in the flume by gate valve on the pump discharge side and a downstream sluice gate. The downstream end of the flume is left opened with an adjustable tailgate for controlling the water level in the channel downstream. To ensure properly developed flow at the breach location, perforated screens and 60 mm long honeycomb are places at the flume inlet. A point gauge is used to measure water surface in channel with an accuracy of 0.1 mm on vernier scale. A 10 MHz 3D nortek acoustic doppler velocimeter (ADV) is used to measure 3D velocity and turbulent components of flow fields.

For the purpose of conducting erosion tests for cohesive soils, procedures and test preparation are considered as in Salaheldin et al. (2004), who presented methodology for prediction of erosion resistance and pattern for cohesive soils. Their methodology operates on both undisturbed and remolded samples; they suggested sample size and procedure for sample placement and onset of erosion of sediment sample. Their procedure extends the erosion resistance determination by scaling it up to prototype scale, yielding very similar results thereby indicating that the methodology is not scale dependent. A false bottom is constructed over the bed of the existing flume with a height of 8 cm above flume bottom and with width less than the flume width by 1 cm to allow for placement in lab flume. The false bottom is intended to enable placing a 20 $\times$ 20 cm steel box flushed with the false floor for placement of cohesive soil samples in. The sample box has sides completely flushed with the false floor and placed at 8 m from the inlet of the flume to have fully developed flow. The soil was initially trimmed flush with the bottom of the flume at the location of the sample. Flow is gradually increased for steps and each flow is maintained for a sufficient period of time to cause erosion and, if erosion does not occur, the flow is further increased for the same amount of time until the erosion is observed and water becomes muddy; this time is taken to be 2 h (Salaheldin et al. 2004). The initiation of erosion is determined visually and in conjunction with the formation of a small cloud of mud in water, the erosion always starts from the upstream side of the sample.

Unlike the velocity measurements that needed only 2,000 sample to be accurately determined, the turbulence parameters of flow need at least 5,000–10,000 sample. Critical bed shear stress at the onset of erosion of cohesive sediment sample is determined in this study using the covariance method (COV). The COV is considered to give unbiased estimates of τ among the different methods available to estimate τ from velocity measurements. Instantaneous velocities measured by the ADV are used to obtain the instantaneous random fluctuation terms such that $u' = U - \bar{U}$, $v' = V - \bar{V}$, $w' = W - \bar{W}$. The local mean covariance, $\overline{u'w'}$, $\overline{u'v'}$, $\overline{v'w'}$, is called Reynolds stress and used for fully turbulent flow to calculate the near bed shear stress through $\tau = -\rho\left(\overline{u'w'}\right)$.

Undisturbed field soil samples that are used are collected from several locations covering all various types of cohesive soils in Egypt from north to south. Samples were collected from Aswan, Kafr ElSheikh, Shark El Tafreea, and Fayoum. These samples are manually collected in big chunks, the surface layer is removed since it is disturbed and contains debris, and then each chunk is cut into pieces of 40 × 40 cm and soil is shoveled to a depth of 20 cm and scooped from the bottom. Samples are protected with plastic and in lab are cut with sharp knives to fit in the flume groove. From each location, various samples are collected and tested. Tests include grain size distribution, compaction tests, water content; and to determine the mineral composition of soil samples, X-Ray Diffraction analysis was performed.

4 Results and Discussions

To ensure repeatability of experiments and determine error in velocity measurements, the experiment was run for several times and velocity components at a selected location have been recorded; it was found that experiments are repeatable and error in velocity measurements was less than 1 %. On the other hand, bed shear stress as calculated by the COV method has been recorded for various soil samples and ranged between 0.9 and 2.6 Pascal. Table 2 shows the critical shear stress range obtained from this experiment versus that obtained by others for cohesive soils. X-Ray Diffraction (XRD) analysis was performed on all samples to determine the mineral composition of selected cohesive soils in Egypt. The cohesive soils had colors from grey to reddish brown according to the location and had considerable variations in mineral contents, as shown in Fig. 4.

Thus, for each sample, we have the mineral composition—in percentages of various minerals as obtained from the XRD—and the corresponding critical shear

Table 2 Critical shear stress as obtained by various researchers for clay

Study	$\tau_{critical}$(Pa)
Dunn (1959)	2–25
Enger et al. (1968)	15–100
Lyle and Smerdon (1965)	0.35–2.25
Smerdon and Beasley (1959)	0.75–5
Arulanandan et al. (1975)	0.1–4
Arulanandan (1975)	0.2–2.7
Kelly and Gularte (1981)	0.02–0.4
Current study	0.9–2.6

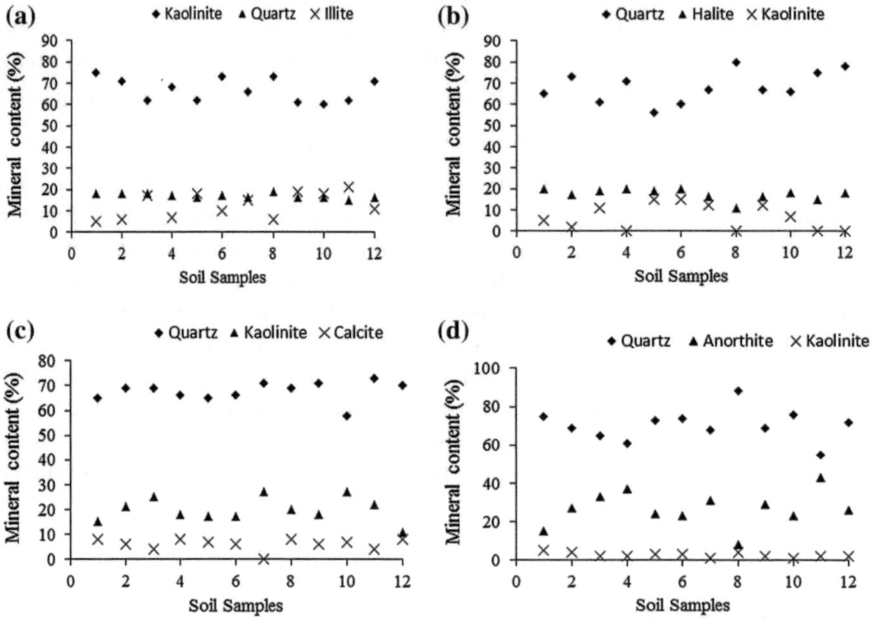

Fig. 4 Mineral composition of selected soil samples at **a** Aswan, **b** Shark El tafreea, **c** Fayoum, **d** Kafr ElSheikh

stress at the onset of initiation of erosion. As mentioned before, this study is concerned about the mineral content of the clay and thus minerals are the only parameters assumed to affect the initiation of erosion. Other parameters, such as grain size, water content, and density were measured and variation is not observed to be significant amongst all samples. The critical shear stress for clay samples can be written as follow;

$$\tau_{critical} = f\{Kaolinite, Quartz, Halite, Illite, Calcite, Chlorite, Montmorillonnite\} \tag{1}$$

To apply the GEP, we need to define the learning environment using a fitness function, as suggested by Ferreira (2001). Previous experience with the GEP suggests the choice of root relative squared error (RRSE) as a fitness function; it helps an efficient evolution for the model and allows it to travel fitness landscape until it finds an optimal solution for the given problem. This function computes the total squared error and normalizes it by the same dimensions as the quantity being predicted; the root relative square error E_i of an individual program i is defined by the following equation;

$$E_i = \sqrt{\frac{\sum_{j=1}^{n}\left(P_{(ij)} - T_j\right)^2}{\sum_{j=1}^{n}\left(T_j - \bar{T}\right)^2}} \tag{2}$$

Where $P_{(ij)}$ is the value predicted by the program i for fitness case j; T_j is the target value for fitness case j; and $\bar{T} = 1/n \sum_{j=1}^{n} T_j$. For a perfect fit $E_i = 0$, and thus the index of RRSE ranges from 0 to infinity, with zero corresponding to the ideal. Thus, the fitness of an individual model f_i can be calculated from the following equation which ranges from 0 to 1,000, with 1,000 corresponding to perfect fit;

$$f_i = 1000.\frac{1}{1 + E_i} \tag{3}$$

The second step is to choose the chromosome architecture; this includes determining the number of chromosomes, head size and number of genes. Each gene is composed of a head and a tail; the head contains symbols that represent a chosen function and the tail contains terminals. The length of the head h is chosen such that it controls the size of the formed expression trees; the tail t is a function of the head size and of the number of functions chosen n, such as $t = h\ (n - 1) + 1$. Several trials with various head sizes need to be attempted to reach for the best fitness and an optimum head. From practice with various problems, a head size of three to five will always give the best models with respect to size and predictability.

The GEP chromosomes are composed of one or more genes of equal length. Each gene codes for a sub-expression tree and each of the sub expression trees interact with each other, forming more complex expression trees. Choosing one gene leads to a complex expression tree, while multigenic chromosomes are always preferred in complex problems with nonlinear relations, where each gene codes for a smaller and simpler building block. The gene number is determined by trial runs and in general two to six genes will be more than sufficient in problems with various complexities.

The third step is choosing the set of functions that will create the chromosomes. These functions are the essence of evolution of the GEP; they allow modifications without restrictions leading to compact correct programs for a specific function. The choice of an appropriate function set is not the same for every problem and depends mainly on the program performance with some chosen arguments. If the evolution is not satisfactory, one can use a wider set of functions until optimum fitness is achieved. However, a professional approach would be to initially use the

Table 3 Optimal parameter settings for the GEP algorithms

Parameters	Settings-GEPI	Settings-GEPII
Number of generations	100,000	100,000
Number of chromosomes	30	30
Number of genes	3, 4	3, 4
Head size	3, 5, 8	3, 5, 8
Linking function	Addition	Addition
Fitness function error type	RRSE	RRSE
Mutation rate	0.05, 0.005	0.05, 0.005
Inversion rate	0.1	0.1
One point recombination rate	0.3	0.3
Two Point recombination rate	0.3	0.3
Gene recombination rate	0.1	0.1
Gene transposition rate	0.1	0.1
Function set	+, −, ×, /, $\sqrt{}$, e, In	+, −, ×, /, $\sqrt{}$, e, In, cos, sin, arctan, 1/

basic mathematical operators (+, −, ×, /) to allow for production of simple models. A second run of GEP is performed using a different set of functions as shown in Table 3.

The fourth step is to set the linking function, which is the interaction between all sub-expression trees of the model, these linking functions can be addition, subtraction, division, and multiplication. Obviously, it works only in case of multigenic models. The choice of linking functions depends on the complexity of the problem and the experience of the model user and for simpler models for a certain problem, addition or subtraction would be appropriate.

The fifth step is to set the values controlling various genetic operations controlling the evolutionary process of GEP. The most efficient operator in GEP is the mutation, which causes populations of individuals to adapt very efficiently, allowing for the evolution of good solutions to all problems. Ferraire (2001) recommends using a mutation rate equivalent to two one-point mutations per chromosome. Mutations cause the expression trees to drastically change in size with no constraints in kind of mutation and number of mutations in a chromosome. Other genetic operations, such as inversion, transposition, and other operators shown in the above flow chart, are of less importance and all depend on the value assigned for mutation rate. Thus, default values are always assigned for all genetic operations, as shown in Table 3, and the value of the mutation rate changes from 0.05 to 0.005 to test the evolution of the program.

To test the performance of the developed model, the mean square error, MSE, mean absolute error, MAE, and relative squared error, RSE, were used as indicators, as calculated from the following equations, respectively

$$MSE_i = \frac{1}{n}\sum_{j=1}^{n}\left(P_{(ij)} - T_j\right)^2 \quad (4)$$

Table 4 Performance of produced functions for critical shear stress

Performance indicators	GEPI	GEPII
Fitness	780	756
MSE	0.0139	0.0158
MAE	0.0946	0.0946
RSE	0.09131	0.1036
R-Square	0.942	0.89

$$MAE_i = \frac{1}{n}\sum_{j=1}^{n}\left|\frac{P_{(ij)} - T_j}{T_j}\right| \tag{5}$$

$$RSE_i = \frac{\sum_{j=1}^{n}\left(P_{(ij)} - T_j\right)^2}{\sum_{j=1}^{n}\left(T_j - \bar{T}\right)^2} \tag{6}$$

Results from both models; GEPI and GEPII, are shown in Table 4. According to Table 4, GEPI model with chosen set of functions produced the best results with a fitness value of 780 and an R-Square value of 0.94. On the other hand, GEPII model with sin, cos and arctan, did not produce results with same accuracy. Both models evolved with four genes and thus each function produced has four expression trees. All expression sub trees are linked by addition and the gene mutation was run using random constants for each gene. The best individual of all generations for GEPI has the explicit formulation as follows

$$\begin{aligned}\tau_{critical} = {} & e^{\left\{Cl^3/(Kt\times Q - Cl)\right\}^9} - 0.85819\sqrt{Il} + 2M + 0.795014 + Kt^2 \times Q^2 - Kt \\ & - 137.684Q \times M(Il - Ca) - Ca \\ & + H^3\left[(Kt - Q)^2 + 64.444 + 16.0557(Kt - Q)\right]\end{aligned} \tag{7}$$

where all minerals are in percentages and $\tau_{critical}$is in Pascal; Q = quartz, Kt = kaolinite, Cl = chlorite, Il = ilite, M = montmorillonite, H = halite, and Ca = calcite.

On the other hand, simpler equations can be obtained using the GEP when only one gene is specified as the main chromosome architecture with rest of parameters as in GEPI; however, the fitness of the model decreased to 745 with MAE of 0.105 and R-squared of 0.89; the following figure shows the architecture of the single genetic expression tree;

According to the ET in Fig. 5, the equation is further simplified to

$$\begin{aligned}\tau_{critical} = {} & M + Q - Kt + 0.44451Ca - 0.884949 - Ca \times Q \\ & + [M - Il + 5.947266Q]^3\end{aligned} \tag{8}$$

where Q = quartz, Kt = kaolinite, Cl = chlorite, Il = ilite, M = montmorillonite, H = halite, and Ca = calcite.

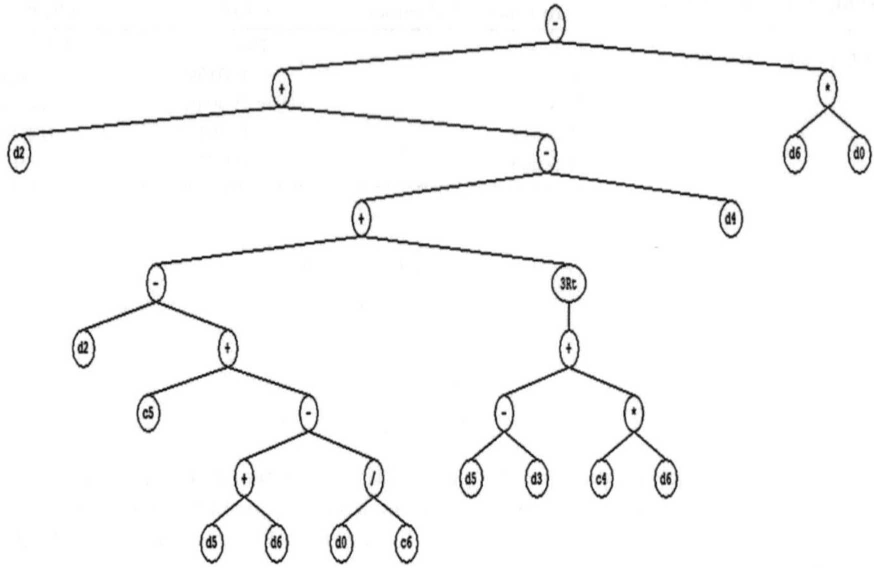

Fig. 5 Expression tree produced by the GEPI, 1 gene used

As interpreted from the simplified equation, the critical shear stress for clay depends on all minerals in the clay samples and thus its increase cannot be attributed alone to the increase or decrease in a certain mineral. It is found that the clay samples of Shark El Tafreea had the highest critical shear stress amongst all other samples with an average value of 2 Pa. The main constituent of this sample is quartz, followed by halite and kaolinite. On the other hand, the lowest shear stress with an average value of 0.9 Pa was in the samples collected from Aswan having kaolinite as major constituent followed by quartz and illite. Other soil parameters are assumed to be constant such as water content and density.

5 Conclusions

The results show that gene expression programming (GEP) is capable of mapping data into a high dimensional feature space with variety of methods to find relations and trends in data. An equation linking the critical shear stress in clays with the mineral composition has been presented and MAE reached less than 0.09. Despite the fact that none of the other parameters, such as water content, percent passing sieve 200, density, swelling, etc., has been included, the relation can be used to provide estimates for critical shear stress versus soil type in Egypt, according to soil mineral composition and location of samples.

Acknowledgments This work has been carried out under the financial support of Egyptian Science and Technology Development Fund (STDF), Egyptian State Ministry for Scientific Research, Project ID39.

References

Arulanandan K (1975) Fundamental aspects of erosion in cohesive soils. J Hydr Div 101(5):635–639 (ASCE)

Azamathulla H, Ghani AA, Zakaria N, Guven A (2009) Genetic programming to predict bridge pier scour. J Hydraul Eng 136(3) (ASCE)

Cao Y, Wang J, Briaud JL, Chen HC, Li Y, Nurtjahyo P (2002) EFA tests and the influence of various factors on the erodibility of cohesive soils. In: Proceedings of the first international conference on scour of foundations, Texas A and M University, Department of Civil Engineering, College Station

Dunn IS (1959) Tractive resistance to cohesive channels. J Soil Mech Found Div 85(3):1–24 (ASCE)

Enger PF, Smerdon ET, Masch FD (1968) Erosion of cohesive soils. J Hydr Div 94(4):1017–1049 (ASCE)

Federal Highway Association FHWA (2001) Evaluation scour at bridges. Tech Circular 4th edn Pub. # NHI 01-001, US Department of Transportation

Ferreira C (2001) Gene expression programming: a new adaptive algorithm for solving problems. Complex Syst 13(2):87–129

Ghani AA, Azamathulla H (2011) Gene expression programming for sediment transport in sewer pipe systems. J Pipeline Syst Eng Pract 2(3):102–106

Guven A, Gunai M (2008) Genetic programming approach for prediction of local scour downstream of hydraulic structures. J Irrig Drainage Eng 134(2):241–249

Guven A, Aytek A, Yuce MI, Aksoy H (2009) Genetic programming based empirical model for daily reference evapotranspiration estimation. Clean: Soil Air Water 36(10–11):905–912

Independent Levee Investigation Team (2006) Investigation of the performance of the new orleans flood protection systems in hurricane katrina on August 29, 2005. Vol I. Main Tests and Executive Summary

Kelly EK, Gularte RC (1981) Erosion resistance of cohesive soils. J Hydr Div 107(10): 1211–1224 (ASCE)

Lyle WM, Smerdon ET (1965) Relation of compaction and other soil properties to erosion and resistance of soils. Trans Am Soc Agric Eng 8(3). St Joseph Michigan

Richardson EV, Davis SR (1995) Evaluating scour at bridges. Rep. No. FHWAIP—90-017 (HEC 18), Federal Highway Administration, Washington

Salheldein T, Imran J, Chaudhry H (2004) Numerical modeling of three dimensional flow field around circular piers. J Hydraul Eng 130(2) (ASCE)

Smerdon ET, Beasley RP (1959) Tractive force theory applied to stability of open channels in cohesive soils. Res Bull No 715, Agricultural Experiment Station, University of Missouri, Columbia

Correlation Measures for Solute Transport Model Identification and Evaluation

Fred Sonnenwald, Virginia Stovin and Ian Guymer

Abstract Correlation measures are used in a range of applications to quantify the similarity between time-series, often between model output and observed data. A software tool implemented by the authors uses optimisation to identify a system's Residence Time Distribution (RTD) from noisy solute transport laboratory data. As part of the further development of the tool, an investigation has been undertaken to determine the most suitable correlation measures, both for solute transport model identification as an optimisation constraint and as an objective means of solute transport model evaluation. Correlation measures potentially suitable for use with solute transport data were selected for evaluation. The evaluation was carried out by manipulating synthetic dye traces in ways that reflect common solute transport model discrepancies. The conditions tested include change in number of sample points (discretisation/series length), transformation (scaling, etc.), transformation magnitude, and noise. BLC, χ^2, FFCBS, R^2, RMSD, R_t^2, ISE, and APE show favourable characteristics for use in model identification. Of these, R^2, R_t^2 and APE are non-dimensional and so are also suitable for model evaluation.

F. Sonnenwald (✉) · V. Stovin
Department of Civil and Structural Engineering, The University of Sheffield,
Mappin St., Sheffield S1 3JD, UK
e-mail: f.sonnenwald@sheffield.ac.uk

V. Stovin
e-mail: v.stovin@sheffield.ac.uk

I. Guymer
School of Engineering, University of Warwick, Coventry CV4 7AL, UK
e-mail: i.guymer@warwick.ac.uk

P. Rowiński (ed.), *Experimental and Computational Solutions of Hydraulic Problems*, GeoPlanet: Earth and Planetary Sciences, DOI: 10.1007/978-3-642-30209-1_28,

1 Introduction

Correlation values express the similarity between either two separate time-series or two segments of the same time-series. They are frequently used to evaluate the goodness of fit for models. Within solute transport and environmental engineering, typically the R^2 (Nash and Sutcliffe 1970), R_t^2 (Young et al. 1980), and RMSD (Anderson and Woessner 1992) correlation measures are used. The conventional understanding is that a larger/smaller correlation value is better, but the reasons for the use of a specific correlation measure are typically not discussed. For example, with reference to solute transport model fitting, Fischer (1967) only specified that "the best possible agreement" be obtained with no further guidance on how agreement is defined or assessed.

Stovin et al. (2010) described a new software tool, based on deconvolution, which identifies a Residence Time Distribution (RTD) from experimentally-obtained upstream and downstream solute traces. This software, which was originally developed for pharmacological applications, utilised the χ^2 (Greenwood and Nikulin 1996) parameter as a goodness-of-fit constraint in the model identification process. Simultaneously, the software tool generated large numbers of candidate models which were evaluated with R_t^2. More information on both χ^2 and R_t^2 was necessary to determine if they were the correct correlation measures to be using, or if other correlation measures might prove to be better. The aim of this research, therefore, was to undertake an objective comparison of candidate correlation measures to determine which would be most suitable for use in the deconvolution software for model identification and for the subsequent evaluation of the identified solute transport models.

1.1 Correlation Measures

Overall, there are many different types of correlation/similarity measures/coefficients. They range from simple formulaic measures, which include R^2 and R_t^2, to algorithms, e.g. Dynamic Time Warping (Berndt and Clifford 1994), which process the data in discrete steps. The most common correlation measures are based on Euclidean geometry. Euclidean based measures derive their correlation value from the Cartesian coordinate differences of the two time-series. There are also statistical and energy based measures. Statistical measures are derived from assumed probability density functions or other statistical functions. Energy based measures treat time-series data as analogous to energy states and apply energy transformations or functions developed in an energy context.

While this paper focuses on solute transport data, which is one-dimensional, correlation measures have applications in many fields, from economic forecasting (Miskiewicz 2010) to voice analysis (Vlachos et al. 2002). Two-dimensional correlation measures (Chen et al. 2005) are used when each time step has two associated values, e.g. $t(1) = a$ and $t(1) = b$. Similarly, while solute transport data are linear, there are also non-linear correlation measures (Wang et al. 2005), in which the exact ordering of the data points is irrelevant. Correlation measures which operate on different time intervals or segments of the same time-series are called cross-correlation or auto-correlation measures (Movahed et al 2006). Auto-correlation measures highlight trends or repeating patterns in a time-series and as such are often used with frequency or wave form data. They can also appear in complex number forms (Boudraa et al. 2008).

Ignoring correlation measures that fall into the two-dimensional, non-linear, algorithmic, and cross/auto-correlation categories, twelve correlation measures have been identified as potentially suitable for use with solute transport data. These are summarised in Table 1. The measures contain a mix of non-dimensional and dimensional correlation measures. Non-dimensional correlation measures produce correlation values that have a defined meaning within a fixed range. R_t^2 is a typical examples of this, where the values between 0 and 1 are defined. 1 indicates a perfect correlation, and 0 no correlation. Values below 0 are undefined, but can be considered to indicate no correlation.

Moriasi et al. (2007) conducted an extensive literature review compiling correlation values reported from various model fitting studies. The study provides a guideline for the use of the correlation measures found to be commonly used for watershed simulations. Through analysis of the measures used and values obtained they recommended minimum satisfactory correlation values. However, these specific values are only of direct relevance in the context of watershed simulation performance evaluation. Unfortunately, the study does not provide much additional understanding of the relative sensitivity of the correlation measures.

Table 1 List of correlation measures

Name	Symbol	Type	Reference
Burnham-Liard criterion	BLC	S	George et al. (1998)
χ^2 test	χ^2	S	Greenwood and Nikulin (1996) Hattersley et al. (2008)
Furthest fitting cost based similarity	FFCBS	E	Ye et al. (2004) *equal weighting applied*
Nash-sutcliffe model efficiency coefficient	R^2	E, D	Nash and Sutcliffe (1970)
Pearson's product moment correlation coefficient	PMCC	S, D	Rodgers and Nicewander (1988)
Root mean square deviation	RMSD	S	Anderson and Woessner (1992)
Coefficient of determination	R_t^2	E, D	Young et al. (1980)
SimilB	SimilB	J, D	Boudraa et al. (2008)
Young information criterion	YIC	S	Cox and Boucher (1989)
2-D correlation coefficient	CORR2	E, D	The MathWorks Inc. (2011)
Integral of squared error	ISE	E	Ghosh (2007)
Average percent error	APE	S, D	Kashefipour and Falconer (2000)

(E)uclidean, non-(d)imensional, (s)tatistical, (J)energy

For the purposes of using a correlation measure in solute transport model identification and evaluation, certain characteristics are desirable. In model identification, several models using different model parameters are compared to the same experimental data. In this scenario, all models have the same number of sample points, so any sensitivity to this parameter can be ignored. However, when evaluating models, data is often of different lengths, e.g. different experiments, so a correlation measure should not be sensitive to number of sample points. In both cases, the correlation measure should be sensitive to transformation and transformation magnitude, e.g. the correlation value obtained should indicate that the overall shape of the model is a poor fit when this is the case. The correlation measures however should not be sensitive to the noise present in natural systems. This paper aims to systematically evaluate the 12 correlation measures in Table 1 against four trace characteristics: number of sample points, transformation, transformation intensity, and noise.

2 Methodology

Three synthetic dye traces have been created, each representing a different solute transport scenario. The synthetic traces were discretised and then modified to reflect different conditions by introducing additional information through transformations and noise. The correlation measures were then used to compare the unmodified traces with their modified counterparts.

2.1 Generating the Synthetic Traces

The three synthetic dye traces are shown in Fig. 1. The first trace (Trace 1) consists of a normal distribution covering ± 3 standard deviations, or 99% of the data as generated by the `pdf` function (The MathWorks Inc. 2011). This represents a typical response to an instantaneous input in pipe flow. Trace 2 represents a fully mixed response to a step input. It is calculated from an exponential increase and decrease. Trace 3 represents a non-ideal mixing response to an instantaneous input in the form of a 2-cell Aggregated Dead-Zone (ADZ) model (Rutherford 1994). It is interesting to note that Trace 1 is symmetric while Traces 2 and 3 have more common asymmetric profiles.

2.2 Generating the Test Data

The test data was generated from the synthetic traces by varying the four parameters. The first step was to discretise the synthetic traces according to the number of sample points parameter. This resulted in a total of thirty traces, three

Fig. 1 The synthetic dye traces

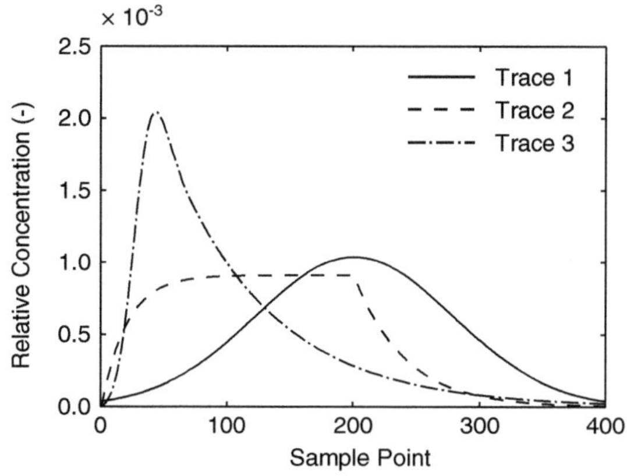

each at 20, 40, 80, 100, 200, 400, 800, 1200, 1600, and 2000 uniform sample points. The range of number of sample points represents common series lengths. Each trace at each number of sample points was normalised to have unit area. This ensures changes due to variation in number of sample points is independent of total solute volume.

From the discretised traces, related traces were generated using five transformations in combination with three transformation magnitudes, 0.05, 0.10, and 0.20. Each transformation was applied in two directions, up/down or left/right as appropriate at magnitude × value (y-axis) or magnitude × sample points (x-axis) as appropriate, e.g. a shift or stretch of $0.20 \times 400 = 80$ sample points. Any additional sample points that were generated as a result of transformation were ignored, and in order to retain a constant time-series length, zeros were inserted when the number of sample points was reduced. Figure 2a–e to illustrate the five transformations, outlined below:

- **Scaling**, Fig. 2a, a linear change in value that may occur when the model has errors in scale or due to poor calibration of measuring equipment;
- **Shifting**, Fig. 2b, a temporal change that may occur when a modelled trace has errors in its advection term;
- **Truncation**, Fig. 2c, a loss of data that could be representative of not monitoring the entire trace;
- **Stretching**, Fig. 2d, an elongation of trace profile that could be representative of an error in sampling frequency or an overestimation of dispersion coefficient;
- **Squeezing**, Fig. 2e, the opposite of stretching.

Noise, e.g. Fig. 2f, was applied to the modified traces to represent the variation common in recorded data due to both instrument noise and natural systems. Noise was generated through the superposition of a random number [5, 15] of sine waves

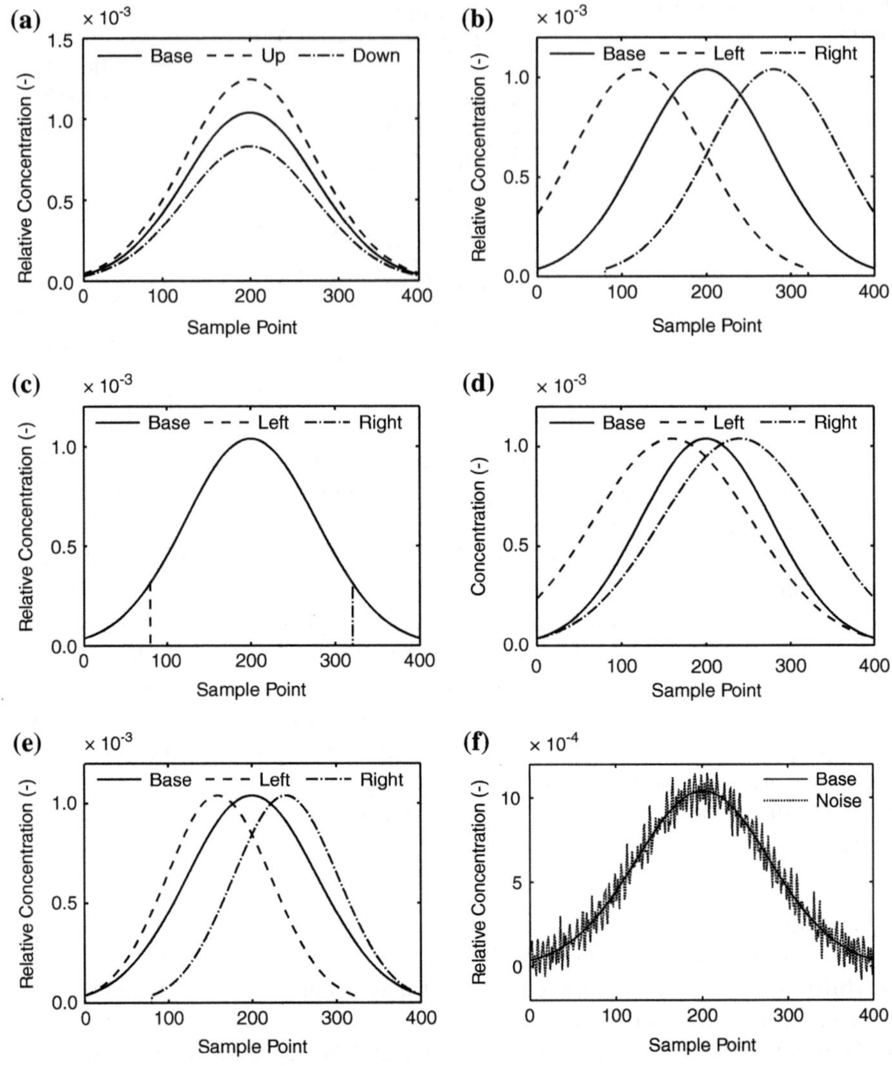

Fig. 2 Examples of generated data, with transformation applied at a magnitude of 0.20 and each displayed using a temporal discretisation of 400 sample points. **a** Scaling. b Shifting. **c** Truncation. **d** Stretching. **e** Squeezing. **f** Noise

with random amplitude [-1, 1], random phase [0, π], and random frequency [0, length/2]. This was found to have a better zero mean when sub-sampling to lower numbers of sample points, and therefore better reflect natural noise than purely random computer generated noise. The noise was scaled to 0, 5, 10, and 20 % ratios of noise peak to discretised trace peak.

2.3 Comparing Correlation Values

Not all the correlation measures tested are non-dimensional, so a method for directly comparing all correlation values was required. Expressing the correlation value as a ratio of the perfect correlation value was rejected as for some measures a perfect correlation is indicated by 0, which would result in division by zero. Instead, it was decided to normalise correlation values into a similar range. Equation (1) is used to do this, where r is the correlation value to be normalised, $r_{perfect}$ is the value of a perfect correlation between the trace and itself prior to transformation and/or addition of noise, r_{max} and r_{min} are respectively the numerically largest and smallest correlation values obtained, each either the perfect or worst correlation. Applying (1) results in a correlation value that is in context for a specific trace, number of sample points, and correlation measure, with 1 the best and 0 the worst correlation value obtained. It should be noted that normalised correlation values are specific to this dataset, so have limited meaning outside the present analysis.

$$r_{normalised} = 1 - \left| \frac{r_{perfect} - r}{r_{max} - r_{min}} \right| \tag{1}$$

The easiest way to directly compare correlation measures is graphically, and complete plots showing the trends in correlation coefficient in response to all four parameters (number of sample points, transformation, transformation magnitude and noise) have been generated and considered. Space does not permit these to be presented in full, but the data for R_t^2 will be presented and discussed. Additionally, a more easily interpreted plot has been generated. Sensitivity of a correlation measure to a given parameter can be expressed as the standard deviation of correlation values with respect to that parameter. That is, for example, the standard deviation of the vector produced by (2), where i, j, k and l are indices for the four parameters. Comparing standard deviations for each parameter and measure gives a good indication of how well the different correlation measures perform.

$$r_{normalised\,vector\,i} = \frac{1}{n_j \times n_k \times n_l} \sum_{j=1}^{n_j} \sum_{k=1}^{n_k} \sum_{l=1}^{n_l} r_{normalised\,i,j,k,l} \tag{2}$$

3 Results and Discussion

There were three stages to analysing the obtained data. The first was to examine the obtained results in detail. Second was to examine trends and overall performance of the correlation measures and discard those correlation measures judged unsuitable for solute transport model identification. Finally, the correlation measures suitable for solute transport model evaluation have been examined.

3.1 Detailed Examination

The complete plots are made up of multiple sub-plots, each of which shows the correlation values for one of the synthetic traces and one of the transformations. There are 3 traces and 5 transformations so there are 15 sub-plots in total. Figure 3 shows an example sub-plot before and after normalisation for R_t^2. Figure 4 presents the complete normalised plot.

Figure 3a shows the raw R_t^2 correlation values for the shifting transformation applied to Trace 2, where 1 is a perfect correlation. The lines are nearly horizontal, which indicates that noise has very little influence on the correlation value obtained when Trace 2 is transformed by shifting. The greater vertical spacing between the dashed lines compared with the solid lines shows that right shifting consistently has more impact than left shifting, which is expected for this asymmetric trace. The 2,000 sample point symbols are close to their lines, but the 20 sample point symbols diverge from their lines as noise increases—noise appears to have greater impact on the shifted Trace 2 when there are fewer sample points to characterise the curve. Figure 3b is the normalised equivalent of Figure 3a. The normalised sub-plot in this case effectively expands the y-axis values. The occurrence of normalised values of 0 shows that shifting resulted in the lowest correlation across all transformations applied to Trace 2.

Figure 4 is a complete plot for the normalised R_t^2 correlation measure. It is evident that shifting also has the greatest impact on Trace 1 and that shifting is about equal with squeezing in Trace 3. For Trace 2, truncation has about half the impact of shifting, and squeezing has an influence somewhere between truncation and shifting. Stretching has the second least impact and scaling has the least impact. Sensitivity to noise is universally low. Stratification of the lines is similar

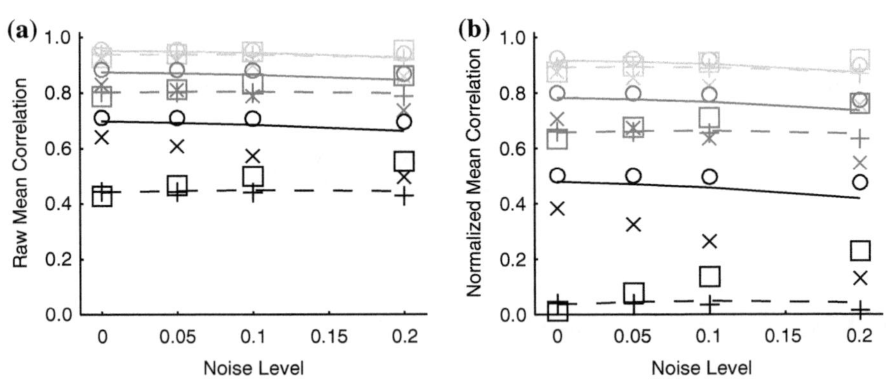

Fig. 3 The shifted trace 2 sub-plot. *Solid lines* (mean), *crosses* (20 sample points), and *circles* (2,000 sample points) show the *up/left* transformation. *Dashed lines* (mean), *squares* (20 sample points) and *pluses* (2,000 sample points) show the *down/right* transformation. Line intensity indicates transformation magnitude. **a** Raw R_t^2correlation values. **b** Normalised R_t^2correlation values

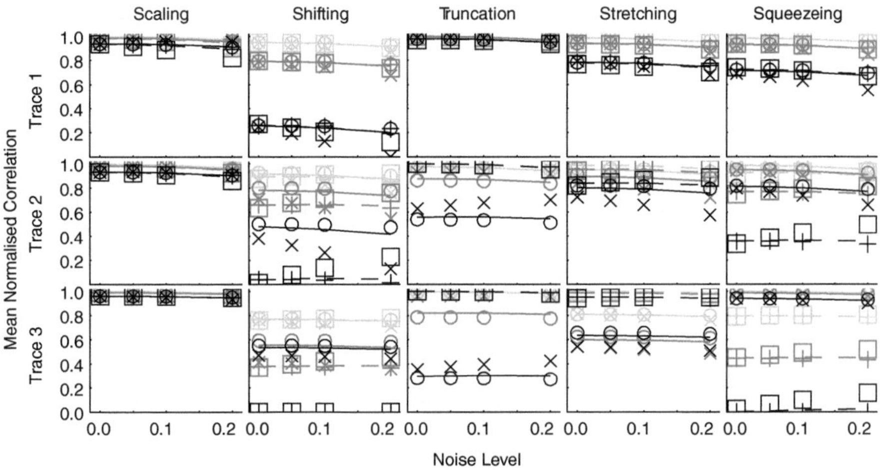

Fig. 4 Mean normalised correlation values for R_t^2. *Horizontally*, different transformations are shown, while the different traces are shown *vertically*. Sub-plots have same key as Fig. 3

across all traces for the scaling transformation, which indicates any traces scaled to the same degree with R_t^2 will have a similar change in correlation value, regardless of trace profile. The stratification of transformation intensity lines otherwise varies significantly. Trace 1 shows the least variation, with left/right transformations being identical and stretching and squeezing being very similar as well. Truncation of Trace 1 shows the least distinction of all the transformation/trace combinations, which implies that R_t^2 will not deal as well with identifying discrepancies associated with long tails. The long tail effect can also be seen in the right truncation of Trace 3, where again there is very little stratification. Conversely, there is a very large impact from left truncation, where the majority of the peak of Trace 3 is cut-off.

3.2 Correlation Measure Comparison

Many of the trends exhibited by R_t^2 apply to other correlation measures as well. The Trace 1 correlation values obtained for the up/left and down/right transformations produce nearly identical results. Increasing transformation magnitude results in decreased correlation. In most cases there is little variation with noise and small variation with number of sample points. This is shown in Fig. 5, generated from (2). A higher value indicates that a correlation measure is more sensitive to that parameter.

The correlation measures fall into three distinct groups. Group 1 comprises BLC, FFCBS, R^2, PMCC, RMSD, R_t^2, CORR2, and APE, which all exhibit high

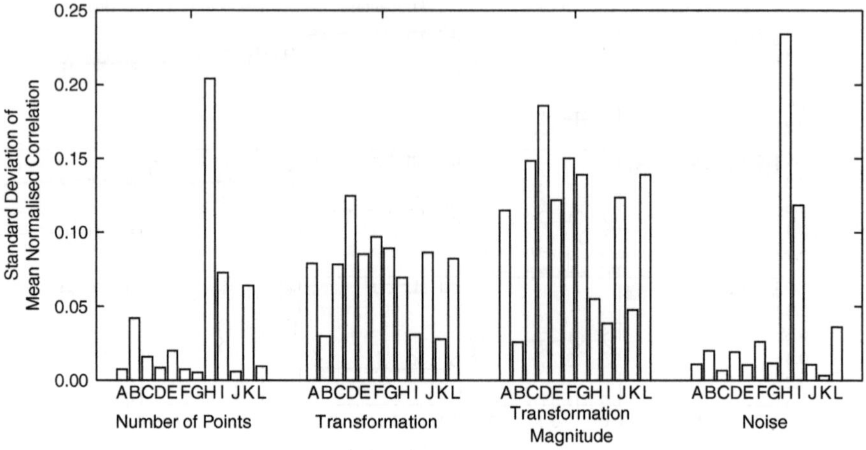

Fig. 5 Standard deviation plot showing sensitivity of the correlation measures. A–BLC, B–χ^2, C–FFCBS, D–R^2, E–PMCC, F–RMSD, G–R_t^2, H–SimilB, I–YIC, J–CORR2, K–ISE, L–APE

sensitivity to transformation/transformation magnitude and low sensitivity to number of sample points/noise. Within this group R^2 has the greatest sensitivity to transformation (change in the underlying shape), whilst APE shows a slightly higher sensitivity to noise. All Group 1 measures appear suitable for model identification, however when examining the complete plots, PMCC and CORR2 show extremely limited sensitivity to scaling. This is attributed to these measures relying on a difference in mean value which will only change slightly when all values in a trace are scaled. Both of the Group 2 measures, χ^2 and ISE, appear potentially suitable despite high sensitivity to number of sample points. The lower sensitivity to transformation/transformation magnitude is still acceptable due to the insensitivity to noise. SimilB and YIC make up Group 3, with a high sensitivity to both number of sample points and noise. This, combined with the lower sensitivity to transformation/transformation magnitude, make all Group 3 measure unsuitable.

Overall there are eight measures which show the characteristics previously mentioned to be desirable for model identification. These six measures of Group 1 (BLC, FFCBS, R^2, RMSD, R_t^2, and APE) and the Group 2 measures (χ^2 and ISE) could all be used for model identification, e.g. within the deconvolution tool. To narrow the choice of measure further requires testing for interaction with other elements of the deconvolution process, which is ongoing research. For solute transport model evaluation, only the non-dimensional measures of Group 1 (R^2, R_t^2 and APE) can be recommended as these are not influenced by number of sample points.

3.3 Application to Model Evaluation

To further investigate the correlation measures for evaluation and provide guidance on choice of correlation measure, R^2, R_t^2, and APE have been applied to the scenarios shown in Fig. 6. Associated (raw) correlation values are presented in Table 2. Note that for APE, a lower value represents a better correlation, with 0 being the best and 100 the worst possible. For R^2 and R_t^2, 1 is the best and 0 the worst possible correlation value.

In Fig. 6a, Model A1 can be visually described as the best fit. All correlation measures correctly identify this. Model C1 is identified as the next best fit, and while R_t^2 shows a somewhat lower correlation, R^2 and APE both show significantly lower correlations. Model B1 is identified as the worst fit, with R^2 indicating that there is no correlation at all when B1 is clearly a temporal shift of A1. This is unique to R^2, and indicates it may actually be unsuitable for model identification. In comparison, any dimensional correlation measure would still indicate some relation.

Figure 6b presents a scenario in which the measured data exhibits a secondary peak representative of recirculation effects or transient storage. This scenario is common and it is essential that correlation measures indicate whether or not the model fits correctly. Here, Model B2 can be visually described as the best as it clearly includes the secondary peak. In this case, the R^2 and R_t^2 correlation values imply an extremely good fit for both models which is typical of these measures. In contrast, APE clearly differentiates between Model A2 and Model B2. Dimensional correlation measures would also show B2 to be the better fit, but like the correlation values for R^2 and R_t^2 would also be difficult to distinguish.

Each of the three correlation measures might be useful in different scenarios. R_t^2 might be most useful when model fits are expected to generally be poor as it will still indicate correlation for a wide range of transformation magnitude. APE exaggerates small differences in trace profile, and so might be useful in some

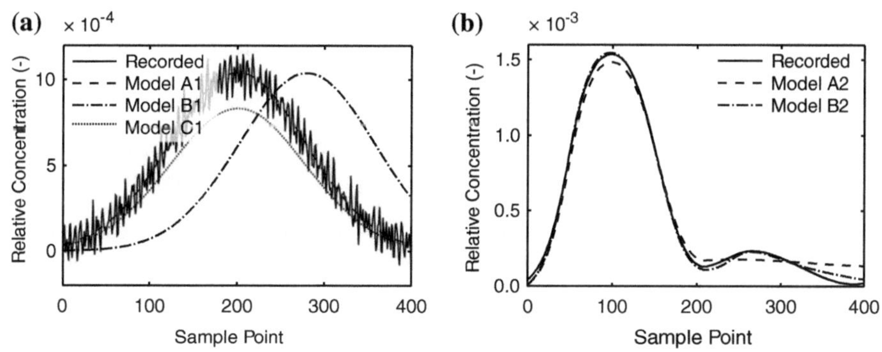

Fig. 6 Models fit to theoretical 'recorded' data for examining correlation measures. **a** Trace 1 type situation. **b** Secondary peak situation

Table 2 Correlation values for the models presented in Fig. 6

Model	A1	B1	C1	A2	B2	% Change
R^2	0.9561	0.0000	0.8386	0.9881	0.9982	1.010
R_t^2	0.9853	0.5457	0.9459	0.9938	0.9990	0.520
APE	12.19	73.32	23.24	9.719	3.745	5.974

% Change shows the change in correlation value between models A2 and B2 as a percentage of the total range

situations, while R^2 exaggerates large differences in trace profile. Dimensional correlation measures may perform similarly to any of these three measures, but lack the finite scale to quantify and compare correlation values. Of the three correlation measures, both R^2 and R_t^2 have been used extensively in civil engineering and therefore the values obtained can be more easily compared to existing research.

4 Conclusion

This paper has presented a systematic comparison of twelve different correlation measures using three realistic solute trace profiles to identify measures that show high sensitivity to profile shape but limited sensitivity to noise. Eight measures (BLC, χ^2, FFCBS, R^2, RMSD, R_t^2, ISE, and APE) have been found to be potentially suitable for model identification. Of these, the three non-dimensional measures, R^2, R_t^2, and APE, were examined in greater detail and found to be suitable for use in model evaluation.

References

Anderson M, Woessner W (1992) Applied groundwater modeling: simulation of flow and advective transport. Academic, London

Berndt DJ, Clifford J (1994) Using dynamic time warping to find patterns in time series. In: Technical report WS-94-03, The AAAI Press, Melno Park, California, pp 359–370

Boudraa AO, Cexus JC, Groussat M, Brunagel P (2008) An energy-based similarity measure for time series. Eurasip J Adv Signal Process

Chen L, Ozsu MT, Oria V (2005) Robust and fast similarity search for moving object trajectories. In: Widom J, Ozcan F, Chirkova R (eds) Proceedings of the ACM SIGMOD international conference on management of data, Baltimore, MD, pp 491–502

Cox CS, Boucher AR (1989) Data based models: an automatic method for model structure determination. In: IEE colloquium model validation control system design simulation, London, UK, pp 2, 1–2, 4

Fischer HB (1967) The mechanics of dispersion in natural streams. J Hydraul Div ASCE 93(6):187–215

George SC, Burnham KJ, Mahtani JL, Inst Elect Engineers L (1998) Modelling and simulation of hydraulic components for vehicle applications–a precursor to control system design. In:

International conference on simulation '98, 457. Institute of Electrical Engineers, London, pp 126–132
Ghosh AK (2007) Introduction to linear and digital control systems. Prentice-Hall, India
Greenwood P, Nikulin M (1996) A guide to chi-squared testing, vol 280. Wiley-Interscience
Hattersley JG, Evans ND, Hutchison C, Cockwell P, Mead G, Bradwell AR, Chappell MJ (2008) Nonparametric prediction of free-lightchain generation in multiple myelomapatients. In: 17th International federation of automatic control world congress (IFAC), Seoul, Korea, pp 8091–8096
Kashefipour S, Falconer R (2000) An improved model for predicting sediment fluxes in estuarine waters. Proceedings of the fourth international hydroinformatics Conference, Iowa, USA,
Miskiewicz J (2010) Entropy correlation distance method. The euro introduction effect on the consumer price index. Phys A 389(8):1677–1687
Moriasi DN, Arnold JG, Van Liew MW, Bingner RL, Harmel RD, Veith TL (2007) Model evaluation guidelines for systematic quantification of accuracy in watershed simulations. Trans ASABE 50(3):885–900
Movahed MS, Jafari GR, Ghasemi F, Rahvar S, Tabar MRR (2006) Multifractal detrended fluctuation analysis of sunspot time series. Theory Exp J Stat Mech
Nash JE, Sutcliffe JV (1970) River flow forecasting through conceptual models part i–a discussion of principles. J Hydrol 10(3):282–290
Rodgers JL, Nicewander WA (1988) Thirteen ways to look at the correlation coefficient. Am Stat 42(1):59–66
Rutherford JC (1994) River mixing. Wiley, Chichester, England
Stovin VR, Guymer I, Chappell MJ, Hattersley JG (2010) The use of deconvolution techniques to identify the fundamental mixing characteristics of urban drainage structures. Water Sci Technol 61(8):2075–2081
The MathWorks Inc (2011) MATLAB R2011a. Natick, MA
Vlachos M, Kollios G, Gunopulos D (2002) Discovering similar multidimensional trajectories. In: Agrawal R (ed) 18th International conference on data engineering. Proceedings, IEEE computer society, pp 673–684
Wang Q, Shen Y, Zhang JQ (2005) A nonlinear correlation measure for multivariable data set. Phys D 200(3–4):287–295
Ye JC, Tang Y, Peng H, Zheng QL, ieee (2004) Ffcbs: a simple similarity measurement for time series. In: Proceedings of the 2004 international conference on intelligent mechatronics and automation, pp 392–396
Young P, Jakeman A, McMurtrie R (1980) An instrumental variable method for model order identification. Automatica 16(3):281–294

Flow Simulations in the Porąbka Lake with the FESWMS Model

Magdalena Trzewik and Paweł S. Hachaj

Abstract The chapter presents an application of a two-dimensional depth averaged numerical model FESWMS to calculate flow in the Porąbka retention reservoir. The lake in question has complex flow characteristics due to the presence of an outlet of a pumped-storage hydro power plant. Results of sample steady-state and dynamic simulations are shown and discussed.

1 Introduction

The main goal of the research presented was a simulation analysis how the work of the Porąbka-Żar pumped-storage hydroelectric power station influences water flow patterns in the Porąbka retention reservoir. Simulations have been conducted using the Finite-Element Surface-Water Modeling System (FESWMS) model included in The Surface-water Modeling Solution comprehensive environment provided by the AQUAVEO Company (AQUAVEO (online) Aquaveo water modeling solutions, http://aquaveo.com). The secondary goal was to check the limitations of the model: will a tool created mainly for rivers be able to work on such a compound hydraulic system as a mountain retention lake which acts also as the lower tank of a pumped-storage power plant.

M. Trzewik (✉) · P. S. Hachaj
Institute of Water Engineering and Water Management, Cracow University of Technology, Kraków, Poland
e-mail: Magdalena.trzewik@iigw.pl

P. S. Hachaj
e-mail: pawel.hachaj@iigw.pl

P. Rowiński (ed.), *Experimental and Computational Solutions of Hydraulic Problems*, GeoPlanet: Earth and Planetary Sciences, DOI: 10.1007/978-3-642-30209-1_29,

2 Characteristics of the Soła Cascade and the Porąbka Lake

Soła is a river in southern Poland, a right-hand tributary of the Vistula River. The total area of the river basin is 1,390.60 km^2 and its length is 88.90 km. soła originates in the Western Beskids mountain range near the Poland-Slovakia border. It is made up of the confluence of several small creeks at the village of Rajcza. It runs downhill north to the towns of Żywiec and Kęty, forming the border between the Silesian and the Żywiec Beskids. After 89 km, just below the town of Oświęcim, Soła flows into Vistula.

One can distinguish two sections of the Soła River. The upper section starts from the sources and ends in the Tresna reservoir's backflow profile. The lower section starts from Czaniec reservoir's dam and ends at the mouth of the river. There is a system of three reservoirs called "Soła Cascade" (IMGW RZGW Kraków 2007) located between the upper and lower sections.

The Soła Cascade consists of three reservoirs: Tresna, Porąbka and Czaniec. One of the main functions of this system is to provide drinking water for Silesia, Bielsko-Biała, Oświęcim and Kęty, and to provide water to factories and fish farms. The reservoirs are also used to produce hydroelectric power and to mitigate exceptional floods. Reservoirs can reduce the maximum flood flows from 1,469 to 650 m^3/s. Hydroelectric power stations located on Tresna and Porąbka generate 28 million and 25 million kWh per year. Moreover, a hydroelectric power facility located east to the Porąbka reservoir, called Porąbka-Żar, works as a pumped-storage power plant. Its task is stabilization of the power system during power grid load changes.

The reservoirs have also become very attractive for tourists, because of their localization near the mountains (Fig. 1).

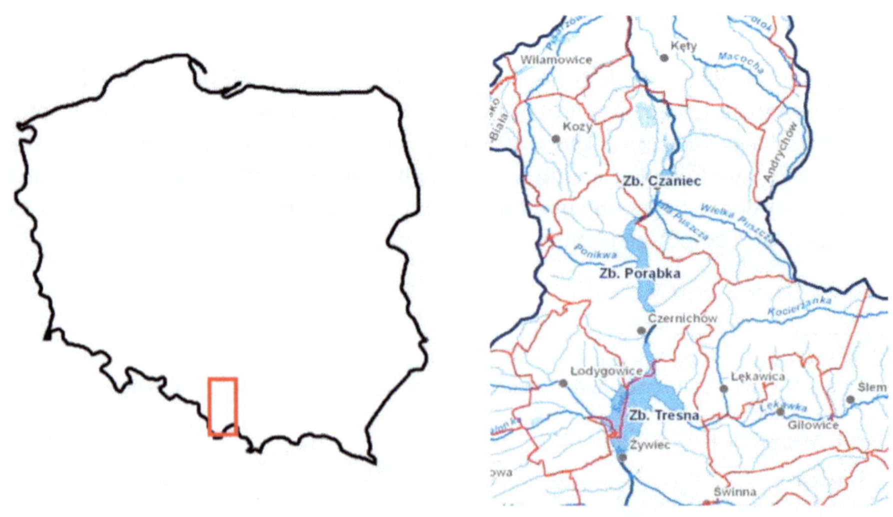

Fig. 1 *Left*: the region on the map of Poland; *right*: the Soła Cascade and its surroundings

The Porąbka reservoir, the second one in the system, is also called the Międzybrodzkie Lake. It's the oldest reservoir in the system (built in 1936). The dam is located on the 32.3 km of the Soła River. Surface of the catchment area is 1,082.0 km^2. The reservoir is long but not very wide.

The Porąbka's reservoir has two hydroelectric power plants. The first one is located on the dam. The second one is located on the eastern shore. This is the second biggest pumped-storage hydroelectricity in Poland. Porąbka's reservoir is its lower tank. The upper tank is located on the top of the Żar hill. The upper and lower tanks are connected with tunnels carved into the rock. Water level difference between the upper and lower tanks is 440 m. There is also a special basin and a dyke built inside the lake to protect the outlet of the power station (Table 1).

At times of low electrical demand, excess grid generation capacity is used to pump water into the higher reservoir. When the demand is high, water is released back into the lower reservoir through turbines, generating electricity. Although the losses during the pumping process makes the plant a consumer of energy overall, the system increases revenue by selling more electricity during periods of peak demand, when electricity prices are higher. The 500 MW turbines systems can work both as a water pump and as an energy generator.

3 The SMS Package and the FESWMS Model

The Surface-water Modeling Solution (SMS) is a comprehensive environment for hydrodynamic modeling (AQUAVEO (online) Aquaveo water modeling solutions, http://aquaveo.com). It allows engineers and scientists to visualize, analyze and understand available data better. The numerical models supported in SMS are able to compute a variety of information applicable to surface water modeling. Primary applications of the models include calculation of water surface elevations and flow velocities for steady-state and dynamic conditions. Particular applications include the modeling of: rivers behavior, contaminate transport, sediment transport, rural and urban flooding, estuarine, coastal circulation, inlet and wave modeling.

Table 1 Porąbka reservoir parameters

Volume [mln m^3]	Total	27.19
	Flood	4.58
	Standard	19.47
	Minimal for hydroelectric power	6.5
	Minimal	3.14
WSL (Water surface level) [m above mean sea level]	Max WSL	321.49
	Normal WSL for summer	320.09
	Min WSL for hydroelectric power	313.69
	Min WSL	311.09

Among the advantages of SMS are flexible modeling approaches. This environment allows building conceptual models by using Geographic Information System (GIS) familiar objects like points, arcs and polygons. The conceptual models are high-level, not relying on meshes. They only need input such as bathymetric data, flow rates, boundary conditions or water surface elevations.

On the other hand, computational models integrated in SMS rely on meshes. Because of that, SMS also has tools that help building them including their automatic generation and optimization as well as provides tools for result analysis and visualization. This makes understanding of the results easier and allows involved people to focus on the merits of the problem. Thanks to SMS's support for many computational models, it gives a possibility of cross-comparison of the simulation results of different models.

The FESWMS model (Froelich 2003) is a program that allows simulations of water and non-cohesive sediment movement in rivers, estuaries or coastal waters. This model is based on Finite Element Method and can be used to provide a solution both steady-state and time-dependent systems. The main limitation of this model is that it uses depth-averaged, two dimensional equations. This means that it should be used only when vertical movements, velocities and accelerations are infinitesimal in comparison to the horizontal ones. FESWMS is especially used for modeling situations where a river crosses a highway, where one-dimensional analysis cannot provide satisfying results. It has been also used to simulate water levels and flows distribution around islands, into and out of channels or even at confluences of large rivers.

FESWMS solves differential equations using Gelarkin Finite Element Method. The first step of the solving procedure is dividing physical region into subregions called elements. Elements have either triangular or quadrilateral shapes. They are defined by nodes placed both on the element edges and in their interior. State variables inside the elements can be interpolated using values in element's nodes. The model uses quadratic functions to interpolate flow rates (basing on all nodes) and linear functions to interpolate water level (basing only on values in the vertex nodes). These mixed interpolations stabilize the numerical solution.

Modeled system is very complicated and contains many variables and parameters. Therefore, some simplifications are introduced to reduce the complexity of the model which results in easier and faster calculations. The most important simplification is the neglect of water flow in the vertical direction. Depth is the parameter that the flow velocity in the xy-plane is dependent on.

FESWMS model equations for the flow of water in the reservoirs have been derived based on the principle of mass and momentum conservation.

Simplified dynamic water flow equation along the x direction has the following form:

$$\frac{\partial q_x}{\partial t} + gH\frac{\partial z_0}{\partial x} + \frac{\partial}{\partial x}\left(\frac{q_x^2}{H} + \frac{gH^2}{2}\right) + \frac{\partial}{\partial y}\left(\frac{q_x q_y}{H}\right) + \frac{1}{\rho}\left(\tau_{\mathrm{bx}} - \frac{\partial H\tau_{\mathrm{xx}}}{\partial x} - \frac{\partial H\tau_{\mathrm{xy}}}{\partial y}\right) = 0. \tag{1}$$

where H is the water depth at a given point; $q_x = V_x H$ is the unitary flow rate in the x direction; $q_y = V_y H$ is the unit flow rate in the y direction; z_0 is the bed elevation; g is the gravitational acceleration; ρ is the density of water; τ_{bx}, τ_{by} are bed shear-stresses in the x and y direction, respectively; τ_{xx}, τ_{xy}, τ_{yx}, τ_{yy} are directional components of lateral shear stress caused by turbulence.

The equation for the y direction is analogous. See e.g. (Froelich 2003; Hachaj 2007).

This model has been designed primarily to simulate flow fields in rivers. In this case it was used for a retention reservoir. It had been used before for Soła's Cascade (Trzewik 2011) (however without taking into account dynamic solutions connected with the pumped-storage power plant influence) as well as for the Dobczyce (e.g. (Hachaj 2006; Hachaj 2007)) and Goczałkowice's reservoirs (http://www.zizozap.pl). This is an innovative application of the model, yet it worked well in each of those cases.

4 Simulations

The simulations presented in this chapter were performed for different hydraulic conditions. First of all the total flow through the lake in the absence of auxiliary Porąbka-Żar outflow has been set (according to (IMGW RZGW Kraków 2007)) to two different values:

- 22 m^3/s divided into 20 m^3/s from Soła and two left bank tributaries, 1 m^3/s each. This setup is close to the yearly average flow in the lake that appears in nature during moderately wet seasons;
- 9.1 m^3/s divided into 8.7 m^3/s from Soła and two left bank tributaries, 0.2 m^3/s each. This setup is close to the most probable flow in the lake.

Then the following scenarios were taken into account:

- The power plant does not work;
- the power plant generates energy injecting additional 14 m^3/s into the lake;
- the power plant pumps water into its upper tank at the rate of -12 m^3/s.

The computational mesh has been built for water surface level of 320 m above the mean sea level. The map digitization, as well as original mesh construction along with all the parameters is described in (Trzewik 2011). For the purpose of this study, the original mesh has been slightly modified near the power plant outlet. It contains over 7,500 elements and over 25,000 computational nodes.

There were two kinds of simulations performed for these conditions:

- Steady state, where the total inflow and outflow are equal and the water level does not change. Such simulations are easier to compute and interpret but they are further from real lake states. Yet they still cover physically possible flows;
- Dynamic, taking into account time-dependence of the flow field.

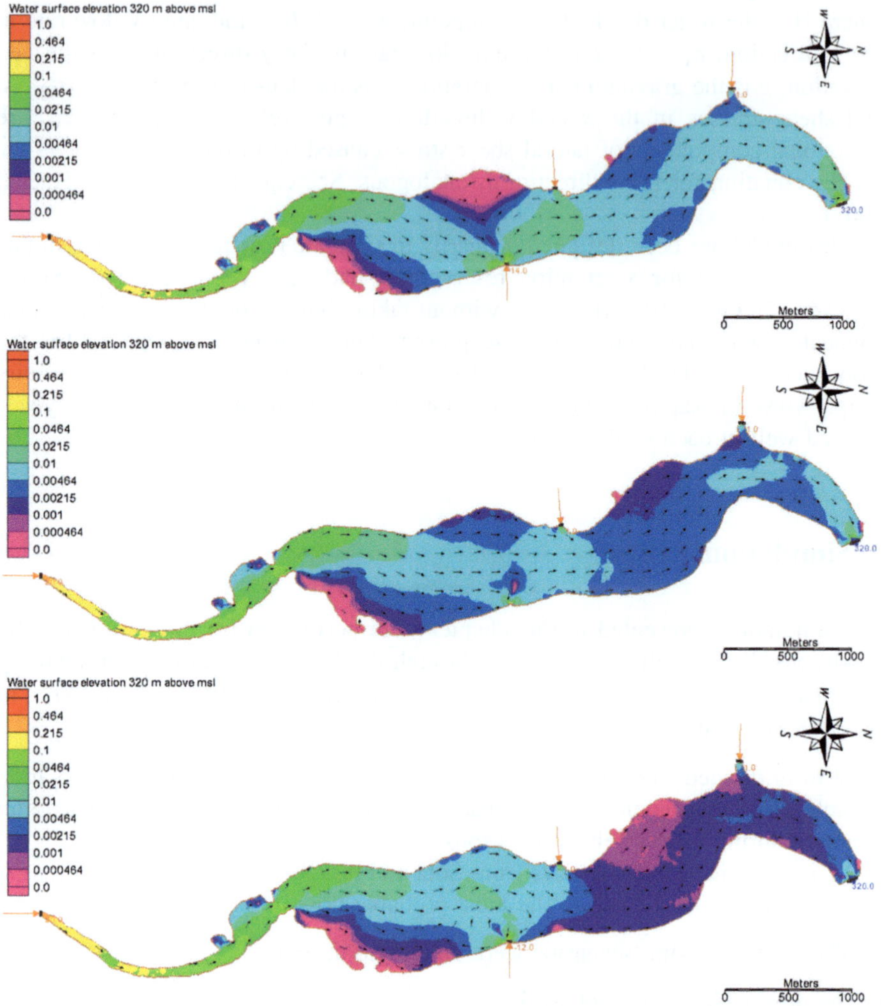

Fig. 2 Planar velocity fields resulting from steady state simulations for 20 m^3/s main inflow: the power plant damps water into the reservoir at 14 m^3/s (*upper* frame); the power plant does not work (*middle* frame); the power plant takes water from the lake at 12 m^3/s (*lower* frame)

4.1 Steady-State Simulations

The first steady-state simulation was carried out for 20 m^3/s rate main inflow and two left bank 1 m^3/s rate inflows. The 20 m^3/s rate main inflow is close to yearly average inflow for the Soła river. The results are shown in Fig. 2.

In Figs. 2, 3, 6, and 7, the leftmost arrow indicates the main inflow (Soła), the two upper arrows denotes the inflows of the tributaries, the lower arrow indicates

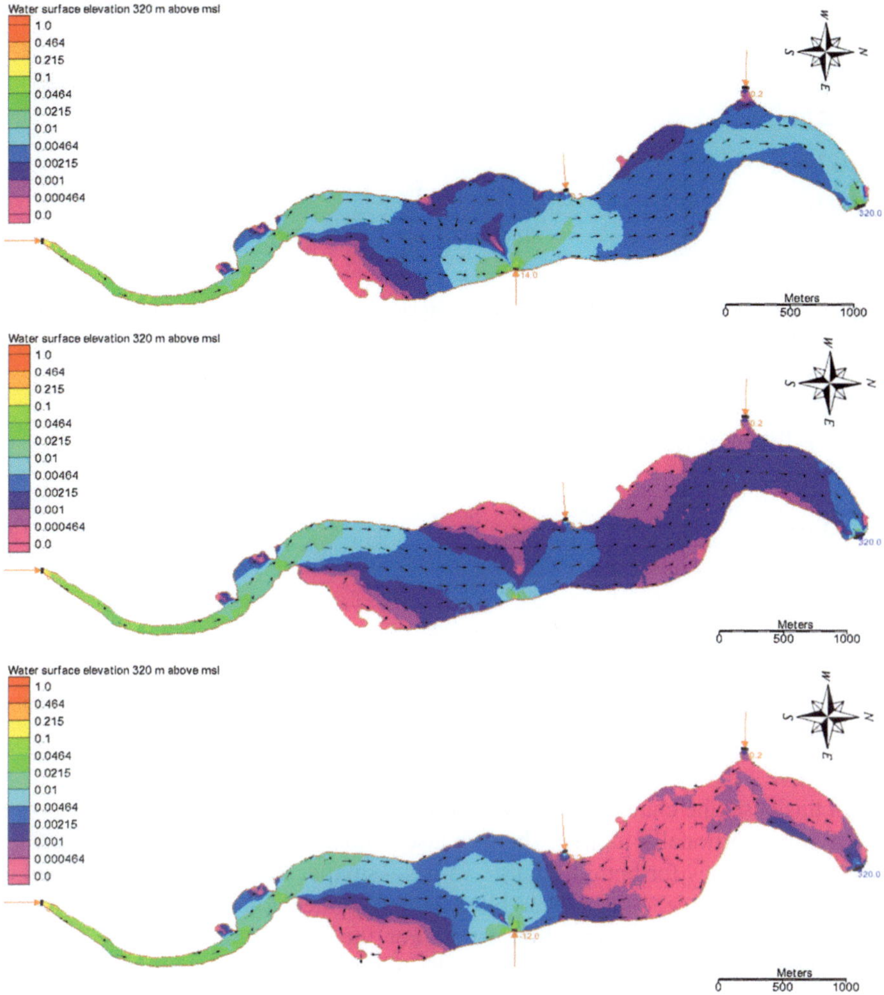

Fig. 3 Planar velocity fields resulting from steady state simulations for 8.7 m^3/s main inflow: the power plant damps water into the reservoir at 14 m^3/s (*upper* frame); the power plant does not work (*middle* frame); the power plant takes water from the lake at 12 m^3/s (*lower* frame)

the power plant inflow or outflow (if present) and the black line on the right, with WSE denoted, is the outflow of the lake.

As it can be seen, while the flow in the upper part of the lake is mostly unaffected by the power plant activity, the flow pattern in the lower part changes dramatically. This suits the steady-state assumption.

The second steady-state simulation was carried out for 8.7 m^3/s rate main inflow and two left bank 0.2 m^3/s rate inflows. This is the most probable flow through this lake. The results are shown in Fig. 3.

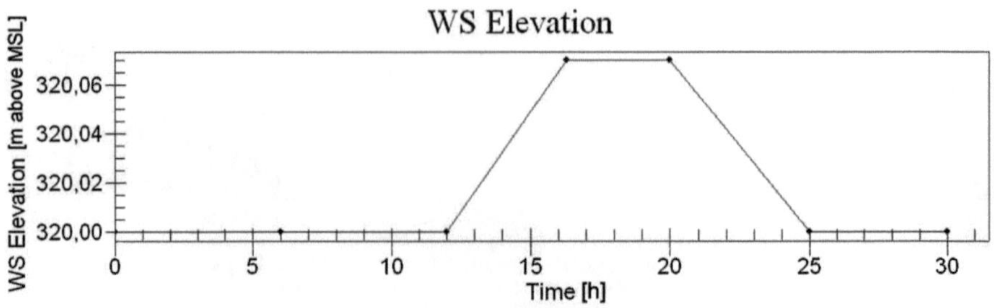

Fig. 4 Water surface boundary condition for dynamic simulations. Damping water from the upper tank causes a 7-cm WSE rise in the lower tank

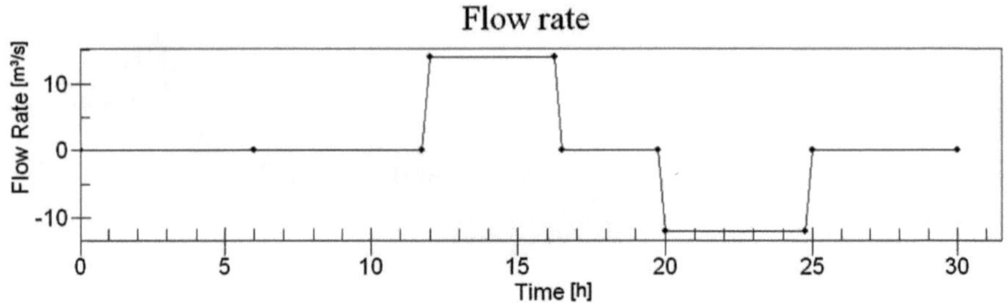

Fig. 5 Power plant outlet discharge rate boundary condition for dynamic simulations. Pumping water upwards takes a longer time as the discharge value is lower

In this case, the changes are much more visible as the contribution of the power plant is proportionally higher. There is also a noteworthy difference between the "inactive" frames for both discharge values—smaller flows fit into the old bed of Soła which runs near the power plant outlet, while large discharges use wider fraction of the lake cross-section there. For the first frame in Fig. 3, where a substantial part of the flow is taken by the power plant, the current fits in the old bed again.

4.2 Dynamic Simulations

The dynamic simulations have been conducted for 30 h of the model time. The first six hours have been used to let the model find a stable solution, while the next 24 h represent the daily cycle of the lake:

- the power plant is inactive from noon to 6 PM, from 10:30 PM to 2 AM, and from 7 AM to noon;
- the power plant generates electricity from 6 PM to 10:30 PM;
- the power plant stores water from 2 AM to 7 AM.

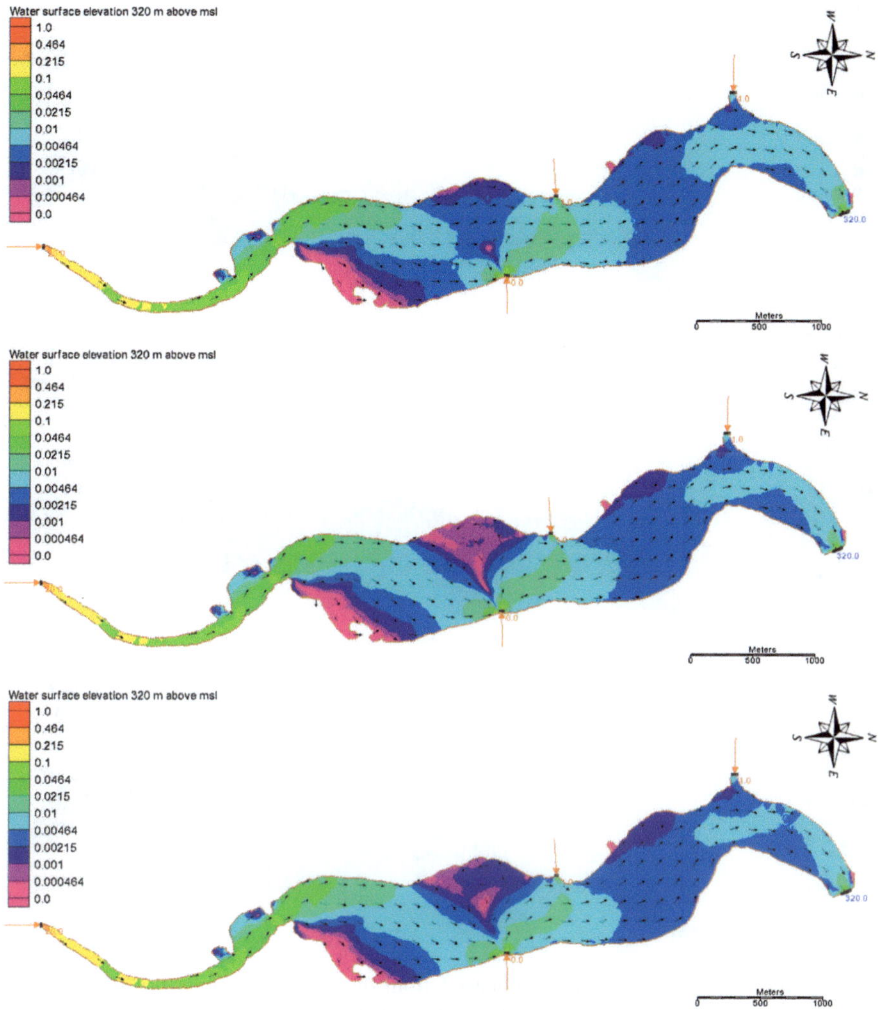

Fig. 6 Sample planar velocity fields obtained from dynamic simulations for 20 m^3/s main inflow: 0.5 h after the power plant started damping water to the lake (*upper* frame); 2 h after the power plant started damping water to the lake (*middle* frame); just when the power plant ended dumping water to the lake (*lower* frame)

This is reflected in appropriate dynamic boundary conditions imposed on water level and right bank discharge, as shown in Figs. 4 and 5.

The first dynamic simulation has been done for 20 m^3/s rate main inflow and two left bank 1 m^3/s rate inflows. Time step size has been chosen to be 5 min and results were saved every 2nd time step, i.e. every 10 min of the model time, which gives 145 result sets for time running from 6 h (noon) to 30 h (noon next day). The calculations took about 10 h of computation time.

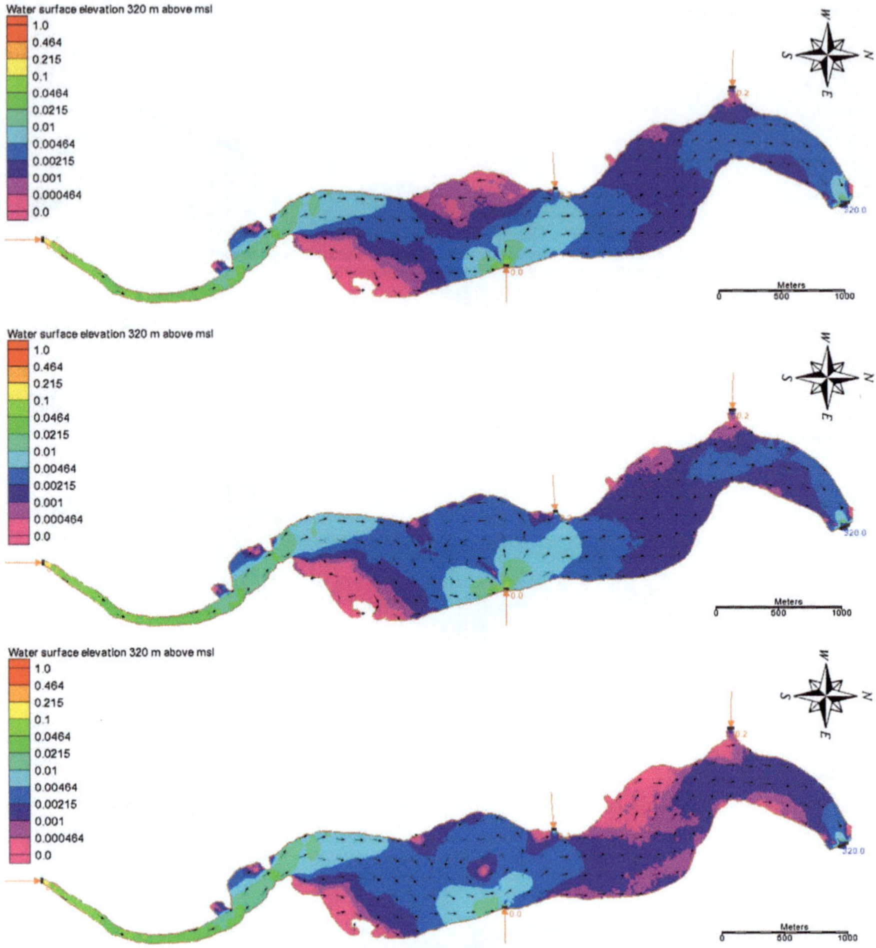

Fig. 7 Sample planar velocity fields obtained from dynamic simulations for 8.7 m^3/s main inflow: 0.5 h after the power plant started damping water to the lake (*upper* frame); 2 h after the power plant started damping water to the lake (*middle* frame); just when the power plant ended dumping water to the lake (*lower* frame)

The second dynamic simulation has been done for 8.7 m^3/s rate main inflow and two left bank 0.2 m^3/s rate inflows. Time step size has been 5 min and results were saved every 2nd time step, which gives 145 result sets for 24 h. Sample results are shown in Fig. 7.

The picture is much different from that for the steady-state simulations. As the water surface elevation rises and the main current in the lower part of the lake becomes disturbed, stagnant areas appear there. The velocity field in the middle part of the reservoir is altered most. However, similarly to the steady-state solution in the upper part of the lake the changes in the velocity field are still small.

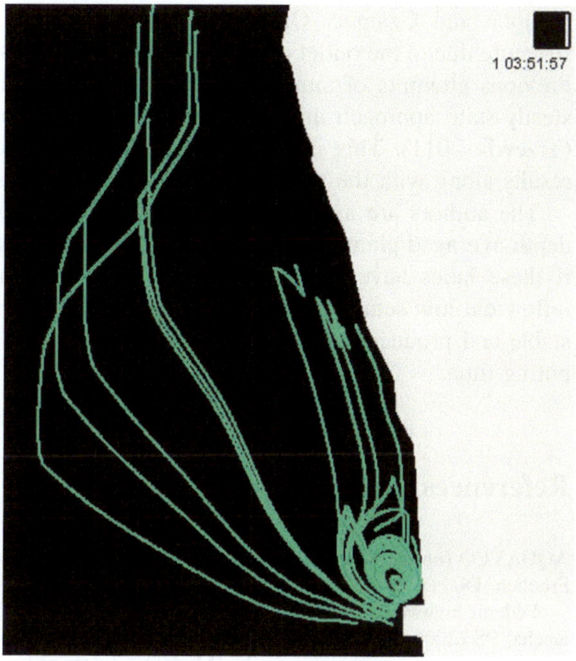

Fig. 8 Paths of several simulated particles tracked by the SMS system. Simulation starts when the power plant starts dumping water to the lake. Some of the particles are being caught inside the protective rim around the power plant outlet. Others run out and flow along their current lines that change in time. Firstly they go along the lines of the current starting at the outlet. The first rapid change of particle movement appears when the extra flow stops. The second one is caused by the power plant starting pumping water up again. Finally the third one appears when the *upper* tank is full and there is no extra flow again

Comparing all 145 result frames computed for a 0–24 h period in a consecutive order leads to a conclusion that both for the yearly average and most probable flows in the lake, turning the Porąbka-Żar power facility on and off causes changes in the velocity field and that this field stabilizes again after about 2 h in its new form. Such a behavior is difficult to show in a paper using static pictures only. Those interested in the results are welcome to contact the authors of this article to obtain animations depicting these phenomena.

The SMS package provides an interesting tool to visualize dynamic planar velocity field and not just drawing velocity maps: it is the particle tracer. This utility allows a user to draw paths of virtual particles moving along current lines even if these lines change in time. It does not include any form of movement dispersion or diffusion and as such it should be used with care. Nevertheless it provides valuable and intuitive visualizations of the velocity field as the one shown in Fig. 8.

5 Final Remarks and the Conclusion

Although designed mainly to cover the hydraulics of rivers and estuaries, the FESWMS model has been previously successfully used by the authors of this paper to simulate flows in retention reservoirs: Dobczyce, Goczałkowice, Tresna,

Porąbka and Czaniec. Out of them, Porąbka has the most complicated flow structure due to the outlet of pumped-storage power station located inside the lake. Previous attempts of simulating the velocity field in that lake contained only a steady-state approach and its results were published only in an engineer thesis (Trzewik 2011). This is the first external publication in English covering these results along with the new ones obtained from dynamic simulations.

The authors are assured that the FESWMS model is a good tool to simulate depth-averaged planar flows (a.k.a. 2.5-dimensional flows) in retention lakes even if these lakes have complex bathymetry as well as unusual and fast-changing inflow/outflow setup. With properly designed computational meshes the model is stable and produces reliable and physically justified results in a reasonable computing time.

References

AQUAVEO (online) Aquaveo water modeling solutions, http://aquaveo.com

Froelich DC (2003) Two-dimensional depth-averaged flow and sediment transport model. Federal highway administration, Virginia

Hachaj PS (2006) Modelowanie pola prędkości wody w zbiorniku dobczyckim—budowa siatki obliczeniowej i wstępne wyniki (Modeling water velocity field In the Dobczyce Lake—construction of computational mesh and preliminary results), Czasopismo Techniczne 15Ś/2007 p 99 (in Polish)

Hachaj PS (2007) Modelling of a two-dimensional velocity field for the water flow in the lake of Dobczyce, in transport phenomena in hydraulics. Publs Inst Geophys Pol Acad Sci, E-7(401):87

IMGW RZGW Kraków (2007) Kaskada rzeki Soły zbiorniki Tresna, Porąbka, Czaniec (The Soła river cascade, reservoirs: Tresna, Porąbka, Czaniec), (in Polish)

Trzewik M (2011) Aplikacja modelu FESWMS na zbiornikach kaskady soły (The FESWMS model application on the Soła Cascade reservoirs), engineer degree thesis, Politechnika Krakowska, (in Polish)

On the Estimation of Solute Transport Parameters for Rivers

S. G. Wallis, M. Osuch, J. R. Manson, R. Romanowicz and B. O. L. Demars

Abstract The modelling of solute transport in rivers is usually based on simulating the physical processes of advection, dispersion and transient storage, which requires the modeller to specify values of corresponding model parameters for the particular river reach under study. In recent years it has become popular to combine a numerical solution scheme of the governing transport equations with a parameter optimisation technique. However, there are several numerical schemes and optimisation techniques to choose from. The chapter addresses a very simple question, namely, do we get the same, or do we get different, parameter values from the application of two independent solute transport models/parameter optimisation techniques to the same data? Results from seven different cases of observed solute transport suggest the latter, which implies that parameter values cannot be transferred between modelling systems.

S. G. Wallis (✉)
Heriot-Watt University, Edinburgh, UK
e-mail: s.g.wallis@hw.ac.uk

M. Osuch · R. Romanowicz
Institute of Geophysics, Warsaw, Poland

J. R. Manson
Richard Stockton College, New Jersey, USA

B. O. L. Demars
James Hutton Institute, Aberdeen, UK

P. Rowiński (ed.), *Experimental and Computational Solutions of Hydraulic Problems*, GeoPlanet: Earth and Planetary Sciences, DOI: 10.1007/978-3-642-30209-1_30,

1 Introduction

Environmental engineers and scientists frequently use mathematical models that simulate the physical processes of solute transport in rivers. Typical applications include assessing the impact of pollution incidents and assessing issues relating to stream ecology. In order to ensure reliable predictions from such models it is preferable to calibrate and validate them using observations of solute transport over a wide range of flow rates for the river of interest. Having achieved this, model parameters can be evaluated for the hydraulic conditions pertaining to the issue of concern, which should result in a reliable prediction. Often, however, the data required to calibrate and validate the model are not available, e.g., in the case where a pollution incident takes place in a river that has not been studied before, in which case parameter values need to be evaluated in another way. In such a situation the only realistic approach is to transfer parameter values from similar rivers, often using empirical equations that have been established using tracer experiments.

During the last 50 years, several methods of analysing tracer data in order to evaluate solute transport parameters have been proposed, e.g., the method of moments (Fischer 1967), routing procedures (Fischer 1968; Singh and Beck 2003) and inverse modelling (Wagner and Gorelick 1986; Scott et al. 2003; Romanowicz et al. 2013). Clearly, if these parameter values are to be used to establish empirical equations the parameter values so estimated need to be reliable, i.e., they need to be an accurate reflection of the physical processes being modelled. If not, e.g., estimated parameter values may be biased away from their true values, then significant inaccuracies in empirical equations that rely on these values will ensue.

In this chapter we focus on the third of the above approaches because it is increasingly replacing the other two. The essential nature of inverse modelling is that it involves optimising objectively the parameters of a solute transport model such that a best fit is obtained between simulated and observed solute concentration data. Alternatively, but not considered herein, properties of observed concentration profiles may be objectively estimated using moment matching techniques (Seo and Cheong 2001; Worman and Wachniew 2007).

Errors in the parameter values obtained by inverse modelling may come from several sources including: inapplicability of the mathematical model to the physical situation; limitations of the numerical scheme used to implement the model; identification of locally optimised rather than globally optimised parameters; non-uniqueness of the optimised parameters; and noisy tracer data. Of particular interest here is the case where an empirical equation might be derived from parameter values obtained from several different sources, in each of which different numerical schemes and/or optimisation algorithms have been used.

This particular situation is explored in this chapter, with the aim of comparing optimised parameter values obtained from a range of solute concentration data using two different numerical scheme/optimisation approaches. A very simple question is posed: do we get the same, or do we get different, parameter values from the application of two independent solute transport models/parameter optimisation techniques to the same data?

2 Solute Transport Modelling

The overwhelming majority of all solute transport modelling in rivers has been undertaken using one of two one-dimensional models, namely the advection–dispersion equation (ADE) and the transient storage model (TSM). The ADE predates the TSM and describes solute transport in terms of advection and longitudinal dispersion. In simple terms, solute is carried downstream at a velocity determined by location within the cross-section (moving faster in the channel centre and moving slower at the channel sides) and solute is continually mixed (by transverse turbulent diffusion and secondary currents) within the cross-section. The differential velocities act to stretch a cloud of solute longitudinally and the cross-sectional mixing acts to reduce the longitudinal stretching. Once a solute cloud has been evolving for long enough in such a flow field, the ADE applies to the cross-sectional average solute concentration (Taylor 1954; Fischer 1967; Rutherford 1994). Often, but not always, observed tracer concentration–time profiles are more skewed than those simulated by the ADE. Typically, observed profiles have more steeply rising limbs and less steeply falling limbs, sometimes with more elevated tails, than simulated ones.

It is recognised that the ADE does not explicitly include the role played by storage areas in rivers, of which there are several types. For example, dead zones are parts of the flow in which no net downstream of water occurs but into which solute migrates, moving back into the main flow sometime later. Thus there is a transient storage of solute in them. The water in some of these dead zones is stagnant but in some others it circulates. Solute transfer between main flow and dead zone and back again occurs by diffusion. In another type of dead zone there is exchange of water and solute with the main flow, in which case solute exchange occurs by a combination of diffusion and local advection. Indeed, some of these more dynamic zones might be considered to be part of the main flow. Generally, but not exclusively, dead zones are located around the channel sides. However, similar, but larger and more dynamic, semi-isolated storage areas exist in river bends, at river junctions, around obstructions such as bridge piers and fallen trees, in pool-riffle structures and amongst in-river vegetation. The hyporheic zone (where there is interaction between the river and the surrounding groundwater) also provides locations where solute can be dynamically stored.

As a result of including transient storage in the ADE, converting it to the TSM, simulated concentration–time profiles often, but not always, mimic observed concentration–time profiles much better. The TSM is described by the following two equations (Bencala and Walters 1983):

$$\frac{\partial c}{\partial t}+\frac{Q}{A}\frac{\partial c}{\partial x}=\frac{1}{A}\frac{\partial}{\partial x}\left(AD\frac{\partial c}{\partial x}\right)+\alpha(s-c) \tag{1}$$

$$\frac{\partial s}{\partial t}=-\alpha\frac{A}{A_s}(s-c) \tag{2}$$

where c is the solute concentration in the main flow, t is time, Q is the flow rate, A is the cross-sectional area of the main flow, x is the longitudinal co-ordinate, D is the dispersion coefficient, α is the exchange rate between the main channel and the storage zones, s is the solute concentration in the storage zones, and A_s is the cross-sectional area of the storage zones. In the absence of transient storage, the ADE is recovered by setting the exchange rate to zero (Eq. 2 becoming redundant).

Note that although Eq. (1) allows for spatially variable parameters, the TSM model is usually implemented in a reach-by-reach basis with parameters being constant within an individual reach. Steady flow is assumed, also.

Although the TSM was originally formulated to include the effect of stagnant dead zones in the ADE, it is often now used to include the effects of all the transient storage processes present in a river reach, which may result in a more complicated formulation and/or interpretation of the transient storage terms in recognition of the wide range of transient storage time scales that may exist (see, e.g., Deng et al. 2010; Bottacin-Busolin et al. 2011). In the current work, two TSM models were used: these are referred to as the OTIS model and the UK model.

2.1 The OTIS Model

This was based on an implicit Eulerian finite difference numerical solution of Eqs. (1) and (2), as described in Runkel and Broshears (1991) and Runkel and Chapra (1993). Overall, the scheme is of the Crank-Nicolson type in which the first three terms in Eq. (1) are approximated by implicit central differences, in time or space, as appropriate, and the final term is treated in a central implicit fashion. Equation (2) is also approximated in the Crank-Nicolson way. The upstream boundary condition, applied at the upstream end of the reach of interest, was provided by an observed main channel concentration–time profile, a zero dispersive flux downstream boundary condition, applied further downstream than the downstream end of the reach of interest, was used and the initial condition was zero solute concentration in the main channel and in the storage zones.

The model parameters (D, A, As and α) were estimated by optimising the numerical solution to an observed concentration–time profile at the downstream end of the reach of interest, which involved minimising an objective function based on the residuals between simulated and observed solute concentrations. This was achieved using a global optimisation approach based on the differential evolution (DE) technique, which is an example of an evolutionary algorithm (Storn and Price 1997; Price et al. 2005). In this an initial population of N randomly selected individuals (N-dimensional vectors of parameters) evolves as a result of mutation, recombination and selection. Mutation involves adding the weighted difference between two random vectors to a third vector, and recombination is a crossover between the new vector and its parents. Then the performance of the old and new generations is compared, and only the better one survives to the next generation.

Differential evolution is an effective, robust and simple global optimisation technique which has only a few control parameters: the population size, the scaling factor used for calculating the weighted differences and the crossover parameter, which govern the probability that a particular element from the parent is passed to the offspring. The following values were assumed for these parameters, respectively: 40, 0.85 and 0.80. Ranges of parameter values for the DE method were chosen as follows: (0, 10,000) for A, D and A_S and (0, 0.001) for α.

2.2 The UK Model

This was based on a different formulation of the model equations, a different numerical solution scheme and a different optimisation algorithm. The model equations are:

$$\frac{\partial(cA)}{\partial t}+\frac{\partial(cAU)}{\partial x}=\frac{\partial}{\partial x}\left(AD\frac{\partial c}{\partial x}\right)+k_1A(s-c) \tag{3}$$

$$\frac{\partial(sA_s)}{\partial t}=-\alpha A(s-c) \tag{4}$$

where U is the cross-sectional average flow velocity in the main channel (=Q/A), k_1 is a solute exchange rate parameter and the other symbols are as previously defined. Clearly, k_1 in Eq. (3) is equivalent to α in Eq. (1). When implemented in the usual steady flow, reach constant parameter form all the terms in Eq. (3) are equivalent to the corresponding terms in Eqs. (1), and (4) can be re-written as

$$\frac{\partial s}{\partial t}=-k_2(s-c) \tag{5}$$

where k_2 is equivalent to the term $\alpha A/A_s$ in Eq. (2). The numerical solution of Eqs. (3) and (5) was based on DISCUS (Manson et al. 2001), a semi-Lagrangian finite volume numerical solution. In DISCUS the (explicit) high-order QUICKEST scheme for advection is implemented in a semi-Lagrangian fashion and is combined with implicit backward finite differencing of the dispersion and transient storage. The boundary conditions and initial conditions were the same as for the OTIS model.

The model parameters (D, U, k_1 and k_2) were estimated by optimising the numerical solution to an observed concentration–time profile at the downstream end of the reach of interest using a global optimisation technique based on a genetic algorithm (GA) approach. In general, GA techniques are similar to DE techniques. However, in the GA technique used here the objective function was defined as the inverse of the sum of the squared residuals between observed and simulated downstream concentrations. The objective function was then maximised by adjusting the parameters according to the micro-genetic algorithm described in Yang et al. (1998). The control parameters of population size, mutation probability and crossover rate were defined as 5, 0.2 and 0.5, respectively. 2,500 generations were allowed to evolve, yielding a parameter space of 33554432^4 potential solutions. Ranges of parameter values for the GA method were chosen as follows: (0, 2) for U, (0, 2,000) for D, (0, 0.01) for k_1 and (0, 0.1) for k_2.

3 Data Preparation and Analysis

The data required for parameter estimation consisted of solute concentration–time profiles at the upstream and downstream ends of a river reach, the corresponding reach length and the corresponding river flow rate. The parameters of both models were optimised by minimising or maximising an objective function as previously described. Seven cases were analysed, taken from a variety of sources (several tracer experiments in streams and rivers and one analytical solution). To ensure that exactly the same concentration–time data were used with both models, both upstream and downstream concentration–time profiles for a particular case were supplied at the same, constant, sampling interval. This removed the need to interpolate the data, which might have been done in different ways by each modelling group. However, the time step varied between the cases. The cases are summarised in Table 1.

For both models, a river reach was treated as a uniform channel so that all model parameters were constants. To investigate the role of spatial resolution, which has been rarely even considered in previous work, each case was analyzed four times using space steps of L/200, L/100, L/50 and L/25, where L is the reach length. In addition, some cases were analysed over a wider range of space step.

Table 1 Summary of cases

Case	River	Reach length (m)	Flow rate (l/s)	Time step (s)
1	Murray Burn	184	144.8	30
2	Rhine	47,150	1,169,000	600
3	Rhine	73,800	1,169,000	1,200
4	Cairn burn	91	10.2	10
5	Uvas creek	152	13.6	600
6	Uvas creek	186	14.0	600
7	Analytical	250	1,000	1

The treatment of the downstream boundary condition was also investigated briefly. As with the spatial resolution issue, there is little published guidance on where the downstream boundary should be located—far enough so as not to influence the solution in the reach of interest being the best advice available, see, e.g., Runkel and Broshears (1991). So for one or two cases results were compared from a series of runs in which the downstream boundary was located progressively further away from the downstream end of the reach of interest.

4 Results and Discussion

Figure 1 shows the variation of the optimised parameters from Case 7 over a wide range of space step values, and shows how the optimised parameter values converge as the magnitude of the space step decreases. In the figure, the parameter values are expressed as a percentage of the value found for 500 space steps and the results are from the OTIS model. Similar behavior was found in the UK model and for other Cases. It is particularly evident that A_s is the parameter that is most sensitive to this issue. There is no published guidance on what degree of spatial resolution is required for these sorts of analyses. Indeed, it is not clear that the issue has been studied before. The most likely reason for this is that, because the optimisation uses temporal data, the issue has been overlooked. Nevertheless, as one might expect, Fig. 1 suggests that the spatial resolution needs to be sufficiently good to prevent it being a source of error. Of course, what really matters is the number of space steps over which the spatial concentration profile, corresponding to the problem in hand, is resolved rather than how well the reach length is resolved. This can be evaluated by dividing the length of the spatial concentration profile (estimated here as the product of the centroid velocity and the time difference between the trailing and leading edges of the upstream concentration–time profile) by the space step. For Case 7 this gives 392, 196, 98 and 49 for reach resolutions of L/200, L/100, L/50 and L/25, respectively, suggesting that resolving the spatial concentration profile over about 100 space steps is sufficient for parameters A, D and α, but maybe as many as about 400 space steps are required for A_S. For the other six Cases, the number of space steps covering the spatial concentration profile length for the L/200 reach resolution varied between about 100 and 1,800. So based on the analysis of Case 7, the other Cases seem to have been adequately resolved in space.

Results from runs with both models where the location of the application of the downstream boundary condition was varied showed that only for the lowest spatial resolution case (L/25) was there any significant sensitivity to this issue. This seems reasonable because the zero dispersive flux boundary condition is more consistent with shallow spatial concentration gradients than with steep ones, with the former and latter being associated, respectively, with high and low spatial resolutions.

Table 2 compares the optimised parameter values from the two models for all seven Cases. In view of the discussion above, it seems reasonable to base the

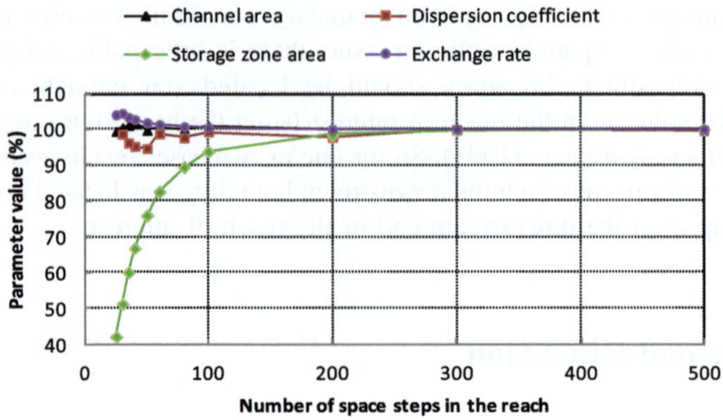

Fig. 1 Sensitivity of optimised parameters to spatial resolution (Case 7; OTIS model)

Table 2 Optimised parameter values from the two models (reach resolution of L/200)

Case	Model	A (m^2)	D (m^2/s)	α (1/s)	A_s (m^2)	DaI (–)
1	OTIS	0.688	0.361	0.00100	0.083	>5
	UK	0.737	0.564	0.00025	0.039	>5
2	OTIS	1,418	22.5	0.00002	86.3	>5
	UK	1,367	46.0	0.00062	112.8	>5
3	OTIS	1,284	299	0.00001	380	>1, < 5
	UK	1,336	743	0.00013	16.8	>5
4	OTIS	0.137	0.187	0.00018	0.017	>1, <5
	UK	0.145	0.235	0.00031	0.661	>0.2, <1
5	OTIS	0.451	0.204	0.00003	0.919	>0.2, <1
	UK	0.460	0.250	0.00003	1.892	>0.2, <1
6	OTIS	0.513	0.153	0.00005	4.710	>0.2, <1
	UK	0.694	1.383	0.00075	0.053	>5
7	OTIS	2.020	1.110	0.00035	3.722	>0.2, <1
	UK	2.019	1.097	0.00038	3.230	>0.2, <1

comparison on the results obtained with the highest reach resolution, i.e., 200 space steps. It is clear that there are only a few examples of close agreement between the models and there are several examples of very poor agreement between them. Clearly, this is a worrying outcome since it suggests that the parameter values are not independent of the model/optimisation system used, indicating that they are not transferable. If this state of affairs were found to be generally true it would seriously threaten the credibility of empirical equations based on published parameter values that had been derived in the way used in this chapter.

Of the four parameters, A shows the best agreement between the models with percentage differences being <10 % for all cases. The percentage errors in the other parameters vary widely (even exceeding 100 %), and show no apparent pattern.

The good agreement for A is consistent with the idea that the response of a solute cloud to being transported through a river reach is most sensitive to A because it controls both the main channel travel time and the time available for dispersion and transient storage to occur. It is also consistent with the results of, e.g., Mrokowska and Osuch (2011) and Wagener et al. (2002). In these studies, the results of Monte Carlo simulations showed that A was the most easily identifiable parameter of the TSM.

Indeed, these and other studies highlight the problem of equifinality with the TSM, namely that many different combinations of the parameters can give equally good simulations. A corollary of this is that we cannot have much confidence in the values of the optimised parameters. This is often expressed in terms of estimates of uncertainty in the parameter values (Romanowicz et al. 2013) or in terms of the identifiability of the parameters (Wagner and Harvey 1997; Wagener et al. 2002). Clearly, if there isn't a unique parameter set that gives an optimum fit to observations, then the differences between the results of the OTIS and UK models may simply be a reflection of differences between the two optimisation methods.

Several authors have discussed the importance of the Damkohler number, DaI, in relation to the identifiability of the TSM parameters (e.g., Wagner and Harvey 1997; Scott et al. 2003; Worman and Wachniew 2007). DaI expresses a dimensionless ratio of main channel to storage zone residence times and is defined as

$$DaI = \frac{\alpha(1 + A/A_s)L}{U} \tag{6}$$

When DaI is small, the solute exchange between the main channel and the storage zones takes place so slowly that observed concentration–time profiles do not reflect the interaction between them. When DaI is large, the solute exchange between the main channel and the storage zones takes place so quickly that it is impossible to differentiate between dispersion caused by the main channel shear flow and by the transient storage. Wagner and Harvey (1997) suggest that reliable parameters are only identifiable for $0.1 < \mathrm{DaI} < 1.0$. Few of the Cases analysed herein fall within this range (see Table 2) so this may be contributing to the differences between the results of the OTIS and UK models. However, there is great uncertainty in the values of DaI because they are derived from the optimised parameter values. What we really need, of course, are the true DaI values, i.e., those corresponding to the physical situations, but these remain elusive unless we can be certain that the optimised parameter values are robust.

Irrespective of the above, it is interesting to note that Case 7 shows the best agreement between the two models. The data used were derived from an analytical solution to the problem (Hart 1995) so these data (unlike the rest) are not subject to experimental errors in field work, laboratory analysis or data analysis. Indeed a comparison of the optimised parameter values and the values used to generate the data ($A = 2\ m^2$, $D = 1\ m^2/s$, $\alpha = 0.00025$ 1/s, $A_s = 0.1\ m^2$) is very encouraging for A and D, even if less so for α and A_s.

Similarly, notwithstanding the equifinality and identifiability issues, the extent to which other issues might be contributing to the differences between the parameter values from the two models is considered below. Four obvious issues stand out. Firstly, the equations on which the two models are based are not identical, secondly the numerical schemes used to solve the equations are different, thirdly different optimisation algorithms are used and fourthly different parameters are optimised.

In regard to the model equations, the OTIS model is based on a so-called non-conservative formulation whereas the UK model is based on a so-called conservative one. The distinction between these formulations is only usually important if there are discontinuities in parameter values or solutions of the equations. So for the steady, uniform flows in question here, in which all parameters are constants, we would not expect this to be a serious issue. The different numerical schemes can be expected to be an influence because they introduce different errors, although we might expect a reduction in this effect as the spatial resolution of the problem increases. Of course, numerical errors might also be introduced through the temporal resolution of the problem, but here this has been fixed at an identical level in both models and has not been studied. Both models employ a global optimisation approach (although the techniques are different), so it is unlikely that either is getting stuck in local minima. Finally, it is not known what differences might be introduced by optimising A, D, α and A_s directly in the OTIS model whilst optimising U, D, k_1 (=α) and k_2 in the UK model and then calculating A and A_s from these values. Although U and A are simply related through Q (which is known), the calculation of A_s in the UK model relies on using two optimised parameter values (k_1 and k_2), which may amplify errors in A_s.

5 Conclusions

The application of two different versions of the TSM to the same solute concentration data has yielded different optimum parameter values. Although it is comforting that the main channel cross-sectional area shows the best agreement between the two models, the differences in the values of the other three parameters are a source of concern. Indeed, if optimised parameter values are not independent of the modeling/optimisation systems being used, this raises serious concerns over the robustness and transferability of the values. It has not been possible to establish the cause of these differences but it is likely to be mainly associated with equifinality and identifiability issues of the problem under study. Nevertheless, various modeling errors derived from the form of the governing equations used, the numerical schemes employed and the combination of parameters being optimised cannot be ruled out. The study has posed more questions than it has answered.

References

Bencala KE, Walters RA (1983) Simulation of solute transport in a mountain pool-and-rifle stream: a transient storage model. Water Resour Res 19:718–724

Bottacin-Busolin A, Marion A, Musner T, Tregnaghi M, Zaramella M (2011) Evidence of distinct contaminant transport patterns in rivers using tracer tests and a multiple domain retention model. Adv Water Resour 34:737–746

Deng Z-Q, Jung H-S, Ghimire B (2010) Effect of channel size on solute residence time distributions in rivers. Adv Water Resour 33:1118–1127

Fischer HB (1967) The mechanics of dispersion in natural streams. J Hydraul Div Proc Am Soc Civ Eng 93:187–216

Fischer HB (1968) Dispersion predictions in natural streams. J Sanit Eng Div Proc Am Soc Civ Eng 94:927–943

Hart DR (1995) Parameter estimation and stochastic interpretation of the transient storage model for solute transport. Water Resour Res 31:323–328

Manson JR, Wallis SG, Hope D (2001) A conservative semi-Lagrangian transport model for rivers with transient storage zones. Water Resour Res 37:3321–3330

Mrokowska MM, Osuch M (2011) Assessing validity of the dead zone model to characterize transport of contaminants in the River Wkra. In: Rowinski P (ed) Experimental methods in hydraulic research. Springer-Verlag, Berlin, pp 235–245

Price KH, Storn RM, Lampinen JA (2005) Differential evolution: a practical approach to global optimization. Springer-Verlag, Berlin

Romanowicz RJ, Osuch M, Wallis SG (2013) Modelling of solute transport in rivers under different flow rates: a case study without transient storage. Acta Geophys (In Press)

Runkel RL, Broshears RE (1991) One-dimensional transport with inflow and storage (OTIS): a solute transport model for small streams. CADSWES Department of Civil, Environmental and Architectural Engineering, University of Colorado

Runkel RL, Chapra SC (1993) An efficient numerical solution of the transient storage equations for solute transport in small streams. Water Resour Res 29:211–215

Rutherford JC (1994) River mixing. Wiley, Chichester

Scott DT, Gooseff MN, Bencala KE, Runkel RL (2003) Automated calibration of a stream solute transport model: implications for interpretation of biogeochemical parameters. J N Am Benthol Soc 22:492–510

Seo IW, Cheong TS (2001) Moment-based calculation of parameters for the storage zone model for river dispersion. J Hydrol Eng Am Soc Civ Eng 127:453–465

Singh SK, Beck MB (2003) Dispersion coefficient of streams from tracer experiment data. J Environ Eng Proc Am Soc Civ Eng 129:539–546

Storn RM, Price KH (1997) Differential evolution—a simple and efficient heuristic for global optimization over continuous spaces. J Global Optim 11:41–359

Taylor GI (1954) The dispersion of matter in turbulent flow through a pipe. Proc R Soc Lond A 233:446–468

Wagener T, Camacho LA, Wheater HS (2002) Dynamic identifiability analysis of the transient storage model for solute transport in rivers. J Hydroinform 4:199–211

Wagner BJ, Gorelick SM (1986) A statistical methodology for estimating transport parameters: theory and applications to one-dimensional advective-dispersive systems. Water Resour Res 22:1303–1315

Wagner BJ, Harvey JW (1997) Experimental design for estimating parameters of rate-limited mass transfer: analysis of stream tracer studies. Water Resour Res 33:1731–1741

Worman A, Wachniew P (2007) Reach scale and evaluation methods as limitations for transient storage properties in streams and rivers. Water Resour Res 43:W10405. doi:10.1029/2006WR005808

Yang G, Reinstein LE, Pai S, Xu Z, Carroll DL (1998) A new genetic algorithm technique in optimization of permanent 125-I prostate implants. Med Phys 25:2308–2315

MIX
Papier aus verantwortungsvollen Quellen
Paper from responsible sources
FSC® C105338

If you have any concerns about our products, you can contact us on
ProductSafety@springernature.com

In case Publisher is established outside the EU, the EU authorized representative is:
Springer Nature Customer Service Center GmbH
Europaplatz 3, 69115 Heidelberg, Germany

Printed by Libri Plureos GmbH
in Hamburg, Germany